Hyperbolic Equations and Related Topics

Proceedings of the Taniguchi International Symposium
Katata and Kyoto, 1984

Sigeru Mizohata, editor

1986

ACADEMIC PRESS, INC.

Harcourt Brace Jovanovich, Publishers

Boston Orlando San Diego
New York Austin London Sydney
Tokyo Toronto

Published in the United States by Academic Press, Inc., Orlando, Florida 32887

Library of Congress Catalog Card Number 86–72475

ISBN 0–12–501658–1

PRINTED IN JAPAN
By Tokyo Press Co., Ltd.

Preface

In the theory of partial differential equations, the hyperbolic differential equation has been and will be one of the most attractive subjects. Many important notions and methods in mathematical analysis sprang from the problems for hyperbolic equations.

Several pioneering works in the 60th have prepared the rapid progress since 1970. In particular, since then, many papers concerning the characterization of hyperbolic operators appeared and the hyperbolic initial-boundary value problem was studied. Moreover the propagation of singularities of solutions was intensively studied; on one hand, by the use of pseudo-differential and Fourier integral operators, and on the other hand, by the use of complex analysis initiated by J. Leray.

For the purpose of reviewing these activities we held the International Workshop on Hyperbolic Equations and Related Topics at Katata from August 27 through August 31, 1984 under the auspices of Taniguchi Foundation. Sixteen mathematicians, seven from abroad and nine from Japan, participated in the Workshop. As an extension of the Workshop we held the International Symposium on Hyperbolic Equations and Related Topics at Kyoto on September 3 through September 5 with about 150 participants. The papers contributed to the Workshop at Katata and the Symposium at Kyoto are included in these Proceedings.

For the workshop and the symposium Taniguchi Foundation kindly gave us a full financial support and provided us with facilities at Kyūzesō, Katata. We would like to express our sincere gratitude to Mr. Taniguchi and Professor K. Itô in this respect. We hope that this symposium has contributed not only to promotion of scientific activities but also to international mutual understanding, the ideal of Mr. Taniguchi in supporting us. Also we would like to thank the Research Institute for Mathematical Sciences at Kyoto University for providing us with lecture hall and lobby.

Professors Y. Ohya and M. Ikawa made painstaking efforts to organize the workshop and the symposium, and edit these proceedings. Mr. S. Okuda, Taniguchi Foundation, effectively made administrative business. Mr. S. Okada, Kinokuniya Company Ltd., eagerly cooperated with us to publish these Proceedings. We are very grateful to all of them.

October, 1985

S. Mizohata

CONTENTS

CONTENTS

Comments on the Development of Hyperbolic Analysis

Looking at the contents of the elaborate contributed papers, we thought that it would be useful to give a brief survey on the development of researches on hyperbolic partial differential equations. Of course, it is a very difficult and almost impossible task to give an impartial account of this vast subject. For instance, we mentioned almost nothing about the works on the initial-boundary value problems. In making up this article, we received precious assistance from our colleagues Y. Hamada, N. Iwasaki, T. Nishitani and S. Tarama. To all of them we express our thanks.

Chapter 1. Cauchy Problem (I)

The systematic treatment of the Cauchy problem for hyperbolic equations began with the famous treatise of J. Hadamard [1]. In this book, the fundamental solutions for general second order hyperbolic equations with analytic coefficients are constructed explicitly. On the other hand, K. O. Friedrichs-H. Lewy [1] and J. Schauder [1] introduced the L^2-norm (energy method) and solved the Cauchy problem for general second order hyperbolic equations with smooth coefficients.

The Cauchy problem for general hyperbolic partial differential equations and systems was studied by I. G. Petrowsky [1, 2]. After then, J. Leray [1], L. Gårding [1], K. O. Friedrichs [1], clarified the work of Petrowsky and extended it from their view-points.

Now we explain it more concretely. We are concerned with general higher order single equations, and the Cauchy problem for them:

$$(1) \qquad \begin{cases} P(t, x; D_t, D_x)u(t, x)=f(t, x), & \text{for } (t, x) \in [0, T]\times \boldsymbol{R}^l=\Omega, \\ D_t^j u|_{t=t_0}=u_j(x), \quad 0\leqq j\leqq m-1, & \text{for } t_0 \in [0, T), \end{cases}$$

where

$$(2) \qquad p(t, x; \tau, \xi)=\tau^m+\sum_{j=1}^{m} a_j(t, x; \xi)\tau^{m-j}, \quad \text{and}$$

$$\text{order } a_j\leqq j \quad \text{(Kowalewskian)}.$$

We say that the Cauchy problem for (1) is C^∞-wellposed, if for any $f(t, x) \in C^\infty([t_0, T]\times \boldsymbol{R}^l)$, and for any initial data $u_j(x) \in C^\infty$, there exists a unique solution $u(t, x) \in C^\infty([t_0, T]\times \boldsymbol{R}^l)$. Further we say that (1) is uniformly C^∞-wellposed in Ω, when the above problem is well posed for any $t_0 \in [0, T)$.

Let us explain Petrowsky's result. In a word, he proved that (1) is C^∞-wellposed under the assumption that P is strictly hyperbolic. The strict hyperbolicity means the following. Let

$$(3) \qquad P_m(t, x; \tau, \xi) = \prod_{j=1}^{m} (\tau - \lambda_j(t, x; \xi)).$$

P is called strictly hyperbolic if 1) $\lambda_j(t, x; \xi)$ are all real for any $(t, x; \xi)$; 2) they are distinct. More precisely, there exists some positive δ such that

$$(4) \qquad |\lambda_i(t, x; \xi) - \lambda_j(t, x; \xi)| \geqq \delta |\xi|, \qquad \text{for all } (t, x; \xi) \in \Omega \times \boldsymbol{R}^l.$$

In 1957, the work of P. D. Lax [1] appeared. In this paper he proposed a prototype of Fourier integral operator and also a new viewpoint. We could say that new approach began with this work. This work was followed by S. Mizohata [1] and he proposed a method which could be called a method of micro-localization of nowadays. They claimed that in order that (1) be C^∞-wellposed, it is necessary that the characteristic roots $\lambda_j(t, x; \xi)$ be real for all $(t, x; \xi)$ irrespective of their multiplicities. Although this result looks like trivial today, we could say that these works offered efficient methods to later researches on hyperbolic equations.

The next step was to consider non-strictly hyperbolic equations and systems as far as the Cauchy problem concerned. Observe that, when the multiplicity of characteristic root is constant, (3) becomes

$$(5) \qquad P_m = \prod_{j=1}^{k} (\tau - \lambda_j(t, x; \xi))^{m_j},$$

and when the multiplicity is not constant (variable multiplicity), the situation becomes, in general, fairly complicated. Famous examples in this case are

$$L_0 u = \partial^2 u(t, x)/\partial t^2 - t^{2k} a(t, x) \partial^2 u/\partial x^2 + t^l b(t, x) \partial u/\partial x = f,$$
$$L_1 u = \partial^2 u(t, x)/\partial t^2 - x^{2k} a(t, x) \partial^2 u/\partial x^2 + x^l b(t, x) \partial u/\partial x = f,$$

where $a(t, x) > 0$, and both equations are considered in a neighborhood of the origin.

(I) Case of constant multiplicity.

i) *C^∞-wellposedness.*

After pioneering works of A. Lax [1] and of M. Yamaguti [1], S. Mizohata-Y. Ohya [1, 2] treated the case $m_j \leqq 2$ in (5), inspired by E. E. Levi [1]. This gave a necessary and sufficient condition for C^∞-well-posedness. That condition can be stated in the following form: Subprincipal symbol of P, denoted by P'_{m-1}, vanishes on double characteristic sets, namely

$$(6)\quad P'_{m-1}(t, x; \lambda_j(t, x; \xi), \xi) \equiv \left(P_{m-1} - \frac{1}{2i}\sum_{i=0}^{l}\frac{\partial^2}{\partial x_i \partial \xi_i} P_m\right)(t, x; \lambda_j(t, x; \xi), \xi) \equiv 0 \text{ for all } (t, x; \xi),$$

where $x_0 = t$, and $\xi_0 = \tau$. Remark that the importance of this condition was pointed out, independently of Mizohata-Ohya, by J. Vaillant [1] too.

After this work, H. Flashka-G. Strang [1] gave a necessary condition for general multiplicity m_j, and J. Chazarain [1] showed that the condition is sufficient using the thoery of Fourier integral operators (see also V. Ya. Ivrii-V. M. Petkov [1]).

ii) $\gamma^{(s)}$-*wellposedness.*

Hereafter we denote by $\gamma^{(s)}$ the space of all functions of Gevrey class s. The systematic treatment of the Cauchy problem in Gevrey class began with the work of Y. Ohya [1], and this result was immediately extended by J. Leray-Y. Ohya [1, 2]. Y. Ohya [1] proved its $\gamma^{(s)}$-wellposedness for $1 < s \leqq m/(m-1)$, and J. Leray-Y. Ohya [1, 2] showed the same result for $s \leqq r/q$, where r is the maximum multiplicity of P_m and $m - r + q$ $(q \geqq 0)$ is the minimum order of $P - \prod_{j=1}^{r} a_j(t, x; D_t, D_x)$, where a_j is defined as an arbitrary strictly hyperbolic operator.

Leray-Ohya [1] gave precious subjects of researches on (formally) hyperbolic equations with constant multiplicity. S. Matsuura [1] proved that $P_m(t, x; \tau, \xi)$ can be factorized by polynomials, $P_m = \prod_{j=1}^{k} q_j(t, x; \tau, \xi)^{d_j}$, where $q_1 \cdots q_k$ is a strictly hyperbolic polynomial. J. C. De Paris [1] has shown a kind of factorization by differential polynomials and defined "bien décomposable" operators which are now known to be equivalent to satisfying the Levi condition. H. Komatsu [1], analyzing the above works, showed a general factorization theorem, and defined the notion of irregularity through this factorization. Concerning the precise meaning of Gevrey indices, we quote H. Komatsu [1, 2], and S. Mizohata [9] which discusses, after a factorization of operator, its necessity too (see also T. Nishitani [1], and S. Mizohata [6]).

Let us observe that the (perfect) factorization of P gives clear image to the results obtained above. Let $P_m = \prod_{j=1}^{k} (\tau - \lambda_j)^{m_j}$, then

$$P = P_k \circ P_{k-1} \circ \cdots \circ P_1 + R,$$

where 1) $P_j = (D_t - \lambda_j)^{m_j} + a_{1j}(t, x; D)(D_t - \lambda_j)^{m_j - 1} + \cdots + a_{ij}(D_t - \lambda_j)^{m_j - i} + \cdots + a_{m_j j}$.
order $a_{ij} \leqq i$.

2) $R=\sum_{j=1}^{n} r_j(t, x; D)D_t^{m-j}$, $r_j(t, x; \xi)$ being regularizing symbols together with $D_t^k r_j$ $(k=1, 2, \cdots)$.

Then we can reformulate above conditions as follows:

Levi condition. order $a_{ij}(t, x; \xi) \leqq 0$.

Index of Gevrey s. $\rho = \max_{1\leqq j\leqq k} \max_{1\leqq i\leqq m_j} \text{order } a_{ij}/i \leqq 1/s$.

(II) Case of variable multiplicity.

i) *C^∞-wellposedness.*

After several pioneering works (for instance M. Y. Chi [1], A. B. Nersesyan [1]), O. A. Oleinik [1] appeared. This paper oriented the researches on this subject. She considers

$$\partial_t^2 u(t, x) - \sum_{i,j=1}^{n} \partial_i(a_{ij}(t, x)\partial_j u) + b_0 \partial_t u + \sum_{j=1}^{n} b_j(t, x)\partial_j u + cu = f,$$

assuming $\sum a_{ij}\xi_i\xi_j \geqq 0$. The Cauchy problem concerns in the space $\Omega = \{(t, x) \in [0, T]\times \boldsymbol{R}^n\}$, with initial data at $t=0$. The following condition was presented as a sufficient condition for C^∞-wellposedness. If we take two positive constants α and A conveniently, it holds that $\alpha t(\sum_i b_j\xi_j)^2 \leqq A\sum a_{ij}\xi_i\xi_j + \sum (a_{ij})'_t\xi_i\xi_j$ for all $(t, x; \xi) \in \Omega\times \boldsymbol{R}^n$.

Next, the research on the necessity has been done in V. Ya. Ivrii-V. M. Petkov [1]. In it the following result is proved: Suppose that the multiplicity of characteristic roots at $(\hat{t}, \hat{x})$ is at most r; more precisely, if there exist positive rational numbers p and q $(p\geqq q)$ such that

$$\left(\frac{\partial}{\partial \tau}\right)^r P_m(\hat{t}, \hat{x}; 0, \xi) \neq 0$$

and

$$\left(\frac{\partial}{\partial \tau}\right)^{\alpha_0}\left(\frac{\partial}{\partial \xi}\right)^{\alpha}\left(\frac{\partial}{\partial t}\right)^{\beta_0}\left(\frac{\partial}{\partial x}\right)^{\beta} P_m(\hat{t}, \hat{x}; 0, \xi) = 0 \qquad (\xi \in R_\xi^n)$$

for $\alpha_0 + |\alpha| + p\beta_0 + q|\beta| < r$, then it is necessary for the C^∞-wellposedness of Cauchy problem that

$$(6) \qquad \left(\frac{\partial}{\partial \tau}\right)^{\alpha_0}\left(\frac{\partial}{\partial \xi}\right)^{\alpha}\left(\frac{\partial}{\partial t}\right)^{\beta_0}\left(\frac{\partial}{\partial x}\right)^{\beta} P_{m-s}(\hat{t}, \hat{x}; 0, \xi) = 0 \qquad (\xi \in \boldsymbol{R}_\xi^n)$$

for $\alpha_0 + |\alpha| + p\beta_0 + q|\beta| + s(1+p) < r$ are satisfied, where P_{m-s} $(1\leqq s\leqq m)$ are the homogeneous lower order terms of order $m-s$ of P.

Observe that they obtained more precise results (necessary conditions) in the case where the multiplicity is not greater than 3. These results gave rise to the important notion of effectively hyperbolic operators (nominated by L. Hörmander [4]), namely the effective hyperbolicity is necessary for the strong hyperbolicity in the case of single equations. Here, the strong hyperbolicity means the stability of wellposedness under any addition of lower order terms, which they called regularly hyperbolic operator. Recently N. Iwasaki [1–3] proved the sufficient part. (See also T. Nishitani [6].) We could conclude the equivalence of the effective hyperbolicity and the strong hyperbolicity for single equations. The effective hyperbolicity means that, at any critical point of $P_m=0$, the fundamental matrix

(7) $$F=\begin{bmatrix} P_{m,\xi x} & P_{m,\xi\xi} \\ -P_{m,xx} & -{}^tP_{m,\xi x} \end{bmatrix}$$ has non zero real eigenvalues.

In order to explain this notion roughly, suppose that we have a smooth decomposition $P_m=qr$, where q and r are strictly hyperbolic polynomials. If we have

$$\{q, r\}=\sum_{\alpha=0}^{l}\left(\frac{\partial q}{\partial \xi_\alpha}\frac{\partial r}{\partial x_\alpha}-\frac{\partial q}{\partial x_\alpha}\frac{\partial r}{\partial \xi_\alpha}\right)\neq 0,$$

on $q=r=0$, then P is effectively hyperbolic. Especially, strictly hyperbolic operators are effectively hyperbolic. The fundamental matrix (7) was introduced without assuming the smooth decomposition.

In non effectively-hyperbolic cases, L. Hörmander [4] precised the necessary condition of Ivrii-Petkov. Namely this condition says that, at any critical point, the subprincipal symbol P'_{m-1} is real and moreover

$$|P'_{m-1}(t, x; \lambda, \xi)|\leqq \tfrac{1}{2}\operatorname{Tr} F,$$

where $\operatorname{Tr} F$ denotes the sum of positive eigenvalues of iF.

On the sufficiency, Ivrii [1] and Hörmander discussed also some cases, where the inequality of A. Melin [1] and related one (Gårding type inequality) can be applied. Complete treatment is still open. Observe that

$$-\tau^2+(t\xi_2+\xi_1)^2+\nu x_1^2\xi_2^2 \qquad \text{on } \boldsymbol{R}^3$$

is an example of opreators with double characteristic. This is effectively hyperbolic if $0\leqq\nu<1$, and this presents two types of other cases according to $\nu=1$ or $\nu>1$.

There are various works on conditions on lower terms connected with the Levi condition; for instance, on sufficiency, A. Menikoff [1], Y. Ohya [3], S. Tarama [1], R. Sakamoto [1, 2], K. Yamamoto [1], and on the necessity, T. Mandai [1]. Many of these treat the cases where the critical points appear at special positions.

ii) $\gamma^{(s)}$-*wellposedness.*

There is a fundamental work of J. M. Bony-P. Schapira [1], which asserts the global existence of analytic solutions assuming P (with analytic coefficients) formally hyperbolic, namely, assuming all the characteristic roots $\lambda_j(t, x; \xi)$ real for all $(t, x; \xi)$. This is a kind of global version of the Cauchy-Kowalewski theorem. Inspired by this result, denoting by r the maximal multiplicity of P_m, V. Ya. Ivrii [1] proved it for $1 < s < (2r-2)/(2r-3)$, and J.-M. Trépreau [1] for $1 < s < \min(r/(r-1), 3/2)$ under the assumptions of analytic principal symbol. Finally, M. D. Bronštein [1, 2] succeeded in proving it for $1 < s < r/(r-1)$. Concerning the admissible data of Gevrey class, see Ivrii [3]. It is shown that, as a result, for operator $P = D_t^2 - t^{2k}a(t, x)D_x^2 - t^m b(t, x)D_x$, $(a > 0, b \neq 0)$, the admissible index s of Gevrey is $s \leqq (2k-m)/(k-1-m)$, if $m < k-1$. There are several works assuming some conditions on the behaviors of double characteristics. See for instance, K. Igari [1].

Some comments should be added. In the Cauchy problem (1) for the operator P, the roles of t and x are not symmetric, as far as we consider it as purely evolution equation. Indeed, in the treatment of Gevrey class mentioned above, we know that, as far as it is concerned only with the existence theorem of the Cauchy problem, it is enough to assume a finite number of differentiability of coefficients, with respect to t. In this direction, we have the results with coefficients depending only on t by F. Colombini-E. De Giorgi-S. Spagnolo [1] for strictly hyperbolic equations, and F. Colombini-E. Jannelli-S. Spagnolo [1] for non strictly hyperbolic case. T. Nishitani [4, 5] showed it also with less regularity with respect to t by the energy method, and Y. Ohya-S. Tarama [1] extended their results to general cases. Moreover, F. Colombini-S. Spagnolo [1] and F. Colombini-E. Jannelli-S. Spagnolo [1] gave the example whose Cauchy problem is $\gamma^{(s)}$-wellposed for any s, but is not C^∞-wellposed.

(III) Systems.

Compared to single equations, we do not have yet the general result for hyperbolic systems of equations, except for the cases of symmetric, more generally, symmetrizable systems proposed by K. O. Friedrichs. Roughly speaking, there may be only three following methods to study the non symmetric systems; the first method is the reduction to symmetrizable system; the second one is the reduction to "bien décomposable" system; the third is the construction of parametrix by solving the transport equations.

For the hyperbolic systems which admit only characteristic roots with constant multiplicity and whose principal part has the stable structure on the characteristics, the above methods work very well. For instance, we quote H. Yamahara [1] for the first method, R. Berzin-J. Vaillant [1] and D. Gourdin [1] for the second, K. Kajitani [1] and V. M. Petkov [3] for the third.

For the general case, the problem is very much complicated (see W. Matsumoto [1]), we leave the reader to consult the following papers; for $\gamma^{(s)}$-wellposedness, J. Leray-Y. Ohya [1, 2], K. Kajitani [2]; for C^∞-wellposedness, Y. Demay [1], H. Uryu [1], D. Gourdin [2].

Chapter 2. Cauchy Problem (II)

We are concerned with the Cauchy probelm of linear partial differential equations with holomorphic coefficients in complex domains. The Cauchy-Kowalewski theorem states that there exists locally a unique holomorphic solution if the initial surface is not characteristic and the data are holomorphic. This local solution determines the function element. Then the problem is to study the analytic continuation of this function element and its singularities. Contrary to the case of ordinary differential equations, this problem is not easy, and up to now, is far from complete. Strictly speaking, we should distinguish the considerations of partial differential operators with real analytic coefficients in $\boldsymbol{R}^n$ from those in $\boldsymbol{C}^n$ with holomorphic coefficients. In particular, in complex domains, the hyperbolicity loses its meaning. However, the notions of constant multiplicity, and of perfect factorizations, Levi condition etc. stated in Chapter 1 keep their meanings in complex domains.

The systematic study of the Cauchy problem for linear partial differential equations in complex domains began with the work of J. Leray [2]. In it he proposed to describe the singularities of solutions and stated the proposition that the singular support of solutions of the Cauchy problem is contained in the characteristics tangent to the initial surface or issuing from the singular support of the initial data. He described the singular support of solution in the case where the initial surface has characteristic points (see also L. Gårding-T. Kotake-J. Leray [1]). Observe that this article contains fundamental ideas and methods for the treatment of general partial differential equations in complex domains. He published a series of articles on the Cauchy problem and on the integrals on complex analytic variety ([3–6]).

Solutions with singularities (or wave form) have been considered by the method of asymptotic expansions on fairly general settings. The works P. D. Lax [1], D. Ludwig [1], S. Mizohata [2], [3], and Gårding-Kotake-Leray [1], have been done until the first half of the 60th. In [3], S. Mizohata confirmed by direct estimates the convergence of formal asymptotic expressions in the case of holomorphic coefficients, and showed the existence of null solutions and non analytic solutions.

After above pioneering works, Y. Hamada [1] appeared, and this article accelerated later development. Let us explain it. We are concerned with the following local Cauchy problem.

$$
(1) \qquad \begin{cases} P(t, x; D_t, D_x)u(t, x)=0 \\ D_t^j u(0, x)=w_j(x)/(\xi^0\cdot x)^{p_j} \qquad (0\leqq j\leqq m-1), \end{cases}
$$

where we assume (2) in Chapter 1, assuming a_j holomorphic in a neighborhood of the origin. Concerning the initial data, we assume $w_j(x)$ to be holomorphic, and ξ^0 $(\neq 0)\in \boldsymbol{C}^n$, and p_j positive integer. Concerning (3) in Chap. 1, we assume $\lambda_j\neq\lambda_k$ $(j\neq k)$ for $\xi=\xi^0$. Then by defining the phase functions $\phi_\mu(t, x)$ $(1\leqq\mu\leqq m)$,

$$
\begin{cases} \dfrac{\partial}{\partial t}\phi_\mu(t, x)=\lambda_\mu(t, x; \operatorname{grad}_x \phi_\mu), \\ \phi_\mu(0, x)=\xi^0\cdot x, \end{cases}
$$

it is proved that the solution u of (1) is represented uniquely in the form

$$
(2) \qquad u(t, x)=\sum_{\mu=1}^{m}\sum_{-\infty<j<\infty} f_j(\phi_\mu(t, x))a_{j,\mu}(t, x),
$$

where $\{f_j(s)\}_{-\infty<j<\infty}$ are defined by $f_0(s)=\log s$, $f_j(s)=f'_{j+1}(s)$; for $j\geqq 1$, $f_j(s)=(1/j!)\{s^j\log s-s^j(1+1/2+\cdots+1/j)\}$.

By denoting (t, x) by $x=(x_0, x_1, \cdots, x_n)=(x_0, x')$, this expression can be written in the form

$$
(3) \qquad u(x)=\sum_{\mu=1}^{m}\left\{\sum_{k=1}^{\infty} F_{k,\mu}(x)/\phi_\mu(x)^k+G_\mu(x)\log\phi_\mu(x)\right\}+H(x),
$$

where $F_{k,\mu}$, G_μ and H are holomorphic in a common complex neighborhood of the origin. The convergence is uniform on any compact set far from the characteristic surfaces $K=\bigcup_{\mu=1}^{m}\{x; \phi_\mu(x)=0\}$. Of course, the sum over k is finite. In the second paper in [2], he proved that the same result holds if one assumes the Levi condition $P'_{m-1}=0$ on double characteristic sets. Observe that the expression (2) opened the way to the construction of the fundamental solutions for general hyperbolic equations in an easier and natural way.

These results and methods were extended and clarified by C. Wagschal [1], [2], [3], J. C. De Paris [2], Hamada [1], and finally Hamada-Leray-Wagschal [1]. Wagschal formulated an elegant method of classical majorant function in [1] and gave a precise definition of ramified Cauchy problem in [3]. J. C. De Paris [2] showed the validity of the above expression under the assumption of (general) Levi condition. Observe that in this case, the sum over k in (3) is finite. Hamada [1] showed its validity without assuming the Levi condition. Observe that, in this case, the sum over k in general extends to infinity. Finally, Hamada-Leray-Wagschal [1] completed the researches in this direction. S. Ōuchi [1] studied asymptotic behaviors of singularities of some specific solutions. Recently S. Mizohata [7] showed that the converse of the result of De Paris is true; In a word, meromorphic propagation property implies Levi condition.

In the case of operators with characteristics of variable multiplicity, assuming involutive, there are the results by G. Nakamura [1], and by D. Schiltz-J. Vaillant-C. Wagschal [1]. J. Vaillant studies the integrals on complex variety, by developing the idea in J. Leray [6]. Thus the studies of singularities of solutions are closely related with interesting problems in the theory of several complex variables, and they will be discussed from various view-points by classical and modern methods, and so in connection with hyperbolic equations and theory of analytic functions.

Finally we add two related topics.

1) Goursat problem. The importance of this problem will be clear if one looks at the treatise of J. Hadamard [1]. There are many works on it. Y. Hasegawa [1] and M. S. Baouendi-C. Goulaouic [1] are already classical. The notion of Fuchsian type in partial differential equations sprang from the previous paper of Hasegawa. For the extension to C^∞-class, see Y. Hasegawa [2], T. Nishitani [6].

2) We know that the Cauchy-Kowalewski theorem can be applied to Leray-Volevič systems. However, up to now, we don't know clearly to what wider class of systems this theorem can be applied. S. Mizohata [4] treats this problem (see also M. Miyake [1]). Concerning single equations, S. Mizohata proves that the converse of that theorem is true [5]. Namely for any non Kowalewskian equation, the Cauchy-Kowalewski theorem fails to hold.

Chapter 3. Micro-local Analysis

(I) Propagation of singularities.

Entering into the 70th, many works on the investigation of the singularities of solutions for general pseudo-differential operators have been done from the view-point of micro-local analysis, and gained fruitful success.

For operators of principal type with real principal symbols, the C^∞-wave front set WF of solutions is invariant under the translation along the bicharacteristics as far as they do not intersect the wave front set of the right hand side. This basic fact was shown in L. Hörmander [1], [2]. For the analytic wave front set, this was shown in M. Sato-T. Kawai-M. Kashiwara [1], and L. Hörmander [3].

In the case of complex principal symbol, these bicharacteristics can be 2 dimensional manifold, see J. J. Duistermaat-L. Hörmander [1]. In the case of analytic wave front set WF_a, the situation becomes simple for real bicharacteristics, see N. Hanges [2], J. Sjöstrand [5]. J. Chazarain [2] has obtained similar results for operators whose principal symbols are product of powers of real symbols of principal type and whose lower order terms satisfy the Levi condition. For operators not necessarily of principal type, and in analytic category, there is an important result of M. Kashiwara-T. Kawai [1] for propagation of analytic wave front set for micro-hyperbolic pseudo-differential operators. For other results, see J. Sjöstrand [5].

Concerning the case of operators with characteristic of variable multiplicity, we should mention the works of V. Ya. Ivrii. In [7], for a class of effectively hyperbolic operators, and in [8] for a class of non effectively hyperbolic operators, C^∞-wave front set was investigated. [6] is devoted to the study of wave front set of solutions of symmetric hyperbolic systems. For general effectively hyperbolic operators, the wave front set of solutions was studied in R. B. Melrose [10] and T. Nishitani [4]. These works rely on the local estimates of wave front set for hyperbolic pseudo-differential operators with double characteristics, and these estimates are based on energy method. See also R. Lascar [1] and R. B. Melrose [1]. Global version of propagation of wave front set based on local eatimates was discussed in S. Wakabayashi [7, 8], and R. B. Melrose [12].

On the other hand, from view-point of parametrices, there are many mutually related works. S. Alinhac [1–5], A. Yoshikawa [1, 2] constructed parametricies for a class of effectively hyperbolic operators. H. Kumano-go considered wave front sets of solutions for first order hyperbolic systems whose principal parts are diagonal matrices with smooth symbols, by investigating structures of multi-products of Fourier integral operators. See, H. Kumano-go [1], Chapter 10, and its references.

N. Hanges [1] studied branching of wave front set for operators with non involutive characteristics. G. A. Uhlmann [1, 2] has constructed parametrices for operators with involutive characteristics satisfying the Levi condition. J.-M. Bony [1] introduced para-differential operators to study the wave front sets for non-linear differential operators.

In R. Lascar [1], and B. Lascar-R. Lascar [2] and R. B. Melrose-

G. A. Uhlmann [1], they studied some version of the theory of Fourier integral operators adapted to operators with characteristics of variable multiplicity.

On Gevrey wave front set WF_s, there are many works. If we assume the coefficients of operator to be analytic, there may be no specific difficulty. However, in order to discuss WF_s under the assumption that the coefficients belong to $\gamma^{(s)}$, careful consideration will be required. Concerning it, see S. Mizohata [8], K. Taniguchi [1].

There are many other results, see references.

(II) Singularities of solutions for boundary-value problems.

In geometric optics, an incident ray in a direction ξ ($\xi \in \boldsymbol{R}^d$, $|\xi|=1$) hittting transversally a mirror Γ at a point a is reflected in a direction

$$\eta=\xi-2(n\cdot\xi)n, \tag{1}$$

where n is the unit outer normal of Γ at a. This fact corresponds to the propagation of singularities of the boundary-value problem of the wave equation

$$\begin{cases} \dfrac{\partial^2}{\partial t^2}u-\displaystyle\sum_{i=1}^{d}\dfrac{\partial^2}{\partial x_i^2}u=0 & \text{in } \Omega\times\boldsymbol{R}, \\ u=0 & \text{on } \Gamma\times\boldsymbol{R}, \end{cases} \tag{2}$$

where Ω is a domain in $\boldsymbol{R}^d$ with smooth boundary Γ. Suppose that

$$WF(u)\cap\{t=t_0\}=(x_0, t_0, \xi, 1)\in T^*\omega, \tag{3}$$

where $\omega=\Omega\times\boldsymbol{R}$, $|\xi|=1$. We assume for simplicity

$$\begin{cases} x_0+(t-t_0)\xi\in\Omega, & \text{for } t<t_1 \\ x_0+(t_1-t_0)\xi=a, & \\ a+(t-t_1)\eta\in\Omega & \text{for } t>t_1, \end{cases} \tag{4}$$

where η is the one defined by (1). Set

$$\gamma(t)=\begin{cases} x_0+(t-t_0)\xi & \text{for } t\leq t_1 \\ a+(t-t_1)\eta & \text{for } t>t_1. \end{cases}$$

It is easy to show that

$$WF(u)\cap T^*\omega=L^-\cup L^+, \tag{5}$$

where

$$L^- = \{(\gamma(t), \xi, 1);\ t<t_1\}, \qquad L^+ = \{(\gamma(t), \eta, 1);\ t>t_1\}.$$

This fact means that the wave front sets of solution (2) propagate according to the law of geometric optics. In the case where the curve $\gamma(t)$ is tangent to Γ at the point a, that is $(\xi, n)=0$, the propagation of singularities is related to the phenomenon of diffraction or to the propagation on the boundary. In this case the propagation of singularities was not enough studied until fairly recent time. In this case, assuming

$$(6) \qquad \mathcal{O} = \boldsymbol{R}^d \setminus \Omega \quad \text{is strictly convex,}$$

D. Ludwig [2] constructed an asymptotic solution by using the Airy function, which aimed at the study of diffraction (see also C. Morawetz-D. Ludwig [1]). Later F. G. Friedlander [1], F. G. Friedlander-R. B. Melrose [1], M. E. Taylor [2] studied the propagation of singularities of solutions for a class of boundary-value problems which include the case (2), assuming (6). These results show that even in the case $(\xi, n)=0$, (5) holds for the wave front set of C^∞-category.

In order to study the problem for general domains, Melrose [7] developed the theory of pseudo-differential operators, Fourier integral operators, canonical transformations and wave front sets WF_b which are attached to boundary-value problems. By using these results, Melrose [3], R. B. Melrose-J. Sjöstrand [1] studied WF for second order operators with L^2-wellposed boundary conditions. They showed that, roughly speaking, WF propagates only to points connected by generalized bicharacteristic curves. This implies that the singularities in C^∞-class do not propagate into the shadow region.

About the propagation of singularities in the analytic class, for example for the analytic wave front sets WF_a, a similar result holds. Namely, suppose that Γ is analytic and consider (2). If

$$(3') \qquad WF_a(u) \cap \{t=t_0\} = (x_0, t_0, \xi, 1) \in T^*\omega,$$

and (4) is satisfied, then

$$(5') \qquad WF_a(u) \cap T^*\omega = L^- \cup L^+.$$

But in the case $(\xi, n)=0$, the phenomenon is very different from that of C^∞-case. For example, in the diffractive case, WF_a propagates into the shadow region (Friedlander-Melrose [1], J. Rauch [1]). J. Sjöstrand [1–3] studied the general theory of the propagation of WF_a for second order operators. On the other hand, G. Lebeau [2] studies the propagation of Gevrey singularities WF_s in the shadow region. Related studies on analytic singularities are found in K. Kataoka [1–2], H. Komatsu-T. Kawai [1] and P. Schapira [1, 2].

Concerning the reflection of singularities for higher order equations, L. Nirenberg [1] showed a fundamental relationship between reflected wave front sets and the wave front sets (in the boundary) of the traces of solutions. Further studies about the problem were made by M. E. Taylor [1], A. Majda-S. Osher [1], and J. Chazarain [4].

As the reader might observe, the researches explained up to now, rely more or less on the basic researches on the mixed problems (initial-boundary value problems) for hyperbolic equations. Concerning this research, we cannot afford to describe its development. For this matter, see for instance the monograph by R. Sakamoto [3], and the references there.

For the mixed problems which are not necessarily L^2-wellposed, even in the case where a singularity hits a boundary transversally, not only reflections but also the propagation in the boundary may take place, and we have not yet any general theory. For operators with constant coefficients in a half space, see G. F. D. Duff [1], S. Wakabayashi [1–4].

Chapter 4. Applications

The methods and the notions developed in hyperbolic analysis are giving impact on several neighboring fields, especially on the spectral theory. To illustrate it, we review a part of recent developments of the scattering theory.

Decay of local energy. Concerning the behavior as t tends to infinity of solutions of the wave equation in the exterior domain Ω of R^d ($d\geqq 3$, odd),

$$(1)\qquad \begin{cases} \left(\dfrac{\partial^2}{\partial t^2}-\Delta\right)u(t,x)=0 \quad \text{in } \boldsymbol{R}\times\Omega, \quad u(t,x)=0 \quad \text{on } \boldsymbol{R}\times\Gamma, \\ u(0,x)=f_1(x), \qquad u(0,x)=f_2(x), \end{cases}$$

the uniform decay of the local energy is one of the fundamental problems. Namely, we are concerned with the characterization of the shape of the obstacle $\mathcal{O}$ $(=\boldsymbol{R}^d\backslash\Omega)$ for which there exists a function $p(t)$ such that

$$p(t)\longrightarrow 0 \qquad \text{as } t\longrightarrow\infty,$$
$$E(u,t;R)\leqq p(t)E(u,0;R) \qquad \text{for all } f_1,f_2\in C_0^\infty(\Omega_R),$$

where $\Omega_R=\Omega\cap\{x;\,|x|\leqq R\}$ and

$$E(u,t;R)=\frac{1}{2}\int_{\Omega_R}|\nabla_x u(t,x)|^2+\left|\frac{\partial}{\partial t}u(t,x)\right|^2dx.$$

Remark that P. D. Lax-C. Morawetz-R. S. Phillips [1] showed the following: If such a $p(t)$ exists, we can take $p(t)$ in the form $\beta e^{-\alpha t}$ with some positive constants α and β. For this problem, Lax-Phillips [1] gave the following conjecture: A necessary and sufficient condition for the existence of $p(t)$ is that the obstacle $\mathcal{O}$ is non-trapping. Here non-trapping means that for any $R>0$, every ray starting from a point in Ω_R goes out from Ω_R within a fixed time T_R.

The necessity of this condition is proved by J. Ralston [1]. The sufficiency was proved by C. Morawetz-J. Ralston-W. A. Strauss [1] under some additional assumption. Later, R. B. Melrose [4] proved its sufficiency by using the theory of propagation of singularities. One might say that the above articles give a suggestion to a closed relation between the local decay of energy and the propagation of singularities. Namely, one can argue in the following way. Since it is shown that the wave front sets of solution propagate along the generalized bicharacteristic strips, when $\mathcal{O}$ is non-trapping, all solution with initial data in Ω_R has no singularities in Ω_R for $t \geqq T_R$. This fact implies that Lax-Phillips' semi-group $Z(t)$ (see Lax-Phillips [1], p. 151) becomes a compact operator for large t. Then by Theorems 5.2 and 5.3 in it, the exponential decay of local energy is derived.

Poles of the scattering matrix. Concerning the relationship between the geometry of $\mathcal{O}$ and the poles of $\mathcal{S}(z)$, Lax-Phillips [1] gave the following conjecture: For non-trapping obstacle, $\mathcal{S}(z)$ is free from poles in $\{z;\ \mathrm{Im}\, z \leqq \alpha\}$ with some $\alpha>0$, and for a trapping obstacle, $\mathcal{S}(z)$ has a sequence of poles with imaginary parts converging to zero (see, p. 158). Moreover it showed that the existence of a $p(t)$ mentioned above implies that $\mathcal{S}(z)$ is free from poles in the region of the form $\{z;\ \mathrm{Im}\, z \leqq a \log(|z|+1)+b\}$, with some positive constnats a and b.

Now, J. Bardos-J.-C. Guillot-J. Ralston [1] proved that when $\mathcal{O}$ consists of two strictly convex obstacles, there exists an infinite number of poles in $\{z;\ \mathrm{Im}\, z \leqq \varepsilon \log(|z|+1)\}$ for any $\varepsilon>0$. V. M. Petkov extended this result to the case where there are many strictly convex obstacles ([5, 6]). These results exhibit a different distribution of poles of $\mathcal{S}(z)$ from that of non-trapping obstacles. However, for any trapping obstacle, whether the set of poles of $\mathcal{S}(z)$ has the above property or not is still open question.

On the other hand, in the case of two strictly convex obstacles, it is known that there exists $c_0>0$ such that $\{z;\ |\mathrm{Im}\, z-c_0|<\varepsilon\}$ contains infinitely many poles for any $\varepsilon>0$, and there are a finite number of poles in $\{z;\ \mathrm{Im}\, z \leqq c_0-\varepsilon\}$ (see M. Ikawa [1]). This shows that the above conjecture of Lax-Phillips fails to hold in general. To know what kind of trapping obstacles satisfy the above conjecture is an interesting problem. Concerning this, M. Ikawa [2] shows such an example.

Concerning the poles for non-trapping obstacles, assuming that $\mathcal{O}$ is non-trapping and Γ is analytic, G. Lebeau obtained precise informations on the distribution of poles, from the Gevrey singularities in the shadow region.

Inverse problem. Although it is well-known that the scattering operator or the scattering matrix determines the scatterer, concrete relationships between geometries of obstacles and the scattering operators are not fully considered. However, the problem to determine the shape of obstacle from the scattering operator is deeply related with the propagation of singularities. A. Majda [1] found a good representation of the scattering kernel, and determined the convex hull of the obstacle from the singularities of the scattering kernel by combining the representation formula with the propagation of singularities. Concerning this problem, for instance there are works of A. Majda-M. E. Taylor [1], H. Soga [1].

Scattering phase. Since the scattering matrix $\mathcal{S}(z)$ is unitary for $z=\lambda$ real,

$$s(\lambda)=\frac{1}{2\pi i}\log\det\mathcal{S}(\lambda)$$

is real-valued function on R. We call it the scattering phase. J. Ralston [2], A. Majda-J. Ralston [1], and A. Jenssen-T. Kato [1] showed that an asymptotic formula

$$s(\lambda)=\omega_d\operatorname{vol}(\mathcal{O})\lambda^d+o(\lambda^d)\qquad\text{for }\lambda\longrightarrow\infty$$

holds for obstacles in some class. We see that the scattering phase is an analogue to $N(\lambda)$ (counting function of eigenvalues) for the interior domain.

V. M. Petkov-G. St. Popov [2] proved the following more precise formula

$$s(\lambda)=\omega_d\operatorname{vol}(\mathcal{O})\lambda^d+\omega_{d-1}\operatorname{vol}(\Gamma)\lambda^{d-1}+O(\lambda^{d-2}),$$

for non-trapping obstacles. In the proof the following trace formula is the starting point:

$$\begin{aligned}2\operatorname{tr}\int_{-\infty}^{\infty}&\rho(t)(\cos t\sqrt{-\Delta}\oplus 0-\cos t\sqrt{-\Delta_0})dt\\ &=\int_{-\infty}^{\infty}\frac{d}{d\lambda}\hat{\rho}(\lambda)s(\lambda)d\lambda\qquad\text{for all }\rho\in C_0^\infty(R).\end{aligned}$$

On the other hand a fairly delicate argument shows that

$$\mathscr{F}\left[t\int_{R^d}(E(t, x, x)-E_0(t, x, x))dx\right](\lambda)=s(\lambda),$$

where $E(t, x, y)(E_0(t, x, y))$ is the Green function for the initial-boundary value problem (Cauchy problem) (1) with initial data $f_1=\delta(x-y)$, $f_2=0$. This indicates that the propagation of singularities plays an essential role. The non-trapping assumption on $\mathcal{O}$ implies that

$$\int_{R^d}(E(t, x, x)-E_0(t, x, x))dx$$

is smooth for $|t|>0$. Thus it suffices to consider the singularities at $t=0$. In the case of trapping obstacles we have no general result. M. Ikawa [3] gives an example which shows a different behavior of $s(\lambda)$ caused by the existence of periodic rays.

Comments. Finally we add some comments concerning the applications of hyperbolic analysis to other problems in the spectral theory. We mention among them, 1) asymptotic distributions of eigen-values, 2) Poisson's summation formula for compact Riemannian manifolds, and further corresponding formula for exterior domains. They appeared already at the fairly early stage, and became the underlying ideas of scattering theory. For these, see references.

Organizing Committee

S. Mizohata
Y. Ohya
M. Ikawa

Bibliography

S. Alinhac: [1] Parametrix et propagation des singularités pour un problème de Cauchy à multiplicité variable. Astérisque, **34–35** (1976), 3–26.

—— [2] Parametrix pour un système hyperbolique à multiplicité variable, Comm. Partial Differential Equations, **2** (1977), 251–296.

—— [3] Solution explicite du problème de Cauchy pour des opérateurs effectivement hyperboliques, Duke Math. J., **45** (1978), 225–258.

—— [4] Branching of singularities for a class of hyperbolic operators, Indiana Univ. Math. J., **27** (1978), 1027–1037.

K. Amano-G. Nakamura: [1] Branching of singularities for degenerate hyperbolic operators, Publ. RIMS. Kyoto Univ., **20** (1984), 225–275.

M. S. Baouendi-C. Goulaouic: [1] Cauchy problems with characteristic initial hypersurfaces, Comm. Pure Appl. Math., **26** (1973), 455–475.

J. Bardos-J.-C. Guillot-J. Ralston: [1] La relation de Poisson pour l'équation des ondes dans un ouvert non borné, Application à la théorie de la diffusion, Comm. Partial Differential Equations, **7** (1982), 905–958.

R. Berzin-J. Vaillant: [1] Systèmes hyperboliques à caractéristiques multiples, J. Math. Pures Appl., **58** (1979), 165–216.

J.-M. Bony: [1] Calcul symbolique et propagation des singularités pour les équations aux dérivées partielles non linéaires, Ann. Sci. Ecole Norm. Sup., 4ᵉ sér. **14** (1981), 209–246.

J.-M. Bony-P. Schapira: [1] Solutions hyperfonctions du problème de Cauchy, Lecture Notes in Math. No. 287, Springer, 1973, 82–98.

L. Boutet de Monvel-P. Krée: [1] Pseudo-differential operators and Gevrey classes, Ann. Inst. Fourier, **17** (1967), 295–323.

M. D. Bronštein: [1] Smoothness of roots of polynomials depending on parameters, Sibirsk, Mat. Ž., **20** (1979, 493–501 (Siberian Math. J., **20** (1979), 347–352).

—— [2] The Cauchy problem for hyperbolic operators with characteristics of variable multiplicity, Trudy Moskov, Mat. Obšč., **41** (1980), 87–103 (Trans. Moscow Math. Soc., **1** (1982), 87–103).

J. Chazarain: [1] Opérateurs hyperboliques à caractéristiques de multiplicité constante, Ann. Inst. Fourier, **24** (1974), 173–202.

—— [2] Propagation des singularité pour une classe d'opérateurs à caractéristiques multiples et resolubilité locale, Ann. Inst. Fourier, **24** (1974), 203–223.

—— [3] Formule de Poisson pour les variétés riemanniennes, Invent. Math., **24** (1974), 65–82.

—— [4] Reflection of C^∞ singularities for a class of operators with multiple characteristics, Publ. RIMS. Kyoto Univ., **12** suppl. (1977), 39–52.

M. Y. Chi: [1] On the Cauchy problem for a class of hyperbolic equations with data given on the degenerate parabolic line, Acta Math. Sinica, **8** (1958), 521–529.

F. Colombini-E. De Giorgi-S. Spagnolo: [1] Sur les équations hyperboliques avec des coefficients qui ne dépendent que du temps, Ann. Scuola Norm. Sup. Pisa, **4** (1979), 511–559.

F. Colombini-E. Jannelli-S. Spagnolo: [1] Well posedness in the Gevrey class of the Cauchy problem for a non strictly hyperbolic equation with coefficients depending on time, Ann. Scuola Norm Sup. Pisa, **10** (1983), 291–312.

F. Colombini-S. Spagnolo: [1] An example of a weakly hyperbolic Cauchy problem not well posed in C^∞, Acta Math., **148** (1982), 243–253.

Y. Demay: [1] Parametrix pour des systèmes hyperboliques du premier ordre à multiplicité constante, J. Math. Pures Appl., **56** (1977), 393–422.

J-C. De Paris: [1] Problème de Cauchy oscillatoire pour un opérateur différentiel à caractéristiques multiples; Lien avec l'hyperbolicité, J. Math. Pures Appl., **51** (1972), 231–256.

—— [2] Problème de Cauchy analytique à données singulières pour un opérateur différentiel bien décomposable, J. Math. Pures Appl., **51** (1972), 465–488.

G. F. D. Duff: [1] Hyperbolic differential equations and waves, Proc. NATO ASI (ed. H.G. Garnir), Reidel Publ., Dordrecht, 1976, 27–155.

J. J. Duistermaat-V. Guillemin: [1] The spectrum of positive elliptic operators and periodic geodesics, Invent. Math., **29** (1975), 39–79.

J. J. Duistermaat L.-Hörmander: [1] Fourier integral operators II, Acta Math., **128** (1972), 183–269.

G. Eskin: [1] A parametrix for mixed problems for strictly hyperbolic equations of arbitray order, Comm. Partial Differential Equations, **1** (1976), 521–560.

—— [2] Parametrix and propagations of singularities for the interior mixed problem, J. Analyse Math., **32** (1977), 17–62.

H. Flashka-G. Strang: [1] The correctness of the Cauchy problem, Adv. in Math., **6** (1971), 347–379.

F. G. Friedlander: [1] The wavefront set of a simple initial-boundary value problem with glancing rays, Math. Proc. Cambridge Philos. Soc., **79** (1976), 145–159.

—— [2] The wave equation on a curved space-time, Cambridge Univ. Press, 1975.

F. G. Friedlander-R. B. Melrose: [1] The wave front set of the solution of a simple initial-boundary value problem with glancing rays, II, Math. Proc. Cambridge Philos. Soc., **81** (1977), 97–120.

K. O. Friedrichs: [1] Symmetric hyperbolic system of linear differential equations, Comm. Pure Appl. Math., **7** (1954), 345–392.

K. O. Friedrichs-H. Lewy: [1] Über die Eindeutigkeit und das Abhängigkeitsgebiet der Lösungen beim Anfangswertproblem linearer hyperbolisher Differentialgleichungen, Math. Ann., **98** (1927), 192–204.

L. Gårding: [1] Cauchy's problem for hyperbolic equations, Lecture Notes, Univ. Chicago, 1957.

L. Gårding-T. Kotake-J. Leray: [1] Uniformisation et développement asymptotique de la solution du problème de Cauchy linéaire à données holomorphes; analogie avec la théorie des ondes asymptotiques et approchées, Bull. Soc. Math. France, **92** (1964), 263–361.

M. Gevrey: [1] Oeuvres de Maurice Gevrey, Editions du C.N.R.S., 1970.

D. Gourdin: [1] Les opérateurs faiblement hyperboliques matricielles à caractéristiques de multiplicité constante, bien décomposables et le problème de Cauchy non caractéristique associé, J. Math. Kyoto Univ., **17** (1977), 539–566.

—— [2] Problème de Cauchy non caractéristique pour les systèmes hyperboliques à caractéristiques de multiplicité variable, domaine de dépendance, Comm. Partial Differential Equations, **4** (1979), 447–507.

V. Guillemin-R. B. Melrose: [1] The Poisson summation formula for manifolds with boundary, Adv. in Math., **32** (1979), 204–232.

—— [2] An inverse spectral result for elliptical regions in R^2, Adv. in Math., **32** (1979), 128–148.

J. Hadamard: [1] Le problème de Cauchy et les équations aux dérivées partielles linéaires hyperboliques, Hermann, Paris, 1932.

Y. Hamada: [1] The singularities of the solutions of the Cauchy problem, Publ. RIMS. Kyoto Univ., **5** (1969), 21–40.

—— [2] On the propagation of singularities of the solution of the Cauchy problem, Ibid., **6** (1970), 357–384.

—— [3] Problème analytique de Cauchy à caractéristiques multiples dont les données de Cauchy ont des singularités polaires, C.R. Acad. Sci. Paris **276**,

Ser. A (1973), 1681–1684.
Y. Hamada-J. Leray-C. Wagschal: [1] Systèmes d'équations aux dérivées partielles à caractéristiques multiples: problème de Cauchy ramifié; hyperbolicité partielle, J. Math. Pures Appl., **55** (1976), 297–352.
N. Hanges: [1] Parametrices and propagation of singularities for operators with non-involutive characteristics, Indiana Univ. Math. J., **28** (1979), 87–97.
—— [2] Propagation of analyticity along real bicharacteristics, Duke Math. J., **48** (1981), 269–277.
Y. Hasegawa: [1] On the initial-value problems with data on a double characteristic, J. Math. Kyoto Univ., **11** (1971), 357–372.
—— [2] On the C^∞-Goursat problem for the equations with constant coefficients, J. Math. Kyoto Univ., **19** (1979), 135–169.
J. W. Helton-J. Ralston: [1] The first variation of the scattering matrix, J. Differential Equations, **21** (1976), 378–399.—An addendum, Ibid., **28** (1978), 155–162.
L. Hörmander: [1] On the existence and regularity of solutions of linear pseudo-differential equations, Enseign. Math., **17** (1971), 99–163.
—— [2] Fourier integral operators I, Acta Math., **127** (1971), 79–183.
—— [3] Uniqueness theorems and wave front sets for solutions of linear differential equations with analytic coefficients, Comm. Pure Appl. Math., **24** (1971), 671–704.
—— [4] The Cauchy problem for differential equations with double characteristics, J. Analyse Math., **32** (1977), 118–196.
K. Igari: [1] An admissible data class of the Cauchy problem for non-strictly hyperbolic operators, J. Math. Kyoto Univ., **21** (1981), 351–373.
M. Ikawa: [1] On the poles of the scattering matrix for two strictly convex obstacles, J. Math. Kyoto Univ., **23** (1983), 127–194.—An addendum, Ibid., **23** (1983), 795–802.
—— [2] Trapping obstacles with a sequence of poles of the scattering matrix converging to the real axis, to appear in Osaka J. Math. .
—— [3] On the scattering matrix for two convex obstacles, Proc. Taniguchi Intern. Sympos. on Hyperbolic Equations and Related Topics, 1984 (these Proceedings).
M. Imai-T. Shirota: [1] On a parametrix for the hyperbolic mixed problem with diffractive lateral boundary, Hokkaido Math. J. **7** (1978), 339–352.
V. Ya. Ivrii: [1] Correctness of the Cauchy problem in Gevrey classes for non strictly hyperbolic operators, Mat. Sb. **96** (1975), 390–413 (Math. USSR-Sb., **25** (1975), 365–387).
—— [2] Sufficient conditions for regular and completely regular hyperbolicity, Trudy Moskov. Mat. Obšč., **33** (1976), Trans. Moscow Math. Soc., **1** (1978), 1–65).
—— [3] Cauchy problem conditions for hyperbolic operators with characteristics of variable multiplicity for Gevrey classes, Sibirsk. Mat. Ž., **17** (1976), 1256–1270. (Siberian Math., J., **17** (1976), 921–931).
—— [4] The well posedness of the Cauchy problem for non strictly hyperbolic operators, Trudy Moskov. Mat. Obšč., **34** (1977), 151–170. (Trans. Moscow Math. Soc., **2** (1978), 149–168).
—— [5] Wave fronts of solutions of boundary-value problems for a symmetric hyperbolic systems, I, II, III, Sibirsk. Mat Ž., **20** (1979), 741–751, 1022–1033, **21** (1980), 74–81. (Siberian Math. J., **20** (1979, 516–523, 722–734, **21** (1980), 54–60).
—— [6] Wave fronts of solutions of symmetric pseudodifferential systems, Sibirsk. Mat. Ž., **20** (1979), 557–578 (Siberian Math. J., **20** (1980), 390–405).
—— [7] Wave fronts of solutions of certain pseudodifferential equations, Trudy

Moskov. Mat. Obšč., **39** (1979), 49–82 (Trans. Moscow Math. Soc., **1** (1981), 49–86).

V. Ya. Ivrii: [8] Wave fronts of solutions of certain hyperbolic pseudodifferential equations, Trudy Moskov. Mat. Obšč., **39** (1979), 83–112 (Trans. Moscow Math. Soc., **1** (1981), 87–119).

—— [9] Wave fronts for solutions of boundary-value problems for a class of symmetric hyperbolic systems, Sibirsk. Mat. Ž., 21 (1980), 62–71 (Siberian Math. J., **21** (1980), 527–534).

—— [10] Second term of the spectral asymptotic expansion of the Laplace-Beltrami operator on manifold with boundary, Funkcional. Anal. i Priložen., **14** (1980), 25–34 (Functional Anal. Appl., **14** (1980), 98–106).

— [11] Precise Spectral Asymptotics for Elliptic Operators, Lecture Notes in Math., No. 1100, Springer-Verlag, 1984.

V. Ya. Ivrii-V. M. Petkov: [1] Necessary conditions for the Cauchy problem for non strictly hyperbolic equations to be well posed, Uspehi Mat. Nauk, **29** (1974), 3–70 (Russian Math. Surveys, **29** (1974), 1–70).

C. Iwasaki-Y. Morimoto: [1] Propagation of singularities of solutions for a hyperbolic system with nilpotent characteristics, I, II, Comm. Partial Differential Equations, **7** (1982), 743–793, **9** (1984), 1407–1436.

N. Iwasaki: [1] The Cauchy problem for effectively hyperbolic equations (a special case), J. Math. Kyoto Univ., **23** (1983), 503–562.; (a standard type), Publ. RIMS. Kyoto Univ., **20** (1984), 551–592.

—— [2] The Cauchy problem for effectively hyperbolic equations (general cases), RIMS. preprint 468, RIMS. Kyoto Univ., 1984, to appear in J. Math. Kyoto Univ..

—— [3] The Cauchy problem for effectively hyperbolic equations (Remarks), Proc. Taniguchi Intern. Sympos. on Hyperbolic Equations and Related Topics 1984, (these Proceedings).

A. Jensen-T. Kato: [1] Asymptotic behavior of the scattering phase for exterior domains, Comm. Partial Differential Equations, **3** (1978), 1165–1195.

K. Kajitani: [1] Cauchy problem for non strictly hyperbolic systems, Publ. RIMS. Kyoto Univ., **15** (1979), 519–550.

—— [2] Cauchy problem for non strictly hyperbolic systems in Gevrey classes, J. Math. Kyoto Univ., **23** (1983), 599–616.

M. Kashiwara-T. Kawai: [1] Micro-hyperbolic pseudo-differential operators I, J. Math. Soc. Japan, **27** (1975), 359–404.

K. Kataoka: [1] Micro-local theory of boundary value problems I and II (Theorem on regularity up to the boundary for reflective and diffractive operators), J. Fac. Sci. Univ. Tokyo, Sect. IA **27** (1980), 355–399, **28** (1981), 31–56.

—— [2] Microlocal analysis of boundary value problems with application to diffraction, Proc. NATO ASI (ed. H.G. Garnir), Reidel Publ., Dordrecht, 1976, 121–131.

K. Kitagawa-T. Sadamatsu: [1] A necessary condition of Cauchy-Kowalewski's theorem, Publ. RIMS. Kyoto Univ., **11** (1976), 523–534.

H. Komatsu: [1] Irregularity of characteristic elements and hyperbolicity, Publ. RIMS. Kyoto Univ., **12** (1977), 233–245.

—— [2] Linear hyperbolic equations with Gevrey coefficients, J. Math. Pures Appl., **59** (1980), 145–185.

H. Komatsu-T. Kawai: [1] Boundary values of hyperfunction solutions of linear partial differential equations, Publ. RIMS. Kyoto Univ., **7** (1971), 95–104.

K. Kubota: [1] A microlocal parametrix for an exterior mixed problem for symmetric hyperbolic systems, Hokkaido Math. J., **10** (1981), 264–298.

H. Kumano-go: [1] Pseudo-differential operators, MIT Press, Cambridge, 1981.

H. Kumano-go-K. Taniguchi: [1] Fourier integral operators of multi-phase and the fundamental solution for a hyperbolic system, Funkcial. Ekvac., **22** (1979),

161–196.
H. Kumano-go-K. Taniguchi-Y. Tozaki: [1] Multi-products of phase functions for Fourier integral operators with an application, Comm. Partial Differential Equations, **3** (1978), 349–380.
B. Lascar: [1] Propagation des singularités pour des équations hyperboliques à caractéristique de multiplicité au plus double et singularités Masloviennes, Amer. J. Math., **104** (1980), 227–285.
B. Lascar-R. Lascar: [1] Propagation des singularités pour des équations hyperboliques à caractéristiques de multiplicité au plus double et singularités Masloviennes II, J. Analyse Math., **41** (1982), 1–38.
B. Lascar-J. Sjöstrand: [1] Equations de Schrödinger et propagation des singularités pour des opérateurs pseudo-différentiels à caractéristiques réelles de multiplicité variable, Astérisque, **95** (1982), 167–207.
R. Lascar: [1] Propagation des singularités pour une classe d'opérateurs pseudo-différentiels à caractéristiques de multiplicité variables, C.R. Acad. Sci., Paris, **283** (1976), 341–343.
A. Lax: [1] On Cauchy's problem for partial differential equations with multiple characteristics, Comm. Pure Appl. Math., **9** (1956), 135–169.
P. D. Lax: [1] Asymptotic solutions of oscillatory initial value problems, Duke Math. J., **24** (1957), 627–646.
P. D. Lax-C. S. Morawetz-R. S. Phillips: [1] Exponential decay of solutions of the wave equation in the exterior of a star-shaped obstacle, Comm. Pure Appl. Math., **16** (1963), 447–486.
P. D. Lax-R.S. Phillips: [1] Scattering Theory, Academic press, New York, 1967.
—— [2] A logarithmic bound on the location of the poles of the scattering matrix, Arch. Rational Mech. Anal., **40** (1971), 268–280.
G. Lebeau: [1] Non-holonomie dans un problème de diffraction, Séminaire Goulaouic-Schwartz, 1979–1980, n°17.
—— [2] Régularité Gevrey 3 pour la diffraction, Comm. Partial differential Equations, **9** (1984), 1437–1494.
J. Leray: [1] Hyperbolic differential equations, Inst. Adv. Study, Princeton, 1953. (Notes miméographiées).
—— [2] Uniformisation de la solution du problème linéaire analytique de Cauchy de la variété qui porte les données de Cauchy, Bull. Soc. Math. France, **85** (1957), 389–429.
—— [3] La solution unitaire d'un opérateur différentiel linéaire, Ibid., **86** (1958), 75–96.
—— [4] Le calcul différentiel et intégral sur une variété analytique complexe, Ibid., **87** (1957), 81–180.
—— [5] Un prolongement de la transformation de Laplace qui transforme la solution unitaire d'un opérateur hyperbolique en sa solution élémentaire, Ibid., **90** (1962), 39–156.
—— [6] Un complément au théorème de N. Nilsson sur les intégrales de formes différentielles à support singulier algébrique, Bull. Soc. Math. France, **95** (1967), 313–374.
J. Leray-Y. Ohya: [1] Systèmes linéaires, hyperboliques non stricts, Colloque de Liège, 1964, C.B.R.M., 105–144.
—— [2] Equations et systèmes non-linéaires, hyperboliques non-stricts, Math. Ann., **170** (1967), 167–205.
E. E. Levi: [1] Caracteristics multiple e problema di Cauchy, Ann. Mat. Pura Appl., **16** (1909), 109–127.
D. Ludwig: [1] Exact and asymptotic solutions of the Cauchy problem, Comm. Pure Appl. Math., **13** (1960), 473–508.
—— [2] Uniform asymptotic expansions at a caustic, Comm. Pure Appl. Math., **19** (1966), 215–250.
—— [3] Uniform asymptotic expansions of the field scattered by a convex object

at high frequencies, Comm. Pure Appl. Math., **20** (1967), 103–138.
A. Majda: [1] A representation formula for the scattering operator and the inverse problem for arbitrary bodies, Comm. Pure Appl. Math., **30** (1977), 165–194.
A. Majda-S. Osher: [1] Reflection of singularities at the boundary, Comm. Pure Appl. Math., **28** (1975), 479–499.
A. Majda-J. Ralston: [1] An analogue of Weyl's theorem for unbounded domains I, II, III, Duke Math. J., **45** (1978), 183–196, **45** (1978), 513–536, **46** (1979), 725–731.
A. Majda-M. E. Taylor: [1] The asymptotic behavior of the diffractive peak in classical scattering, Comm. Pure Appl. Math., **30** (1977), 639–669.
T. Mandai: [1] A necessary and sufficient condition for the well-posedness of some weakly hyperbolic Cauchy problem, Comm. Partial Differential Equations, **8** (1983), 735–771.
W. Matsumoto: [1] On the conditions for the hyperbolicity of systems with double characteristic routs, I & II, J. Math. Kyoto Univ., **21** (1981), 47–84 & 251–271.
S. Matsuura: [1] On non strict hyperbolicity, Proc. Funct. Anal. Related Topics, 1969, Univ. Tokyo Press, 1970, 171–176.
A. Melin: [1] Lower bounds for pseudo-differential operators, Ark. Mat., **9** (1971), 117–140.
R. B. Melrose: [1] Normal self-intersections of the characteristic variety, Bull. Amer. Math. Soc., **81** (1975), 939–940.
—— [2] Microlocal parametrices for diffractive boundary value problems, Duke Math. J., **42** (1975), 605–635.
—— [3] Differential boundary value problems of principal type, Seminar on Singularities of Solutions, Princeton Univ. Press, 1978, 81–112.
—— [4] Singularities and energy decay in acoustical scattering, Duke Math. J., **46** (1979), 43–59.
—— [5] Forward scattering by a convex obstacle, Comm. Pure Appl. Math., **33** (1980), 461–500.
—— [6] Hypoelliptic operators with characteristic variety of codimension two and the wave equation, Séminaire Goulaouic-Schwartz, 1980.
—— [7] Transformation of boundary value problems, Acta Math., **147** (1981), 149–236.
—— [8] Scattering theory and the trace of the wave group, J. Funct. Anal., **45** (1982), 29–40.
—— [9] Polynomial bound on the number of scattering poles, J. Funct. Anal., **53** (1983), 287–303.
—— [10] The Cauchy problem for effectively hyperbolic operators, Hokkaido Math. J., **12** (1983), 371–391.
—— [11] Polynomial bound on the distribution of poles in scattering by an obstacle, Journées Equations aux Dérivées Partielles, St. Jean de Monts, 1984.
—— [12] The trace of the wave group, Contemporary Mathematics, **27**, 1984, 127–167.
R. B. Melrose-J. Sjöstrand: [1] Singularities of boundary value problems, I, II, Comm. Pure Appl. Math., **37** (1978), 593–617, **35** (1982), 129–168.
R. B. Melrose-G. A. Uhlmann: [1] Lagrangian intersection and the Cauchy problem, Comm. Pure Appl. Math., **32** (1979), 483–519.
—— [2] Microlocal structure of involutive conical refraction, Duke Math. J., **46** (1979), 571–582.
A. Menikoff: [1] The Cauchy problem for weakly hyperbolic equations, Amer. J. Math., **97** (1975), 548–558.
M. Miyake: [1] On Cauchy-Kowalevski's theorem for general systems, Publ. RIMS. Kyoto Univ., **15** (1979), 315–337.
S. Mizohata: [1] Some remarks on the Cauchy problem, J. Math. Kyoto Univ.,

1 (1961), 109–127.

S. Mizohata: [2] Analyticity of the fundamental solutions of hyperbolic systems, J. Math. Kyoto Univ., **1** (1962), 327–355.

—— [3] Solutions nulles et solutions non analytiques, J. Math. Kyoto Univ., **1** (1962), 271–302.

—— [4] On Kowalewskian systems, Uspehi Mat. Nauk, **29** (1974), 216–227.

—— [5] On the Cauchy-Kowalewski theorem, Math. Analysis & Appl., Adv. in Math., Suppl. Studies, Vol. **7B** (1981), Acad. Press, New York, 617–652.

—— [6] On the hyperbolicity in the domain of real analytic functions and Gevrey classes, Hokkaido Math. J., **12** (1983), 298–310.

—— [7] On the meromorphic propagation of singularities and the Levi condition, Astérisque, **131** (Colloque de L. Schwartz) (1985), 127–135.

—— [8] Propagation de régularité au sens de Gevrey pour les opérateurs dif férentiels à multiplicité constante, Séminaire sur les Équations aux Dérivées Partielles Hyperboliques et Holomorphes (J. Vaillant), Hermann, Paris, 1984, 106–133.

—— [9] On the Cauchy problem for hyperbolic equations and related problems — micro-local energy method —, Proc. Taniguchi Intern. Sympos. on Hyperbolic Equations and Related Topics 1984, (these Proceedings).

S. Mizohata-Y. Ohya: [1] Sur la condition de E.E. Levi concernant des équations hyperboliques, Publ. RIMS. Kyoto Univ., **4** (1968/69), 511–526.

—— [2] Sur la condition d'hyperbolicité pour les équations à caractéristiques multiples II, Japan. J. Math., **40** (1971), 63–104.

C. S. Morawetz-D. Ludwig: [1] The generalized Huygens' principle for reflecting bodies, Comm. Pure Appl. Math., **22** (1969), 189–205.

C. S. Morawetz-J. Ralston-W. A. Strauss: [1] Decay of solutions of the wave equation outside nontrapping obstacles, Comm. Pure Appl. Math., **30** (1977), 447–508.

Y. Morimoto: [1] On the propagation of the wave front set of a solution for a hyperbolic system, Math. Japon., **27** (1982), 501–508.

G. Nakamura: [1] The singularities of solutions of the Cauchy problem for systems whose characteristic roots are non uniform multiple, Publ. RIMS. Kyoto Univ., **13** (1977), 255–275.

G. Nakamura-H. Uryu: [1] Parametrix of certain weakly hyperbolic operators, Comm. Partial Differential Equations **5** (1980), 837–896.

S. Nakane: [1] Propagation of singularities and uniqueness in the Cauchy problem at a class of doubly characteristic points, Comm. Partial Differential Equations, **6** (1981), 917–927.

A. B. Nersesyan: [1] Cauchy problem for degenerate second order hyperbolic equations, Izv. Akad. Nauk Armjan. SSR Mat., **3** (1968), 79–100.

L. Nirenberg: [1] Lectures on linear partial differential equations, Regional Conf. Ser. in Math. n° 17, AMS, 1973.

T. Nishitani: [1] On the Lax-Mizohata theorem in the analytic and Gevrey classes, J. Math. Kyoto Univ., **18** (1979), 501–512.

—— [2] On the $\mathscr{E}$-wellposedness for the Goursat problem with constant coefficients, J. Math. Kyoto Univ., **20** (1980), 179–190.

—— [3] On wave front sets of solutions for effectively hyperbolic operators, Sci. Rep. College Gen. Ed. Osaka Univ., **32**, 2 (1983), 1–7. Note on some non effectively hyperbolic operators, Ibid., 9–17.

—— [4] Sur les équations hyperboliques à coefficients hölderiens en t et de classe de Gevrey en x, Bull. Sci. Math., **107** (1983), 113–138.

—— [5] Energy inequality for non strictly hyperbolic operators in the Gevrey class, J. Math. Kyoto Univ., **23** (1983), 739–773.

—— [6] Local energy integrals for effectively hyperbolic operators, I & II, J. Math. Kyoto Univ., **24** (1984), 623–658, 659–666.

J. C. Nosmas: [1] Parametrix du problème de Cauchy pour une classe de systèmes hyperboliques symétrisables à caractéristiques involutives de multiplicités variables, Comm. Partial Differential Equations, **5** (1980), 1–22.

Y. Ohya: [1] Le problème de Cauchy pour les équations hyperboliques à caractéristique multiple, J. Math. Soc. Japan, **16** (1964), 268–286.

—— [2] On E.E. Levi's functions for hyperbolic equations with triple characteristics, Comm. Pure Appl. Math., **25** (1972), 257–263.

—— [3] Le problème de Cauchy à caractéristiques multiples, Ann. Scuola Norm. Sup. Pisa, **4** (1977), 755–805.

Y. Ohya-S. Tarama: [1] Le problème de Cauchy à caractéristiques multiples -coefficients hölderiens en *t*-, Proc. Taniguchi Intern. Sympos. on Hyperbolic Equations and Related Topics 1984, (these Proceedings).

O. A. Oleinik: [1] On the Cauchy problem for weakly hyperbolic equations, Comm. Pure Appl. Math., **23** (1970), 569–586.

T. Ōshima: [1] On analytic equivalence of glancing hypersurfaces, Sci. Papers College Gen. Ed. Univ. Tokyo, **28** (1978), 51–57.

S. Ōuchi: [1] Asymptotic behaviour of singular solutions of linear partial differential equations in the complex domain, J. Fac. Sci. Univ. Tokyo, Sect. IA **27** (1980), 1–36.

V. M. Petkov: [1] Propagation des singularités pour une classe des systèmes hyperboliques à caractéristiques de multiplicité variable, Serdica, **3** (1977), 189–197.

—— [2] Propagation des singularités pour des systèmes hyperboliques non symétrisables à caractéristiques de multiplicité constante, Serdica, **3** (1977), 152–158.

—— [3] The parametrix of the Cauchy problem for non symmetrizable hyperbolic systems with characteristics of constant multiplicity, Trudy Moskov. Mat. Obšč., **37** (1978), 3–48 (Trans. Moscow Math. Soc., **1** (1980), 1–47).

—— [4] Propagation of singularities and inverse scattering problem for transparent obstacles, J. Math. Pures Appl., **61** (1982), 65–90.

—— [5] Note on the distribution of poles of the scattering matrix, J. Math. Anal. Appl., **101** (1984), 582–587.

—— [6] Poisson relation for manifolds with boundary, Proc. Taniguchi Intern. Sympos. on Hyperbolic Equations and Related Topics 1984, (these Proceedings).

V. M. Petkov-G. St. Popov: [1] Propagation des singularités pour des systèmes hyperboliques non symétrisables, Serdica, **2** (1976), 283–294.

—— [2] Asymptotic behavior of the scattering phase for non trapping obstacles, Ann. Inst. Fourier, **32** (1982), 114–149.

I. G. Petrowsky: [1] Über das Cauchysche Problem für Systeme von partiellen Differentialgleichungen, Mat. Sb., **2** (1937), 815–870.

—— [2] Über das Cauchysche Problem für ein System linearer partieller Differentialgleichungen in Gebiete der nicht analytischen Funktionen, Bull. Univ. Moscou, Sér. Intern, **7** (1938), 1–72.

J. Ralston: [1] Solutions of the wave equation with localized energy, Comm. Pure Appl. Math., **22** (1969), 807–823.

—— [2] Propagation of singularities and the scattering matrix, Proc. NATO ASI (ed. H.G. Garnir), Reidel Publ., Dordrecht, 1981, 169–184.

J. Rauch: [1] The leading wave front for hyperbolic mixed problems, Bull. Soc. Roy. Sci. Liège, **46** (1977), 156–161.

J. Rauch-J. Sjöstrand: [1] Propagation of analytic singularities along diffracted rays, Indiana Univ. J. Math., **30** (1981), 381–401.

R. Sakamoto: [1] On a sufficient condition for well-posedness of weakly hyperbolic Cauchy problem, Publ. RIMS. Kyoto Univ., **15** (1979), 469–518.

—— [2] Some remarks on degenerate hyperoblic Cauchy problems, Comm. Pure Appl. Math., **33** (1980), 817–830.

R. Sakamoto: [3] Hyperbolic boundary value problems, Cambridge Univ. Press,

Cambridge, 1982.
M. Sato-T. Kawai-M. Kashiwara: [1] Hyperfunctions and pseudodifferential equations, Lecture Notes in Math., No. 287, Springer, 1973, 265–529.
P. Schapira: [1] Propagation au bord et réflexion des singularités analytiques des solutions des équations aux dérivées partielles II, Séminaire Goulaouic-Schwartz, 1976–77, n°9.
—— [2] Propagation at the boundary of analytic singularities, Proc. NATO ASI (ed. H.G. Garnir), Reidel Publ., Dordrecht, 1981, 185–212.
J. Schauder: [1] Das Anfangswert Problem einer quasilinearer hyperbolischen Differentialgleichung zweiter Ordnung in beliebiger Anzahl von unabhängigen Veränderlichen, Fund. Math., **24** (1935), 213–246.
D. Schiltz-J. Vaillant-C. Wagschal: [1] Problème de Cauchy ramifié: racine caractéristique double ou triple en involution, J. Math. Pures Appl., **4** (1982), 423–443.
K. Shinkai: [1] On the fundamental solution for a degenerate hyperbolic system, Osaka J. Math., **18** (1981), 257–288.
—— [2] Wave front sets of solutions for a degenerate hyperbolic system, Math. Japon., **29** (1984), 945–959.
J. Sjöstrand: [1] Propagation of analytic singularities for second order Dirichlet problems, Comm. Partial Differential Equations, **5** (1980), 49–91. II, Ibid., **5** (1980), 187–207.
—— [2] Analytic singularities and micro-hyperbolic boundary value problem, Math. Ann., **254** (1980), 211–256.
—— [3] Analytic singularities of solutions of boundary value problems, Proc. NATO ASI (ed. H.G. Garnir), Reidel Publ., Dordrecht, 1981, 235–270.
—— [4] Singularités analytiques microlocales, Astérisque, **95** (1982), 1–166.
H. Soga: [1] Singularities of scattering kernel for convex obstacles, J. Math. Kyoto Univ., **22** (1983), 729–765.
H. Tahara: [1] Singular hyperbolic systems, I, J. Fac. Sci. Univ. Tokyo, Sect. IA **26** (1979), 213–238, II, Ibid., **26** (1979), 391–412, III, Ibid., **27** (1980), 465–507.
K. Taniguchi: [1] Fourier integral operators in Gevrey class on $\boldsymbol{R}^n$ and the fundamental solution for a hyperbolic operator, Publ. RIMS. Kyoto Univ., **20** (1984), 491–542.
K. Taniguchi-Y. Tozaki: [1] A hyperbolic equation with double characteristics which has a solution with branching singularities, Math. Japon., **25** (1980), 279–300.
S. Tarama: [1] Sur le problème de Cauchy pour une classe des opérateurs différentiels du type faiblement hyperbolique, J. Math. Kyoto Univ., **22** (1982), 333–368.
M. E. Taylor: [1] Reflection of singularities of solutions to systems of differential equations, Comm. Pure Appl. Math., **28** (1975), 457–478.
—— [2] Grazing rays and reflection of singularities of solutions to wave equations, Comm. Pure Appl. Math., **29** (1976), 1–37, II (systems), Ibid., **29** (1976), 463–481.
—— [3] Pseudo-differential Operators, Princeton Univ. Press, 1981.
J.-M. Trépreau: [1] Le problème de Cauchy hyperbolique dans les classes d'ultrafonctions et d'ultradistributions, Comm. Partial Differential Equations, **4** (1979), 339–387.
G. A. Uhlmann: [1] Pseudo-differential operators with double characteristics, Comm. Partial Differential Equations, **2** (1977), 713–779.
—— [2] Parametrices for operators with multiple involutive characteristics, Comm. Partial Differential Equations, **4** (1979), 739–767.
H. Uryu: [1] The Cauchy problem for weakly hyperbolic systems with variable multiple characteristics, Publ. RIMS. Kyoto Univ., **15** (1979), 719–739.
J. Vaillant: [1] Données de Cauchy portées par une caractéristique double, Dans

le cas d'un système linéaire d'équations aux dérivées partielles, Rôle des bicaractéristiques., J. Math. Pures Appl., **47** (1968), 1–40.
—— [2] Ramifications d'intégrales holomorphes, Proc. Taniguchi Intern. Sympos. on Hyperbolic Equations and Related Topics 1984, (these Proceedings).
S. Wakabayashi: [1] Singularities of the Riemann functions of hyperbolic mixed problems in a quarter space, Publ. RIMS. Kyoto Univ., **11** (1976), 417–440.
—— [2] Analytic wave front sets of the Riemann functions of hyperbolic mixed problems in a quarter space, Publ. RIMS. Kyoto Univ., **11** (1976), 758–807.
—— [3] Microlocal parametrices for hyperbolic mixed problems in the case where boundary waves appear, Publ. RIMS. Kyoto Univ., **14** (1978), 283–307.
—— [4] Propagation of singularities of the fundamental solutions of hyperbolic mixed problems, Publ. RIMS. Kyoto Univ., **15** (1979), 653–678.
—— [5] Singularities of solutions of the Cauchy problems for operators with nearly constant coefficient hyperbolic principal part, Comm. Partial Differential Equations, **8** (1983), 347–406.
—— [6] Singularities of solutions of the Cauchy problem for symmetric hyperbolic systems, Comm. Partial Differential Equations, **9** (1984), 1147–1177.
—— [7] Singularities of solutions of the Cauchy problem for hyperbolic systems in Gevrey classes, Japan. J. Math., **11** (1985), 157–201.
C. Wagschal: [1] Problème de Cauchy analytique, à données méromorphes, J. Math. Pures Appl., **51** (1972), 375–397.
—— [2] Une généralisation du problème de Goursat pour des systèmes d'équations intégro-différentielles holomorphes ou partiellement holomorphes, J. Math. Pures Appl., **53** (1974), 99–134.
—— [3] Sur le problème de Cauchy ramifié, J. Math. Pures Appl., **53** (1974), 147–164.
M. Yamaguti: [1] Le problème de Cauchy et les opérateurs d'intégrale singulière, Mem. College Sci. Univ. Kyoto Ser. A, **32** (1959), 121–151.
H. Yamahara: [1] On the Cauchy problem for weakly hyperbolic systems, Publ. RIMS. Kyoto Univ., **12** (1976), 493–512.
K. Yamamoto: [1] The Cauchy problem for some class of hyperbolic differential operators with variable multiple characteristics, J. Math. Soc. Japan, **31** (1979), 481–502.
A. Yoshikawa: [1] Construction of a parametrix for the Cauchy problem of some weakly hyperbolic equations, I, II, III, Hokkaido Math. J., **6** (1977), 313–344, **7** (1978), 1–26, **7** (1978), 127–141.
—— [2] Parametrices for a class of effectively hyperbolic operators, Comm. Partial Differential Equations, **5** (1980), 1073–1151.

Taniguchi Symp. HERT
Katata 1984, pp. 1–9

Complex Vector Fields, Holomorphic Extension of CR Functions and Related Problems

M. S. BAOUENDI

We present in this lecture some recent results on microlocal hypo-analyticity of functions annihilated by complex vector fields and some applications to holomorphic extension of CR functions. This talkis based on several joint articles with C. H. Chang, L. P. Rothschild and F. Treves ([2], [3], [4]).

§ 1. Hypo-analytic structures

Let Ω be an open set of $\boldsymbol{R}^N$, $N=n+m$, $n\geqslant 0$, $m>0$, and $L_1, \cdots, L_n$ be n complex vector fields with smooth coefficients defined in Ω i.e.

$$L_j=\sum_{k=1}^{N} a_{j,k}(x)\frac{\partial}{\partial x_k},$$

where the complex valued functions $a_{j,k}$ are smooth in Ω.

We assume throughout this section the following conditinos:

(1.1) *$L_1, \cdots, L_n$ are $\boldsymbol{C}$-linearly independent, i.e. the matrix $(a_{j,k}(x))$ is of rank n at every point $x \in \Omega$.*

(1.2) *There exist m complex valued smooth functions $Z_1, \cdots, Z_m$ defined in Ω, satisfying*

$$L_j Z_k=0 \quad \text{in } \Omega, \quad 1\leqslant j\leqslant n, \quad 1\leqslant k\leqslant m,$$

and the differentials $dZ_1, \cdots, dZ_m$ are $\boldsymbol{C}$-linearly independent at each point of Ω.

We shall think of $Z=(Z_1, \cdots, Z_m)$ as a map from Ω into $\boldsymbol{C}^m$. Denote by $\boldsymbol{L}$ the set of all complex vector fields in Ω of the form

$$\sum_{j=1}^{n} a_j(x)L_j, \qquad a_j \in C^\infty(\Omega).$$

The *integrability* condition (1.2) implies that

$$[L, L] \subset L,$$

i.e. if $L, M \in L$ then the Lie bracket $[L, M] \in L$. We shall say that $(L_1, \cdots, L_n)$ is a basis of the Lie algebra L and $Z_1, \cdots, Z_m$ are first integrals of L.

If (1.1) and (1.2) are satisfied we shall say that Ω is equipped with a *hypo-analytic structure* of codimension n. Of course we do not change the structure if we change basis in L, nor if we replace Z by $\tilde{Z} = H \circ Z$, where H is a holomorphism defined in an open neighborhood of $Z(\Omega)$ in C^m.

We are interested in the study of the solutions of the overdetermined system

$$(1.3) \qquad L_j h = 0, \qquad 1 \leqslant j \leqslant n,$$

where $h \in \mathscr{D}'(\Omega)$ (distribution in Ω).

Since all the results summarized here are local, we will consider a fixed point of Ω (say the origin of R^N), and will freely replace Ω by a smaller neighborhood of 0 if needed.

Definition 1.1. A distribution h, solution of (1.3) is said to be hypo-analytic at $\omega_0 \in \Omega$ iff there exist an open set Ω', $\omega_0 \in \Omega' \subset \Omega$, and a holomorhic function F defined in a neighborhood of $Z(\Omega')$ in C^m, such that

$$h = F \circ Z \qquad \text{in } \Omega'.$$

The complement of the set where h is hypo-analytic will be called the *hypo-analytic singular support* of h. If the first integrals Z_j's are (real) analytic, then hypo-analyticity coincides with analyticity.

We shall give some sufficient conditions to insure that any solution of (1.3) is hypo-analytic at ω_0.

We need to "microlocalize" Definition 1.1. We first do so when the codimension n of the hypo-analytic structure is zero.

a) Hypo-analytic structures of codimension 0

In this subsection we assume $0 \in \Omega \subset R^m$ and $Z_1, \cdots, Z_m \in C^\infty(\Omega)$ with independent differentials (Here $n=0$, there are first integrals but no vector fields!). After shrinking Ω about 0 and changing Z we may assume

$$(1.4) \qquad Z(x) = x + i\phi(x),$$

with $\phi \in C^\infty(\Omega, R^m)$, $\phi(0)=0$, $\phi'(0)=0$. The map Z is then a diffeomorphism of Ω onto a totally real submanifold $Z(\Omega)$ of C^m. If $u \in \mathscr{E}'(\Omega)$

(distribution with compact suport) we define its FBI (Fourier-Bros-Iagolnitzer) transform by

$$(1.5)\qquad F_\kappa(u, t, \zeta)=\int_\Omega \exp(-iZ(x)\cdot\zeta-\kappa\langle\zeta\rangle\langle Z(x)-t\rangle^2)u(y)dZ(x),$$

where $t, \zeta \in \boldsymbol{C}^m$, $|\operatorname{Im}\zeta|<|\operatorname{Re}\zeta|$, $\langle\zeta\rangle=(\sum\zeta_j^2)^{1/2}$, and $\kappa>0$, $dZ(x)=dZ_1(x)\wedge\cdots\wedge dZ_m(x)$. (When $\phi\equiv 0$ in (1.4), the integral (1.5) has been used in [6], see also Sjöstrand [10].)

Let $\xi^0 \in T_0^*\Omega\setminus 0$. Consider the following property:

(1.6) There are an open neighborhood $\mathcal{O}$ of 0 in $\boldsymbol{C}^m$, an open cone $\Gamma\subset\boldsymbol{C}^m\setminus\{0\}$, $\xi^0 \in \Gamma$, and a constant $C>0$ such that

$$|F_\kappa(u, t, \zeta)|\leqslant Ce^{-|\zeta|/C}, \quad t \in \mathcal{O}, \quad \zeta \in \Gamma.$$

It is shown in [2] that u is hypo-analytic at 0 iff (1.6) holds for all $\zeta^0 \in T_0^*\Omega\setminus 0$, provided κ is large enough.

We shall say that u is *hypo-analytic* at $(0, \xi^0)$ if (1.6) holds with κ sufficiently large. This *microlocal hypo-analyticity* could also be defined using boundary values of holomorphic functions in certain "conoids" in $\boldsymbol{C}^m$ well situated with respect to ξ^0, as in the usual microlocal analyticity defined by Sato [9].

The set of covectors in $T^*\Omega\setminus 0$ at which a distribution $h \in \mathscr{D}'(\Omega)$ is not hypo-analytic is called the *hypo-analytic wave front set* of h. It coincides with the usual analytic *wave front set* of Sato when Z is analytic. Its base projection on Ω is the hypo-analytic singular support of h.

b) General hypo-analytic structures

We assume in this subsection that Ω is equipped with a hypo-analytic structure with positive codimension, i.e. $n>0$.

Let X be an m dimensional real submanifold of Ω. We say that X is *maximally real* if the restriction of the first integrals $Z_1, \cdots, Z_m$ to X have linearly independent differentials. If X is maximally real then the functions $Z_j|_X$, $1\leqslant j\leqslant m$, define a hypo-analytic structure on X of codimension zero. If X is such a manifold and $\omega_0 \in X\subset\Omega$, we can find coordinates x, y near ω_0 in Ω, vanishing at ω_0, $x \in \boldsymbol{R}^m$, $y \in \boldsymbol{R}^n$, such that after changing Z by a holomorphism we have

$$Z(x, y)=x+i\phi(x, y),$$

where ϕ is smooth and real valued, $\phi(0)=0$, $\phi'_x(0)=0$, and X is defined by $y=0$.

A distribution h, solution of (1.3) is *hypo-analytic* at $(\omega_0, \xi^0) \in T^*\Omega \setminus 0$ if there exists a maximally real manifold containing ω_0 such that $h|_X$ is hypo-analytic at the canonical projection $\pi_X(\omega_0, \xi^0)$ $(\pi_X : T^*\Omega \to T^*X)$. The hypo-analytic wave front set of h is then the set of covectors in $T^*\Omega \setminus 0$ at which h is not hypo-analytic. The base projection of $WF_{\mathrm{ha}}h$ coincides with the hypo-analytic singular support of h.

A covector $(\omega_0, \xi^0) \in T^*\Omega$ is called *characteristic* if

$$\langle \xi^0, L_{\omega_0} \rangle = 0, \qquad L \in \boldsymbol{L}.$$

The first basic result deals with non-characteristic covectors.

Theorem 1.1 ([2]). *If* $(\omega_0, \xi^0) \in T^*\Omega$ *is non-characteristic, then any solution of* (1.3) *is hypo-analytic at* (ω_0, ξ^0).

The follownig criterion using the Levi form deals with characteristic coverctors.

Theorem 1.2 ([2]). *Let* (ω_0, ξ^0) *be a characteristic covertor. Assume there exists* $L \in \boldsymbol{L}$ *such that*

$$\frac{1}{\sqrt{-1}} \langle \xi^0, [L, \bar{L}] \rangle < 0,$$

then any distribution solution of (1.3) *is hypo-analytic at* (ω_0, ξ^0).

Define the Levi matrix at a characteristic point (ω_0, ξ^0) by

$$C_{j,k}(\omega_0, \xi^0) = \frac{1}{\sqrt{-1}} \langle \xi^0, [L_j, \bar{L}_k]_{\omega_0} \rangle$$

where $L_1, \cdots, L_n$ is a basis of $\boldsymbol{L}$.

We have an immediate corollary of Th. 1.1 and Th. 1.2.

Corollary. *Let* $\omega_0 \in \Omega$. *If for every characteristic vector* $\xi \in T^*_{\omega_0}\Omega \setminus 0$, *the Levi matrix at* (ω_0, ξ) *has at least one positive eigenvalue, and at least one negative eigenvalue, then any distribution solution of* (1.3) *is hypo-analytic at* ω_0.

§ 2. C. R. Structures

In this section, in addition to (1.1) and (1.2) we shall assume:

(2.1) $L_1, \cdots, L_n, \bar{L}_1, \cdots, \bar{L}_n$ *are linearly independent*.

This implies that $N \geqslant 2n$. We take

$$N=2n+l, \qquad m=n+l$$

$(n>0, l\geqslant 0)$. $M=Z(\Omega)$ is a C. R. generic real manifold of codimension l in $\boldsymbol{C}^{n+l}$, The map Z is a local diffeomorphism. Solutions of (1.3) can be viewed as CR distributions defined on M (distributions annihilated by the induced Cauchy-Riemann operator on M). The characteristic set of $\boldsymbol{L}$ is now an l dimensional vector subbundle of $T^*\Omega$. A distribution solution of (1.3) is hypo-analytic at ω_0 iff its push-forward (CR distribution on M) extends holomorphically in a neighborhood of $Z(\omega_0)$ in $\boldsymbol{C}^{n+l}$. In this context, Corollary 1.1 can be viewed as a generalization of the classical result of H. Lewy [7] for hypersurfaces in $\boldsymbol{C}^{n+1}$. Using "analytic discs" method Boggess and Polking [5] obtained a different proof of Corollary 1.1 in the CR case.

Assume (1.1), (1.2) and (2.1). A point $(\omega_0, \xi^0)\in T^*\Omega\setminus 0$ is of *finite type*, if there exist $M_1, \cdots, M_m \in \boldsymbol{L}\oplus\bar{\boldsymbol{L}}$ such that

$$\langle \xi^0, [M_1, [M_2, [\cdots [M_{m-1}, M_m]]]]_{\omega_0}\rangle \neq 0.$$

The smallest such positive integer $m=m(\omega_0, \xi^0)$ is called the *type* at (ω_0, ξ^0). A point $\omega_0\in\Omega$ is of finite type if (ω_0, ξ) is of finite type for all $\xi\in T^*_{\omega_0}\Omega\setminus 0$.

a) Hypersurfaces of $\boldsymbol{C}^{n+1}$

We assume in this subsection that $l=1$. Therefore $M=Z(\Omega)$ is a hypersurface of $\boldsymbol{C}^{n+1}$.

We have the following result.

Theorem 2.1 ([4]). *Let M be a smooth real hypersurface of $\boldsymbol{C}^{n+1}$. Let $\omega_0\in M$ be a point of finite type, then any CR distribution h defined on M extends holomorphically to at least one side of M in a neighborhood of ω_0 in $\boldsymbol{C}^{n+1}$.*

Note that in the hypersurface case, the fiber dimension of the characteristic set is one. Holomorphic extendability to one side of M in a neighborhood of ω_0 is equivalent to the existence of a characteristic covector $\xi^0\in T^*_{\omega_0}\Omega\setminus 0$ such that any solution of (1.3) is hypo-analytic at (ω_0, ξ^0). Theorem 2.1 is therefore a microlocal statement about solutions of (1.3). It follows from a more general result that we shall now describe.

Assume (1.1), (1.2) and (2.1), with $l=1$. Let $\omega_0\in\Omega$ be a point of type m. We can find coordinates $x, y\in\boldsymbol{R}^n$, $s\in\boldsymbol{R}$ vanishing at ω_0, and first integrals Z_j, $1\leqslant j\leqslant n+1$, such that

$$(2.2)\qquad \begin{aligned} &Z_j=x_j+iy_j=z_j, \qquad 1\leqslant j\leqslant n,\\ &Z_{n+1}=s+i\phi(x, y, s), \end{aligned}$$

with ϕ real, $\phi(0)=0$, $\phi'(0)=0$. Furthermore, if we write

$$\phi(x, y, 0)=\phi_0(z, \bar{z}) \qquad (z \in \boldsymbol{C}^n),$$

then

$$\phi_0(z, \bar{z})=P_m(z)+O(|z|^{m+1}) \tag{2.3}$$

where P_m is a *non-pluriharmonic* real homogeneous polynomial of degree m.

Note that in the coordinates (x, y, s) the characteristic covectors at 0 are of the form λds, $\lambda \in \boldsymbol{R}$.

We say that P_m satisfies the *line sector property* if there exist a real pluriharmonic polynomial of degree m, $h_m(z)$, a complex line $\mathscr{L}$ passing through 0 in $\boldsymbol{C}^n$ and a sector $\mathscr{S}$ in $\mathscr{L}$ satisfying:

$$\begin{aligned} &P_m(z)+h_m(z)|_{\mathscr{S}}<0, \\ &\text{angle } \mathscr{S}>\pi/m. \end{aligned} \tag{2.4}$$

Theorem 2.2 ([4]). *Assume* (2.1), (2.3) *and* (2.4) *then any solution of* (1.3) *is hypo-analytic at* $(0, ds)$.

A coordinate free version of Th. 2.2 (using commutators of the L_j's and $\bar{L}_j$'s) can be found in [4].

Finally note that when the type of ω_0 is odd then any CR distribution extends to a full neighborhood of $Z(\omega_0)$ in $\boldsymbol{C}^{n+1}$.

b) Higher codimension-rigid CR structures

Results of this subsection are part of the joint work [3]. Details will appear elsewhere.

Let Ω be an open set of $\boldsymbol{R}^{2n+l}$ $(n>0, l>0)$, and $\mathscr{V}\subset \boldsymbol{C}T\Omega$ be a complex vector subbundle of the compexified tangent bundle of Ω. Assume that the (complex) fibre dimension of $\mathscr{V}$ is n. Denote by $\boldsymbol{L}$ the smooth sections of $\mathscr{V}$

$$\boldsymbol{L}=C^\infty(\Omega, \mathscr{V}).$$

We assume there exists an l dimensional real vector space π of (real) vector fields in Ω

$$\pi\subset C^\infty(\Omega, T\Omega)$$

such that

$$\mathscr{V}_\omega+\bar{\mathscr{V}}_\omega+\boldsymbol{C}\pi_\omega=\boldsymbol{C}T_\omega\Omega, \qquad \forall\omega \in \Omega. \tag{2.5}$$

We also make the following assumptions:

(2.6) $$[\mathscr{V}, \mathscr{V}]\subset\mathscr{V} \qquad \text{(i.e. } [\boldsymbol{L}, \boldsymbol{L}]\subset\boldsymbol{L}),$$

(2.7) $$[\boldsymbol{L}, \boldsymbol{\pi}]\subset\boldsymbol{L},$$

(2.8) $$[\boldsymbol{\pi}, \boldsymbol{\pi}]=0.$$

Making use of Newlander-Nirenberg Theorem [8] we can show that conditions (2.5)-(2.8) imply local integrability (i.e. (1.2) where the L_j's are a basis of $\boldsymbol{L}$). In fact given $\omega_0 \in \Omega$ there exist coordinates x, y, s ($x, y \in \boldsymbol{R}^n$, $s \in \boldsymbol{R}^l$) near ω_0, vanishing at ω_0, and $n+l$ first integrals $Z_1, \cdots, Z_n, W_1, \cdots, W_l$ of the form

(2.9) $$\begin{aligned} Z_j&=x_j+iy_j, & 1\leqslant j\leqslant n,\\ W_k&=s_k+i\phi_k(x, y), & 1\leqslant k\leqslant l, \end{aligned}$$

where ϕ_k is real valued, $\phi_k(0)=0$ and $\phi_k'(0)=0$.

Denote by (z, w) the coordinates in $\boldsymbol{C}^{n+l}$ ($z \in \boldsymbol{C}^n$, $w \in \boldsymbol{C}^l$). The manifold M defined in a neighborhood of the origin in $\boldsymbol{C}^{n+l}$ by

(2.10) $$\operatorname{Im} w=\phi(\operatorname{Re} z, \operatorname{Im} z),$$

($\phi=(\phi_1, \cdots, \phi_l)$) is a CR generic submanifold of $\boldsymbol{C}^{n+l}$ of (real) codimension l. The map (Z, W): $\Omega\to\boldsymbol{C}^{n+l}$ defined by (2.9) is a diffeomorphism from Ω onto M. We refer to the hypo-analytic structure on Ω with first integrals (2.9) as a *rigid* CR structure, and we say that M is a *rigid* CR manifold. Note that we can choose a local basis of $\boldsymbol{L}$ of the form

$$L_j=\frac{\partial}{\partial\bar{z}_j}-i\sum_{k=1}^{l}\frac{\partial\phi_k}{\partial\bar{z}_j}(z, \bar{z})\frac{\partial}{\partial s_k}, \qquad 1\leqslant j\leqslant n,$$

here $z=x+iy \in \boldsymbol{C}^n$, and a basis of the vector space $\boldsymbol{\pi}$ of the form

$$T_k=\frac{\partial}{\partial s_k}, \qquad 1\leqslant k\leqslant l.$$

Also note that the characteristic covectors at the origin are of the form

$$\xi=\sum_{j=1}^{l}\lambda_j ds_j, \qquad \lambda_j \in \boldsymbol{R}.$$

Let Γ be a convex proper open cone of $\boldsymbol{R}^l$, and $\mathcal{O}$ an open neighborhood of the origin in $\boldsymbol{C}^{n+l}$. We will need certain sets in $\boldsymbol{C}^{n+l}$ of the form

(2.11) $$\mathscr{W}=\{(z,w)\in\mathcal{O};\ \operatorname{Im} w-\phi(\operatorname{Re} z, \operatorname{Im} z)\in\Gamma\}.$$

$\mathscr{W}$ is an open subset of $\boldsymbol{C}^{n+l}$ and $M\cap\mathcal{O}$ is contained in the boundary of $\mathscr{W}$. We will refer to sets of the form (2.11) as *wedges.*

A holomorphic function f in $\mathscr{W}$ is called with *tempered growth* if the following holds:

$$|f(z,w)|\leqslant C \text{ dist. } ((z,w);M)^{-N}, \qquad (z,w)\in\mathscr{W},$$

with some constants C and N. Such a holomorphic function has a boundary value on $M\cap\mathcal{O}$, bf, which is a CR distribution.

We can now state the following results:

Theorem 2.3 ([3]). *Let M be given by* (2.10) *and h be a CR distribution defined on M. There exist a finite number of wedges $\mathscr{W}_j$, $1\leqslant j\leqslant r$, of the form* (2.11) (*with different cones $\Gamma_j\subset\boldsymbol{R}^l$*) *and holomorphic functions with tempered growth, f_j, defined in $\mathscr{W}_j$ such that*

$$h=\sum_{j=1}^{r} bf_j,$$

in a neighborhood of 0 *in M.*

Theorem 2.4 [3]. *Let M be given by* (2.10) *and assume that* 0 *is of finite type. Then for any CR distribution h defined on M there exist a single wedge $\mathscr{W}$ of the form* (2.11) *amd a holomorphic function f defined in $\mathscr{W}$ with tempered growth such that*

$$h=bf$$

in a neighborhood of 0 *in M.*

In the case $l=1$ (hypersurface), Th. 2.3 was proved in Andreotti-Hill [1] ($r=2$, the wedges are both sides of the hypersurface), and Th. 2.4 follows from Th. 2.1 above.

Theorem 2.4 is a consequence of a more general microlocal result that we shall describe.

Let $\xi^0=\sum_{j=1}^{l}\lambda_j ds_j$, $\lambda_j\in\boldsymbol{R}$, be a characteristic covector at the origin. We say that ξ^0 satisfies the *curve sector property* if there exists a holomorphic curve $\gamma: D\to\boldsymbol{C}^n$, D being a neighborhood of 0 in $\boldsymbol{C}$, $\gamma(0)=0$, such that

$$\sum_{j=1}^{l}\lambda_j\phi_j(\operatorname{Re}\gamma(c), \operatorname{Im}\gamma(c))=P_m(c)+O(|c|^{m+1}).$$

where P_m is a homogeneous real polynomial of degree m. In addition we assume there exist $\mu \in \boldsymbol{C}$ and a sector $\mathscr{S}$ in the complex plane such that

$$P_m(c) + \mathrm{Re}\,\mu c^m|_{\mathscr{S}} < 0$$

and

$$\text{angle } \mathscr{S} > \pi/m.$$

Theorem 2.5 ([3]). *Let M be defined by* (2.10) *and* ξ^0 *be a characteristic covector at the origin satisfying the curve sector property then any CR distribution defined on M is hypo-analytic at* $(0, \xi^0)$.

When 0 is of finite type, making use of Theorem 2.5 we can show that if h is a CR distribution defined on M, then its hypo-analytic wave front set is contained in a strictly convex closed cone. This fact will then imply the extendability result of Th. 2.4.

Let us mention that the curve sector property could be stated in an invariant way (i.e. coordinate free) using commutators of vector fields $L \in \boldsymbol{L}$ and their complex conjugates.

Finally the "rigidity" assumption in Theorem 2.4 could be weakened. Such results will appear elsewhere.

References

[1] A. Andreotti and C. D. Hill, E.E. Levi convexity and the Hans Lewy problem I and II, Ann. Scuola Norm. Sup. Pisa **26** (1972), 325–363, 747–806.

[2] M. S. Baouendi, C. H. Chang and F. Treves, Microlocal hypo-analyticity and extension of CR functions, J. Differential Geom. **18** (1983), 331–391.

[3] M. S. Baouendi, L. P. Rothschild and F. Treves, CR structures with group action and extendability of CR functions, to appear.

[4] M. S. Baouendi and F. Treves, About the holomorphic extension of CR functions on real hypersurfaces in complex space, Duke Math. J. **51** (1984), 77–107.

[5] A. Boggess and J. Polking, Holomorphic extension of CR functions, Duke Math. J. **49** (1982), 757–784.

[6] J. Bros et D. Iagolnitzer, Support essentiel et structure analytique des distributions, Séminaire Goulaouic-Lions, Schwartz, 1975–76, Exp. 18.

[7] H. Lewy, On the local character of the solution of an atypical differential equation in three variables and a related problem for regular functions of two complex variables, Ann. of Math. **64** (1956), 514–522.

[8] A. Newlander and L. Nirenberg, Complex analytic coordinates in almost complex manifolds, Ann. of Math. **65** (1957), 391–404.

[9] M. Sato, Hyperfunctions and partial differential equations, Proc. Internat. Conf. Funct. Anal. Tokyo (1969), 91–94.

[10] J. Sjöstrand, Singularités analytiques microlocales, Astérisque, Soc. Math. France **95** (1982), 1–166.

Department of Mathematics
Purdue University
West Lafayette, Indiana 47907
U.S.A.

Taniguchi Symp. HERT
Katata 1984, pp. 11–49

Second Microlocalization and Propagation of Singularities for Semi-Linear Hyperbolic Equations

Jean-Michel BONY

§ 0. Introduction

0.1. Propagation of singularities

Let u be a solution belonging to H^s, in an open set of $\boldsymbol{R}^3$ containing 0, of

$$\frac{\partial^2 u}{\partial t^2} - \frac{\partial^2 u}{\partial x^2} - \frac{\partial^2 u}{\partial y^2} = f(t, x, y, u). \tag{0.1}$$

We are interested in the following problem: is it possible to determine the singularities of u for $t>0$, knowing its singularities for $t<0$?

We shall have to assume $s>3/2$. For small values of s, it is possible to find solutions u belonging to C^∞ in the past, and singular for $t>0$ (see [8]).

If we want to know whether u belongs to $H^{s'}$ or not, locally or microlocally, for $s' \leqslant 2s-3/2$, a complete answer is known—and is valid for any nonlinear strictly hyperbolic equation—singularities travel along bicharacteristics.

For $s'>2s-3/2$, nonlinear phenomena appear: self-spreading [1] and interaction, which we will study assuming that u, in the past, has only good (i.e. conormal, see § 7 below) singularities along a finite number of characteristic surfaces $\Sigma_1, \Sigma_2, \cdots$.

For two singularities, the result is known: u will be regular outside $\Sigma_1 \cup \Sigma_2$ in the future and will have conormal singularities near Σ_1 and Σ_2 ([3]). This result is valid for equations of order 2 in any dimension. For higher order equations, the singularities are localized on the characteristic hypersurfaces starting from $\Sigma_1 \cap \Sigma_2$ ([4]). We refer to [5] for a survey and more references.

In this paper, we shall study the interaction of 3 singularities along $\Sigma_1, \Sigma_2, \Sigma_3$. The main result (Corollary 7.5) says that, in the future, u is regular outside $\Sigma_1, \Sigma_2, \Sigma_3$ and the half light cone starting from the point $\Sigma_1 \cap \Sigma_2 \cap \Sigma_3$. This can be pictured as follows

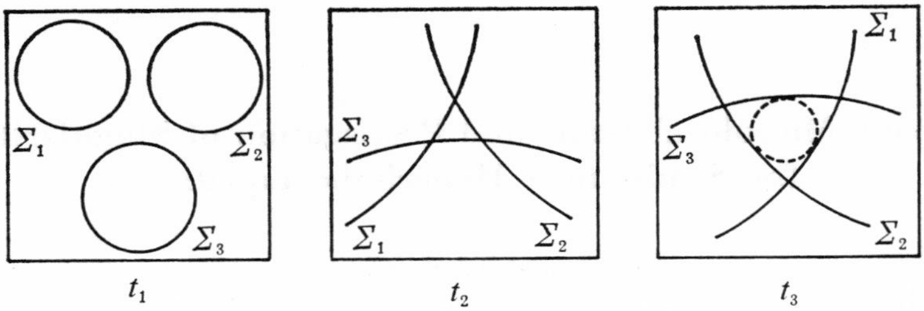

Between t_1 and t_2, interactions two by two create no new singularity. Between t_2 and t_3, interaction creates a new singularity less strong than the incident ones, on the dotted circle.

An example showing that the new singularity can appear is due to Rauch and Reed [11]. A first proof of our result had been briefly given in [6]. Independently, and by different methods, Melrose and Ritter [10] proved a result less precise than Corollary 7.5, but whose geometrical meaning is the same.

Using the same methods, we can prove an analogous result for Cauchy problem. If the Cauchy data on $t=0$ of a solution u of (0.1) are conormal distributions along a finite number of curves, then u will be regular outside the characteristic surfaces starting from the curves and the light cones starting from the points of intersection of the curves (see [6], Theorem A′).

0.2. Second microlocalization

We shall have to use a new symbolic calculus, analogous to that developed by Y. Laurent [9] in the analytic case. Definitions and results are summed up in § 1. Given a Lagrangean submanifold Λ of $T^*\boldsymbol{R}^n$, one first associates spaces $H_\Lambda^{s,s'}$ of distributions belonging (for $s'>0$) to H^s microlocally near Λ, to $H^{s+s'}$ microlocally outside Λ, with a "control" of what happens when one approaches Λ. A family of operators (2-microdifferential operators) acting on these spaces is described. There exists a good symbolic calculus for these operators, the symbols being singular on Λ. This allows us to define the second wave front of type $H^{s,s'}$, which coincides with the usual $H^{s+s'}$ wave front outside Λ, but contains more information near Λ.

In §§ 2 and 3, the symbolic calculus is constructed when Λ is the conormal of 0 in $\boldsymbol{R}^n$. In that case, one can construct, using Littlewood-Paley theory, an isomorphism between spaces $H^{s,s'}$ and weighted Sobolev spaces in $\boldsymbol{R}^n\setminus\{0\}$ (operators Π and Pf). To each (convenient) symbol p singular at 0, it is then possible to associate a properly supported (in $\boldsymbol{R}^n\setminus\{0\}$) pseudo-differential operator, mapping weighted Sobolev spaces into themselves. Via the above isomorphism, one can associate to p an operator on spaces $H^{s,s'}$, which will be a 2-microdifferential operator of symbol p.

In §4, we show that this construction is invariant under Fourier integral operators (and even under more general transformations). This proves that second microlocalization can be made along any Lagrangean submanifold. In §5, we study the propagation of 2-microlocal singularities for operators of principal type, when a bicharacteristic intersects Λ. As usual, one can be reduced to study $\partial/\partial x_1$+lower order term (i.e. a term of bi-order $(+1, -1)$). Actually, one needs to prove 3 different theorems to describe what happens when a singularity travels across Λ.

In §6, we prove the only results of this paper which are actually non-linear. There exist 2-microlocal operators Z ("singular vector fields") whose symbol is linear with respect to ξ, and a Leibniz type formula for $Z(u\,v)$ is valid for these opreators.

It is then possible to prove our result on equation (0.1). For singular vector fields (Z_i) which are tangent to Σ_1, Σ_2, Σ_3 and to the light cone, we will show inductively that $Z_i u \in H^{s,s'}$, $Z_i Z_j u \in H^{s,s'}$, $\cdots$.

Thanks to Leibniz formula, the spaces of functions satisfying these conditions are algebras. It remains to prove that vector-valued functions $U_1=(Z_i u)$, $U_2=(Z_i Z_j u)$, $\cdots$ satisfy equations $\square U+\mathscr{R}U=F$, which is a theorem on commutators. Then the results of §5 on propagation of 2-microlocal singularities will give the regularity of U.

It should be noted that this proof would not work with usual (C^∞) vector fields instead of singular vector fields. That vector fields would be too flat at 0, and commutation relations leading to equations for U would have been false.

§ 1. Second microlocalization along a Lagrangean submanifold (Definitions and statements)

1.1. Notations. Let Λ be a conic Lagrangean submanifold of $T^*\boldsymbol{R}^n$. It can be defined by equations

$$m_1(x, \xi)=\cdots=m_n(x, \xi)=0$$

where m_j are homogeneous of degree 1, such that the Poisson brackets $\{m_i, m_j\}$ vanish on Λ. We shall denote by M_j pseudo-differential operators with principal symbol m_j.

1.2. Definition (2-microlocal Sobolev space). *Given $s \in \boldsymbol{R}$ and $k \in \boldsymbol{N}$, we shall say that $u \in H^{s,k}_{\Lambda,\mathrm{loc}}$ if*

$$M^I u \in H^s_{\mathrm{loc}} \qquad \textit{for } |I|\leqslant k$$

(with standard notations: $M^I=M_{i_1}\circ\cdots\circ M_{i_l}$; $|I|=l$).

For $s \in \boldsymbol{R}$, $s' \in \boldsymbol{R}$, spaces $H^{s,s'}_{\Lambda,\mathrm{loc}}$ are defined by duality and interpolation.

It makes sense to say that $u \in H^{s,s'}_{\Lambda}$ locally near x_0 or microlocally near (x_0, ξ_0). One can also define spaces $H^{0,s'}$ as domains of fractional powers of a self-adjoint extension of $\sum M_i^* M_i$. It is clear that the local spaces depend only on Λ and not on the choice of the M_j.

1.3. Symbols. We shall denote by $\mathscr{H}_0$ the space of vector fields in $T^*\boldsymbol{R}^n$ which are homogeneous of degree 0 (they are generated by $\partial/\partial x_j$ and $\xi_j(\partial/\partial\xi_k)$). We shall denote by $\mathscr{H}_1$ the space of vector fields belonging to $\mathscr{H}_0$ and tangent to Λ. We shall call $d_1(x, \xi)$ the "distance to Λ":

$$d_1(x, \xi) = (1 + \sum |m_j(x, \xi)|^2)^{1/2}$$
$$\langle \xi \rangle = (1 + |\xi|^2)^{1/2}.$$

1.4. Definition (symbols of 2-microlocal operators). *We shall denote by $\Sigma^{m,m'}_{\Lambda}$, the space of functions $a(x, \xi) \in C^\infty(\boldsymbol{R}^{2n})$ such that*

$$|H^I K^J a(x, \xi)| \leqslant C \langle \xi \rangle^{m+|I|} d_1(x, \xi)^{m'-|I|}$$

for finite families of vector fields $H_i \in \mathscr{H}_0$ and $K_j \in \mathscr{H}_1$.

It is clear that $\Sigma^{m,m'}_{\Lambda}$ coincides with the space of classical symbols $S^{m+m'}_{1,0}$ in $T^*\boldsymbol{R}^n \setminus \Lambda$, and that one has $S^m_{1,0} \subset \Sigma^{m,0}_{\Lambda}$ for any Λ. It is easy to see (using for instance invariance by canonical transformations) that, for $a \in \Sigma^{m_1, m_1'}$ and $b \in \Sigma^{m_2, m_2'}$, one has

$$ab \in \Sigma^{m,m'}_{\Lambda} \quad \{a, b\} \in \Sigma^{m,m'-1}_{\Lambda}$$

with $m = m_1 + m_2$ and $m' = m_1' + m_2'$.

1.5. Definition (2-microlocal operators). *Let A be a properly supported operator mapping $C^\infty(\boldsymbol{R}^n)$ into $\mathscr{D}'(\boldsymbol{R}^n)$. We will say that A belongs to* $\mathrm{Op}(\Sigma^{m,m'}_{\Lambda})$ *if*

$$\mathrm{Ad}\, P_1 \circ \cdots \circ \mathrm{Ad}\, P_k \circ \mathrm{Ad}\, M_{i_1} \circ \cdots \circ \mathrm{Ad}\, M_{i_l} \cdot A \text{ maps}$$
$$H^{s,s'}_{\Lambda} \text{ into } H^{s-m-\Sigma p_i,\, s'-m'+k}_{\Lambda} \text{ for } s, s' \in \boldsymbol{R}$$

where P_i are classical pseudo-differential operators of order p_i, where M_j are defined in 1.1, *and where* $\mathrm{Ad}\, B \cdot A = [B, A]$.

Using R. Beals' characterization (see [7]), it is easy to see that, microlocally outside Λ, $\mathrm{Op}(\Sigma^{m,m'}_{\Lambda})$ coincides with $\mathrm{Op}(S^{m+m'}_{1,0})$.

1.6. Theorem (Fundamental properties of 2-microlocal calculus).
There exists one unique family of continuous linear applications $\sigma_{m,m'}$ *(principal symbol) from* $\mathrm{Op}(\Sigma_\Lambda^{m,m'})$ *onto* $\Sigma_\Lambda^{m,m'}/\Sigma_\Lambda^{m,m'-1}$ *such that*

a) *For* $A \in \mathrm{Op}(\Sigma_\Lambda^{m_1,m_1'})$ *and* $B \in \mathrm{Op}(\Sigma_\Lambda^{m_2,m_2'})$ *one has, with* $m=m_1+m_2$, $m'=m_1'+m_2'$:

$$\sigma_{m,m'}(A \circ B)=\sigma_{m_1,m_1'}(A)\cdot\sigma_{m_2,m_2'}(B)$$

$$\sigma_{m,m'-1}([A, B])=\frac{1}{i}\{\sigma_{m_1,m_1'}(A),\ \sigma_{m_2,m_2'}(B)\}.$$

b) *The kernel of* $\sigma_{m,m'}$ *is* $\mathrm{Op}(\Sigma_\Lambda^{m,m'-1})$.

c) $\sigma_{m,m'}(A^*)=\overline{\sigma_{m,m'}(A)}$.

d) *If* $P \in Op(S_{1,0}^m)$, *then* $\sigma_{m,0}(P)\equiv p_m(x, \xi)$ $(mod\ \Sigma^{m,-1})$.

Here, the continuity of $\sigma_{m,m'}$ means that it is induced by a continuous (for their natural topology of Frechet spaces) linear map from $\mathrm{Op}(\Sigma_\Lambda^{m,m'})$ into $\Sigma_\Lambda^{m,m'}$.

This theorem will be proved first in the case $\Lambda=T_0^*(\boldsymbol{R}^n)=\{(x, \xi)\,|\,x=0\}$ (§ 3) and then extended to general Λ using invariance properties (§ 4). It is evident that, in the whole theory, $\boldsymbol{R}^n$ can be replaced by any n-dimensional manifold.

1.7. *Remark.* These operators are not microlocal. However, given $a \in \Sigma_\Lambda^{m,m'}$, we can always choose an $A \in \mathrm{Op}\,\Sigma_\Lambda^{m,m'}$, with $\sigma_{m,m'}(A)=a$, which is "as microlocal as one wants". We have just to take any $\tilde{A}$ with $\sigma_{m,m'}(\tilde{A})=a$, and then put $A=\sum \psi_i(x, D)\tilde{A}\varphi_i(x, D)$ with φ_i, $\psi_i(x, \xi) \in S_{1,0}^m$; $1=\sum \varphi_i$; $\psi_i=1$ near $\mathrm{Supp}\,(\varphi_i)$; $\mathrm{Supp}\,(\psi_i)$ small.

1.8. *2-microlocal ellipticity.* Let (x_0, ξ_0) belong to Λ and let $(\delta x_0, \delta\xi_0)$ be a vector tangent to $T^*\boldsymbol{R}^n$ at the point (x_0, ξ_0) and not tangent to Λ at that point. If $A \in \mathrm{Op}(\Sigma_\Lambda^{m,m'})$, we will say that A is *elliptic* at $(x_0, \xi_0, \delta x_0, \delta\xi_0)$ if one has

$$|\sigma_{m,m'}(A)(x_0+\varepsilon\delta x_0,\ \lambda(\xi_0+\varepsilon\delta\xi_0)|\geqslant C\lambda^m(\lambda\varepsilon)^{m'}$$

with $C>0$, for $0<\varepsilon\leqslant\varepsilon_0$ and $\lambda\varepsilon\geqslant\mu_0$.

It is easy to see that this definition does not depend on the choice (modulo $\Sigma^{m,m'-1}$) of $\sigma_{m,m'}(A)$, and that it depends only on the class of $(\delta x_0, \delta\xi_0)$ in $T_\Lambda(T^*\boldsymbol{R}^n)_{(x_0,\xi_0)}=T_{(x_0,\xi_0)}(T^*\boldsymbol{R}^n)/T_{(x_0,\xi_0)}\Lambda$. It is also easy to see that, if A is elliptic at $(x_0, \xi_0; \delta x_0, \delta\xi_0)$, it is elliptic at $(x, \xi; \delta x, \delta\xi)$, for (x, ξ) belonging to Λ sufficiently close to (x_0, ξ_0) and $(\delta x, \delta\xi)$ in a neighbourhood of $(\delta x_0, \delta\xi_0)$.

1.9. Definition (Second wave front).

a) Let $(x_0, \xi_0) \in \Lambda$ and let u belong to $H_\Lambda^{s,-\infty}(=\bigcup H_\Lambda^{s,s'})$ microlocally at (x_0, ξ_0). We shall say that u belongs to $H_\Lambda^{s,s'}$ 2-microlocally at $(x_0, \xi_0; \delta x_0, \delta\xi_0)$ if there exists $A \in \mathrm{Op}(\Sigma^{0,0})$, elliptic at that point, concentrated near (x_0, ξ_0), such that $Au \in H_\Lambda^{s,s'}$.

b) If $u \in H_\Lambda^{s,-\infty}$, we can define its $H^{s,s'}$-wave front as a closed subset of $(T^*\boldsymbol{R}^n \setminus \Lambda) \cup T_\Lambda(T^*\boldsymbol{R}^n)$ equipped with its natural topology of blow-up: it consists of points $(x, \xi) \in T^*\boldsymbol{R}^n \setminus \Lambda$ such that u does not belong to $H^{s+s'}$ microlocally at (x, ξ) and of points $(x, \xi; \delta x, \delta\xi)$ in $T_\Lambda(T^*\boldsymbol{R}^n)$ such that u does not belong to $H^{s,s'}$ 2-microlocally at that point.

One should pay attention to the fact that, before talking about its $H^{s,s'}$-2-microlocal regularity, one should know that u belongs microlocally to $H^{s,-\infty}$. The same observation is valid for the following theorem, which is a standard consequence of Theorem 1.6.

1.10. Theorem. *Let A belong to* $\mathrm{Op}\,\Sigma^{m,m'}$.

a) *Assume u belongs to $H^{s,-\infty}$ microlocally near Λ and to $H^{s,s'}$ 2-microlocally at $(x, \xi, \delta x, \delta\xi)$. Then Au belongs to $H^{s-m,s'-m'}$ 2-microlocally at that point.*

b) *Assume now that u belongs to $H^{s,-\infty}$ microlocally near Λ that $Au \in H^{s-m,s'-m'}$ 2-microlocally at $(x, \xi, \delta x, \delta\xi)$, and that A is elliptic at that point. Then, one has $u \in H^{s,s'}$ 2-microlocally at the same point.*

§ 2. Littlewood-Paley theory and 2-microlocal Sobolev spaces

2.1. Littlewood-Paley decomposition. Let $\varphi(\xi)$ be a C^∞ function with support in the ring $C_0=\{\xi \in \boldsymbol{R}^n | k^{-1} \leqslant |\xi| \leqslant 2k\}$, $k>1$, such that $\sum_{-\infty}^{+\infty} \varphi(2^{-j}\xi) =1$. We shall write $\psi(\xi)=\sum_{-\infty}^{-1} \varphi(2^{-j}\xi)$, and $C_j=2^j C_0$. The Littlewood-Paley decomposition of $u \in \mathscr{S}'(\boldsymbol{R}^n)$ is defined by

$$u=u_{-1}+\sum_0^\infty u_j=\psi(D)u+\sum_0^\infty \varphi(2^{-j}D)u.$$

It is well known (see [7]) that if u belongs to H^s, (the norm being noted $|\cdot|_s$), one has

$$(2.1) \qquad 2^{js}|u_j|_0 \leqslant c_j \quad \text{with} \quad \sum c_j^2 < \infty.$$

Conversely, if a family of functions u_j satisfies (2.1) and if the spectrum (=support of Fourier transform) of each u_j is contained in C_j, one has $\sum u_j \in H^s$.

2.2. Proposition. *Assume u belongs to H^s and* $\mathrm{Supp}(u) \subset F$. *Then one has, for each $N>0$, an estimate:*

(2.2) $$|2^{js}(1+2^j d(x, F))^N u_j(x)|_0 \leqslant c_{jN};\ \sum_j c_{jN}^2 < \infty$$

where $d(x, F)$ is the euclidean distance to the closed set F.

It is an easy consequence of

(2.3) $$u_j = \varphi(2^{-j}D)u = u_*[2^{nj}h(2^j x)]$$

where h (the inverse Fourier transform of φ) belongs to $\mathscr{S}$ and satisfies $\int P(t)h(t)dt = 0$ for each polynomial P.

This proposition says that L.P. decomposition is as local as allowed by the uncertainty principle. It will allow us in §§2 and 3, where the theory is local near 0, to consider only (sometimes implicitly) distributions with support in, for instance, the unit ball $B(0, 1)$. The following theorem gives more precise informations on the non-local character of L.P. decomposition.

2.3. Theorem. *Let T_{pq} the operators defined, for $p \leqslant q$, by*

(2.4) $$T_{pq} = \varphi(2^p x)\varphi(2^{-q}D);\ T_{pq}^* = \varphi(2^{-q}D)\varphi(2^p x).$$

Then there exist N_0 and constants C_k such that

(2.5) $$\| T_{p',q'} \circ T_{pq}^* \| \leqslant C_k 2^{-k(|p-p'|+|q-q'|)}$$

(2.6) $$\| T_{p'q'}^* \circ T_{pq} \| \leqslant C_k 2^{-k(|p-p'|+|q-q'|)}.$$

Where $\|\cdot\|$ denotes the operator norm in L^2. These estimates are also valid for $T_{pq} = \varphi_1(2^p x)\varphi_2(2^{-q}D)$, if φ_1 and φ_2 are supported in some ring.

The estimate (2.5) is trivial for $|p-p'|$ and $|q-q'|$ bounded by some fixed N_0. We have to estimate

$$(T_{p'q'} \circ T_{pq}^* u, v) = (T_{pq}^* u, T_{pq}^* v) \qquad \text{for } |u|_0 = |v|_0 = 1.$$

This scalar product is 0 for $|q-q'|$ larger than some convenient N_0, because the Fourier transforms have disjoint supports. For $|q-q'| \leqslant N_0$, it follows immediately from (2.2), applied to $\varphi(2^p x)u(x)$, that the scalar product can be estimated by any negative power of 2^q times the distance between the supports of $\varphi(2^p x)$ and $\varphi(2^{p'}x)$. For $|p-p'| > N_0$, this distance is $\geqslant c2^{-\mathrm{Min}(p,p')}$ and thus

(2.7) $$\| T_{p'q'} \circ T_{pq}^* \| \leqslant C_k 2^{-k[q-\mathrm{Min}(p,p')]}.$$

Using $p \leqslant q$ and $p' \leqslant q'$, we have then proved (2.5) in all cases.

The proof of (2.6) is along the same lines. The left hand side is 0 for $|p-p'|>N_0$ and one has

$$\|T^*_{p'q'}\circ T_{pq}\|\leqslant C_k 2^{-k(\mathrm{Max}(q,q')-p)} \qquad \text{for } |p-p'|\leqslant N_0,\ |q-q'|\geqslant N_0. \tag{2.8}$$

2.4. Theorem (Characterization of 2-microlocal Sobolev spaces). *When Λ is the Lagrangean manifold conormal to 0 in $\boldsymbol{R}^n$ ($\Lambda=T^*_0\boldsymbol{R}^n$), we shall note $H^{s,s'}_0$ (instead of $H^{s,s'}_\Lambda$) the spaces defined in n° 2.1. A distribution u belongs to $H^{s,s'}_0$ if and only if one has*

$$|2^{js}(1+2^j|x|)^{s'}u_j(x)|_0\leqslant c_j; \qquad \sum c_j^2<\infty \tag{2.9}$$

or

$$\begin{cases}|T_{pq}u|_0\leqslant c_{pq}2^{-qs}2^{-(q-p)s'}; & \sum\sum c_{pq}^2<\infty\\ |\psi(2^q x)\varphi(2^{-q}D)u|_0\leqslant c'_q 2^{-qs}; & \sum c'^2_q<\infty.\end{cases} \tag{2.10}$$

It is clear that (2.9) and (2.10) are equivalent. Consider now the case $s'=k\in\boldsymbol{N}$. The operators x_jD_k will play the part of operators M_i in Definition (1.2), and then

$$u\in H^{s,k}_0 \Longleftrightarrow x^\alpha u\in H^{s+|\alpha|}, \qquad |\alpha|\leqslant k. \tag{2.11}$$

Then, using (2.1) for functions $x^\alpha u$, and commuting $\varphi(2^{-q}D)$ and multiplications by x_j, one obtains easily the equivalence with (2.9).

It is evident that spaces defined by (2.9) for (s, s') and for $(-s, -s')$ are dual spaces, and that the family of these spaces is stable by complex interpolation. So (2.9) characterizes $H^{s,s'}_0$ in all cases.

2.5. Definition (Weighted Sobolev spaces). Let v belong to $\mathscr{D}'(\boldsymbol{R}^n\setminus\{0\})$, vanishing outside $B(0, 1)$. We will say that v belongs to $SP(s, s')$ if

$$|\varphi(x)u(2^{-j}x)|_{s+s'}\leqslant c_j 2^{-j(s-n/2)}; \qquad \sum c_j^2<\infty. \tag{2.12}$$

It is equivalent to say that

$$\begin{cases}|T^*_{pq}v|_0\leqslant c_{pq}2^{-qs}2^{-(q-p)s'}; & \sum\sum c_{pq}^2<\infty\\ |\psi(2^{-q}D)[\varphi(2^q x)v]|_0\leqslant c'_q 2^{-qs}; & \sum c'^2_q<\infty.\end{cases} \tag{2.13}$$

For $s+s'\in\boldsymbol{N}$, (2.12) is equivalent to

$$|x|^{-s+|\lambda|}D^\lambda u\in L^2 \qquad \text{for } 0\leqslant|\lambda|\leqslant s+s'. \tag{2.14}$$

These spaces are actually quite classical weighted Sobolev spaces. Our indexation by (s, s') will be convenient for us, but is not the usual one.

2.6. Definition (Flattening and finite part operators). The operator Π is defined for $u \in \mathscr{D}'(\mathbf{R}^n)$ (vanishing outside $B(0, 1)$ for instance by

(2.15)
$$\Pi u = \sum_{p \leqslant q}\sum T_{pq}u = \sum_{p \leqslant q}\sum \varphi(2^p x)\varphi(2^{-q}D)u.$$

The operator Pf is defined for $v \in \bigcup SP(s, s')$ (extendable distributions in $\mathbf{R}^n \setminus \{0\}$) by

(2.16)
$$Pf\, v = \sum_{p \leqslant q}\sum T^*_{pq}v = \sum_{p \leqslant q}\sum \varphi(2^{-q}D)\varphi(2^p x)v.$$

2.7. Theorem. *For $s \geqslant 0$, $s+s' \geqslant 0$, $H_0^{s,s'}$ and $SP(s, s')$ are subspaces of $L^2(\mathbf{R}^n) = L^2(\mathbf{R}^n \setminus \{0\})$ and one has*

a) $SP(s, s') \subset H_0^{s,s'}$; $SP(s, s') \cap H_0^{s,\infty} = SP(s, \infty)$

b) *Π maps $H^{s,s'}$ into $SP(s, s')$*

c) *Pf maps $SP(s, s')$ into $H_0^{s,s'}$*

d) *$E = I - \Pi$ maps $H_0^{s,s'}$ into $H_0^{s,\infty}$ and $F = I - Pf$ maps $SP(s, s')$ into $H_0^{s,\infty}$.*

Using interpolation, it is sufficient to prove this theorem when $s \in \mathbf{N}$ and $s+s' \in \mathbf{N}$, which we will assume now. Recall that $u \in H^{s,-k}$, with $k \in \mathbf{N}$, if and ony if one has

(2.17)
$$u = \sum_{|\alpha| \leqslant k} x^\alpha v_\alpha; \qquad v_\alpha \in H^{s-|\alpha|}.$$

a) The two properties are evident when $s' \geqslant 0$, because $u \in H_0^{s,s'}$ is equivalent to $x^\mu D^\lambda u \in L^2$ for $\lambda \leqslant s+s'$ and $|\mu| \geqslant \mathrm{Sup}(0, |\lambda|-s)$, (see (2.11)), while $u \in SP(s, s')$ is equivalent to the same proprety for $|\mu| \geqslant |\lambda|-s$ (see (2.14)).

When $u \in SP(s, -k)$, $k \in \mathbf{N}$, one can write

(2.18)
$$u = |x|^{-2k}\left(\sum x_i^2\right)^k u = \sum_{|\alpha|=k}\sum_{|\beta|=k} C_{\alpha\beta}x^\alpha \frac{x^\beta u(x)}{|x|^{2k}}.$$

It is clear that $|x|^{-2k}x^\beta u(x)$ belongs to $SP(s-k, 0)$, and $u \in H_0^{s,-k}$ is a consequence of (2.17) and (2.18).

b) In view of Theorem 2.3 and of Cotlar lemma, Π is bounded on L^2. We shall prove that Π maps $H^{1,-1}$ into $SP(1, 1)$ and $H^{0,1}$ into $SP(0, 1)$, the general case will follow easily by induction.

If $u \in H_0^{1,-1}$, one has $u = u_0 + \sum x_i u_i$, with $u_0 \in H^1$ and $u_i \in L^2$. We have

$$\frac{1}{|x|}\Pi u=\left[\sum 2^{p-q}\frac{\varphi(2^p x)}{2^p|x|}\frac{\varphi(2^{-q}D)}{2^{-q}|D|}\right]|D|u_0+\left[\sum\frac{2^p x_i\varphi(2^p x)}{2^p|x|}\varphi(2^{-q}D)\right]u_i$$
$$+\left[\sum 2^{p-q}\frac{\varphi(2^p x)}{2^p|x|}\varphi_i'(2^{-q}D)\right]u_i.$$

The brackets above are bounded operators in L^2, in view of Cotlar lemma and Theorem 2.3. Hence, $|x|^{-1}\Pi u\in L^2$ which means that $u\in SP(1,-1)$.

Assume now $u\in H_0^{0,1}=SP(0,1)$ which means that u and $x_k D_l u$ belong to L^2. We have just to prove that $[x_k D_l, \Pi]$ is bounded in L^2, which follows from Theorem 2.3 and from

$$[x_k D_l, \Pi]=[\sum(2^p x_k)\varphi_l'(2^p x)\varphi(2^{-q}D)]+[\sum\varphi(2^p x)(2^{-q}D_l)\varphi_k'(2^{-q}D)].$$

c) The proof is along the same lines, Pf being bounded on L^2 by Theorem 2.3.

Assume first $v\in SP(+1,-1)$, we have

$$Pfv=\sum_i x_i[\sum\varphi(2^{-q}D)\varphi(2^p x)]\left(\frac{x_i v}{|x|^2}\right)+\sum_i[\sum 2^{-q}\varphi_i'(2^{-q}D)\varphi(2^p x)]\left(\frac{x_i v}{|x|^2}\right).$$

The first terms are products of x_i by functions of L^2, while the last ones belong to H^1, because their derivatives

$$[\sum(2^{-q}D_l)\varphi_i'(2^{-q}D)\varphi(2^p x)]\left(\frac{x_i v}{|x|^2}\right)$$

belong to L^2 by Theorem 2.3. We have Thus proved that $Pfv\in H_0^{1,-1}$.

Assume now $v\in SP(0,1)=H_0^{0,1}$. As above, we can prove easily that $[x_k D_l, Pf]$ is bounded on L^2 and thus that Pfv belong to $H^{0,1}$. The general case will follow by induction.

Part d) will follow from the following lemma.

2.8. Lemma. *Let φ_1 belong to $C_0^\infty(C_0)$ and $\psi_1\in\mathscr{S}(R^n)$. Then the operators*

$$E_1=\sum_{p=0}^{\infty}\psi_1(2^p x)\varphi_1(2^{-p}D)$$

and

$$F_1=\sum_{p=0}^{\infty}\varphi_1(2^{-p}D)\psi_1(2^p x)$$

are bounded on L^2.

Put $T_p=\psi_1(2^p x)\varphi_1(2^{-p}D)$. It is clear that one has $T_pT_q^*=0$ for $|q-p|$ larger than some N_0. We are going to prove now that $\|T_p^*T_q\|\leqslant C2^{-|p-q|}$, and it is sufficient to prove this in the following two particular cases.

1st case: Assume the support of $\hat{\psi}$ is contained in a small neighbourhood of 0. Then the support of $\widehat{T_pu}$ is contained in a ring slightly larger than C_p. We have then $(T_pu\,|\,T_qv)=0$ for $|p-q|$ larger than some N_0 and hence $T_p^*T_q=0$ for $|p-q|\geqslant N_0$.

2nd case: Assume $\psi(0)=0$, and, for instance $q\geqslant p$,

$$T_p^*T_q=\varphi_1(2^{-p}D)\psi_1(2^p x)\psi_1(2^q x)\varphi_1(2^{-q}D).$$

$$\begin{aligned}|\psi_1(2^p x)\psi_1(2^q x)|&\leqslant C2^p|x|\,|\psi_1(2^q x)|\leqslant C2^{p-q}|2^q x\psi_1(2^q x)|\\&\leqslant C'2^{p-q}.\end{aligned}$$

We have then $\|T_p^*T_q\|\leqslant C''2^{p-q}$.

Any function ψ_1 being the sum of functions as above, we have proved in all cases that $\|T_p^*T_q\|$ and $\|T_pT_q^*\|$ are bounded by $C2^{-|p-q|}$ and it follows immediately from Cotlar Lemma that $\sum T_p$ and $\sum T_p^*$ are bounded on L^2.

Proof of part d) *of Theorem* 2.7.: We have

$$E=I-\Pi=\sum\psi(2^p x)\varphi(2^{-p}D)$$
$$F=I-Pf=\sum\varphi(2^{-p}D)\psi(2^p x)$$

where the function ψ is defined in n° 2.1.

We can first note that $x^\alpha D^\beta E$ and $x^\alpha D^\beta F$, for $|\alpha|=|\beta|$ are as described in Lemma 2.8. For instance

$$x_kD_lE=\sum(2^p x_k)\psi(2^p x)(2^{-p}D_l)\varphi(2^{-p}D)+\sum(2^p x_k)\psi_l'(2^p x)\varphi(2^{-p}D).$$

It remains to prove that operators E_1 and F_1 in Lemma 2.8 map $H_0^{s,-s}$ (resp. $SP(s,-s)$) into $H_0^{s,-s}$, which is true for $s=0$. If $u\in H_0^{1,-1}$, we have $u=u_0+\sum x_iu_i$ as in proof of part b). We have then $E_1u_0\in H^1$ because

$$D_lE_1u_0=E_1D_lu_0+\left[\sum\psi_{1,l}'(2^p x)\frac{\varphi_1(2^{-p}D)}{2^{-p}|D|}\right](Du_0).$$

A similar argument shows that $E_1(x_iu_i)$ belongs to $x_i\cdot L^2+H^1$. This proves that E_1u belongs to $H_0^{1,-1}$. The general case follows by induction, and the proof for F is along the same lines.

2.9. Theorem. *For* $s, s'\in \mathbf{R}$, *one has*

a) Π *maps* $H_0^{s,s'}$ *into* $SP(s,s')$

b) *Pf maps* $SP(s, s')$ *into* $H_0^{s,s'}$
c) $I-\Pi\circ Pf$ *maps* $SP(s, s')$ *into* $SP(s, \infty)$
d) $I-Pf\circ\Pi$ *maps* $H_0^{s,s'}$ *into* $H_0^{s,\infty}$.

From Theorem 2.7, we know that Π maps $H_0^{s,s'}$ into $SP(s, s')$ for $s\geqslant 0$, $s'\geqslant 0$, and thus $Pf=\Pi^*$ maps $SP(-s, -s')$ into $H_0^{-s,-s'}$. For the same reason, Pf maps $SP(s, s')$ into $H_0^{s,s'}$ and thus $\Pi=Pf^*$ maps $H_0^{-s,-s'}$ into $SP(-s, -s')$ We deduce then parts a) and b) of the theorem by complex interpolation.

For $s\geqslant 0$, $s'\geqslant 0$, we have:

$$(I-\Pi\circ Pf)u=E\circ Pfu+Fu.$$

For $u\in SP(s, s')$, we have $Pf\,u\in H_0^{s,s'}$, $E\circ Pf\,u\in H_0^{s,\infty}$, $Fu\in H_0^{s,\infty}$, and then $(I-\Pi\circ Pf)u\in H_0^{s,s'}$. We have obviously $(I-\Pi\circ Pf)u\in SP(s, s')$ and then, using Theorem 2.7. a), $(I-\Pi\circ Pf)u\in SP(s, \infty)$.

We have also $(I-Pf\circ\Pi)u=F\circ\Pi u+Eu$. For $u\in H_0^{s,s'}$, we have $\Pi u\in SP(s, s')$, $F\circ\Pi u\in H_0^{s,\infty}$, $Eu\in H_0^{s,\infty}$, and then $(I-Pf\circ\Pi)u\in H_0^{s,\infty}$.

Take now $s\in \boldsymbol{R}$, and $a\geqslant|s|$. We have that $(I-Pf\circ\Pi)$ maps $SP(s+a, 0)$ into $SP(s+a, 2N)$, and (it is the adjoint of $I-\Pi\circ Pf$) maps $SP(s-a, -2N)$ into $SP(s-a, 0)$. By complex interpolation, one gets that it maps $SP(s, -N)$ into $SP(s, +N)$ which proves part d) of the theorem. A similar argument gives part c).

2.10. *Remark.* We have thus proved that Π and Pf induce inverse isomorphims:

$$H_0^{s,s'}/H_0^{s,\infty}\xrightarrow{\ \sim\ } SP(s, s')/SP(s, \infty).$$

It is easy to see that this isomorphism does not depend on the choice of the function φ used to define Π and Pf. Actually, if $\tilde{\Pi}$ and $\widetilde{Pf}$ come from another choice, the proof of Theorem 2.9 gives that $I-\tilde{\Pi}\circ Pf$, $I-\Pi\circ\widetilde{Pf}$ are regularizing operators.

It is also possible to define $\tilde{\Pi}$ and $\widetilde{Pf}$ in the following manner. Let $\chi(t)$ be a C^∞ function, nonnegative, equal to 1 for $t\geqslant 2k$ and to 0 for $t\leqslant k/2$, and put

$$\widetilde{Pf}^*u=\tilde{\Pi}u=\int e^{ix\xi}\chi(|x||\xi|)\hat{u}(\xi)d\xi/(2\pi)^n.$$

It is not difficult to prove that $\Pi-\tilde{\Pi}$ maps $H_0^{s,s'}$ into $SP(s, \infty)$ and that $Pf-\widetilde{Pf}$ maps $SP(s, s')$ into $H_0^{s,\infty}$.

§ 3. Quantization of 2-microlaoal Symbols (Related to $\Lambda=T_0^*(\boldsymbol{R}^n)$)

3.1. 2-microlocal symbols. According to Definition 1.3, where the

M_i become the $x_j D_k$, and $\mathscr{H}_1$ is the space of vector fields generated by $\xi_j(\partial/\partial\xi_k)$ and $x_j(\partial/\partial x_k)$, one has

$$(3.1)\quad a(x, \xi) \in \Sigma_0^{m,m'} \Longleftrightarrow |D_\xi^\alpha D_x^\beta a(x, \xi)| \leqslant C_{\alpha\beta}\langle\xi\rangle^{m-|\alpha|+|\beta|}(1+|x|\,|\xi|)^{m'-|\beta|}.$$

In this section, we shall associate to such symbols operators defined on spaces $H_0^{s,s'}$. We shall first associate operators defined on spaces $SP(s, s')$, and then we shall use isomorphisms given by Theorem 2.9.

We shall always assume implicitly that symbols and operators are 0 outside the unit ball, and shall not write the associated cut off functions.

3.2. Definition (Special symbols). *We shall say that a C^∞ function $a(x, \xi)$ defined for $x \neq 0$ belongs to $S\Sigma_0^{m,m'}$ if*

a) *$a(x, \xi)$ is flat in the following sense:*

$$(3.2)\qquad |D_\xi^\alpha D_x^\beta a(x, \xi)| \leqslant C_{\alpha\beta}|x|^{-m+|\alpha|-|\beta|}(1+|x|\,|\xi|)^{m+m'-|\alpha|}.$$

b) *$a(x, D)$ is properly supported, in a strong sense, in $\boldsymbol{R}^n \setminus \{\theta\}$:*

$$(3.3)\qquad \operatorname{Supp}(\hat{a}^2(x, x-y)) \subset \left\{(x, y) \,\middle|\, k^{-1} \leqslant \frac{|y|}{|x|} < k\right\}$$

for some $k > 1$.

Here, $\hat{a}^2$ means the Fourier transform with respect to ξ.

3.3. Theorem. *Let $\sigma(t)$ be a C^∞ function, equal to 1 for $|t| \leqslant 1/4$ and to 0 for $|t| \geqslant 1/2$. Then, the map*

$$a(x, \xi) \rightarrow [1-\sigma(|x|\,|\xi|)]a(x, \xi)$$

induces isomorphisms independent of σ):

$$S\Sigma_0^{m,m'}/S\Sigma_0^{m,-\infty} \xrightarrow{\ \sim\ } \Sigma_0^{m,m'}/\Sigma_0^{m,-\infty}.$$

Estimates (3.1) and (3.3) are equivalent for $|x||\xi| \geqslant 1/4$, and it is evident that the quotient map is well defined, does not depend on σ, and is injective. We have to prove that it is surjective.

Let $a(x, \xi) \in \Sigma_0^{m,m'}$, and put

$$b(x, \xi) = a(x, \xi)[1-\sigma(|x|\,|\xi|)].$$

$$c(x, \xi) = \int e^{i\xi z}\hat{b}^2(x, z)\sigma\left(\frac{|z|}{|x|}\right)dz/(2\pi)^n.$$

Property (3.3) is then valid for c. One has:

$$c(x, \xi) = \int b\left(x, \xi - \frac{\theta}{|x|}\right) h(\theta) d\theta,$$

where $h(\theta)$ is the inverse Fourier transform of $\sigma(|z|)$. Integrating first in the domain $|\theta|/|x| \leqslant |\xi|/2$, we obtain an estimate by $|x|^{-m}(|x||\xi|)^{m+m'}$. For $|\theta|/|x| \geqslant |\xi|/2$ one has $|\xi - \theta/|x|| \leqslant 3|\theta|/|x|$ and

$$\left| b\left(x, \xi - \frac{\theta}{|x|}\right) \right| \leqslant c|x|^{-m}|\theta|^{|m+m'|}(1+|0|)^{-M},$$

and

$$\int_{|\theta| \geqslant |x||\xi|/2} b\left(x, \xi - \frac{\theta}{|x|}\right) h(\theta) d\theta \leqslant c|x|^{-m}(1+|x||\xi|)^{-N}$$

for every $N > 0$. The same method works for derivatives and estimate (3.2) is proved for $c(x, \xi)$.

We have now to prove that $b(x, \xi) - c(x, \xi)$ satisfies estimates (3.2) (or 3.1) for $|x||\xi| \geqslant 1/4$, with $m' \to -\infty$. Using

$$\int h(\theta) d\theta = 1 \quad \text{and} \quad \int h(\theta)\theta^{\lambda} d\theta = 0$$

(h is the inverse Fourier transform of $\sigma(|z|)$), one gets for each M:

$$c(x, \xi) - b(x, \xi) = \int \left[b\left(x, \xi - \frac{\theta}{|x|}\right) - \sum_{|\lambda| < M} \frac{1}{\lambda!} D_{\xi}^{\lambda} b(\xi) \left(-\frac{\theta}{|x|}\right)^{\lambda} \right] h(\theta) d\theta.$$

Integration over $|\theta|/|x| \leqslant |\xi|/2$ gives an estimate by

$$|x|^{-m}(|x||\xi|)^{m+m'-M} \int |\theta|^M h(\theta) d\theta,$$

while the integral over $|\theta|/|x| \geqslant |\xi|/2$ can be estimated as above.

The same method works for $D_{\xi}^{\alpha} D_{x}^{\beta}(c-b)$ and Theorem 3.3 is proved.

3.4. Theorem. *An operator A (in $C_0^\infty(\boldsymbol{R}^n \setminus \{0\})$, vanishing outside the unit ball can be written $A = a(x, D)$ with $a \in S\Sigma^{m, m'}$, if and only if one has:*

(3.4) $$A = \sum_{p=0}^{\infty} 2^{pm} \Theta_p \mathscr{A}_p \Theta_{-p}$$

where θ_p is the dilatation operator: $\Theta_p u(x) = u(2^p x)$, and $\mathscr{A}_p$ is a ***bounded*** *family of pseudo-differential operators of order $m + m'$, with support in a* ***fixed ring****:*

$$k_0^{-1} \leqslant |x| \leqslant 2k_0.$$

Let $A = a(x, D)$ with $a \in S\Sigma^{m,m'}$. With the function φ defined in 2.1, one has:

$$A = \sum_{p=0}^{\infty} \varphi(2^p x) a(x, D)$$

and it is clear, using (3.3), that the pseudo-differential operator $\varphi(2^p x) a(x, D)$ is supported in a ring $k_0^{-1} 2^{-p} \leqslant |x| \leqslant 2k_0 2^{-p}$, with k_0 independent of p. Put

$$\mathscr{A}_p = 2^{-pm} \Theta_{-p} \varphi(2^p x) a(x, D) \Theta_p,$$

one has (3.4), and the $\mathscr{A}_p$ are supported by a fixed ring. The symbol σ_p of $\mathscr{A}_p$ is given by

$$\sigma_p(x, \xi) = 2^{-pm} \varphi(x) a(2^{-p} x, 2^p \xi).$$

From estimate (3.2), one gets:

$$\text{(3.5)} \qquad |D_\xi^\alpha D_x^\beta \sigma_p(x, \xi)| \leqslant C_{\alpha\beta} |x|^{-m+|\alpha|-|\beta|} (1 + |x|\,|\xi|)^{m+m'-|\alpha|}$$

but, $|x|$ and $|x|^{-1}$ being bounded on the support of σ_p, one gets:

$$\text{(3.6)} \qquad |D_\xi^\alpha D_x^\beta \sigma_p(x, \xi)| \leqslant C'_{\alpha\beta} (1 + |\xi|)^{m+m'-|\alpha|}$$

with constants $C'_{\alpha\beta}$ independent of p.

Conversely, let $\mathscr{A}_p$ be such a bounded family, and define A by (3.4). One has:

$$a(x, \xi) = \sum 2^{pm} \sigma_p(2^p x, 2^{-p} \xi)$$

where the sum is finite ($2^p \in [k_0^{-1}/|x|, 2k_0/|x|]$) for each x.

Property (3.3) comes from the fact that $\hat{\sigma}_p^2(x, x-y)$ has its support in a fixed ring, while estimate (3.2) comes from (3.5) (equivalent to (3.6)).

3.5. Definition. *We shall say that an operator (in $\boldsymbol{R}^n \setminus \{0\}$) is* ***negligible of order*** *m if it maps $SP(s, s')$ into $SP(s-m, +\infty)$ for each $s, s' \in \boldsymbol{R}$.*

We shall say that an operator A belongs to $\mathrm{Op}(S\Sigma_0^{m,m'})$ *if $A = a(x, D) + R$ with $a(x, \xi) \in S\Sigma_0^{m,m'}$ and R negligible of order m.*

3.6. Theorem. *An operator A (in $C_0^\infty(\boldsymbol{R}^n \setminus \{0\})$, vanishing outside the unit ball) belongs to* $\mathrm{Od}(S\Sigma_0^{m,m'})$ *if and only if the commutators*

$$(\mathrm{ab}\, D)^\beta (\mathrm{ad}\, M_x)^\alpha A$$

maps $Sp(s, s')$ *into* $SP(s-m-|\beta|+|\alpha|, s'-m'+|\beta|)$. *Here* M_x *denote the family of operators* M_{x_j}: *multiplication by* x_j.

In view of (3.4) and (2.12) it is clear that if A belongs to $\mathrm{Op}(S\Sigma^{m,m'})$, A maps $SP(s, s')$ into $SP(s-m, s'-m')$. It is also evident that $(\mathrm{ad}\, D)^\beta(\mathrm{ad}\, M_x)^\alpha A$ belongs to $\mathrm{Op}(S\Sigma^{m+|\beta|-|\alpha|, m'-|\beta|})$, and the first part of the theorem is proved.

Conversely, assume now A is an operator whose commutators satisfy the above property. The following properties are easy to prove:

(3.7) $$\|\Theta_p u\|_{SP(s,s')} \leqslant 2^{p(s-n/2)}\|u\|_{SP(s,s')}.$$

(3.8) $$|||\Theta_{-p}B\Theta_p||| = 2^{pm}|||B|||,$$

if $|||\cdot|||$ is the operator norm from $SP(s, s')$ into $SP(s-m, s'-m')$.

(3.9) $$\mathrm{ad}\, D_l\cdot(\Theta_{-p}B\Theta_p) = 2^p\Theta_{-p}(\mathrm{ad}\, D_l\cdot B)\Theta_p.$$

(3.10) $$\mathrm{ad}\, M_{x_k}\cdot(\Theta_{-p}B\Theta_p) = 2^{-p}\Theta_{-p}(\mathrm{ad}\, M_{x_k}\cdot B)\Theta_p.$$

A direct consequence of these properties is that

(3.11) $$(\mathrm{ad}\, D)^\beta(\mathrm{ad}\, M_x)^\alpha\cdot(\Theta_{-p}A\Theta_p)$$

is bounded from $SP(s, s')$ into $SP(s-m, s'-m')$ with a norm $\leqslant C2^{pm}$, with C independent of p.

Consider now $\mathscr{A}_p = 2^{-pm}\Theta_{-p}A\Theta_p\varphi(x)$. It is an operator from $C^\infty(\boldsymbol{R}^n\setminus\{0\})$ into itself. Its kernel is supported in $\{(x, y)\,|\,k^{-1}\leqslant y\leqslant 2k;\ |x|\leqslant 2^p\}$, and one has:

$$A = \sum 2^{pm}\Theta_p\mathscr{A}_p\Theta_{-p}.$$

Put:

$$\sigma_p(x, \xi) = e^{-ix\xi}\mathscr{A}_p(e^{ix\xi}).$$

For every $s\in\boldsymbol{R}$, one has:

$$\begin{aligned}\|\sigma_p(x, \xi)e^{ix\xi}\|_{SP(s,-s)} &\leqslant C\|\varphi(x)e^{ix\xi}\|_{SP(s+m,-s+m')}\\ &\leqslant C'|\varphi(x)e^{ix\xi}|_{m+m'}\leqslant C''(1+|\xi|)^{m+m'}.\end{aligned}$$

Recall now (2.14) that

$$v\in SP(s, -s)\Longleftrightarrow |x|^{-s}v\in L^2.$$

Then

$$\||x|^{-s}\sigma_p(x, \xi)\|_0 \leqslant C''(1+|\xi|)^{m+m'}.$$

Using the same argument for commutators (3.11), one obtains:

$$\||x|^{-s}D_\xi^\alpha D_x^\beta \sigma_p(x, \xi)\|_0 \leqslant C_{s,\alpha,\beta}(1+|\xi|)^{m+m'-|\alpha|}$$

and, by Sobolev theorem

$$\operatorname*{Sup}_x \||x|^{-s}D_\xi^\alpha D_x^\beta \sigma_p(x, \xi)| \leqslant C'_{s,\alpha,\beta}(1+|\xi|)^{m+m'-|\alpha|}.$$

We have then proved that $\mathscr{A}_p$ is a bounded family of pseudo-differential operators of order $m+m'$, and that the same is true for the $|x|^s\mathscr{A}_p$, for each $s \in \boldsymbol{R}$.

Let now $\varphi_1(x)$ be a C^∞ function, equal to 1 in a neighbourhood of the support of φ, and with support in a ring: $k_1^{-1} \leqslant |x| \leqslant 2k_1$. Put

$$\mathscr{A}'_p = \varphi_1(x)\mathscr{A}_p; \qquad \mathscr{R}_p = \mathscr{A}_p - \mathscr{A}'_p.$$

$(\mathscr{A}'_p)$ is a bounded family of pseudo-differential operators of order $m+m'$ supported in a fixed ring, and by Theorem 3.4, one has $\sum 2^{pm}\Theta_p\mathscr{A}'_p\Theta_{-p} = a(x, D)$ with $a \in S\Sigma_0^{m,m'}$. For any $s \in \boldsymbol{R}$ and $N>0$, $(|x|^s R_p)$ is a bounded family of pseudo-differential operators of order $-N$, and one deduces easily from (2.12) that $\sum 2^{pm}\Theta_p R_p \Theta_{-p}$ is a negligible operator of order m.

3.7. Corollary. *The application $a(x, \xi) \to a(x, D)$ from $S\Sigma_0^{m,m'}$ into* Op *$Op(S\Sigma_0^{m,m'})$ induces an isomorphism of the bigraded $*$-algebra $S\Sigma_0^{m,m'}/S\Sigma_0^{m,-\infty}$ (for usual operations $a\# b$ and a^* for symbols) onto the bigraded $*$-algebra* $\mathrm{Op}(S\Sigma_0^{m,m'})/\mathrm{Op}(S\Sigma_0^{m,-\infty})$ *(for composition and adjoint).*

Everything is just a consequence of usual calculus of properly supported operators in $\boldsymbol{R}^n\setminus\{0\}$, except the fact that if $a(x, \xi) \in S\Sigma_0^{m,m'}$, and if $a(x, D)$ is negligible, then $a(x, \xi) \in S\Sigma^{m,-\infty}$. This last point is a direct consequence of Theorem 3.6.

3.8. Theorem. *If P is a differential operator of order m, it belongs to* $\mathrm{Op}(S\Sigma^{m,0})$, *and one has*

(3.12) $\quad P(x, D) - Pf \circ P(x, D) \circ \Pi$ *maps* $H^{s,s'}$ *into* $H^{s-m,+\infty}$.

Moreover, for multiplication M_{x_k} by x_k, one has

(3.13) $\quad M_{x_k} - Pf \circ M_{x_k} \circ \Pi$ *maps* $H^{s,s'}$ *into* $H^{s+1,\infty}$.

It is clear that the symbol $p(x, \xi)$ belongs to $S\Sigma^{m,0}$, and that $x_k \in S\Sigma^{-1,1}$. The operator $p(x, D)$ with its usual meaning is then defined as an element of $\mathrm{Op}(S\Sigma^{m,0})$.

Assume first $u \in H^{s,s'}$, with $s \geqslant m$, $s' \geqslant 0$. We have $u - \Pi u \in H^{s,\infty}$ (Theorem 2.7) and then:

$$Pu - P\Pi u \in H^{s-m,\infty}. \tag{3.14}$$

We have $\Pi u \in SP(s, s')$, hence $P\Pi u \in SP(s-m, s')$, and then (Theorem 2.7):

$$P\Pi u - PfP\Pi u \in H^{s-m,\infty}. \tag{3.15}$$

(3.12) is then a consequence, for $s \geqslant m$, $s' \geqslant 0$, of (3.14) and (3.15). The same proof is valid for (3.13) using $x_k(u - \Pi u) \in H^{s+1,\infty}$, and $x_k \Pi u \in SP(s+1, s'-1)$.

For general s, s', it is evident that $P - Pf \circ P \circ \Pi$ maps $H^{s,s'}$ into $H^{s-m,s'}$. Interpolating between this result (for large negative values of s, s') and (3.12) (for large positive values of s, s'), one gets (3.12) in full generality. The same argument is valid for (3.13).

3.9. Definition (2-microlocal operators related to $T_0^* \boldsymbol{R}^n$). *An operator A (in $C_0^\infty(\boldsymbol{R}^n)$ vanishing outside the unit ball) belongs to* $\mathrm{Op}(\Sigma_0^{m,m'})$ *if the commutators*

$$(\operatorname{ad} D)^\beta (\operatorname{ad} M_x)^\alpha A$$

map $H_0^{s,s'}$ into $H^{s-m-|\beta|+\alpha,\, s'-m'+|\beta|}$.

3.10. Theorem (Symbolic calculus). *There exists a canonical isomorphism from the bigraded $*$-algebra* $\mathrm{Op}(\Sigma_0^{m,m'})/\mathrm{Op}(\Sigma^{m,-\infty})$ *(for composition and adjoint) onto the bigraded $*$-algebra $\Sigma_0^{m,m'}/\Sigma_0^{m,-\infty}$ (for usual operations $a \# b$ and a^* for symbols).*

For $A \in$ $\mathrm{Op}(\Sigma_0^{m,m'})$, *we shall note $\sigma(A)$ its canonical image in $\Sigma_0^{m,m'}/\Sigma_0^{m,-\infty}$ (full symbol of A).*

Consider the map $A \to \tilde{A} = \Pi \circ A \circ Pf$. For $A \in \mathrm{Op}(\Sigma^{m,m'})$ and $B \in \mathrm{Op}(\Sigma^{m_1,m_1'})$ one has, using Theorem 2.9:

$$\tilde{A} \text{ maps } SP(s, s') \text{ into } SP(s-m, s'-m'). \tag{3.16}$$

$$\widetilde{AB} - \tilde{A}\tilde{B} \text{ maps } SP(s, s') \text{ into } SP(s-m-m_1, +\infty). \tag{3.17}$$

These properties, combined with Theorem 3.8 show that $(\operatorname{ad} D)^\beta \times (\operatorname{ad} M_x)^\alpha \tilde{A}$ maps $SP(s, s')$ into $SP(s-m-|\beta|+|\alpha|, s'-m'+|\beta|)$ and that $\tilde{A} \in \mathrm{Op}(S\Sigma^{m,m'})$ by Theorem 3.6.

By similar arguments, one gets that $\tilde{A} \to Pf \circ \tilde{A} \circ \Pi$ is a left and right inverse (modulo negligible operators), and commuting (3.17) with D and

M_x, that these maps are homomorphisms of bigraded algebras. Lastly, $(\Pi \circ A \circ Pf)^* = \Pi \circ A^* \circ Pf$ is a consequence of $\Pi^* = Pf$.

We shall now define the map σ as the composition

$$\mathrm{Op}(\Sigma^{m,m'})/\mathrm{Op}(\Sigma^{m,-\infty}) \xrightarrow{\sim} \mathrm{Op}(S\Sigma^{m,m'})/\mathrm{Op}(S\Sigma^{m,-\infty}) \xrightarrow{\sim}$$
$$\xrightarrow{\sim} S\Sigma^{m,m'}/S\Sigma^{m,-\infty} \xrightarrow{\sim} \Sigma^{m,m'}/\Sigma^{m,-\infty},$$

where the first map is $A \to \Pi \circ A \circ Pf$, the second one is defined in Corollary 3.7, and the third one in Theorem 3.3.

3.11. Theorem. *Let P be a classical pseudo-differential operator of order m. Then $P \in \mathrm{Op}(\Sigma^{m,0})$, and $\sigma(P)$ is equal to its ordinary symbol (modulo $\Sigma^{m,-\infty}$).*

Let us first prove that P maps $H_0^{s,s'}$ into $H_0^{s-m,s'}$. This is evident for $s' \in N$, using (2.11), and can be extended to general s' by duality and interpolation.

The operator $(\mathrm{ad}\, D)^\beta (\mathrm{ad}\, M_x)^\alpha P$ is of order $m-|\alpha|$ and, then it maps $H^{s,s'}$ into $H^{s-m+|\alpha|,s'} \subset H^{s-m+|\alpha|-|\beta|,s'+|\beta|}$. It follows that $P \in \mathrm{Op}(\Sigma_0^{m,0})$. It remains to prove that its ordinary symbol $p(x, \xi)$ differs from that defined in theorem 3.10 by an element of $\Sigma^{m,-\infty}$.

Assuming (for simplicity of notations) $m=0$, consider

$$\Pi \circ P \circ Pf = \sum_{p' \leqslant q';\; p \leqslant q} \sum \sum \sum \varphi(2^p x)\varphi(2^{-q}D)P\varphi(2^{-q'}D)\varphi(2^{p'}x) \tag{3.18}$$

one knows already that, modulo a negligible operator (in $(\mathrm{Op}(S\Sigma^{0,-\infty})$, its kernel $k(x, y)$ vanishes outside $k^{-1} \leqslant |x/y| \leqslant k$. We are then allowed to keep in (3.18) only terms with $|p-p'| \leqslant N_0$, choosing N_0 such that

$$\sum_{\nu=-N_0}^{N_0} \varphi(2^\nu x) = 1 \qquad \text{on Supp}\,(\varphi).$$

$$\Pi \circ P \circ Pf \simeq \sum_{\nu=-N_0}^{N_0} \sum_p \varphi(2^p x)[1-\psi(2^{-p}D)]P[1-\psi(2^{-p-\nu}D)]\varphi(2^{p+\nu}x). \tag{3.19}$$

The right hand side in (3.19) is equal to

$$\Sigma \Theta_p \mathscr{A}_p \Theta_{-p} \tag{3.20}$$

with $\mathscr{A}_p = \sum_\nu \varphi(x)[1-\psi(D)]P(2^{-p}x, 2^p D)[1-\psi(2^{-\nu}D]\varphi(2^\nu x)$.

Using classical symbolic calculus in a fixed ring, we get

$$\sigma(\mathscr{A}_p) = \varphi(x)p(2^{-p}x, 2^p\xi) + r_p(x, \xi)$$

where $r_p(x, \xi)$ is a bounded family in $S_{1,0}^{-N}$ for each N. The symbol of

(3.20) is equal to $p(x, \xi)+r(x, \xi)$ where $r(x, \xi)=\sum r_p(2^p x, 2^{-p}\xi)$ belongs to $S\Sigma_0^{0,-N}$ for each N.

3.12. *Remark.* It is clear, in view of the proofs above, that one can construct linear and continuous (for convenient topologies) maps from $\bigcup \Sigma_0^{m,m'}$ into $\bigcup \mathrm{Op}(\Sigma_0^{m,m'})$ (quantization map) and from $\bigcup \mathrm{Op}(\Sigma_0^{m,m'})$ into $\bigcup \Sigma_0^{m,m'}$ (symbol map) inducing the isomorphism of Theorem 3.10, but there is no reasonable canonical choice for these maps.

§ 4. Invariance Results and Quantization of 2-Microlocal Symbols (General Case).

The following theorem will prove that there exists only one reasonable notion of principal symbol.

4.1. Theorem. *Let T be a linear operator from $\bigcup \Sigma_0^{m,m'}$ into itself, such T maps continuously $\Sigma_0^{m,m'}$ into itself for each $(m, m') \in \boldsymbol{R}^2$. Assume that*

$$T(ab)=Ta\cdot Tb+R(a, b), \tag{4.1}$$

where R maps $\Sigma_0^{m_1,m_1'}\times\Sigma_0^{m_2,m_2'}$ into $\Sigma_0^{m_1+m_2,m_1'+m_2'-1}$.

$$T(\{a, b\})=\{Ta, Tb\}+S(a, b), \tag{4.2}$$

where S maps $\Sigma_0^{m_1,\prime_1}\times\Sigma_0^{m_2,m_2'}$ into $\Sigma^{m_1+m_2,m_1'+m_2'-2}$.

$$T(x_i)=x_i+r_i;\ T(\xi_j)=\xi_j+s_j;\ T(1)=1+t \tag{4.3}$$

with $r_i, \in \Sigma_0^{-1,0}$, $s_j \in \Sigma_0^{1,-1}$, $t \in \Sigma_0^{0,-1}$.

Then, for $a \in \Sigma_0^{m,m'}$, one has $Ta-a \in \Sigma_0^{m,m'-1}$.

One can introduce spaces which are invariant by dilatations in ξ for, say, $0<|x|\leqslant 1$.

$$a \in \tilde{\Sigma}^{m,m'} \Longleftrightarrow |D_\xi^\alpha D_x^\beta a(x, \xi)|\leqslant C_{\alpha\beta}|\xi|^{m+m'-|\alpha|}|x|^{m'-|\beta|}. \tag{4.4}$$

We shall denote by $N(m, m', k; a)$ the sum of the best constants $C_{\alpha\beta}$ in (4.4) for $|\alpha|+|\beta|\leqslant k$. One may assume that T maps $\tilde{\Sigma}_0^{m,m'}$ into itself, using a cutoff near $|x||\xi|=0$.

The following properties of dilatation operators θ_λ are immediate

$$\theta_\lambda a(x, \xi)=a(x, \lambda\xi). \tag{4.5}$$

$$N(m, m', k; \theta_\lambda a)=\lambda^{m+m'}N(m, m', k; a). \tag{4.6}$$

(4.7) $$\lambda \frac{\partial}{\partial \lambda} \theta_\lambda a = \Delta \circ \theta_\lambda a = \theta_\lambda \circ \Delta a,$$

with

(4.8) $$\Delta a = \sum \xi_j \frac{\partial a}{\partial \xi_j}.$$

We shall now prove that $[T, \Delta]$ maps $\tilde{\Sigma}^{m,m'}$ into $\tilde{\Sigma}^{m,m'-1}$. This will result from the fact that T commutes, up to lower order terms, with $\partial/\partial\xi_j$ and multiplication by ξ_j. For $a \in \tilde{\Sigma}^{m,m'}$, one has

$$T(\xi_j a) = T\xi_j \cdot Ta + R(\xi_j, a) = \xi_j Ta + s_j \cdot Ta + R(\xi_j, a)$$

using (4.1) and (4.3) and thus $T\xi_j a - \xi_j Ta \in \tilde{\Sigma}^{m+1,m'-1}$.

$$T\left(\frac{\partial a}{\partial \xi_j}\right) = T(\{x_j, a\}) = \{Tx_j, Ta\} + S(x_j, a)$$

$$= \frac{\partial}{\partial \xi_j} Ta + \{r_j, Ta\} + S(x_j, a)$$

and

$$\left[T, \frac{\partial}{\partial \xi_j}\right] a \in \tilde{\Sigma}^{m-1,m'}.$$

Consider now

(4.9) $$T^\lambda = \theta_\lambda T \theta_{1/\lambda}.$$

(4.10) $$T^\infty = \lim_{\lambda\to\infty} T_\lambda = T + \int_1^\infty \left(\lambda \frac{\partial}{\partial \lambda} T^\lambda\right) \frac{d\lambda}{\lambda}.$$

From (4.7) and (4.6), one gets

$$\lambda \frac{\partial}{\partial \lambda} T^\lambda = \theta_\lambda [T, \Delta] \theta_{1/\lambda}.$$

(4.11) $$\forall k, \exists k', N\left(m, m'-1, k; \lambda \frac{\partial}{\partial \lambda} T^\lambda a\right) \leqslant \frac{C}{\lambda} N(m, m', k'; a).$$

This proves that the integral in (4.10) is convergent and defines a continuous operator from $\Sigma^{m,m'}$ into $\Sigma^{m,m'-1}$. We are going to prove now that $T^\infty = \mathrm{Id}$, which will prove the theorem.

First, one has

(4.12) $$T^{\lambda}(ab)=T^{\lambda}a\cdot T^{\lambda}b+\theta_{\lambda}R(\theta_{-\lambda}a,\theta_{-\lambda}b),$$

and, given $a\in\tilde{\Sigma}^{m_1,m_1'}$ and $b\in\tilde{\Sigma}^{m_2,m_2'}$, any semi-norm in $\tilde{\Sigma}^{m_1+m_2,m_1'+m_2'-1}$ of $\theta_{\lambda}R(\theta_{-\lambda}a,\theta_{-\lambda}b)$ is bounded by a constant time λ^{-1} in view of (4.6). This proves that

(4.13) $$T^{\infty}(a\cdot b)=(T^{\infty}a)(T^{\infty}b).$$

For the same reason, one has

(4.14) $$T^{\infty}x_i=x_i;\quad T^{\infty}\xi_j=\xi_j;\quad T^{\infty}1=1.$$

For instance, $T^{\lambda}x_i=(1/\lambda)\theta_{\lambda}Tx_i=x_i+(1/\lambda)\theta_{\lambda}r_i$, and any semi-norm of $(1/\lambda)\theta_{\lambda}r_i$ in $\tilde{\Sigma}^{-1,0}$ is bounded by C/λ.

It is sufficient to prove that $T^{\infty}\varphi=\varphi$ for any $\varphi\in C_0^{\infty}(\boldsymbol{R}^n\diagdown\{0\}\times\boldsymbol{R}^n\diagdown\{0\})$, because any $a\in\tilde{\Sigma}^{mm'}$ can be approximated by such φ in the space $\tilde{\Sigma}^{-N,-N}+\tilde{\Sigma}^{N,N}$, with its natural topology, for $N>|m|+|m'|$. One can write, for any (x^0,ξ^0):

$$\varphi(x,\xi)=\varphi(x^0,\xi^0)+\sum(x_j-x_j^0)\alpha_j(x,\xi)+\sum(\xi_j-\xi_j^0)\beta_j(x,\xi)$$

with α_j and β_j belonging to $C^{\infty}(\boldsymbol{R}^n)$. One may moreover assume

$$\alpha_j=-\frac{(x_j-x_j^0)\varphi(x^0,\xi_0^0)}{\sum(x_i-x_i^0)^2+\sum(\xi_i-\xi_i^0)^2};\qquad \beta_j=-\frac{(\xi_j-\xi_j^0)\varphi(x^0,\xi^0)}{\sum(x_i-x_i^0)^2+\sum(\xi_i-\xi_i^0)^2}$$

outside a compact set, and hence, α_j and β_j belong to $\Sigma_0^{0,0}$. In view of (4.13) and (4.14), one has

$$T^{\infty}\varphi=\varphi(x^0,\xi^0)+\sum(x_j-x_j^0)T^{\infty}\alpha_j+\sum(\xi_j-\xi_j^0)T^{\infty}\beta_j,$$

and thus $T^{\infty}\varphi(x^0,\xi^0)=\varphi(x^0,\xi^0)$ which proves the theorem.

4.2. *Remark.* A similar, and simpler, argument proves the analogous result for classes $S_{1,0}^m$ of pseudo-differential symbols. The following theorem says that there is only one reasonable notion of principal symbol for $\mathrm{Op}(S_{1,0}^m)$. It can be proved using θ_{λ} ond Δ as above.

4.3. Theorem. *Let T be a linear operator from* $\bigcup S_{1,0}^m$ *into itself. One assume that T maps continuously* $S_{1,0}^m$ *into itself, and that, for* $a\in S_{1,0}^{m_1}$, $b\in S_{1,0}^{m_2}$, $m=m_1+m_2$, *one has*

$$T(ab)-ab\in S_{1,0}^{m-1}$$

$$T(\{a,b\})-\{Ta,Tb\}\in S_{1.0}^{m-2}$$

$$(T1-1);\quad (Tx_j-x_j);\quad |\xi|^{-1}(T\xi_j-\xi_j)\in S^{-1}_{1,0}.$$

Then, $(T-I)$ maps $S^m_{1,0}$ into $S^{m-1}_{1,0}$.

We shall obtain now the invariance of 2-microlocal calculus by Fourier integral operators as a direct consequence of Theorems 3.11 and 4.1.

4.4. Corollary. *Let $(Y(x,\xi),\ H(x,\xi))$ be an invertible canonical map from a neighbourhood of $(0,\xi_0)$ onto a neighbourhood of $(0,\eta_0)$, preserving $T^*_0(\mathbf{R}^n)$ (i.e. $Y(0,\xi)=0$). Let F and G be Fourier integral operators associated to this map and to its inverse, such that $G\circ F-I$ (resp. $F\circ G-I$) are of order $-\infty$ near $(0,\xi_0)$ (resp. $(0,\eta_0)$). Then*

a) *F maps $H^{s,s'}_0$ into $H^{s,s'}_0$*

b) *If $a(y,\eta)$ is a principal symbol of $A\in \mathrm{Op}(\Sigma^{m,m'}_0)$, then $A'=G\circ A\circ F$ belongs to $\mathrm{Op}(\Sigma^{m,m'}_0)$ and $a(Y(x,\xi),\ H(x,\xi))$ is a principal symbol of A' near $(0,\xi_0)$.*

Let $\mathcal{M}_0$ be the space of elements of $\mathrm{Op}(S^1_{1,0})$ whose principal symbol vanishes on $T^*_0(\mathbf{R}^n)$. It is clear that, if $M\in\mathcal{M}_0$, one has $G\circ M\circ F\in\mathcal{M}_0$. For $s'\in N$, the space $H^{s,s'}_0$ can be characterized by $M_1\circ\cdots\circ M_l u\in H^s$ for $M_j\in\mathcal{M}_0$ and $l\leqslant s'$. This proves a) for $s'\in N$, and the general case follows by duality and interpolation.

In view of Theorems 3.10 and 3.11, the Definition 3.9 of $A\in\mathrm{Op}(\Sigma^{m,m}_0)'$ is equivalent to the stronger condition:

$(\mathrm{ad}\,P_1)\circ\cdots\circ(\mathrm{ad}\,P_k)(\mathrm{ad}\,M_1)\cdots(\mathrm{ad}\,M_l)A$ maps H_0 into $H^{t,t'}_0$, with $t=s-m-\sum p_j,\ t'=s'-m'+k$, for $P_j\in\mathrm{Op}(S^{p_j}_{1,0})$ and $M_j\in\mathcal{M}_0$.

This condition is obviously preserved for $A'=G\circ A\circ F$ and A' belongs to $\mathrm{Op}(\Sigma^{m,m'}_0)$. Consider now the composition:

$$a(y,\eta)\longrightarrow b(x,\xi)=a(Y(x,\xi),\ H(x,\xi))\longrightarrow B\longrightarrow\sigma(F\circ B\circ G)$$

where the 2nd arrow and σ stand for a quantization and a symbol map (see Remark 3.12). One deduces easily from Theorem 4.1 that $\sigma(F\circ B\circ G)\equiv a$ $(\mathrm{mod}\ \Sigma^{m,m'-1}_0)$ near $(0,\eta_0)$. Microlocally near $(0,\xi_0)$, one has $A'\equiv B$ modulo a lower order term, and part b) of the theorem is proved.

4.5. *Proof of Theorem* 1.6.

Let $Y(x,\xi),\ H(x,\xi)$ be a canonical map, defined near $(x_0,\xi_0)\in\Lambda$, such that the image of the Lagrangean manifold Λ is $T^*_0\mathbf{R}^n$, and put $\eta_0=H(x_0,\xi_0)$. Let F and G be Fourier integral operators associated to this map and to its inverse, such that $F\circ G$ and $G\circ F$ coincide microlocally with the identity operator.

It is obvious that $b(y,\eta)\to b(Y(x,\xi),\ H(x,\xi))$ defines isomorphisms from $\Sigma_0^{m,m'}$ onto $\Sigma_\Lambda^{m,m'}$ (Definition 1.4), consistent with product and Poisson bracket. Microlocally $u \to Fu$ and $v \to Gv$ define inverse isomorphisms between $H_\Lambda^{s,s'}$ and $H_0^{s,s'}$. For $s' \in \boldsymbol{N}$, this follows from the fact that $M \to G \circ M \circ F$ is an isomorphism between $\mathscr{M}_\Lambda$ (pseudo-differential operators of order 1 whose principal symbol vanishes on Λ) and $\mathscr{M}_0$ defined in n° 4.4.

Let now A belong to $\mathrm{Op}(\Sigma_\Lambda^{m,m'})$. From Definition 1.5, one deduces that $A' = F \circ A \circ G$ satisfies property (4.13) and thus $A' \in \mathrm{Op}(\Sigma_0^{m,m'})$. If $a'(y, \eta)$ is a principal symbol for A', we will put

$$\sigma_{m,m'}(A)(x, \xi) = a'(Y(x, \xi), H(x, \xi)) \quad (+\Sigma_\Lambda^{m,m'-1}).$$

Corollary 4.4 guarantees that this principal symbol does not depend on the choice of the canonical map and of the Fourier integral operators F and G. Parts a) to d) of Theorem 1.6 follow obviously from the fact that they are already proved when $\Lambda = T_0^*\boldsymbol{R}^n$.

4.6. *Remark.* The following property is an immediate consequence of Corollary 3.4.

Let Λ and Λ' be Lagrangean manifolds. Let $Y(x, \xi)$, $H(x, \xi)$ be a canonical map defined in a neighbourhood of $(x_0, \xi_0) \in \Lambda$, and mapping Λ into Λ'. Let F and G be Fourier integral operators associated to this map and to its inverse, such that $F \circ G$ and $G \circ F$ coincide microlocally with I. Then, microlocally, the maps:

$$H_\Lambda^{s,s'} \longrightarrow H_{\Lambda'}^{s,s'}: \quad u \longrightarrow Fu$$

$$\Sigma_{\Lambda'}^{m,m'}/\Sigma_{\Lambda'}^{m,m'-1} \longrightarrow \Sigma_\Lambda^{mm'}/\Sigma_\Lambda^{m,m'-1}: \quad a \longrightarrow a(Y(x, \xi), H(x, \xi))$$

$$\mathrm{Op}(\Sigma_{\Lambda'}^{mm'})/\mathrm{Op}(\Sigma_{\Lambda'}^{m,m'-1}) \longrightarrow \mathrm{Op}(\Sigma^{m,m'})/\mathrm{Op}\,\Sigma^{m,m'-1}: \quad A \longrightarrow GAF$$

are consistent isomorphisms.

4.7. Diffeomorphisms and weighted Sobolev spaces. Let $\chi(x)$ be a homeomorphism of $\boldsymbol{R}^n$ (near 0) into $\boldsymbol{R}^n$, such that $\chi(0) = 0$, χ is a diffeomorphism outside 0, homogeneous of degree 1 (this condition can be relaxed, it is sufficient to assume $D_x^\alpha \chi(x) = O(|x|^{1-|\alpha|})$).

It is easy to see that $v \to v \circ \chi$ maps $SP(s, s')$ into itself, and that $B \to B(v \circ \chi) \circ \chi^{-1}$ maps $\mathrm{Op}(S\Sigma_0^{m,m'})$ into itself. In fact (see theorem 3.4), if

$$B = \sum 2^{pm} \Theta_p \mathscr{B}_p \Theta_{-p}$$

with $(\mathscr{B}_p)$ bounded family of pseudo-differential operators of oder $m + m'$ supported in a fixed ring

$$Cv = B(v \circ \chi) \circ \chi_1 = \sum 2^{pm} \Theta_p \mathscr{C}_p \Theta_{-p}$$

with $\mathscr{C}_p w = B_p(v \circ \chi) \circ \chi_1$. Thus, χ being a diffeomorphism in the ring, $\mathscr{C}_p$ is a bounded family of pseudo-differential operators. Moreover, a principal symbol $\sigma_{m+m'}(\mathscr{C}_p)$ is equal to $\sigma_{m+m'}(\mathscr{B}_p)(\chi^{-1}(x), {}^t\chi'(x)\xi)$. Now, the symbol b of B is equal to

$$b(x, \xi) = \sum 2^{pm}\sigma(\mathscr{B}_p)(2^p x, 2^{-p}\xi).$$

Using the same property for the symbol $c(x, \xi)$ of C, one obtains

$$c(x, \xi) - b(\chi^{-1}(x), {}^t\chi'(x)\xi) \in S\Sigma_0^{m, m'-1}.$$

We shall now obtain operators on $\boldsymbol{R}^n$, using isomorphisms Π and Pf.

4.8. Definition (singular diffeomorphisms). *Let χ be as above. The operator χ^* is defined in $\mathscr{D}'(\boldsymbol{R}^n)$ (near 0) by*

$$\chi^* u = Pf[(\Pi u \circ \chi)].$$

The following theorem is nothing but the translation of results of n° 4.7.

4.9. Theorem.

a) *χ^* maps $H_0^{s, s'}$ into $H_0^{s s'}$. If χ_1 and χ_2 are diffeomorphisms as above, then $(\chi_1 \circ \chi_2)^* - \chi_1^* \circ \chi_2^* \in \mathrm{Op}(\Sigma_0^{0, -\infty})$.*

b) *If $A \in \mathrm{OP}(\Sigma_0^{m, m'})$ with principal symbol a, then $A' = (\chi^{-1})^* \circ A \circ \chi^*$ belongs to $\mathrm{Op}(\Sigma_0^{m, m'})$, and $a(\chi^{-1}(x), {}^t\chi'(x)\xi)$ is a principal symbol for A'.*

4.10. Singular Fourier integral operators

Usual Fourier integral operators, and singular diffeomorphisms are particular cases of a much larger class of transformations, well-behaved with respect to 2-microlocal calculus, which we shall now describe briefly.

Consider first a homeomorphism $Y(x, \xi)$, $H(x, \xi)$ defined near $T_0^*(\boldsymbol{R}^n)$, mapping $T_0^*(\boldsymbol{R}^n)$ onto itself, and such that, outside $x=0$, it is a smooth canonical map, homogeneous with respect to ξ and also with respect to dilatations in $\boldsymbol{R}^n$ (this last condition can be relaxed as in $n°$ 4.7 for diffeomorphisms):

$$Y(\lambda x, \mu\xi) = \lambda Y(x, \xi); \qquad H(\lambda x, \mu\xi) = \mu H(x, \xi).$$

It is clear that, for x belonging to the fixed ring C_0, $Y(x, \xi)$ belongs to a larger ring. We will denote by $\tilde{\varphi}$ a C^∞ function, equal to 1 on $Y(C_0 \times \boldsymbol{R}^n)$, with support in some larger ring $\tilde{C}_0$ (as usual $\varphi \in C_0^\infty(C_0)$ and defines a partition of unity).

Let now (F_p) be a bounded family of Fourier integral operators of order $m+m'$ associated to the given canonical transformation, from C_0 to $\tilde{C}_0$,and put

$$Fu=Pf\left\{\sum_p 2^{pm}\Theta_p\tilde{\varphi}F_p\varphi_1\Theta_{-p}\Pi u\right\}.$$

The sum of such operators and of operators mapping $H_0^{s,s'}$ into $H^{s-m,\infty}$ will be called singular Fourier integral operators of order (m, m') associated to (Y, H).

It is easy to prove that F maps $H_0^{s,s'}$ into $H_0^{s-m,s'-m'}$, and that, if G is associated to the invere canonical map, with $(I-FG)$ and $(I-GF)$ belonging to $\mathrm{Op}(\Sigma^{0,\infty})$, $A\to F\circ A\circ G=A'$ maps $\mathrm{Op}(\Sigma_0^{m,m'})$ into itself with $\sigma_{mm'}(A')(x,\xi)\equiv\sigma_{mm'}(A)[Y(x,\xi), H(x,\xi)]$ (mod $\Sigma^{m,m'-1}$). One is reduced, via Π and Pf, to prove results on $SP(s,s')$ and $\mathrm{Op}(S\Sigma^{m,m'})$, and by Theorem 3.4 to classical results on Fourier integral operators on a fixed compact set. The details are left to the reader.

Using invariance by usual Fourier integral operators, it would not be difficult now to define classes of operators mapping $H_\Lambda^{s,s'}$ into $H_{\Lambda'}^{s-m,s'-m'}$, well behaved with respect to 2-microlocal calculus, associated to canonical transformations from (microlocally) $T^*\boldsymbol{R}^n\diagdown\Lambda$ into $T^*\boldsymbol{R}^n\diagdown\Lambda'$ which are "almost homogeneous" of degree 1 with respect to the distances to the Lagrangean manifolds Λ and Λ'. This is the equivalent of "quantized bicanonical transformations" defined by Y. Laurent [9] in the analytic theory of second microlocalization.

§ 5. Proragation of 2-microlocal Singularities for Operators of Real Principal Type

5.1. Notations. We shall denote by Λ a Lagrangean submanifold of $\boldsymbol{R}^n$, and by P an element of $S_{1,0}^m$, whose principal symbol p_m is homogeneous and real, defined microlocally near $(x_0,\xi_0)\in\Lambda$. We shall assume that the restriction of dp_m to Λ is non degenerate, which means that the Hamiltonian field H_{p_m} is transversal to Λ at (x_0,ξ_0) and $p_m(x_0,\xi_0)=0$.

We shall denote by γ_- and γ_+ the two open half-bicharacteristics of p_m starting from (x_0,ξ_0). We shall still denote by $(x_0,\xi_0; H_{p_m}(x_0,\xi_0))$ the element of $T_\Lambda(T^*\boldsymbol{R}^n)\subset(T^*\boldsymbol{R}^n\diagdown\Lambda)\cup T_\Lambda(T^*\boldsymbol{R}^n)$ canonically associated to $H_{p_m}(x_0,\xi_0)\in T_{(x_0,\xi_0)}(T^*\boldsymbol{R}^n)$ (see n° 1.8). Our main result is the following.

5.2. Theorem. *Let R belong to* $\mathrm{Op}(\Sigma_\Lambda^{m,-1})$ *microlocally near (x_0,ξ_0) and assume that, 2-microlocally near the points $(x_0,\xi_0; \pm H_{p_m}(x_0,\xi_0))$, we have $R\in$* $\mathrm{Op}(\Sigma_\Lambda^{m,-2})+\mathrm{Op}(\Sigma_\Lambda^{m-1,0})$. *Assume moreover*

a) *$u\in H_\Lambda^{s,-\infty}$ microlocally near Λ.*

b) *$(P+R)u\in H_\Lambda^{s-m,1/2}$ microlocally near (x_0,ξ_0).*

c) *$u \in H^{s-1/2}$ microlocally along γ.*

Then one has $u \in H_\Lambda^{s-\varepsilon, -1/2}$ microlocally near (x_0, ξ_0) and hence $u \in H^{s-1/2-\varepsilon}$ along γ_+, for each $\varepsilon > 0$.

We shall write down the proof only in the scalar case, but the extension to the case u vector valued, R matrix valued, but P diagonal, is immediate.

5.3. Reduction to $\partial/\partial x_1 + R$ near 0

One can find a canonical transformation, near (x_0, ξ_0), mapping Λ into $T^*_{\{0\}}\boldsymbol{R}^n$ and $\{p_m = 0\}$ into $\{\xi_1 = 0\}$. Using an associated microlocally elliptic Fourier integral operator, and multiplying P by an elliptic pseudo-differential operator of order $1-m$, we are reduced to the following situation.

We shall denote by G and Γ small conic neighbourhoods of $\delta x_0 = (1, 0)$ in $\boldsymbol{R}_{x_1} \times \boldsymbol{R}^{n-1}_{x'}$ and of $\xi_0 = (0, \xi_0')$ in $\boldsymbol{R}_{\xi_1} \times \boldsymbol{R}^{n-1}_{\xi_1'}$, and by ω a neighbourhood of 0 in $\boldsymbol{R}^n_x$.

(5.1) $$R \in \mathrm{Op}(\Sigma_0^{1,-1}) \quad \text{in } \omega.$$

(5.2) $$R \in \mathrm{Op}(\Sigma_0^{1,-2}) + \mathrm{Op}(\Sigma_0^{0,0}) \quad \text{in } (\omega \cap \pm G) \times \Gamma.$$

(5.3) $$u \in H_0^{s,-N} \text{ locally in } \omega \text{ for some } N.$$

(5.4) $$\frac{\partial u}{\partial x_1} + Ru \in H_{\{0\}}^{s-1,1/2} \text{ microlocally in } \omega \times \Gamma.$$

(5.5) $$u \in H^{s-1/2} \text{ microlocally in } (\omega \cap (-G)) \times \Gamma.$$

We want to show that $u \in H^{s-\varepsilon,-1/2}$ in $\omega \times \Gamma$. We shall first prove that it is sufficient to prove the result when $R=0$, and then, we will have to prove 3 different theorems on propagation of 2-microlocal singularities for $\partial/\partial x_1$.

5.4. Theorem. *Under assumptions* (5.1) *and* (5.2), *for each $N > 0$, and for each small cone G_0 of $\boldsymbol{R}^n_x$, there exists $E \in \mathrm{Op}(\Sigma^{0,0})$, elliptic in $(\omega \cap G_0) \times \Gamma$ such that*

$$E \circ \left(\frac{\partial}{\partial x_1} + R\right) - \frac{\partial}{\partial x_1} \circ E \in \mathrm{Op}(\Sigma^{1,-N}) \qquad \textit{in } (\omega \cap G_0) \times \Gamma.$$

We shall have to consider two different cases.

a) G_0 is a small conic neighbourhood of $\pm(1, 0)$. One can assume $G_0 = G$ for instance. We shall obtain the symbol of E as a sum $e_0 + e_1 + \cdots + e_N$, with $e_j \in \Sigma_0^{0,-j}$ in G. The first equation to solve is

$$(5.6) \qquad \frac{\partial e_0}{\partial x_1} + re_0 = 0.$$

$$(5.7) \qquad e_0 = \exp\left[\int_{x_1}^{a} r(t, x', \xi)dt\right].$$

By integration of the estimate

$$|D_\xi^\alpha D_x^\beta r(t, x', \xi)| \leqslant C\frac{|\xi|^{-|\alpha|+|\beta|}}{(1+a|\xi|)^{|\beta|}} + C\frac{|\xi|^{1-|\alpha|+|\beta|}}{(1+t|\xi|)^{2+|\beta|}}, \qquad 0<t<a,$$

we obtain that the integral in (5.6), and thus e_0 itself, belongs to $\Sigma^{0,0}$ in G. By standard computation, one obtains that e_j should satisfy

$$(5.8) \qquad \frac{\partial e_j}{\partial x_1} + re_j = s_j \in \Sigma^{1,-j-1}$$

and that e_j belongs to $\Sigma^{0,-j}$.

b) G_0 does not contain $(\pm 1, 0)$, and one can assume $|x_1|<C|x'|$ in G_0. For $\sigma(E)=e_0+\cdots+e_N$, one has to solve (5.6) and then (5.8). A solution of (5.6) is given by

$$(5.9) \qquad e_0(x_1, x', \xi) = \exp\left(-\int_{\alpha|x'|}^{x_1} r(t, x', \xi)\,dt\right)$$

assuming for instance G_0 defined by $x' \in G_0'$ and $(\alpha-\varepsilon)|x'|<x_1<(\alpha+\varepsilon)x'$. From

$$|D_\xi^\alpha D_x^\beta r(t, x', \xi)| < C\langle\xi\rangle^{1-|\alpha|+|\beta|}(1+|x'|\langle\xi\rangle)^{-1-|\beta|}$$

one deduces easily that the integral in (5.9), and thus e_0 itself, belong to $\Sigma^{0,0}$. The resolution of transport equations (5.8) is straightforward.

The following lemma shows that estimates on Littlewood-Paley decomposition give a control of 2-microlocal regularity.

5.5. Lemma. *Let u belong to $H^{s,-\infty}$ locally near 0, and $\delta x_0 \in \boldsymbol{R}^n\setminus\{0\}$. The following properties are equivlent:*

a) *For each $\xi \in \boldsymbol{R}^n\setminus\{0\}$, u belongs to $H^{s,s'}$ 2-microlocally at $(0, \xi; \delta x_0, 0)$.*

b) *There exists a small open cone $G_0 \ni \delta x_0$ and a small open set $\omega \ni 0$ such that*

$$(5.10) \qquad \|2^{ps}(1+2^p|x|)^{s'}\varphi(2^{-p}D)u\|_{L^2(G_0\cap\omega)} \leqslant c_p; \qquad \sum c_p^2 < \infty.$$

We have to prove the equivalence assuming $u \in H^{s,\sigma}$ locally. It will be sufficient to prove it, using commutations with D_j and x_jD_k, for $s\geqslant 0$, $\sigma\geqslant 0$.

Let $g(x)$ be homogeneous of degree 0, C^∞ outside 0, with support in a small conic neighbourhood of δx_0, such that $g=1$ near δx_0. Then $Pf \circ M_g \circ \Pi$ belongs to $\mathrm{Op}(\Sigma_0^{0,0})$ and is elliptic at $(0, \xi; \delta x_0, 0)$. Property a) is thus equivalent to $Pf[g \cdot \Pi u] \in H^{s,s'}$ near 0.

For $s \geqslant 0$ and $\sigma \geqslant 0$, using Theorem 2.7, one has

$$u \equiv Pf[g\Pi u] + Pf[(1-g)\Pi u] \qquad (\mathrm{mod}\ H^{s,\infty})$$
$$u \equiv Pf[g\Pi u] + (1-g)\Pi u \qquad (\mathrm{mod}\ H^{s,\infty}).$$

Let now G_0 be an open cone containing δx_0 such that $\overline{G}_0$ is contained in $\{g(x)=1\}$. Using Proposition 2.2, one has

$$\|2^{ps}(1+2^p|x|)^N \varphi(2^{-p}D)(1-g)\Pi u\|_{L^2(G_0)} \leqslant c_p; \qquad \Sigma c_p^2 < \infty$$

and thus b) is a consequence of a).

Conversely, assumption (5.10) gives immediately

$$g(x)\Pi u(x) \in Sp(s, s')$$

and thus $Pf[g\Pi u] \equiv g\Pi u \in SP(s, s') \subset H_0^{s,s'}$, modulo $H^{s,\infty}$, using again Theorem 2.7.

5.6. Theorem (Propagation from incoming bicharacteristics). *Assume* (5,1), (5.2), (5.3) *and, for* $\sigma > -1/2$,

$$\frac{\partial u}{\partial x_1} + Ru \in H^{s-1,\sigma+1} \text{ 2-microlocally at } (0, \xi_0; -\delta x_0, 0)$$

$$u \in H^{s+\sigma} \text{ microlocally at points } (x_1, 0; \xi_0) \text{ with } x_1 < 0.$$

Then $u \in H^{s,\sigma}$ *2-microlocally at* $(0, \xi_0; -\delta x_0, 0)$.

Let $h(D)$ be a pseudo-differential operator of order 0, with symbol concentrated near ξ_0, let E be given by Theorem 5.4, let $\chi(x)$ be a C^∞ function with support near 0 such that $\chi(0) \neq 0$, and put $v(x) = \chi(x)E[h(D)u](x)$. We are reduced to prove the 2-microlocal regularity of v, knowing:

$$\frac{\partial v}{\partial x_1} = w \in H^{s-1,\sigma+1} \text{ 2-microlocally at points } (0, \xi, -\delta x_0, 0)$$

$$v \in H^{s+\sigma} \text{ locally for } x_1 < 0.$$

With notation $v_p = \varphi(2^{-p}D)v$, one has, using Lemma 5.5 and $|x_1| \sim |x|$ in G:

$$v_p(x) = \int_{-\infty}^{x_1} w_p(t, x')\, dt$$

$$\|2^{p(s-1)}(1+2^p|x_1|)^{\sigma+1}w_p(x)\|_{L^2(G)} \leqslant c_p; \qquad \sum c_p^2 < \infty.$$

The following estimate

$$\|2^{ps}(1+2^p|x_1|)^{\sigma}v_p(x)\|_{L^2(G)} \leqslant d_p; \qquad \sum d_p^2 < \infty.$$

which implies bylemma 5.5 the 2-microlocal regularity of v, is an immediate consequence of the classical Hardy type inequality:

$$\left\|(\varepsilon+t)^{\alpha-1}\int_t^\infty (\varepsilon+s)^{-\alpha}f(s)ds\right\|_{L^2(\boldsymbol{R}+)} \leqslant C(\alpha)\|f\|_{L^2(\boldsymbol{R}+)}$$

for $\alpha > 1/2$, with $C(\alpha)$ independent of ε (put $\varepsilon = 2^{-p}$, $\alpha = \sigma+1$, and $f(t) = (\varepsilon+t)^{\sigma+1}w_p(-t, x')$).

5.7. Theorem (Propagation along second bicharacteristics). *Let $\delta x_0' \neq 0$ belong to $\boldsymbol{R}^{n-1}$, and consider the subset $\tilde{\gamma}$ of $T_\Lambda(T^*\boldsymbol{R}^n)$ made of points $(0, \xi_0; \delta x_1, \delta x_0')$ with $a \leqslant \delta x_1 \leqslant b$. Assume* (5.1), (5.3), *and, for any $\sigma \in \boldsymbol{R}$*

$$\frac{\partial u}{\partial x_1} + Ru \in H^{s-1,\sigma+1} \text{ 2-microlocally at points of } \tilde{\gamma}$$

$$u \in H^{s,\sigma} \text{ 2-microlocally at } (0, \xi_0; c, \delta x_0') \text{ for some } c \in [a, b].$$

Then $u \in H^{s,\sigma}$ 2-microlocally at points of $\tilde{\gamma}$.

Shrinking $[a, b]$ if necessary, one may assume the operator E given by Theorem 5.4 is elliptic in a set $(\omega \cap G_0) \times \Gamma$ containing $\tilde{\gamma}$. Putting $v(x) = \chi(x)E(h(D)u)(x)$ as in Theorem 5.6, one obtains

$$\frac{\partial v}{\partial x_1} = w \in H^{s-1,\sigma+1} \text{ microlocally at points } (0, \xi; \delta x) \text{ with } \delta x \in G_0$$

$$v \in H^{s,\sigma} \text{ microlocally at points } (0, \xi; \delta x) \text{ with } \delta x \in G_1$$

where one may assume that

$$G_0 = \{(\delta x_1, \delta x') \mid \delta x' \in G_0', \ \alpha|\delta x'| < \delta x_1 < \beta|\delta x'|\}$$
$$G_1 = \{(\delta x_1, \delta x') \mid \delta x' \in G_0', \ \alpha_1|\delta x'| < \delta x_1 < \beta_1|\delta x'|\}$$

for $a < \alpha_1 < \beta_1 < \beta$, and G_0' a convenient conic neighbourhood of δx_0.

Using Lemma 5.5 and $|x'| \sim |x|$ in G_0, one has

$$\left\|2^{p(s-1)}(1+2^p|x'|)^{\sigma+1}\frac{\partial v_p}{\partial x_1}\right\|_{L^2(G_0)} \leqslant c_p; \qquad \sum c_p^2 < \infty$$

$$\|2^{ps}(1+2^p|x'|)^{\sigma}v_p(x)\|_{L^2(G_1)} \leqslant d_p; \qquad \sum d_p^2 < \infty.$$

The variables x' playing the part of a parameter, we obtain

$$\|2^{ps}(1+2^p|x'|)^\sigma v_p(x)\|_{L^2(G_0)} \leqslant e_p; \qquad \sum e_p^2 < \infty$$

which implies the microlocal regularity of v in G_0.

5.8. Theorem (Propagation on outgoing bichracteristics). *Assume* (5.1), (5.2), (5.3) *and, for* $\sigma < -1/2$

$$\frac{\partial u}{\partial x_1} + Ru \in H^{s-1,\sigma+1} \text{ 2-microlocally, for } \xi = \xi_0,\ \delta x \in G$$

$$u \in H^{s,\sigma} \text{ 2-microlocally for } \xi = \xi_0,\ \delta x \in G \text{ and } \delta x' \neq 0.$$

Then $u \in H^{s,\sigma}$ *2-microlocally for* $\xi = \xi_0$, $\delta x = (1, 0, \cdots, 0)$. *In particular we have* $u \in H^{s+\sigma}$ *microlocally for* $x \in G \cap \omega$ *and* $\xi = \xi_0$.

Putting $v(x) = \chi(x) E(h(D)u)(x)$ as in Theorem 5.6, we obtain

$$\frac{\partial v_p}{\partial x_1} = w_p \tag{5.11}$$

with

$$\|2^{p(s-1)}(1+2^p|x_1|)^{\sigma+1} w_p(x)\|_{L^2(G)} \leqslant c_p \qquad \sum c_p^2 < \infty \tag{5.12}$$

$$\|2^{ps}(1+2^p|x_1|)^\sigma v_p(x)\|_{L^2(G\setminus\tilde{G})} \leqslant d_p \qquad \sum d_p^2 < \infty \tag{5.13}$$

where $\tilde{G}$ is a small conic neighbourhood of $(1, 0, \cdots, 0)$. Let $h_p(x) = h(2^p x_1)$, with $h(t) = 1$ for $t \geqslant 1$ and 0 near 0, and $\chi(x)$ be homogeneous of degree 0, C^∞ outside 0, equal to 1 near $\tilde{G}$, with support in G. We have

$$\left\| 2^{p(s-1)}(1+2^p|x_1|)^{\sigma+1} \frac{\partial}{\partial x_1}(\chi h_p v_p) \right\|_{L^2(G)} \leqslant c'_p \qquad \sum c_p'^2 < \infty \tag{5.14}$$

using (5.3), (5.11), (5.12), (5.13). In order to prove the theorem, it is sufficient to prove,

$$\|2^{ps}(1+2^p|x_1|)^\sigma \chi h_p v_p\|_{L^2} \leqslant e_p, \qquad \Sigma e_p^2 < \infty.$$

This estimate is an easy consequence of (5.14) and of the classical Hardy inequality

$$\left\| (\varepsilon + t)^{-\alpha-1} \int_0^t (\varepsilon + s)^\alpha f(s) ds \right\|_{L^2(R^+)} \leqslant C(\alpha) \|f\|_{L^2(R^+)}$$

with $C(\alpha)$ independent of ε, for $\alpha > -1/2$ (put $\varepsilon = 2^{-p}$, $\alpha = -\sigma - 1$, $f(t) = (\varepsilon + t)^{\sigma+1}(\partial/\partial x_1)(\chi h_p v_p)(t, x')$).

5.9. Corollary. *Under assumptions* (5.1) *to* (5.5), *we have* $u \in H^{s-\varepsilon,-1/2}$ *microlocally in* $\omega \times \Gamma$, *for every* $\varepsilon > 0$.

It is an immediate consequence of Theorems 5.6 to 5.8. The value $-1/2$ and the loss of ε are imposed by conditions $\sigma > -1/2$ in Theorem 5.6 and $\sigma < -1/2$ in theorem 5.8.

5.10. *Reuark.* Theorem 5.2 is nothing but an invariant formulation of Corollary 5.9. It is of course possible to give an invariant definition of the second bicharacteristics (curves in $T_{\Lambda}(T^*\boldsymbol{R}^n)$ corresponding to the $\tilde{\gamma}$ of Theorem 5.7) and to state three invariant theorems for propagation of 2-microlocal singularities. This is left to the reader.

5.11. *Remark.* Assumption a) in Theorem 5.2 is necessary when R is a 2-microdifferential operator, but can be replaced just by $u \in \mathscr{D}'$ when $R \in \mathrm{Op}(S^{m-1}_{1,0})$. In that case, $P+R$ can be reduced to $\partial/\partial x_1$ using only Fourier integral operators, and one needs just one integration to prove Corollary 5.9.

§ 6. Singular Vector Fields and Leibniz Formula

We are going to prove that spaces $H_0^{s,s'}$ are algebras for $s > n/2$ and $s+s' > n/2$. We recall first the following properties (Theorem 2.7) which are valid for $s \geqslant 0$ and $s+s' \geqslant 0$.

(6.1) — Any $u \in H^{s,s'}$ can be written $(u = \Pi u + Eu)$

as a sum of an element of $SP(s, s')$ and of an element of $H_0^{s,\infty}$. This decomposition is unique, up to an element of $SP(s, \infty)$ $(= H_0^{s,\infty} \cap SP(s, s'))$.

(6.2) — For $v \in SP(s, s')$ one has $v - Pfv \in H_0^{s,\infty}$.

6.1. Theorem. *Assume* $s > n/2$ *and* $s+s' > n/2$.

a) $H_0^{s,s'}$ *is an algebra, and* $SP(s, s')$ *is an ideal of* $H_0^{s,s'}$.

b) *If* $u_j \in H_0^{s,s'}$ *and if* F *is a* C^∞ *function of its arguments, then* $F(x, u_1(x), \cdots, u_N(x))$ *belong to* $H_0^{s,s'}$.

c) *For* $u, v, u_1, \cdots, u_N$ *belongings to* $H^{s,s'}$, *we have*

$$E(uv) \equiv Eu \cdot Ev \qquad (mod\ SP(s, \infty)) \tag{6.3}$$

$$E[F(x, u_1(x) \cdots)] \equiv F(x, Eu_1(x), \cdots) \qquad (mod\ SP(s, \infty)). \tag{6.4}$$

a) It is well known that H^s is an algebra, and an easy induction shows that $x^\alpha D^\beta(uv)$, for $|\alpha| = |\beta|$, belong to H^s if the same is true for u and v. This proves that $H^{s,\infty}$ is an algebra. Note that, moreover, for

$w \in H^{s,\infty}$, one has $x^\alpha D^\beta w \in L^\infty$ for $|\alpha|=|\beta|$.

By Definition 2.5, u belongs to $SP(s, s')$ if the norm of $u(2^{-j}x)$ in $H^{s+s'}$ of a fixed ring C_0 is bounded by $2^{-j(s-n/2)} \cdot c_j$ with $\sum c_j^2 < \infty$. For two such functions u and v, one has

$$\|u(2^{-j}x)v(2^{-j}x)\|_{H^{s,s'}(C_0)} \leqslant 2^{-j(2s-n)} c_j c_j' \leqslant d_j 2^{-j(s-n/2)}$$

which proves that $SP(s, s')$ is an algebra.

Moreover, if $w \in H_0^{s,\infty}$, the functions $w(2^{-j}x)$ and their derivatives are uniformly bounded on C_0, and thus $\|u(2^{-j}x)w(2^{-j}x)\|_{H^{s+s'}(C_0)} \leqslant Cc_j 2^{-j(s-n/2)}$. This proves that the product maps $H_0^{s,\infty} \times SP(s, s')$ into $SP(s, s')$.

From

$$uv = Eu \cdot Ev + [Eu \cdot \Pi v + \Pi u \cdot Ev + \Pi u \cdot \Pi v]$$

one gets immediately part a) of the theorem and (6.3).

b) We shall write down the proof of part b) when F is just a function of one variable u. It is clear that $F(Eu(x))$ belong to $H_0^{s,\infty}$. Put

$$U(x) = F(u(x)) - F(Eu(x))$$
$$U(2^{-j}x) = G_j(x, v_j(x))$$

with $G_j(x, v) = F[Eu(2^{-j}x) + v] - F(Eu(x))$

$$v_j(x) = \Pi u(2^{-j}x).$$

The functions $G_j(x, v)$ are contained in a bounded set of $C^\infty(C_0 \times \boldsymbol{R})$ and vanish for $v=0$, while the v_j are contained in a bounded set of $H^{s+s'}(C_0)$.

It is well known that there exists a constant C (depending on the bounded sets) such that

$$\|G_j(x, v_j(x))\|_{H^{s+s'}(C_0)} \leqslant C\|v_j\|_{H^{s+s'}(C_0)} \leqslant Cc_j 2^{-j(s-n/2)}.$$

This proves that U belongs to $SP(s, s')$, and then part b) of the theorem and (6.4).

6.2. Definition. *We shall call* ***singular vector field*** *an elemetn Z of $Op(\Sigma_0^{0,1})$ whose symbol (mod $\Sigma_0^{0,-\infty}$) is as follows:*

$$z(x, \xi) = \sum a_j(x)(i\xi_j)$$
$$|D_x^\alpha a_j(x)| \leqslant C_{j,\alpha} |x|^{1-|\alpha|}.$$

6.3. Theorem (Leibniz formula). *Let Z be a singular vector field, let $u, v, u_1, \cdots, u_N$ belong to $H^{s,s'}$, $s > n/2$, $s+s' > n/2$, and let F be a C^∞*

function. Then there exist operators $M_1, M_2, M_1', \cdots, M_N'$ *in* $\mathrm{Op}(\Sigma_0^{0,0})$ *(depending on* $u, \cdots, u_N$*) such that*

$$Z(u\cdot v)\equiv Zu\cdot v+u\cdot Zv+M_1u+M_2v \qquad (\textit{mod } H^{s,\infty}) \tag{6.5}$$

$$ZF(u_1(x), \cdots, u_N(x))\equiv\sum\frac{\partial F}{\partial u_j}\cdot(Zu_j)+\sum M_j'\left[\frac{\partial F}{\partial u_j}\right] \quad (\textit{mod } H^{s,\infty}). \tag{6.6}$$

Let $\tilde{Z}$ be the usual vector field $\sum a_j(x)\partial/\partial x_j$ in $\boldsymbol{R}^n\backslash\{0\}$. By definition $Z=Pf\circ\tilde{Z}\circ\Pi$ (mod Op $\Sigma^{0,-\infty}$), and in view of (6.2), we have

$$Zu\equiv\tilde{Z}\Pi u \qquad (\text{mod } H^{s,\infty}).$$

$$Z(uv)\equiv\tilde{Z}(uv-E(uv))\equiv\tilde{Z}(uv-Eu\cdot Ev)$$

using (6.3). Then

$$Z(uv)\equiv\tilde{Z}(\Pi u\Pi v+Eu\Pi v+\Pi uEv)$$

$$Z(uv)\equiv\tilde{Z}\Pi u\cdot v+u\cdot\tilde{Z}\Pi v+(\tilde{Z}Eu)\Pi v+(\tilde{Z}Ev)\Pi u.$$

The function $a=\tilde{Z}Eu$ satisfies $|D^\lambda a(x)|\leqslant C|x|^{-\lambda}$. Let M_2 be the operator in $\mathrm{Op}(\Sigma^{0,0})$ with symbol a. One has $M_2v=Pf(a\Pi v)\equiv(\tilde{Z}Eu)\Pi v$. With an analogous definition for M_1, (6.5) is proved.

We shall prove (6.6) when $n=2$. In view of (6.4), we have

$$ZF(u, v)\equiv\tilde{Z}\Pi F(u, v)\equiv\tilde{Z}[F(u, v)-F(Eu, Ev)]$$

$$ZF(u, v)\equiv\frac{\partial F}{\partial u}(u, v)\tilde{Z}\Pi u+\frac{\partial F}{\partial v}(u, v)\tilde{Z}\Pi v+\tilde{Z}Eu\left(\frac{\partial F}{\partial u}(u, v)-\frac{\partial F}{\partial u}(Eu, Ev)\right)$$

$$+\tilde{Z}Ev\left(\frac{\partial F}{\partial v}(u, v)-\frac{\partial F}{\partial v}(Eu, Ev)\right)$$

$$ZF(u, v)\equiv\frac{\partial F}{\partial u}Zu+\frac{\partial F}{\partial v}Zv+(\tilde{Z}Eu)\left(\Pi\frac{\partial F}{\partial u}\right)+(\tilde{Z}Ev)\left(\Pi\frac{\partial F}{\partial v}\right)$$

which proves (6.6).

6.4. Definition (Algebras defined by singular vector fields). *Let* $\mathscr{Z}$ *be a Lie sub-algebra of* $\mathrm{Op}(\Sigma_0^{0,1})$ *generated (over* $\mathrm{Op}(\Sigma_0^{0,0})$*) by* 1 *and a finite number of singular vector fields* $Z_1, \cdots, Z_m$.

We will say, for $k\in\boldsymbol{N}$, *that u belongs to* $H^{s,s'}(\mathscr{Z}, k)$ *if*

$$Z^I u\in H_0^{s,s'}, \qquad \textit{for } |I|\leqslant k$$

where $I=(i_1, \cdots, i_l)$, $|I|=l$, *and* $Z^I=Z_{i_1}\circ\cdots\circ Z_{i_l}$.

The assumption on $\mathscr{Z}$ means that

$$M \in \mathscr{Z} \Longleftrightarrow M = \sum_1^m A_j Z_j + A_0; \qquad A_j \in \mathrm{Op}(\Sigma_0^{00})$$

$$[Z_i, Z_j] = \sum_1^m A_{ijk} Z_j + A_{ij0}; \qquad A_{ijk} \in \mathrm{Op}(\Sigma_0^{0,0}).$$

6.5. Theorem. *Assume $s > n/2$, $s + s' > n/2$. Then $H^{s,s'}(\mathscr{Z}, k)$ is an algebra. Moreover, if F belongs to C^∞, and $u_1, \cdots, u_N$ belong to $H^{s,s'}(\mathscr{Z}, k)$, one has*

$$F(x, u_1(x), \cdots, u_N(x)) \in H^{s,s'}(\mathscr{Z}, k).$$

This is an easy consequence of Leibniz formula and of Theorem 6.1. One has just to verify by induction, that $Z^I F(x, u_1,, \cdots, u_N)$ can be expressed, starting from the $Z^J u_k$, $|J| \leqslant I$, by a formula involving only non-linear functions and operators in $\mathrm{Op}(\Sigma_0^{0,0})$.

§ 7. Interaction of singularities

7.1. Notations. We shall consider a non-linear Klein-Gordon equation in $\boldsymbol{R}^3$ (coordinates x, y, t)

$$\square u = f[x, y, t, u] \tag{7.1}$$

with $\square = \partial^2/\partial t^2 - \partial^2/\partial x^2 - \partial^2/\partial y^2$, and f a real C^∞ function. We shall assume that u is a given real valued solution of (7.1) in a neighbourhood Ω of 0. Put $\Omega^\pm = \Omega \cap \{t \gtrless 0\}$.

Let $\Sigma_1, \Sigma_2, \Sigma_3$ be characteristic surfaces, intersecting transversally at 0, and let Γ be the full light cone, $\Gamma^\pm$ the two half light cones.

Our main result (Corollary 7.5) will say that, if u has only good (i.e. conormal) singularities on $\Sigma_1 \cup \Sigma_2 \cup \Sigma_3$ in the past, then it will have only good singularities on $\Sigma_1 \cup \Sigma_2 \cup \Sigma_3 \cup \Gamma^+$ in the future.

7.2. Definition. *Let Σ be either a smooth submanifold or the union of hypersurfaces intersecting only two by two and transversally. We shall say that u belongs to $H^s(\Sigma, k)$ (space of* ***conormal distributions****) if*

$$X^I u \in H^s \qquad \text{for } |I| \leqslant k$$

where X^I is a product of $|I|$ smooth vector fields tangent to Σ.

7.3. Definition. *Let $\mathscr{Z}$ be the space of singular vector fields (Definition* 6.2) *which are tangent, outside* 0, *to $\Sigma_1, \Sigma_2, \Sigma_3$ and Γ. The spaces $H^{s,s'}(\mathscr{Z}, k)$ are defined according to Definition* 6.4.

It is clear that, if $u \in H^{s,s'}(\mathscr{L}, k)$, one has $u \in H^{s+s'}(\Sigma_1 \cup \Sigma_2 \cup \Sigma_3 \cup \Gamma, k)$ outside the generating lines $\Sigma_j \cap \Gamma$.

7.4. Theorem. *Assume that u is a solution of* (7.1) *belonging to* $H^{3/2+\varepsilon}(\Omega)$, *and that* $u \in H^{s,-1/2}(\mathscr{L}, k)$ *locally in* Ω^-, *with* $s>2$ *and* $k \in \mathbf{N}$. *Then,* $u \in H^{s',-1/2}(\mathscr{L}, k)$ *near* 0, *for each* $s'<s$.

7.5. Corollary. *Assume that u is a solution of* (7.1) *belonging to* $H^{3/2+\varepsilon}(\Omega)$, *and that* $u \in H^{\sigma}(\Sigma_1 \cup \Sigma_2 \cup \Sigma_3, k)$ *in* Ω^-, *for* $\sigma>3/2$ *and* $k \in \mathbf{N}$. *Then, one has near* 0 *in* Ω^+, *and for each* $\sigma'<\sigma$:

a) $u \in H^{\sigma'+k}$ *outside* $\cup \Sigma_j \cup \Gamma^+$

b) $u \in H^{\sigma'}(\Sigma_j, k)$ *near* $\Sigma_j \setminus (\cup_{i \neq j} \Sigma_i \cup \Gamma^+)$

c) $u \in H^{\sigma'}(\Gamma, k)$ near $\Gamma^+ \setminus (\cup \Sigma_j)$.

It is clear that the assumptions imply that $u \in H^{\sigma+1/2,-1/2}(\mathscr{L}, k)$ in Ω^-, while a), b), c) are just simple consequences of $u \in H^{\sigma+1/2,-1/2}(\mathscr{L}, k)$ near 0. In view of Theorem 6.1 of [2], one has moreover $u \in H^{\tau}$ near $\Gamma^+ \setminus (\cup \Sigma_j)$, with $\tau = \mathrm{Min}(2\sigma - n/2 + 1, \sigma + k)$. Actually, it is possible to improve c) up to

$$u \in H^{\sigma'+g}(\Gamma, k-g) \text{ near } \Gamma^+ \setminus (\cup \Sigma_j), \text{ with } g = \mathrm{Min}\left(k, \left[\sigma - \frac{n}{2} + 1\right]\right).$$

The proof of Theorem 7.4 will be a direct consequence of results of § 5 (propagation of singularities), § 6 (algebras defined by singular vector fields), and of Theorem 7.7 below (commutation). The part b) of this theorem, which is analogous to an argument of Melrose and Ritter [10], will allow us to avoid the introduction of a third microlocalization, which we used in our first proof [6].

7.6. Definition. *Let* $\mathscr{M}$ *be the* $\mathrm{Op}(\Sigma^{0,0})$-*module generated by* $\mathscr{L}$. *We shall say that a finite family* (M_i) *of elements of* $\mathscr{M}$ *is a system of generators, if any element M of* $\mathscr{M}$ *can be written*

$$M = \sum_i A_i M_i + A_0, \quad \textit{with} \quad A_0, A_i \in \mathrm{Op}(\Sigma^{0,0}). \tag{7.2}$$

7.7. Theorem. *There exists a system* (M_i) *of generators of* $\mathscr{M}$ *such that*

a) (7.3) $$[\Box, M_i] = \sum A_{ij} M_j + B_i \Box + A_{i0}$$

with $A_{i0}, A_{ij} \in \mathrm{Op}(\Sigma_0^{2,-1})$, $B_i \in \mathrm{Op}(\Sigma_0^{0,0})$.

b) (7.4) $$A_{i0}, A_{ij} \in \mathrm{Op}(\Sigma_0^{2,-2}) + \mathrm{Op}(\Sigma_0^{1,0})$$

and

$$B_i \in \mathrm{Op}(\Sigma_0^{0,-1})+\mathrm{Op}(\Sigma_0^{-1,1})$$

2-microlocally near the set $\tilde{\Gamma}$ *of points* $(0;\ \xi,\ \eta,\ \tau;\ \delta x,\ \delta y,\ \delta t)$ *satisfying:*

$$\tau^2-\xi^2-\eta^2=0, \qquad \frac{\delta x}{\xi}=\frac{\delta y}{\eta}=-\frac{\delta t}{\tau}.$$

1. We shall first show that it is sufficient to prove the existence of such generators 2-microlocally. Actually, it is possible to construct a partition of identity $1=\sum T_{\alpha\beta}$, as refined as we want by operators in $\mathrm{Op}(\Sigma_0^{0,0})$ with symbol $\chi_{\alpha,\beta}(x)h_\alpha(\xi)$, where $\chi_{\alpha\beta}$ and h_α are homogenous of degree 0, such that

(7.5) $\quad [\Box, T_{\alpha\beta}] \in \mathrm{Op}(\Sigma^{2,-2})+\mathrm{Op}(\Sigma^{1,0})$, 2-microlocally near $\tilde{\Gamma}$.

We can first take a partition of identity $1=\sum h_\alpha(D)$ with the h_α supported in small cones γ_α. If γ_α contains characteristic directions, there exist 2 small cones $\pm g_\alpha$ in $\boldsymbol{R}^n$ such that $(0;\ \xi,\ \eta,\ \tau,\ \delta x,\ \delta y,\ \delta t) \in \tilde{\Gamma}$ and $(\xi,\ \eta,\ \tau) \in \gamma_\alpha$ imply $(\delta x,\ \delta y,\ \delta t) \in \pm g_\alpha$. We have just to take two functions $\chi_{\alpha,\beta}$ equal to 1 in a neighbourhood of $\pm g_\alpha$, and the other ones equal to 0 near these cones such that $\sum_\beta \chi_{\alpha\beta}=1$, and (7.5) is obviously satisfied.

If we are able to find, near the support of each $T_{\alpha\beta}$, a family of operators $M_{\alpha\beta i} \in \mathcal{M}$ such that (7.2), (7.3), (7.4) are valid 2-microlocally near this support, then the $T_{\alpha\beta} \circ M_{\alpha\beta i}$ will satisfy (7.2), (7.3) and (7.4) in a fullneighbourhood of 0.

2. It remains to study the situation near each $(0;\ \xi,\ \eta,\ \tau;\ \delta x,\ \delta y,\ \delta t)$ and it turns out that 2-microlocally, $\mathscr{L}$ is generated by the usual (C^∞) vector fields which are tangent to the parts of the Σ_j and Γ closed to $(\delta x,\ \delta y,\ \delta t)$. This is trivial at almost all directions, except perhaps when $(\delta x,\ \delta y,\ \delta t) \in \Gamma \cap \Sigma_j$. In that case, one can find a diffeomorphism of $\boldsymbol{R}^n$ mapping Γ into Γ and Σ_j into the plane $x+t=0$ (in a conic neighbourhood of $\Gamma \cap \Sigma_j$) and the image of $\mathscr{L}$ is generated by $(xD_x+yD_y+tD_t)$; (xD_t+tD_x); $((x+t)D_y+y(D_t-D_x))$. The property (7.3) follows by direct inspection, and (7.4) is obvious for usual vector fields.

One obtains easily by induction the following:

7.8. Corollary. *Let* (M_i) *be given by Theoem 7.7, and note* $M^I=M_{i_1} \circ \cdots \circ M_{i_l}$ *for* $I=(i_1,\ \cdots,\ i_l)$. *Then, there exist* $A_{IJ} \in \mathrm{Op}(\Sigma_0^{2,-1})$, $B_{IK} \in \mathrm{Op}(\Sigma_0^{0,0})$, *satisfying, 2-microlocally near* $\tilde{\Gamma}$:

$$A_{IJ} \in \mathrm{Op}(\Sigma_0^{2,-2})+\mathrm{Op}(\Sigma_0^{1,0})$$
$$B_{IK} \in \mathrm{Op}(\Sigma_0^{0,-1})+\mathrm{Op}(\Sigma_0^{-1,1})$$

such that

$$[\square, M^I] = \sum_{0 \leqslant |J| \leqslant |I|} A_{IJ} M^J + \sum_{0 \leqslant |K| \leqslant |I|-1} B_{IK} M^K \square .$$

7.9. *Proof of Theorem* 7.4 (beginning)

Let U_l be the vector valued function whose components are the $M^I u$, for $0 \leqslant |I| \leqslant l$. We want to prove by induction, for $l \leqslant k$, the following equation, with $\mathscr{R}_l$ matrix-valued:

$$(7.6\ l) \quad \begin{cases} \square U_l + \mathscr{R}_l U_l = F_l \quad \text{with} \\ \mathscr{R}_l \in \mathrm{Op}(\Sigma_0^{2,-1}) \quad \text{and} \quad \mathscr{R}_l \in Op(\Sigma_0^{2,-2}) + Op(\Sigma^{1,0}) \\ \text{2-microlocally near } \tilde{\Gamma} \\ F_l \in H^{s'-2,1/2} \quad \text{for} \quad s' < s. \end{cases}$$

We can first note that $u \in H^{s,-1/2}(\mathscr{L}, k)$ in Ω^- implies that $U_l \in H^{s-1/2}(\Omega^-)$ for $l \leqslant k$.

For $l=0$, we have $U_0 = u$ and

$$\square u = F_0 = f(x, y, t, u).$$

Using for instance the results of [2], we obtain first that u, and thus F_0, belong to $H^{s-1/2}$ near 0. But we can obtain the stronger result $u \in H_0^{s,-1/2}$. This is trivial microlocally at non-characteristic points, and follows from the propagation of 2-microlocal singularities (see Remark 5.11) at characteristic points. We can now start the induction.

7.10. *Proof of* $(u \in H^{s',-1/2}(\mathscr{L}, l)) \Rightarrow (7.6.(l+1))$

Using Corollary (7.8), we have

$$\square U_{l+1} + \mathscr{R}_{l+1} U_{l+1} = F_{l+1}$$

where $\mathscr{R}_{l+1}$ is the collection of A_{IJ}, and F_{l+1} is the collection of $[M^I + \sum B_{IK} M^K] \square u$ for $|I| = l+1$. In view of Theorem 6.5, $\square u = f(t, x, y, u)$ belongs to $H^{s',-1/2}(\mathscr{L}, l)$ and thus $M^K \square u \in H^{s',-1/2}$ for $|K| \leqslant l$. Then

$$F_l \in H^{s',-3/2} \subset H^{s'-2,1/2}.$$

The properties of $\mathscr{R}_{l+1}$ being given by Corollary 7.8, (7.6($l+1$)) is proved.

7.11. *Proof of* $(7.6\ l) \Rightarrow (u \in H^{s',-1/2}(\mathscr{L}, l))$; $l \leqslant k$.

First note that $u \in H^{s,-1/2}$ implies $U_l \in H^{s,-1/2-l}$. At non characteristic points $\square + \mathscr{R}_l$ has a microlocal inverse belonging to $\mathrm{Op}(\Sigma^{-2,0})$, and thus $U_l \in H^{s',1/2}$ microlocally. At characteristic points, all the assumptions of

Theorem 5.2 are satisfied, and thus $U_l \in H^{s', -1/2}$ microlocally. This completes the proof of Theorem 7.4.

Bibliography

[1] M. Beals, Self spreading and strength of singularities for solutions to semilinear wave equations, Ann. of Math., **118** (1983), 187–214.

[2] J.-M. Bony, Calcul symbolique et propagation des singularités pour les équations aux dérivées partielles non linéaires, Ann. Sci. Ecole Norm. Sup., (4) **14** (1981), 209–246.

[3] ——, Interaction des singularités pour les équations aux dérivées partielles
[4] non linéaires, Sém. Goulaouic-Meyer-Schwartz, 1979–80, n° 22 and 1981–82, n° 2.

[5] ——, Propagation et interaction des singularités par les solutions des équations aux dérivées partielles non linéaires, Proc. Intern. Cong. Math., Warszawa, 1983, 1133–1147.

[6] ——, Interaction des singularités pour les équations de Klein-Gordon non linéaires, Sém. Goulaouic-Meyer-Schwartz, 1983–84, n° 10.

[7] R. Coifman et Y. Meyer, Au-delà des opérateurs pseudo-différentiels, Astérisque, Soc. Math. France, vol. 57, 1978.

[8] T. Ishii, T. Kobayashi and G. Nakamura, Singular solutions of non linear partial differential equations, Preprints, Univ. Tokyo.

[9] Y. Laurent, Théorie de la deuxième microlocalisation dans le domaine complexe, Progress in Math., vol. 53, Birkhäuser, 1985.

[10] R. Melrose and N. Ritter, Interaction of non-linear progressing waves, Preprint.

[11] J. Rauch and M. Reed, Singularities produced by the nonlinear interaction of three progressing waves; examples, Comm. Partial Differential Equations **7** (1982), 1117–1133.

Université de Paris-Sud
Mathématiques, Bât, 425
91405 Orsay-France

Taniguchi Symp. HERT
Katata 1984, pp. 51–62

Le Domaine d'Existence et le Prolongement Analytique des Solutions des Problèmes de Goursat et de Cauchy à Données Singulières

Yûsaku HAMADA et Akira TAKEUCHI

Abstract

In the lemma 9.1 of his paper [6], J. Leray has given precise information about the Cauchy-Kowalewski's theorem. He has constructed an existence domain of the solution depending onyl on the domain of Cauchy data and on the principal part of operator.

C. Wagschal [10] has studied the generalized Goursat problem for systems of integro-differential equations.

By following the reasoning of the lemma 9.1 of [6] and that of [10], we construct an existence domain of the solution depending only on the domain of data and on spectral matrices of Goursat problems. Next, we apply these results to the Cauchy problem with singular data for differential operator with constant multiple characteristics. By employing a method of analytic continuation due to J. Leray [6], we give a complement of the results of [11] and [4].

Dans le lemme 9.1 de son article [6], J. Leray a précisé le théorème de Cauchy-Kowalewski. Il a construit un domaine d'existence de la solution ne dépendant que de celui des données de Cauchy et de la partie principale de l'opérateur.

C. Wagschal [10] a étudié le problème de Goursat généralisé pour les systèmes d'équations intégro-différentielles.

En suivant les raisonnements de [6] et [10], nous construisons un domaine d'existence de la solution ne dépendant que de celui des données et de la matrice spectrale du problème de Goursat. Puis nous appliquons ces résultats au problème de Cauchy à données singulières pour l'opérateur différentiel à caractéristiques de multiplicité constante. En employant une méthode du prolongement analytique due à J. Leray [6], nous complétons les résultats de [11] et [4].

§ 1. Un domaine d'existence du problème de Goursat

Dans cette section, en suivant les raisonnements de J. Leray [6] et de C. Wagschal [10], nous étudions le domaine d'existence de la solution du problème de Goursat pour le système d'équations différentielles.

Soit $Z=\{x=(x_1, \cdots, x_n); |x_j|\leq r_j, 1\leq j\leq n\}$ un voisinage de l'origine de $\boldsymbol{C}^n$. Soit $a(x, D)$ un opérateur différentiel d'ordre m, holomorphe sur Z:

$$a(x, D)=\sum_{|\alpha|\leq m} a_\alpha(x)D^\alpha, \quad D^\alpha=D_1^{\alpha_1}\cdots D_n^{\alpha_n}, \quad |\alpha|=\sum_{i=1}^{m}\alpha_i.$$

Le polynôme de degré m, homogène en

$$H(\xi)=\sum_{|\alpha|=m}\sup_{x\in Z}|a_\alpha(x)|\xi^\alpha, \qquad \xi\in(\boldsymbol{R}_+)^n, \tag{1.1}$$

sera appelé fonction spectrale de $a(x, D)$ sur Z.

Nous introduisons un ordre partiel dans le multi-indice à composantes non négatives: pour des multi-indices $\alpha=(\alpha_1, \cdots, \alpha_n)$, $\beta=(\beta_1, \cdots, \beta_n)\in\boldsymbol{N}^n$ et $i\in\{1, \cdots, n\}$, $\beta\prec_i\alpha$ signifie que $\beta_j=0$ $(0\leq j\leq i-1)$, $0\leq\beta_i\leq\alpha_i-1$ *et* $\beta_j=\alpha_j$ $(i\geq j+1)$. C'est impossible si $\alpha_i=0$. On définit la relation d'ordre partiel $\beta\prec\alpha$ s'il existe un entier $i\in\{1, \cdots, n\}$ tel que $\beta\prec_i\alpha$.

Soit $\alpha(\mu)$ $(1\leq\mu\leq N)$ une suite de multi-indices. Nous étudions le problème de Goursat:

$$\begin{cases} D^{\alpha(\mu)}u_\mu(x)=\sum_{\nu=1}^{N} a_\mu^\nu(x, D)u_\nu(x)+v_\mu(x), & 1\leq\mu\leq N, \\ D^{\beta(\mu)}u_\mu(x)-\sum_{\nu=1}^{N} b_\mu^{\nu,\beta(\mu)}(x, D)u_\nu(x)-D^{\beta(\mu)}w_\mu(x)|_{x_i=0}=0, & \\ \qquad\qquad \text{pour } \beta(\mu)\prec_i\alpha(\mu), & \end{cases} \tag{1.2}$$

où $\beta(\mu)$ décrit l'ensemble des multi-indices vérifiant $\beta(\mu)\prec\alpha(\mu)$.

Nous faisons l'

Hypothèse 1.1. (i) *Les opérateurs différentiels a_μ^ν, $b_\mu^{\nu,\beta(\mu)}$ et les données w_μ, v_μ sont holomorphes sur Z.*

(ii) *Il existe une suite $(n_\mu)_{1\leq\mu\leq N}$ de N entiers telle que l'ordre de $a_\mu^\nu\leq|\alpha(\mu)|+n_\mu-n_\nu$ et l'ordre de $b_\mu^{\nu,\beta(\mu)}\leq|\beta(\mu)|+n_\mu-n_\nu$.*

Note. Nous avons convenu que si $|\alpha(\mu)|+n_\mu-n_\nu<0$, $a_\mu^\nu\equiv0$ et que l'opérateur différentiel $a\equiv0$ a l'ordre $-\infty$.

Soient $H_\mu^\nu(\xi)$ et $G_\mu^{\nu,\beta(\mu)}(\xi)$ les fonctions spectrales respectives de $a_\mu^\nu(x, D)$ et $b_\mu^{\nu,\beta(\mu)}(x, D)$ sur Z, étant convenu que ces opérateurs sont d'ordres respectifs $|\alpha(\mu)|+n_\mu-n_\nu$ et $|\beta(\mu)|+n_\mu-n_\nu$.

Posons pour $\xi \in (R_+)^n$:

(1.3) $$S^{\nu}_{\mu}(\xi)=\max_{\beta(\mu):\beta(\mu)\prec\alpha(\mu)}\{H^{\nu}_{\mu}(\xi)/\xi^{\alpha(\mu)},\ G^{\nu,\beta(\mu)}_{\mu}(\xi)/\xi^{\beta(\mu)}\}.$$

La matrice $(S^{\nu}_{\mu}(\xi))$ sera nommée matrice spectrale sur Z, du problème (1.2). Notons $\rho(\xi)$ sa plus grande valeur propre, nommée rayon spectral de (1.2) sur Z; $\rho(\xi)$ est une fonction de ξ positivement homogène de degré 0 en ξ; $\rho(\xi)\geq 0$. Supposons:

Hypothèse 1.2. *Il existe* $\theta=(\theta_1, \cdots, \theta_n)$ *tel que* $\theta_j\geq 1/r_j$ $(1\leq j\leq n)$ *et* $\rho(\theta)<1$.

Nous obtenons alors le:

Théorème 1.1. *Sous les hypothèses* 1.1 *et* 1.2, *le problème de Goursat* (1.2) *admet une unique solution holomorphe sur*

$$\left\{x;\ \sum_{j=1}^{n}\theta_j|x_j|<1-\rho(\theta)\right\}.$$

La démonstration de ce théorème est faite par la méthode des fonctions majorantes ([1]~[4], [6], [8] et [10]). En la rappelant, nous donnerons quelques lemmes.

Soit $\mathcal{O}$ l'anneau des germes de fonctions holomorphes à l'origine de $\boldsymbol{C}^n$. Soient $\{\Omega_\lambda\}_{\lambda\in\Lambda}$ un système fondamental de voisinage ouvert de l'origine et $\mathcal{O}(\Omega_\lambda)$ l'espace des fonctions holomorphes dans Ω_λ, muni de la topologie de la convergence uniforme sur tout compact de Ω_λ: $\mathcal{O}(\Omega_\lambda)$ est un espace de Fréchet. $\mathcal{O}$ est un espace vectoriel topologique mnui la topologie limite inductive des $\mathcal{O}(\Omega_\lambda)$ $(\lambda \in \Lambda)$.

Notons $\mathcal{O}^N_+=\{f=(f_1, \cdots, f_N) \in \mathcal{O}^N; f_i\gg 0,\ 1\leq i\leq N\}$. Soient T une application linéaire de $\mathcal{O}^N$ dans $\mathcal{O}^N$ et S une application positivement linéaire de $\mathcal{O}^N_+$ dans $\mathcal{O}^N_+$: $S(c_1U^{(1)}+c_2U^{(2)})=c_1S(U^{(1)})+c_2S(U^{(2)})$ pour $U^{(i)} \in \mathcal{O}^N_+$, $c_i\geq 0$, $1\leq i\leq 2$. Si on a $T(u)\ll S(U)$ pour tout $u \in \mathcal{O}^N$, $U \in \mathcal{O}^N_+$ vérifiant $u\ll U$, nous disons que S est une application majorante de T: $T\ll S$. On a le

Lemme 1.1. *Soient* $T\ll S$ *et* $\{u_k\}_{k=1,2,\ldots}$ *une suite de* $\mathcal{O}^N(\Omega_\lambda)$, *où* Ω_λ *est un voisinage de l'origine. Supposons que* $\{u_k\}$ *converge vers* $u \in \mathcal{O}^N(\Omega_\lambda)$ *dans* $\mathcal{O}^N(\Omega_\lambda)$.

Alors il existe $\Omega_{\mu(\lambda)}$ *un voisinage de l'origine ne dépendant que de* Ω_λ *tel que* $\{T(u_k)\} \in \mathcal{O}^N(\Omega_{\mu(\lambda)})$ *et* $\{T(u_k)\}$ *converge vers* $T(u) \in \mathcal{O}^N(\Omega_{\mu(\lambda)})$ *dans* $\mathcal{O}^N(\Omega_{\mu(\lambda)})$.

Preuve. Il suffit de démontrer le lemme dans le cas où $u=0$. Il existe $\varphi \in \mathcal{O}^N_+$ vérifiant que, étant donné $\varepsilon>0$, il existe un entier $p\geq 1$ tel que pour tout $k \geq p$ on a $u_k \ll \varepsilon\varphi$. On a donc $T(u_k) \ll \varepsilon S(\varphi)$, $k \geq p$. Cela signifie que $\{T(u_k)\}$ converge uniformément vers 0 dans un voisinage de l'origine. C.Q.F.D.

Note. Ce lemme démontre que si T admet une application majorante S, T est une application continue de $\mathcal{O}^N$ dans $\mathcal{O}^N$; $\mathcal{O}^N$ est un espace vectoriel topologique muni la topologie limite inductive des $\mathcal{O}^N(\Omega_\lambda)$ $(\lambda \in \Lambda)$.

Lemme 1.2. *Soient T une application linéaire de $\mathcal{O}^N$ dans $\mathcal{O}^N$ et S une application majorante de T: $T \ll S$.*

Considérons les équations

$$u = T(u) + f, \qquad f \in \mathcal{O}^N, \tag{1.4}$$

$$U \gg S(U) + F, \qquad F \in \mathcal{O}^N_+. \tag{1.5}$$

Supposons que l'équation majorante (1.5) *admette une solution $U \in \mathcal{O}^N_+$ quelle que soit la donnée $F \in \mathcal{O}^N_+$. Alors l'équation* (1.4) *admet une uinque solution $u \in \mathcal{O}^N$ quelle que soit la donnée $f \in \mathcal{O}^N$ et si $f \ll F$, on a $u \ll U$.*

Preuve. Soit $U \in \mathcal{O}^N_+$ une solution de (1.5) avec $F \gg f$. Par récurrence, on a $T^k(f) \ll S^k(F)$ et $\sum_{l=0}^{k} S^l(F) \ll U$ pour tout $k \geq 0$. Alors $u = \sum_{l=0}^{\infty} T^l(f)$ converge uniformément au voisinage de l'origine; $u \in \mathcal{O}^N$. D'après le lemme 1.1, u est une solution de (1.4) et on a $u \ll \sum_{l=0}^{\infty} S^l(F) \ll U$. On note que $S^k(F)$ $(k \to \infty)$ converge uniformément vers 0 dans un voisinage de l'origine pour tout $F \in \mathcal{O}^N_+$.

Enfin, soit u une solution de (1.4) avec $f = 0$. Prenons $G \in \mathcal{O}^N_+$ vérifiant $u \ll G$. Par récurrence, on a $u = T^k(u) \ll S^k(G)$ pour tout $k \geq 0$. On a noté que $S^k(G)$ $(k \to \infty)$ converge uniformément vers 0 au voisinage de l'origine. On a donc $u = 0$, d'où l'unicité du problème (1.4).

Note. Sous l'hypothèse du lemme 1.2, $\sum_{l=0}^{\infty} S^l(F)$ est une solution d'équation (1.5). Aussi, si la série $\sum_{l=0}^{\infty} S^l(F)$ converge au voisinage de l'origine, elle est une solution de (1.5).

En utilisant ce lemme, on a le

Lemme 1.3. *Considérons le problème de Goursat et le problème de Goursat majorant:*

$$\begin{cases} D^{\alpha(\mu)} u_\mu(x) = \sum_{\nu=1}^{N} p_\mu^\nu(x, D) u_\nu(x) + v_\mu(x), & 1 \leq \nu \leq N, \\ D^{\beta(\mu)} u_\mu(x) - \sum_{\nu=1}^{N} q_\mu^{\nu, \beta(\mu)}(x, D) u_\nu(x) - D^{\beta(\mu)} w_\mu(x)|_{x_i=0} = 0 & \\ \qquad\qquad\qquad\qquad \textit{pour } \beta(\mu) \prec_i \alpha(\mu), \end{cases} \tag{1.6}$$

et

$$(1.7)\quad \begin{cases} D^{\alpha(\mu)}U_\mu(x) \gg \sum_{\nu=1}^{N} P_\mu^\nu(x, D)U_\nu(x) + V_\mu(x), \qquad 1 \leq \mu \leq N, \\ D^{\beta(\mu)}U_\mu(x)|_{x_i=0} \gg \sum_{\nu=1}^{N} Q_\mu^{\nu,\beta(\mu)}(x, D)U_\nu(x) + D^{\beta(\mu)}W_\mu(x)|_{x_i=0} \\ \qquad\qquad\qquad\qquad\qquad\qquad \text{pour } \beta(\mu) \prec_i \alpha(\mu), \end{cases}$$

où $\beta(\mu)$ décrit l'ensemble des multi-indices vérifiant $\beta(\mu) \prec \alpha(\mu)$. p_μ^ν, $q_\mu^{\nu,\beta(\mu)}$ et P_μ^ν, $Q_\mu^{\nu,\beta(\mu)}$ sont des opérateurs différentiels holomorphes au voisinage de l'origine, vérifiant $p_\mu^\nu \ll P_\mu^\nu$ et $q_\mu^{\nu,\beta(\mu)} \ll Q_\mu^{\nu,\beta(\mu)}$, et $(v_\mu(x))$, $(w_\mu(x)) \in \mathcal{O}^N$, $(V_\mu(x))$, $(W_\mu(x)) \in \mathcal{O}_+^N$.

Supposons que le problème (1.7) admette une solution $(U_\mu(x)) \in \mathcal{O}_+^N$ quelles que soient les données $(V_\mu(x))$, $(W_\mu(x)) \in \mathcal{O}_+^N$.

Alors le problème (1.6) admet une unique solution $(u_\mu(x)) \in \mathcal{O}^N$ et on a $(u_\mu(x)) \ll (U_\mu(x))$, où $(U_\mu(x))$ est une solution de (1.7) pour les données $(V_\mu(x))$, $(W_\mu(x))$ vérifiant $(V_\mu(x)) \gg (v_\mu(x))$, $(W_\mu(x)) \gg (w_\mu(x))$.

Remarque 1.1. Nous ne faisons aucune hypothèse sur les ordres de $p_\mu^\nu(x, D)$, $q_\mu^{\nu,\beta(\mu)}(x, D)$.

Preuve. Soient $\hat{u} = (\hat{u}_\mu)_{1 \leq \mu \leq N} \in \mathcal{O}^N$ et $\hat{U} = (\hat{U}_\mu)_{1 \leq \mu \leq N} \in \mathcal{O}_+^N$ les solutions de systèmes d'équations respectives

$$\begin{cases} D^{\alpha(\mu)}\hat{u}_\mu(x) = \sum_{\nu=1}^{N} p_\mu^\nu(x, D)u_\nu(x), \\ D^{\beta(\mu)}\hat{u}_\mu(x) - \sum_{\nu=1}^{N} q_\mu^{\nu,\beta(\mu)}(x, D)u_\nu(x)|_{x_i=0} = 0, \qquad \text{pour } \beta(\mu) \prec_i \alpha(\mu), \end{cases}$$

et

$$\begin{cases} D^{\alpha(\mu)}\hat{U}_\mu(x) = \sum_{\nu=1}^{N} P_\mu^\nu(x, D)U_\nu(x), \\ D^{\beta(\mu)}\hat{U}_\mu(x) - \sum_{\nu=1}^{N} Q_\mu^{\nu,\beta(\mu)}(x, D)U_\nu(x)|_{x_i=0} = 0, \qquad \text{pour } \beta(\mu) \prec_i \alpha(\mu), \end{cases}$$

où $\beta(\mu)$ décrit l'ensemble $\beta(\mu) \prec \alpha(\mu)$, $u = (u_\mu)_{1 \leq \mu \leq N} \in \mathcal{O}^N$ et $U = (U_\mu)_{1 \leq \mu \leq N} \in \mathcal{O}_+^N$.

Définissons les applications T et S par $\hat{u} = T(u)$ et $\hat{U} = S(U)$. T est alors une application linéaire de $\mathcal{O}^N$ dans $\mathcal{O}^N$ et S est son application majorante: $T \ll S$.

Soient $f = (f_\mu)_{1 \leq \mu \leq N} \in \mathcal{O}^N$ et $F = (F_\mu)_{1 \leq \mu \leq N} \in \mathcal{O}_+^N$ les solutions des systèmes d'équations

$$D^{\alpha(\mu)}f_\mu(x) = v_\mu(x), \qquad D^{\beta(\mu)}f_\mu(x)|_{x_i=0} = D^{\beta(\mu)}w_\mu(x)|_{x_i=0}$$

et

$$D^{\alpha(\mu)}F_\mu(x) = V_\mu(x), \qquad D^{\beta(\mu)}F_\mu(x)|_{x_i=0} = D^{\beta(\mu)}W_\mu(x)|_{x_i=0}$$
$$\text{pour } \beta(\mu) \prec_i \alpha(\mu),$$

où $\beta(\mu) \prec \alpha(\mu)$.

Les problèmes (1.6) et (1.7) équivalent alors aux équations

$$u = T(u) + f, \qquad f \in \mathcal{O}^N,$$
$$U \gg S(U) + F, \qquad F \in \mathcal{O}_+^N.$$

L'hypothèse et le lemme 1.2 prouvent le lemme 1.3. C.Q.F.D.

Le lemme 1.3 nous permet de démontrer le théorème 1.1. En effet, nous utilisons l'inégalité suivante: Soient $a(x, D) = \sum_{|\alpha| \leq m} a_\alpha(x) D^\alpha$ un opérateur différentiel, holomorphe sur Z et $H(\xi)$ sa fonction spectrale. Posons $z = \sum_{j=1}^n \theta_j x_j$, $\theta_j \geq 1/r_j$, $1 \leq j \leq n$. Si on a $u(x) \ll U(z)$, on a

$$a(x, D)u(x) \ll \frac{1}{1-z}\{H(\theta)D_z^m + A(\theta, D_z)\}U(z),$$

où $A(\theta, D_z) = \sum_{|\alpha| \leq m-1} \sup_{x \in Z} |a_\alpha(x)| \theta^\alpha D_z^{|\alpha|}$.

Le théorème 1.1 a pour conséquence le

Corollaire 1.1. *Dans les hypothèses* 1.1 *et* 1.2, *supposons que:*

(i) *pour* $\alpha(\mu) = (\alpha(\mu)_1, \cdots, \alpha(\mu)_n) \in N^n$, $1 \leq \mu \leq N$, *on a* $\alpha(\mu)_i = 0$ $(i \geq l+1)$, *l étant un entier* $1 \leq l \leq n$.

(ii) *les opérateurs* a_μ^ν, $b_\mu^{\nu, \beta(\mu)}$ *et les données* v_μ, w_μ *sont holomorphes sur*

(1.8) $$\{x; |x_j| \leq r_j, 1 \leq j \leq n, (x_{l+1}, \cdots, x_n) \in \mathcal{R}(\mathcal{D})\},$$

où $\mathcal{R}(\mathcal{D})$ *est le revêtement universel d'un domaine* $\mathcal{D}$ *de* $\boldsymbol{C}^{n-l}$. *Les parties principales de* a_μ^ν *et* $b_\mu^{\nu, \beta(\mu)}$ *sont bornées sur* (1.8), *étant convenu que ces opérateurs sont d'ordres respectifs* $|\alpha(\mu)| + n_\mu - n_\nu$ *et* $|\beta(\mu)| + n_\mu - n_\nu$.

(iii) *le rayon spectral* $\rho(\xi)$ *de* (1.2) *sur* (1.8) *est indépendant de* $(\xi_{l+1}, \cdots, \xi_n) \in (R_+)^{n-l}$ *et il existe* $\hat{\theta} = (\theta_1, \cdots, \theta_l)$ $(\theta_j \geq 1/r_j, 1 \leq j \leq l)$ *tel que* $\rho(\hat{\theta}) < 1$.

Alors la solution du problème (1.2) *est holomorphe sur*

$$\left\{x; \sum_{j=1}^{l} \theta_j |x_j| < (1 - \rho(\hat{\theta})), (x_{l+1}, \cdots, x_n) \in \mathcal{R}(\mathcal{D})\right\}.$$

§ 2. Un problème de Goursat particulier

Nous appliquons le théorème 1.1 et le corollaire 1.1 au problème de Goursat particulier qui concerne le problème de Cauchy à données singulières.

Nous notons (t, x) $[x = (x_0, x_1, \cdots, x_n)]$ un point de $\boldsymbol{C}^{n+2}$ et b une autre variable complexe. $[x'=(x_1,\cdots,x_n)]$.

Étudions le problème de Goursat, d'inconnue $(u_i(t,x,b))_{1\leq i\leq l}$,

$$(2.1)\quad \begin{cases} D_0^p D_t^m u_i(t,x,b)=A_i u_i(t,x,b)+v_i(t,x,b), \\ D_t^k u_i(\gamma(b),x,b)=w_{i,k}(x,b), \qquad 0\leq k\leq m-1, \\ D_0^q D_t^m u_i(t,x,b)\Big|_{x_0=0}=\sum_{j=1}^{l} B_i^{q,j} u_j(t,x,b)\Big|_{x_0=0}+z_{i,q}(t,x',b), \\ \qquad\qquad 0\leq q\leq p-1. \end{cases}$$

Nous faisons l'

Hypothèse 2.1. (i) *$A_i=A_i(x,D_t,D_x)$ et $B_i^{q,j}=B_i^{q,j}(x,D_t,D_x)$ sont des opérateurs différentiels, d'ordres respectifs $m+p$ et $m+q$, holomorphes sur*

$$\{(t,x);\ |t|<\infty,\ |x_j|\leq r,\ 0\leq j\leq n\}.$$

Les parties principales H_i et $K_i^{q,j}$ de A_i et $B_i^{q,j}$ sont d'ordres respectifs m et $m-1$ en D_t et H_i ne contient pas $D_0^p D_t^m$.

(ii) *Les données v_i, $w_{i,k}$ et $z_{i,q}$ sont holomorphes sur*

$$\{(t,x,b);\ t\in \boldsymbol{C},\ |t-\gamma(b)|<\varepsilon(b),\ |x_j|\leq r,\ 0\leq j\leq n,\ b\in \mathscr{R}(\mathscr{D})\},$$

où $\mathscr{R}(\mathscr{D})$ est le revêtement universel d'un domaine $\mathscr{D}$ de $\boldsymbol{C}$, $\gamma(b)$ est une application holomorphe de $\mathscr{R}(\mathscr{D})$ dans $\boldsymbol{C}$ et $\varepsilon(b)$ (>0) une fonction continue sur $\mathscr{R}(\mathscr{D})$.

Nous avons alors la

Proposition 2.1. *Sous l'hypothèse* 2.1, *il existe une constante C $(0<C<1)$ ne dépendant que de H_i et $K_i^{q,j}$ telle que le problème de Goursat* (2.1) *admette une unique solution holomorphe sur*

$$\{(t,x,b);\ t\in \boldsymbol{C},\ |t-\gamma(b)|<C\min(\varepsilon(b),r),\ |x_j|\leq Cr,\ 0\leq j\leq n,\ b\in \mathscr{R}(\mathscr{D})\}.$$

§ 3. Problème de Cauchy à données singulières pour des opérateurs à caractéristiques de multiplicité constante

Dans cette section, nous appliquons le résultat précédent au problème de Cauchy à données singulières. Soit X un voisinage de l'origine de $\boldsymbol{C}^{n+1}$:

$$X=\{x=(x_0,x');\ x'=(x_1,\cdots,x_n),\ |x_j|\leq r,\ 0\leq j\leq n\}.$$

On considère $a(x,D)$ un opérateur différentiel d'ordre m holomorphe sur X. Son polynôme caractéristique est noté $g(x,\xi)$. Soit S l'hyperplan

$x_0=0$, supposé non caractéristique pour g.

Nous faisons l'

Hypothèse 3.1. *$g(x, \xi)$ est de la forme $g(x, \xi)=\prod_{i=1}^{s} g_i(x, \xi)^{p_i}$, où $g_i(x, \xi)$ $(1\leq i\leq s)$ sont des polynômes de degrés l_i en ξ. Notons $g_0(x, \xi)=\prod_{i=1}^{s} g_i(x, \xi)$. L'équation $g_0(0, \xi)=0$ pour $\xi=(\xi_0, 1, 0, \cdots, 0)$ possède l racines distinctes, où $l=\sum_{i=1}^{s} l_i$.*

Soient $\varphi_i(x)$ $(1\leq i\leq l)$ les solutions des équations: $g_0(x, D_x\varphi_i(x))=0$, $\varphi_i(0, x')=x_1$. Les $\varphi_i(x)$ sont holomorphes sur X, en diminuant X si nécessaire.

Étudions le problème de Cauchy

$$a(x, D)u(x)=0, \quad D_0^k u(0, x')=w_k(x'), \quad 0\leq k\leq m-1, \tag{3.1}$$

où les données w_k satisfont l'hypothèse suivante:

Hypothèse 3.2. (i) *Soit $\mathscr{D}$ un domaine de $\boldsymbol{C}$ tel que $0 \in \bar{\mathscr{D}}$ et $\mathscr{D}\subset \{t \in \boldsymbol{C}; |t|<r\}$. Définissons le domaine $\Delta=\{x \in X\cap S; x_1 \in \mathscr{D}\}$ dans $X\cap S$. Choisissons un point $(\hat{a}, y)$ de $\mathscr{D}\times(X\cap S)$ tel que $\hat{a}=\varphi_i(y)=y_1$: on a $y \in \Delta$.*

(ii) *Soient $w_k(t, x')$ $(0\leq k\leq m-1)$ des fonctions holomorphes au voisinage de $(\hat{a}, y)$ se prolongeant analytiquement sur $\mathscr{R}(\mathscr{D})\times(X\cap S)$. Supposons que $w_k(x')=w_k(t, x')|_{t=x_1}$. Alors $w_k(x')$ sont des fonctions holomorphes au voisinage de y se prolongeant analytiquement sur $\mathscr{R}(\Delta)$.*

Remarque 3.1. Le choix définitif d'un point y sera fait dans le théorème 3.1.

Nous avons alors le

Théorème 3.1. *Sous les hypothèses* 3.1 *et* 3.2, *le problème de Cauchy* (3.1) *admet une unique solution $u(x)$ holomorphe sur le revêtement universel $\mathscr{R}(\Phi)$ de toute composante connexe Φ de l'ouvert $\{x \in \Omega; \varphi_i(x) \in \mathscr{D}, 1\leq i\leq l\}$, telle que $\Phi\cap S$ ne soit pas vide. Ω est un voisinage de l'origine de $\boldsymbol{C}^{n+1}$ ne dépendant que de $g(x, D)$ et X.*

Remarque 3.2. Dans l'hypothèse 3.2, nous choisissons comme un point y un point de $\Phi\cap S$.

§ 4. Preuve du Théorème 3.1

En utilisant un résultat de [4], nous pouvons nous ramener au cas où $g(x, \xi)=g_0(x, \xi)^p$.

Nous cherchons la solution de (3.1) sous la forme:

$$u(x)=\sum_{i=1}^{l} u_i(\varphi_i(x), x),$$

où $u_i(t, x)$ $(1 \leq i \leq l)$ sont des fonctions holomorphes au voisinage de $(\hat{a}, y)$ se prolongeant analytiquement sur $\mathscr{R}(\mathscr{D}_{\omega'}(\hat{a})) \times \Omega$, $\mathscr{D}_{\omega'}(\hat{a})$ étant la composante connexe contenant le point $\hat{a}$ de $\mathscr{D} \cap \{t \in \boldsymbol{C};\, |t| < \omega'\}$ $(\omega' > 0)$.

Posons $u_i(t, x) = D_t^r \hat{u}_i(t, x)$ $(r = m+p-2)$ et $D_t^k \hat{u}_i(\hat{a}, x) = 0$ $(0 \leq k \leq m-2)$.

Pour que $u(x)$ satisfasse (3.1), il suffit que les $\hat{u}_i(t, x)$ $(1 \leq i \leq l)$ vérifient les équations suivante:

$$(4.1) \qquad \begin{cases} D_0^p D_t^{m-1} \hat{u}_i(t, x) = A_i \hat{u}_i(t, x), \\ D_t^k \hat{u}_i(\hat{a}, x) = 0, \quad 0 \leq k \leq m-2, \\ D_0^q D_t^{m-1} \hat{u}_i(t, x)|_{x_0=0} = \sum_{j=1}^{l} B_i^{q,j} \hat{u}_j(t, x)|_{x_0=0} + z_{q,i}(t, x'), \\ \qquad\qquad 0 \leq q \leq p-1, \end{cases}$$

où A_i et $B_i^{q,j}$ sont les opérateurs différentiels vérifiant l'hypothèse 2.1 (i) où m est remplacé par $m-1$, et

$$z_{i,q}(t, x') = \sum_{k=0}^{m-1} \Delta_k(x') D_{t,\hat{a}}^{-k-(p-1)+q} w_k(t, x'), \qquad 0 \leq q \leq p-1,$$

$\Delta_k(x')$ étant des fonctions holomorphes sur X. Pour une fonction $w(t, x')$ holomorphe sur $\mathscr{R}(\mathscr{D}) \times \{x';\, |x_j| \leq r,\, 1 \leq j \leq n\}$ et $p \in \boldsymbol{N}$, l'opérateur $D_{t,\hat{a}}^{-p}$ est défini comme suit:

$$D_{t,\hat{a}}^{-p} w(t, x') = 1/(p-1)! \int_{\hat{a}}^{t} (\underline{t} - \underline{s})^{p-1} w(s, x') ds,$$

où $\underline{t} = \pi(t)$, $t \in \mathscr{R}(\mathscr{D})$, π étant la projection naturelle de $\mathscr{R}(\mathscr{D})$ sur $\mathscr{D}$.

$z_{i,q}(t, x')$ sont donc des fonctions holomorphes au voisinage de $(\hat{a}, y)$ se prolongeant analytiquement sur

$$\{(t, x');\, t \in \mathscr{R}(\mathscr{D}),\, |x_j| \leq r,\, 1 \leq j \leq n\}.$$

D'après la proposition 2.1, le problème (4.1) admet une unique solution holomorphe au voisinage de $(\hat{a}, y)$. Il s'agit d'étudier le prolongement analytique de cette solution.

Pour faire cela, nous considérons les équations:

$$(4.2) \qquad \begin{cases} D_0^p D_t^{m-1} \hat{u}_i(t, x, b) = A_i \hat{u}_i(t, x, b), \\ D_t^k \hat{u}_i(b, x, b) = 0, \quad 0 \leq k \leq m-1, \\ D_0^q D_t^{m-1} \hat{u}_i(t, x, b)|_{x_0=0} = \sum_{j=1}^{l} B_i^{q,j} \hat{u}_j(t, x, b)|_{x_0=0} + z_{i,q}(t, x'), \\ \qquad\qquad 0 \leq q \leq p-1, \end{cases}$$

où $b \in \mathscr{R}(\mathscr{D})$. On a alors $\hat{u}_i(t, x, \hat{a}) = \hat{u}_i(t, x)$.

Les fonctions $z_{i,q}(t, x')$ sont holomorphes sur $\{(t, x, b);\ t \in \mathscr{R}(\mathscr{D}),$ dist. $(t, b) < \rho(b),\ |x_j| \leq r,\ 0 \leq j \leq n,\ b \in \mathscr{R}(\mathscr{D})\}$, où $\rho(b) =$ dist. $(b, \partial\mathscr{R}(\mathscr{D}))$.

On note que ce domaine est homéomorphe au domaine

$$\{(t, x, b);\ t \in \boldsymbol{C},\ \text{dist.}\ (t, b) < \rho(b),\ |x_j| \leq r,\ 0 \leq j \leq n,\ b \in \mathscr{R}(\mathscr{D})\}.$$

D'après la proposition 2.1, on a le

Lemme 4.1 (local). *La solution* $(\hat{u}_i(t, x, b))_{1 \leq i \leq l}$ *est holomorphe sur*

$$\{(t, x, b);\ t \in \mathscr{R}(\mathscr{D}),\ \text{dist.}(t, b) < C \inf(\rho(b), r),\ |x_j| \leq Cr,\ 0 \leq j \leq n,\ b \in \mathscr{R}(\mathscr{D})\},$$

où C $(0 < C < 1)$ *est une constante ne dépendant que de* $g(x, D)$.

Prolongeons analytiquement $(\hat{u}_i(t, x, b))_{1 \leq i \leq l}$ par la formule suivante: pour t voisin de $b \in \mathscr{R}(\mathscr{D})$,

$$\hat{u}_i(t, x, b) = \int_t^b \hat{U}_i(t, x, \sigma) d\sigma, \tag{4.3}$$

où $\hat{U}_i(t, x, b) = D_b \hat{u}_i(t, x, b)$.

Considérons alors les équations:

$$\begin{cases} D_0^p D_t^{m-1} \hat{U}_i(t, x, b) = A_i \hat{U}_i(t, x, b), \\ D_t^k \hat{U}_i(\underline{b}, x, b) = 0, \quad 0 \leq k \leq m-3, \\ D_t^{m-2} \hat{U}_i(\underline{b}, x, b) = -D_t^{m-1} \hat{u}_i(b, x, b), \\ D_0^q D_t^{m-1} \hat{U}_i(t, x, b)|_{x_0=0} = \sum_{j=1}^{l} B_i^{q,j} \hat{U}_j(t, x, b)|_{x_0=0}, \\ \qquad\qquad\qquad 0 \leq q \leq p-1. \end{cases} \tag{4.4}$$

D'après le lemme 4.1, les fonctions $D_t^{m-1} \hat{u}_i(b, x, b)$ $(1 \leq i \leq l)$ sont holomorphes au voisinage de $(x, b) = (y, \hat{a})$ dans $X \times \mathscr{R}(\mathscr{D})$ et se prolongent analytiquement sur $\{(x, b);\ |x_j| \leq Cr,\ 0 \leq j \leq n,\ b \in \mathscr{R}(\mathscr{D})\}$.

D'après la proposition 2.1, le problème (4.4) admet une unique solution $(\hat{U}_i(t, x, b))_{1 \leq i \leq l}$ holomorphe au voisinage de $(t, x, b) = (\hat{a}, y, \hat{a})$. Il s'agit d'un prolongement analytique de $(\hat{U}_i(t, x, b))_{1 \leq i \leq l}$.

En fait, vu la proposition 2.1, on a le

Lemme 4.2 (global). *Les fonctions* $\hat{U}_i(t, x, b)$ $(1 \leq i \leq l)$ *se prolongent analytiquement sur*

$$\{(t, x, b);\ t \in \boldsymbol{C},\ |t - \underline{b}| \leq C^2 r,\ |x_j| \leq C^2 r,\ 0 \leq j \leq n,\ b \in \mathscr{R}(\mathscr{D})\}.$$

Alors vu (4.3) nous avons la

Proposition 4.1. *Les fonctions $\hat{u}_i(t, x, b)$ sont holomorphes sur*

$$\{(t, x, b);\ t \in \mathscr{R}(\mathscr{D}_{\omega_0}(\hat{a})),\ |x_j| \leq C^2 r,\ 0 \leq j \leq n,\ b \in \mathscr{R}(\mathscr{D}_{\omega_0}(\hat{a}))\},$$

où $\omega_0 = C^2 r/2$.

Les fonctions $\hat{u}_i(t, x) = \hat{u}_i(t, x, \hat{a})$ sont donc holomorphes sur

$$\{(t, x);\ t \in \mathscr{R}(\mathscr{D}_{\omega_0}(\hat{a})),\ |x_j| \leq C^2 r,\ 0 \leq j \leq n\}, \qquad \text{pour } |\hat{a}| < \omega_0.$$

Cela prouve le théorème 3.1.

Les démonstrations détaillées des résultats seront données dans une autre publication.

Bibliographie

[1] J.-C. de Paris, Problème de Cauchy analytique à données singulières pour un opérateur différentiel bien décomposable, J. Math. Pures Appl., **51** (1972), 465–488.

[2] L. Gårding, T. Kotake et J. Leray, Uniformisation et développement asymptotique de la solution du problème de Cauchy linéaire à données holomorphes; analogie avec la théorie des ondes asymptotiques et approchées, Bull. Soc. Math. France **92** (1964), 263–361.

[3] L. Gårding, Une variante de la méthode de majoration de Cauchy, Acta Math., **114** (1965), 143–158.

[4] Y. Hamada, J. Leray et C. Wagschal, Systèmes d'équations aux dérivées partielles à caractéristiques multiples: problème de Cauchy ramifié; hyperbolicité partielle, J. Math. Pures Appl., **55** (1976), 297–352.

[5] Y. Hamada et A. Takeuchi, Le domaine d'existence et le prolongement analytique des solutions des problèmes de Goursat et de Cauchy à données singulières, C.R. Acad. Sci., Paris, **276** sér. I, (1982), 377–380.

[6] J. Leray, Uniformisation de la solution du problème linéaire analytique de Cauchy près de la variété qui porte les données de Cauchy, Bull. Soc. Math. France, **85** (1957), 389–429.

[7] ——, Un complément au théorème de N. Nilsson sur les intégrales de formes différentielles à support singulier algébrique, Bull. Soc. Math. France, **95** (1967), 313–374.

[8] J. Persson, A boundary problem for analytic linear systems with data on intersecting hyperplanes, Math. Scand., **14** (1964), 183–208.

[9] D. Schiltz, J. Vaillant et C. Wagschal, Problème de Cauchy ramifié: racine caractéristique double ou triple en involution, J. Math. Pures Appl., **4** (1982), 423–443.

[10] C. Wagschal, Une généralisation du problème de Goursat pour des systèmes d'équations intégro-différentielles holomorphes ou partiellement holomorphes, J. Math. Pures Appl., **53** (1974), 99–132.

[11] ——, Sur le problème de Cauchy ramifié, J. Math. Pures Appl., **53** (1974), 147–164.

Y. HAMADA
DÉPARTEMENT DE MATHÉMATIQUES
UNIVERSITÉ TECHNOLOGIQUE DE KYOTO
KYOTO, 606, JAPON

A. TAKEUCHI
SECTION DE MATHÉMATIQUES
FACULTÉ DES ARTS LIBÉRAUX
UNIVERSITÉ DE KYOTO
KYOTO, 606, JAPON

Taniguchi Symp. HERT
Katata 1984, pp. 63–84

On the Scattering Matrix for Two Convex Obstacles

Dedicated to Professor Sigeru Mizohata on his sixtieth birthday

Mitsuru IKAWA

§ 1. Introduction

Let $\mathcal{O}$ be a bounded open set in $\boldsymbol{R}^3$ with smooth boundary Γ. We set $\Omega=\boldsymbol{R}^3-\bar{\mathcal{O}}$. Suppose that Ω is connected. Consider the following acoustic problem

$$(1.1)\qquad \begin{cases} \Box u(x,t)=\dfrac{\partial^2 u}{\partial t^2}-\sum\limits_{j=1}^{3}\dfrac{\partial^2 u}{\partial x_j^2}=0 & \text{in } \Omega\times(-\infty,\infty) \\ u(x,t)=0 & \text{on } \Gamma\times(-\infty,\infty). \end{cases}$$

Denote by $\mathcal{S}(\sigma)$ the scattering matrix for this problem. About the definition of the scattering matrix, see for example Lax and Phillips [9, page 9]. Note that $\mathcal{S}(\sigma)$ is a unitary operator from $L^2(S^2)$ into itself for all $\sigma\in\boldsymbol{R}$. As its fundamental properties it is well known that "$\mathcal{S}(\sigma)$ extends to an operator valued function $\mathcal{S}(z)$ analytic in $\operatorname{Im} z\leqslant 0$ and meromorphic in the whole complex plane, and the scattering matrix determines the scatterer (Theorems 5.1 and 5.6 of Chapter V of [9])".

A problem on which we like to consider is about relationships between the geometric properties of an obstacle $\mathcal{O}$ and the analytic properties of the scattering matrix $\mathcal{S}(z)$. Here we take up the trapping or nontrapping property of obstacles.[1)]

The purpose of this note is to consider the case of $\mathcal{O}$ consisting of two convex objects as a simplest example of trapping obstacles, and to show several typical aspects of $\mathcal{S}(z)$ which are different from those for nontrapping obstacles.

Concerning a reflexion of these characters to the distributions of the poles of $\mathcal{S}(z)$ Lax and Phillips gave a following conjecture ([9, paeg 158]): for a nontrapping obstacle the scattering matrix $\mathcal{S}(z)$ is free for poles in $\{z;\operatorname{Im} z\leqslant\alpha\}$ for some $\alpha>0$, and for a trapping obstacle $\mathcal{S}(z)$ has a sequence of poles $\{z_j\}_{j=1}^{\infty}$ such that $\operatorname{Im} z_j\to 0$ as $j\to\infty$. As regard this conjecture Morawetz, Ralston and Strauss [18] and Melrose [12] proved that

[1)] For the definition see [12] and [18].

for a nontrapping obstacle $\mathscr{S}(z)$ is free for poles in

$$\{z;\ \operatorname{Im} z \leqslant a \log(1+|z|)+b\} \tag{1.2}$$

for some a, b positive constants, which implies that the part of the conjecture for nontrapping obstacles is correct. On the other hand, Bardos, Guillot and Ralston [1], Petkov [19] show the existence of an infinite number of poles of $\mathscr{S}(z)$ in

$$\{z;\ \operatorname{Im} z \leqslant \varepsilon \log(1+|z|)\} \tag{1.3}$$

for any $\varepsilon>0$ when $\mathcal{O}$ belongs to some class which contains $\mathcal{O}$ consisted of two disjoint strictly convex obstacles. The nonexistence in (1.2) and the existence in (1.3) show a difference of the distributions of poles for trapping obstacles and for nontrapping obstacles.

Here we shall give a more precise information on the distribution of the poles for the case of two strictly convex objects (Theorem 1), and this result indicates that the conjecture of Lax and Phillips for trapping obstacles is not correct in general. In spite of the above result it seems yet very sure that the conjecture remains to be correct for a great part of trapping obstacles. But we have not known even an example of obstacle for which is proved the existence of a sequence of poles converging to the real axis. Now we shall show the existence of a sequence of poles converging to the real axis for a certain pair of disjoint not strictly convex objects (Theorem 2).

Next take up as another analytic property of $\mathscr{S}(z)$ the asymptotic behavior of the scattering phase

$$s(\sigma) = -\frac{1}{2\pi i} \operatorname{Log} \det \mathscr{S}(\sigma).$$

Majda and Ralston [15], Jensen and Kato [11] and Petkov and Popov [19] consider it when $\mathcal{O}$ is nontrapping. Especially [19] shows the following formula

$$\frac{d}{d\sigma} s(\sigma) = \frac{6}{\pi^2} |\mathcal{O}| \sigma^2 + \frac{1}{4\pi} |\Gamma| \sigma + O(1) \qquad \text{as } \sigma \longrightarrow \infty, \tag{1.4}$$

where $|\mathcal{O}|$ and $|\Gamma|$ denote the volume of $\mathcal{O}$ and the area of Γ. For trapping obstacles we have not known any result on asymptotic behavior of $s(\sigma)$. Majda and Ralston [16] conjectured that for some trapping obstacles an asymptotic formula

$$\frac{d}{d\sigma} s(\sigma) = \frac{6}{\pi^2} |\mathcal{O}| \sigma^2 + o(\sigma^2)$$

does not hold. Even though it is very far from their conjecture, we can show for two convex objects an asymptotic behavior of $s(\sigma)$ different from (1.4). This difference is generated by the existence of a periodic ray for $\mathcal{O}=\mathcal{O}_1\cup\mathcal{O}_2$.

The main part of the proofs of these theorems is reduced to considerations of the asymptotic behavior of solutions for oscillatory boundary data, that is,

$$(1.5)\qquad \begin{cases} \square u=0 & \text{in } \Omega\times(-\infty,\infty) \\ u=m(x,t;k) & \text{on } \Gamma\times(-\infty,\infty) \\ \operatorname{supp} u\subset\Omega\times\{t\geqslant 0\} \end{cases}$$

where m is a function of the form

$$m(x,t;k)=e^{ik(\psi(x)-t)}f(x,t),\qquad f\in C_0^\infty(\Gamma\times(0,d/2)).$$

For example, in the case where $\mathcal{O}_1$ and $\mathcal{O}_2$ are strictly convex, u the solution of (1.5) converges exponentially to a periodic type function. The periodicity in a certain sense corresponds to a series of poles of the Laplace transform in t of $u(x,t;k)$, and the poles are linked with those of $\mathcal{S}(z)$.

About Theorems 2 and 3 we shall use trace formulas in [1] and [19], and an estimation of $\operatorname{tr}_{L^2}\int(\cos t\sqrt{-\Delta}\oplus O-\cos t\sqrt{-\Delta_0})\rho(t)dt$ is essential part of the proofs. To do this we represent the kernel distribution of $\cos t\sqrt{-\Delta}$ as a superposition of asymptotic solutions, and pick out the main part of oscillating integral by using an asymptotic form of solutions of (1.5). Comparing Theorems 1 and 2 the difference of the distributions of poles is generated by a difference of asymptotic behavior of solutions of (1.5), which is a consequence of the difference of the curvatures of the wave fronts of asymptotic solutions of (1.5).

In Section 2 we state Theorems and in Sections 3 and 4 we shall give a sketch of proofs.

§ 2. Statement of theorems

Hereafter we suppose that

$$\mathcal{O}=\mathcal{O}_1\cup\mathcal{O}_2$$

where $\mathcal{O}_j$, $j=1,2$, are bounded open convex sets in $\boldsymbol{R}^3$ such that $\bar{\mathcal{O}}_1\cap\bar{\mathcal{O}}_2=\emptyset$. Set

$$d=\operatorname{dis}(\mathcal{O}_1,\mathcal{O}_2)=|a_1-a_2|,\qquad a_j\in\Gamma_j.$$

Theorem 1. *Suppose that $\mathcal{O}_1$ and $\mathcal{O}_2$ are strictly convex, that is, the Gaussian curvatures of the boundary Γ_j of $\mathcal{O}_j$, $j=1, 2$ never vanish. Then there exist positive constants c_0 and c_1 such that*

(i) *for any $\varepsilon>0$ a region*

$$\{z;\ \operatorname{Im} z\leqslant c_0+c_1-\varepsilon\}-\bigcup_{j=-\infty}^{\infty}\{z;\ |z-z_j|\leqslant C(1+|j|)^{-1/2}\}$$

contains only a finite number of poles of $\mathcal{S}(z)$, where

$$z_j=ic_0+\frac{\pi}{d}j,$$

and C is a positive constant independent of ε,

(ii) *for every j*

$$\{z;\ |z-z_j|\leqslant C(1+|j|)^{-1/2}\}$$

contains at least a pole of $\mathcal{S}(z)$.

Remark. The constants c_0 and c_1 are determined by d and the principal curvatures and directions of Γ_j at a_j such that $0<c_1<c_0/2$. $z_j, j=0, \pm 1, \cdots$ are nothing but the pseudo-poles $\alpha_{j,m}$ for $\boldsymbol{m}=0$ of [1]. Then Theorem 1 says that the pseudo-poles $\alpha_{j,m}$ for $\boldsymbol{m}=0$ approximate the actual poles of $\mathcal{S}(z)$.

On the poles of $\mathcal{S}(z)$ in $\{z;\ \operatorname{Im} z\geqslant c_0+c_1\}$ [6] considers the case of Γ_j are umbilial at $a_j, j=1, 2$. Though it is not yet proved precisely, it seems sure that for some $\mathcal{O}_1$ and $\mathcal{O}_2$ $\alpha_{j,m}$ for $|\boldsymbol{m}|=1$ do approximate a series of actual poles. Then it seems that not only the curvatures of Γ_j at a_j but also the derivatives of the curvatures may closely related to the distribution of poles in $\operatorname{Im} z\geqslant c_0+c_1$.

Theorem 2. *Let $\mathcal{O}_j$, $j=1, 2$, be bounded convex open sets in $\boldsymbol{R}^3$. Suppose that the principal curvatures $\kappa_{jl}(x)$, $l=1, 2$, of Γ_j at x satisfy*

$$C|x-a_j|^e\geqslant\kappa_{jl}(x)\geqslant C^{-1}|x-a_j|^e \qquad \text{for all } x\in\Gamma_j$$

for some

$$2\leqslant e<\infty$$

and $C>0$. Then the scattering matrix $\mathcal{S}(z)$ has a sequence of poles $\{z_j\}_{j=1}^{\infty}$ such that

$$\operatorname{Im} z_j\longrightarrow 0 \qquad \text{as } j\longrightarrow\infty.$$

Theorem 3. *Let $\mathcal{O}_j$, $j=1, 2$, be bounded convex open sets in $\boldsymbol{R}^3$. Suppose that*

$$\kappa_{11}(a_1)>0$$

and that κ_{12} and κ_{2l}, $l=1, 2$, satisfy

$$C|x-a_j|^{e_j}\geqslant\kappa_{jl}(x)\geqslant C^{-1}|x-a_j|^{e_j} \qquad \text{for all } x\in\Gamma_j$$

for some

$$2\leqslant e_1,\ e_2<\infty$$

and $C>0$. Then as asymptotic expansion

$$\frac{d}{d\sigma}s(\sigma)=\frac{6}{\pi^2}|\mathcal{O}|\sigma^2+\frac{1}{4\pi}|\Gamma|\sigma+\sigma^{1/2-1/e_0}g(\sigma)$$
$$+O(\sigma^{1/2-3/2e_0}) \qquad \text{for } \sigma\longrightarrow\pm\infty$$

$$g(\sigma)=c\sum_{q=0}^{\infty}\gamma^{2q}\cos 2dq\sigma$$

holds, where c is a non zero real number,

$$e_0=\min(e_1, e_2)+2$$

and γ is a constant $0<\gamma<1$ determined by κ_{jl}, $j, l=1, 2$.

§ 3. Construction of asymptotic solutions

We note some properties of broken rays according to the law of the geometric optics in Ω, which are in close connection with the propagation of singularities of the problem (1.1). First we introduce notations for broken rays. Denote by $n(x)$ the unit outer normal of Γ at x, and set

$$\Sigma_x^+=\{\xi;\ |\xi|=1,\ n(x)\cdot\xi\geqslant 0\}.$$

We denote by $\mathscr{X}(x, \xi)$ the broken ray starting from $x\in\Gamma$ the direction $\xi\in\Sigma_x^+$, by $X_j(x, \xi)$, $j=1, 2, \cdots$, the points of reflection of the broken ray and by $\Xi_j(x, \xi)$ the direction of the ray reflected at $X_j(x, \xi)$. More precisely, if

$$\{x+l\xi;\ l>0\}\cap\Gamma=\emptyset$$

we set $L_0=\{x+l\xi;\ l\geqslant 0\}$. If $\{x+l\xi;\ l>0\}\cap\Gamma\neq\emptyset$, we set

$$l_0(x, \xi)=\inf\{l>0;\ x+l\xi \in \Gamma\}$$
$$L_0(x, \xi)=\{x+l\xi;\ 0\leqslant l\leqslant l_0(x, \xi)\}$$
$$X_1(x, \xi)=x+l_0(x, \xi)\xi$$
$$\Xi_1(x, \xi)=\xi-2(n(X_1(x, \xi), \xi)n(X_1(x, \xi)).$$

When $\{X_1+l\Xi_1;\ l>0\}\cap\Gamma=\varnothing$ we set $L_1(x, \xi)=\{X_1+l\Xi_1;\ l>0\}$. Otherwise we set

$$l_1(x, \xi)=\inf\{l;\ l>0,\ X_1+l\Xi_1 \in \Gamma\}$$
$$L_1(x, \xi)=\{X_1+l\Xi_1;\ 0\leqslant l\leqslant l_1\}$$
$$X_2(x, \xi)=X_1+l_1\Xi_1$$
$$\Xi_2(x, \xi)=\Xi_1-2(n(X_2)\cdot\Xi_1)n(X_2).$$

Thus we define successively $l_j(x, \xi)$, $X_j(x, \xi)$, $\Xi_j(x, \xi)$, $L_j(x, \xi)$ until $\{X_j+l\Xi_j;\ l>0\}\cap\Gamma=\varnothing$. If there exists j_0 such that for $j\leqslant j_0$, l_j, X_j, Ξ_j are defined and $\{X_{j_0}+l\Xi_{j_0};\ l>0\}\cap\Gamma=\varnothing$, then we define

$$\mathscr{X}(x, \xi)=\bigcup_{j=0}^{j_0} L_j(x, \xi),$$
$${}^{\#}\mathscr{X}(x, \xi)=j_0.$$

Otherwise

$$\mathscr{X}(x, \xi)=\bigcup_{j=0}^{\infty} L_j(x, \xi),$$
$${}^{\#}\mathscr{X}(x, \xi)=\infty.$$

Denote by $S_j(\delta)$, $j=1, 2$, for $\delta>0$ the connected component containing a_j of $\Gamma_j\cap\{x;\ \mathrm{dis}\,(x, L)=\delta\}$, where L is the line passing a_1 and a_2. We set

$$S(\delta)=S_1(\delta)\cup S_2(\delta).$$

By using the assumption that the Gaussian curvature does not vanish on Γ_j except at a_j we have the following lemmas.

Lemma 3.1. *For each $\delta_1>0$ there exists a positive integer K such that for $x \in \Gamma-S(\delta_1)$, $\xi \in \Sigma_x^+$, if $\mathscr{X}(x, \xi)\cap S(\delta_1)=\varnothing$ we have*

$${}^{\#}\mathscr{X}(x, \xi)\leqslant K.$$

Lemma 3.2. *For each $\delta_1>0$, if we choose δ_2 as $\delta_2>\delta_1$ and sufficiently close to δ_1, then*

$$^{\#}\mathscr{X}(x, \xi) \leqslant K+1$$

holds for any $x \in S(\delta_2)$, $\xi \in \Sigma_x^+$ *such that* $X_1(x, \xi) \in S(\delta_2) - S(\delta_1)$.

Let $x_0 \in \Gamma$ and Γ be represented in a neighborhood of x_0 as $x = x(\sigma)$, $\sigma = (\sigma_1, \sigma_2) \in I$, I is an open interval in $\boldsymbol{R}^2$. Then we have by Fourier's inversion formula for $h(x, t) \in C_0^\infty(\mathscr{U} \times (0, 1))$ $(\mathscr{U} = \{x(\sigma); \sigma \in I\})$

$$(3.1) \quad \begin{aligned} h(x(\sigma), t) &= w(x(\sigma), t) \int_{R^2} d\eta \int_R d\tau\, e^{i(\eta\cdot\sigma + t\tau)} \hat{h}(\eta, \tau), \\ \hat{h}(\eta, \tau) &= \int d\sigma' \int dt'\, e^{-i(\eta\cdot\sigma' + t'\tau)} h(x(\sigma'), t'), \end{aligned}$$

where $w(x, t) \in C_0^\infty(\mathscr{U} \times (0, 1))$ verifying

$$w = 1 \qquad \text{on } \operatorname{supp} h.$$

Denote by $w(x, t, \eta, \tau)$ the solution of

$$(3.2) \quad \begin{cases} \square u = 0 & \text{in } \Omega \times (-\infty, \infty) \\ u = w(x, t) e^{i(\sigma\cdot\eta + t\tau)} & \text{on } \Gamma \times (-\infty, \infty) \\ \operatorname{supp} u \subset \Omega \times \{t \geqslant 0\}. \end{cases}$$

Then the solution of boundary value problem

$$(3.3) \quad \begin{cases} \square u = 0 & \text{in } \Omega \times (-\infty, \infty) \\ u = h & \text{on } \Gamma \times (-\infty, \infty) \\ \operatorname{supp} u \subset \Omega \times \{t \geqslant 0\} \end{cases}$$

is represented as

$$(3.4) \quad u(x, t) = \int_{R^2} d\eta \int_R d\tau\, w(x, t; \eta, \tau) \hat{h}(\eta, \tau).$$

By using a representation (3.4) of the solution of (3.3) we may take out the behavior of $u(x, t)$ from the behavior of $w(x, t; \eta, \tau)$ for $|\eta| + |\tau|$ large.

Admit the fact that the main part of the solution of (3.2), when $|\eta| + |\tau|$ is large, propagates along the ray of the geometric optics. Then, if the main part of the solution of (3.2) remains near $\mathcal{O}$ for large t, we see from Lemma 3.1 that it must reflect many times on $S(\delta_1)$. Therefore the consideration of the behavior of $w(x, t; \eta, \tau)$ for large t can be reduced to the consideration of solutions for oscillatory boundary data of the form

$$(3.5) \quad m(x, t; k) = f(x, t) e^{ik(\psi(x) - t)}, \qquad f \in C_0^\infty(S_1(\delta_1) \times (0, 1))$$

where $\psi \in C^\infty(S_1(\delta_1))$, $|\nabla_\Gamma \psi| < 1$.

Let $u_0(x, t; k)$ be an asymptotic solution of the form

$$u_0(x, t; k) = e^{ik(\varphi_0(x)-t)} \sum_{j=0}^{N} v_{j,0}(x, t)(ik)^{-j} \tag{3.6$_0$}$$

satisfying

$$\begin{aligned}
&|\nabla \varphi_0| = 1 && \text{in } \Omega \\
&\varphi_0 = \psi, \quad \frac{\partial \varphi_0}{\partial n} > 0 && \text{on } S_1(\delta_1), \\
&T_0 v_{j,0} = \Box v_{j-1,0} && \text{in } \Omega \times \boldsymbol{R}, \quad j = 0, 1, \cdots, N, \\
&v_{0,0} = f, \quad v_{j,0} = 0, \quad j = 1, 2, \cdots, N && \text{on } S_1(\delta_1) \times \boldsymbol{R},
\end{aligned}$$

where $T_0 = 2(\partial/\partial_t) + 2\nabla \varphi_0 \cdot \nabla + \Delta \varphi_0$, and we set $v_{-1,0} \equiv 0$. Then we have

$$\begin{aligned}
\Box u_0 &= (ik)^{-N} e^{ik(\varphi_0 - t)} \Box v_{N,0} && \text{in } \Omega \times \boldsymbol{R} \\
u_0 &= m(x, t; k) && \text{on } \Gamma_1 \times \boldsymbol{R}.
\end{aligned}$$

Let $\upsilon_j \in C_0(S_j(\delta_2))$ such that $\upsilon_j = 1$ on $S_j(\delta_1)$. Define u_1 as follows:

$$u_1(x, t; k) = e^{ik(\varphi_1(x)-t)} \sum_{j=0}^{N} v_{j,1}(x, t)(ik)^{-j}, \tag{3.6$_1$}$$

$$\begin{aligned}
&|\nabla \varphi_1| = 1 && \text{in } \Omega \\
&\varphi_1 = \varphi_0, \quad \frac{\partial \varphi_1}{\partial n} > 0 && \text{on } S_2(\delta_2), \\
&T_1 v_{j,1} = \Box v_{j-1,1} && \text{in } \Omega \times \boldsymbol{R}, \quad j = 0, 1, \cdots, N, \\
&v_{j,1} = \upsilon_2(x) v_{j,0} && \text{on } S_2(\delta_2) \times \boldsymbol{R}, \quad j = 0, 1, \cdots, N,
\end{aligned}$$

where $T_1 = 2(\partial/\partial t) + 2\nabla \varphi_1 \cdot \nabla + \Delta \varphi_1$, and we set $v_{-1,1} \equiv 0$.

Suppose that u_q is defined as

$$u_q(x, t; k) = e^{ik(\varphi_q(x)-t)} \sum_{j=0}^{N} v_{j,q}(x, t)(ik)^{-j}. \tag{3.6$_q$}$$

We define $u_{q+1}(x, t; k)$ as follows:

$$u_{q+1}(x, t; k) = e^{ik(\varphi_{q+1}(x)-t)} \sum_{j=0}^{N} v_{j,q+1}(x, t)(ik)^{-j}, \tag{3.6$_{q+1}$}$$

$$\begin{aligned}
&|\nabla \varphi_{q+1}(x)| = 1 && \text{in } \Omega \\
&\varphi_{q+1}(x) = \varphi_q(x), \quad \frac{\partial \varphi_{q+1}}{\partial n} > 0 && \text{on } S_\varepsilon(\delta_2), \\
&T_{q+1} v_{j,q+1} = \Box v_{j-1,q+1} && \text{in } \Omega \times \boldsymbol{R}, \quad j = 0, 1, \cdots, N, \\
&v_{j,q+1}(x, t) = \upsilon_\varepsilon(x) v_{j,q}(x, t) && \text{on } S_\varepsilon(\delta_2) \times \boldsymbol{R},
\end{aligned}$$

where $T_{q+1}=2(\partial/\partial t)+2\nabla\varphi_{q+1}\cdot\nabla+\Delta\varphi_{q+1}$, $v_{-1,q+1}\equiv 0$, and

$$\varepsilon(q)=\begin{cases}1 & \text{for } q \text{ even}\\ 2 & \text{for } q \text{ odd.}\end{cases}$$

We set

$$u^{(1)}(x,t;k)=\sum_{q=0}^{\infty}(-1)^q u_q(x,t;k). \tag{3.7}$$

Then we have

$$u^{(1)}-m=\sum_{q=0}^{\infty}(1-v_1)u_{2q+1} \qquad \text{on } \Gamma_1\times \boldsymbol{R},$$

$$u^{(1)}=\sum_{q=0}^{\infty}(1-v_2)u_{2q} \qquad \text{on } \Gamma_2\times \boldsymbol{R},$$

and

$$\Box u^{(1)}=(ik)^{-N}\sum_{q=0}^{\infty}e^{ik(\varphi_q-t)}\Box v_{N,q}. \tag{3.8}$$

For each q an asymptotic solution u'_q of

$$\begin{cases}\Box u=0 & \text{in } \Omega\times\boldsymbol{R}\\ u=(1-v_\varepsilon)u_q & \text{on } \Gamma_{\varepsilon(q)}\times\boldsymbol{R}\\ u=0 & \text{on } \Gamma_{\varepsilon(q+1)}\times\boldsymbol{R}\\ \operatorname{supp} u\subset\Omega\times\{t\geqslant 0\}\end{cases}$$

can be constructed easily by the usual method, and we see from Lemma 3.2 that its support leaves from $\mathcal{O}$ within K-times reflections on the boundary. Therefore if we know the asymptotic properties of u_q when q tends to the infinity, we can easily take out the behavior of u'_q. Then

$$u(x,t;k)=u^{(1)}(x,t;k)+u^{(2)}(x,t;k) \tag{3.9}$$

where

$$u^{(2)}=\sum_{q=0}^{\infty}u'_q$$

can be an approximation of the solution for an oscillatory boundary data (3.5).

Remark. Concerning the estimate of the right hand side of (3.8) we have an estimate

$$|D^{\alpha}_{x,t}\square v_{N,q}(x, t)| \leqslant C_{\alpha}|t|^{|\alpha|+N},$$

and in some cases an estimate

$$\sup|\square v_{N,q}(x, t)| \geqslant ct^{N} \qquad (c>0)$$

holds. Therefore generally even if k is very large $\square u^{(1)}$ is small only for $t<k^{1-\varepsilon}$ $(\varepsilon>0)$.

§ 4. Outline of the proofs of theorems

4.1. *Distribution of the poles of $\mathscr{S}(z)$ of the case of two strictly convex objects.*[2)]

Let $g \in C^{\infty}(\Gamma)$ and $\mathrm{Re}\,\mu>0$. Denote by $(U(\mu)g)(x)$ a solution in $\bigcap_{m>0} H^{m}(\Omega)$ of

$$\begin{cases}(\mu^2-\Delta)u=0 & \text{in } \Omega\\ u=g & \text{on } \Gamma.\end{cases}$$

Then $U(\mu)$ is analytic in $\mathrm{Re}\,\mu>0$ as $\mathscr{L}(C^{\infty}(\Gamma), C^{\infty}(\bar{\Omega}))$ valued function, where $\mathscr{L}(E, F)$ denotes the set of all continuous linear mappings from E into F. It is known that $U(\mu)$ is prolonged analytically into the whole complex plane as a meromorphic function (Mizohata [17], Lax and Phillips [9]). Concerning $U(\mu)$ we have the following

Theorem 4.1. *Suppose that $\mathcal{O}_j$, $j=1, 2$, are strictly convex. Then $U(\mu)$ is holomorphic in*

$$\mathscr{D}_{\varepsilon}=\{\mu;\ \mathrm{Re}\,\mu \geqslant -c_0-c_1+\varepsilon\}-\bigcup_{j=-\infty}^{\infty}\{\mu;\ |\mu-\mu_j| \leqslant C(1+|j|)^{-1/2}\}$$
$$-\{\mu;\ |\mu| \leqslant C_{\varepsilon}\}$$

for any $\varepsilon>0$, where

$$\mu_j=-c_0+i\frac{\pi}{d}j \quad (=iz_j).$$

Moreover an estimate

$$\sum_{|\beta|\leqslant m}\sup_{x\in\Omega_R}|D^{\beta}_{x}(U(\mu)g)(x)| \leqslant C_{R,m,\varepsilon}\sum_{j=0}^{m+7}|\mu|^{j}\|g\|_{H^{m-j+7}(\Gamma)}$$

holds for all $\mu \in \mathscr{D}_{\varepsilon}$, where $\Omega_R=\Omega\cap\{x;\ |x|\leqslant R\}$.

---

2) A detailed proof is given in [4] and [5].

Theorem 4.2. *For each j $U(\mu)$ has at least a pole in*

$$\{\mu;\,|\mu-\mu_j|\leqslant C(1+|j|)^{-1/2}\}.$$

Recall the relation of the poles of $\mathscr{S}(z)$ and $U(\mu)$. Theorem 5.1 of Chapter V of [9] shows that $\mathscr{S}(z)$ has a pole at exactly those points z for which $\mu=iz$ is the pole of $U(\mu)$. Then Theorem 1 follows immediately from Theorems 4.1 and 4.2.

We shall give a brief explanation how to prove the above theorems by using asymptotic solutions constructed in the previous section.

Let $g \in C^\infty(\Gamma)$ and $m(t) \in C_0(0, 1)$, and let $u(x, t)$ be a solution of

$$\begin{cases} \Box u=0 & \text{in } \Omega\times \boldsymbol{R} \\ u=g(x)m(t) & \text{on } \Gamma\times \boldsymbol{R} \\ \operatorname{supp} u\subset\Omega\times\{t\geqslant 0\}. \end{cases}$$

Then the Laplace transformation

$$\hat{u}(x, \mu)=\int_{-\infty}^{\infty} e^{-\mu t}u(x, t)dt \tag{4.1}$$

converges in $\operatorname{Re}\mu>0$, and $\hat{u}$ satisfies

$$\begin{cases} (\mu^2-\Delta)\hat{u}(x, \mu)=0 & \text{in } \Omega \\ \hat{u}(x, \mu)=g(x)\hat{m}(\mu) & \text{on } \Gamma. \end{cases}$$

Evidently $\hat{u} \in L^2(\Omega)$. Then if $\hat{m}(\mu)\neq 0$ it follows that

$$(U(\mu)g)(x)=\hat{u}(x, \mu)/\hat{m}(\mu). \tag{4.2}$$

Through (4.2) the consideration of the analytic continuation of $U(\mu)$ can be reduced to an examination of convergence of (4.1) for $\operatorname{Re}\mu\leqslant 0$.

Without loss of generality we may suppose that

$$a_1=(0, 0, 0), \qquad a_2=(0, 0, d),$$

and that $S_1(\delta_2)$, $S_2(\delta_2)$ are represented as $\{y(\sigma)=(\sigma_1, \sigma_2, y_3(\sigma));\ \sigma \in I\}$ and $\{z(\eta)=(\eta_1, \eta_2, z_3(\eta));\ \eta \in I\}$ respectively.

Let $i(\sigma)=(i_1(\sigma), i_2(\sigma), i_3(\sigma))$ be a smooth $\boldsymbol{R}^3$-valued function such that $|i(\sigma)|=1$ for all $\sigma \in I$. Set

$$\begin{cases} \dfrac{\partial i}{\partial\sigma_j}(\sigma)=(\kappa_{j1}(\sigma), \kappa_{j2}(\sigma), \kappa_{j3}(\sigma)),\ j=1, 2, \\ \mathscr{K}(\sigma)=\begin{bmatrix}\kappa_{11}(\sigma) & \kappa_{12}(\sigma)\\ \kappa_{21}(\sigma) & \kappa_{22}(\sigma)\end{bmatrix}. \end{cases} \tag{4.3}$$

We suppose that $\mathscr{K}(\sigma)\geqslant 0$. Let $r(\eta)$ be $\boldsymbol{R}^3$-valued function defined by

$$r(\eta)=i(\sigma)-2(i(\sigma)\cdot m(\eta))m(\eta) \tag{4.4}$$

where $m(\eta)$ is the unit outer normal of Γ_2 at $z(\eta)$, and σ and η are linked by a relation

$$z(\eta)=y(\sigma)+l(\sigma)i(\sigma). \tag{4.5}$$

Set

$$\begin{cases} \dfrac{\partial r}{\partial \eta_p}(\eta)=(\tilde{\kappa}_{p1}(\eta),\ \tilde{\kappa}_{p2}(\eta),\ \tilde{\kappa}_{p3}(\eta)), & p=1, 2,\\ \tilde{\mathscr{K}}(\eta)=\begin{bmatrix}\tilde{\kappa}_{11}(\eta) & \tilde{\kappa}_{12}(\eta)\\ \tilde{\kappa}_{21}(\eta) & \tilde{\kappa}_{22}(\eta)\end{bmatrix}, \end{cases} \tag{4.6}$$

$$\begin{cases} \dfrac{\partial m}{\partial \eta_p}(0)=(k_{p1}^{(2)},\ k_{p2}^{(2)},\ k_{p3}^{(2)}), & p=1, 2,\\ K_2=\begin{bmatrix}k_{11}^{(2)} & k_{12}^{(2)}\\ k_{21}^{(2)} & k_{22}^{(2)}\end{bmatrix}. \end{cases} \tag{4.7}$$

Lemma 4.3. *For $r(\eta)$ defined by* (4.5) *it holds that*

$$\|\tilde{\mathscr{K}}(\eta)-F_2(\mathscr{K}(\sigma))\|\leqslant C\delta_2\|\mathscr{K}(\sigma)\| \qquad \text{for all } \eta\in S_2(\delta_2) \tag{4.8}$$

where

$$F_2(\mathscr{K})=\mathscr{K}(I+d\mathscr{K})^{-1}+2K_2.$$

Let $i(\sigma)$, $j(\sigma)$ be $\boldsymbol{R}^3$-valued functions satisfying $|i(\sigma)|=|j(\sigma)|=1$. Denote by $\mathscr{K}(\sigma)$, $\mathscr{H}(\sigma)$ 2×2 matrices defined by (4.3) for i and j respectively. For $j(\sigma)$ we set

$$s(\eta)=j(\tilde{\sigma})-2(j(\tilde{\sigma})\cdot m(\eta))m(\eta)$$

where η and $\tilde{\sigma}$ are linked by $z(\eta)=y(\tilde{\sigma})+h(\tilde{\sigma})j(\tilde{\sigma})$, $h(\tilde{\sigma})>0$. Denote by $\tilde{\mathscr{H}}(\eta)$ the 2×2 matrix defined by (4.6) for $s(\eta)$. Then we have

Lemma 4.4. *Suppose that $\mathscr{K}(\sigma)$, $\mathscr{H}(\sigma)\geqslant C_0>0$. Then it holds that*

$$\max_{\eta\in S_2(\delta_2)}|s(\eta)-r(\eta)|\leqslant((1+dC_0)^{-1}+C\delta_2)\max_{\sigma\in S_1(\delta_2)}|j(\sigma)-i(\sigma)|, \tag{4.9}$$

$$\begin{aligned}\max_{\eta\in S_2(\delta_2)}\|\tilde{\mathscr{K}}(\eta)-\tilde{\mathscr{H}}(\eta)\|\leqslant{}&((1+dC_0)^{-1}+C\delta_2)\max_{\sigma\in S_1(\delta_2)}\|\mathscr{K}(\sigma)-\mathscr{H}(\sigma)\|\\ &+C(|i|_2+|j|_2+|y|_2+|z|_2)\max_{\sigma\in S_1(\delta_2)}|i(\sigma)-j(\sigma)|.\end{aligned} \tag{4.10}$$

Set

$$i_q(\sigma)=\nabla\varphi_{2q}(y(\sigma)), \qquad r_q(\eta)=\nabla\varphi_{2q+1}(z(\eta)),$$

and denote by $\mathscr{K}_q(\sigma)$ and $\tilde{\mathscr{K}}_q(\eta)$ the matrices defined for i_q and r_q. From the strict convexity of $\mathcal{O}_j$ we have $K_2>0$. Note that the same results as Lemmas 4.3 and 4.4 also hold about reflections on $S_1(\delta_2)$. Then if $\mathscr{K}_0(\sigma)\geqslant 0$ we have

$$\mathscr{K}_q(\sigma)\geqslant 2K_1 \quad \text{and} \quad \tilde{\mathscr{K}}_q(\eta)\geqslant 2K_2 \quad \text{for } q=1, 2, \cdots.$$

By applying (4.9) repeatedly we have for some $0<\alpha<1$

$$\max_{\eta\in S_2(\delta_2)} |r_{q+1}(\eta)-r_q(\eta)|+\max_{\sigma\in S_1(\delta_2)} |i_{q+1}(\sigma)-i_q(\sigma)|\leqslant C\alpha^q.$$

Then there exist $r_\infty(\eta)$ and $i_\infty(\sigma)$ such that

$$\max_{\eta\in S_2(\delta_2)} |r_q(\eta)-r_\infty(\eta)|\leqslant C\alpha^q,$$

$$\max_{\sigma\in S_1(\delta_2)} |i_q(\sigma)-i_\infty(\sigma)|\leqslant C\alpha^q.$$

Let $\varphi_\infty(x)$ and $\tilde{\varphi}_\infty(x)$ be functions such that

$$\nabla\varphi_\infty(y(\sigma))=i_\infty(\sigma), \quad \varphi_\infty(a_1)=0, \quad \nabla\tilde{\varphi}_\infty(z(\eta))=r_\infty(\eta), \quad \tilde{\varphi}_\infty(a_2)=0.$$

Similarly we have by using (4.10)

$$\|\mathscr{K}_q(\sigma)-\mathscr{K}_\infty(\sigma)\|\leqslant C\alpha^q, \qquad \|\tilde{\mathscr{K}}_q(\eta)-\tilde{\mathscr{K}}_\infty(\eta)\|\leqslant C\alpha^q.$$

From (4.8) and the one on $S_1(\delta_2)$ we see that $\mathscr{K}_\infty(0)$ is the fixed point of

$$\mathscr{F}_1(\mathscr{K})=F_1\circ F_2(\mathscr{K})$$

and $\tilde{\mathscr{K}}_\infty(0)$ is that of $\mathscr{F}_2=F_2\circ F_1$, where $F_1(\mathscr{K})=\mathscr{K}(I+d\mathscr{K})^{-1}+2K_1$ and K_1 is a matrix defined for Γ_1 by (4.7) replacing $m(\eta)$ by $n(\sigma)$ the unit outer normal of Γ_1.

By using the above result on the convergence of the gradient of the phase functions we can show the exponential convergence of the broken ray $\mathscr{X}(y, \nabla\varphi_0)$, $y\in S_1(\delta_2)$ such that $X_{2q}(y, \nabla\varphi_0)=x$. Then we have

Proposition 4.5. *It holds that*

$$|\varphi_{2q}(\cdot)-(\varphi_\infty(\cdot)+2dq+d_0)|_m\leqslant C_m\alpha^{2q},$$

$$|\varphi_{2q+1}(\cdot)-(\tilde{\varphi}_\infty(\cdot)+(2q+1)d+\tilde{d}_0)|_m\leqslant C_m\alpha^{2q}$$

for $m=0, 1, \cdots$, *where* $|\cdot|_m$ *denotes a norm of* $\mathscr{B}^m$.

Next consider the convergence of the amplitude functions for $q\to\infty$. Set

$$\begin{cases}\lambda=[\det(I+d\mathscr{K}_\infty(0))]^{-1/2}\\ \tilde{\lambda}=[\det(I+d\tilde{\mathscr{K}}_\infty(0))]^{-1/2}.\end{cases} \tag{4.11}$$

Proposition 4.6. *We have*

$$v_{0,q}=w_q+z_q,$$
$$w_{2q}(x,t)=\lambda^q\tilde{\lambda}^q a(x)bf(A,t-j_\infty(x)-d_\infty-2qd),$$
$$w_{2q+1}(x,t)=\lambda^{q+1}\tilde{\lambda}^q\tilde{a}(x)bf(A,t-\tilde{j}_\infty(x)-\tilde{d}_\infty-(2q+1)d),$$
$$|z_q(x,t)|\leqslant C(\lambda\tilde{\lambda})^q\alpha^q,$$

where $a(x)$, $\tilde{a}(x)$, $j_\infty(x)$, $\tilde{j}_\infty(x)$ are C^∞-functions in $\omega(\delta_2)$ determined by $\mathcal{O}$ only and b, d_∞, $\tilde{d}_\infty$ are constants depending on $\mathcal{O}$ and φ_0, A is a point on $S_1(\delta_2)$.

Keeping in mind (4.1) and (4.2) consider the Laplace transformation of

$$E_0(x,t;k)=\sum_{q=0}^{\infty}e^{ik(\varphi_{2q}(x)-t)}v_{0,2q}(x,t),$$
$$\tilde{E}_0(x,t;k)=\sum_{q=0}^{\infty}e^{ik(\varphi_{2q+1}(x)-t)}v_{0,2q+1}(x,t).$$

Set

$$F(x,t;k)=e^{ik(\varphi_\infty(x)+d_0-t)}a(x)bf(A,t-j_\infty(x)-d_\infty).$$

Then

$$E_0(x,t;k)-\sum_{q=0}^{\infty}(\lambda\tilde{\lambda})^qF(x,t-2dq;k)$$
$$=\sum_{q=0}^{\infty}(e^{ik(\varphi_{2q}-t)}-e^{ik(\varphi_\infty(x)+d_\infty+2dq-t)})v_{0,2q}$$
$$+\sum_{q=0}^{\infty}e^{ik(\varphi_\infty(x)+d_\infty+2dq-t)}z_{2q}=I(x,t)+II(x,t).$$

Putting

$$\lambda\tilde{\lambda}=e^{-2dc_0},\qquad \alpha=e^{-2dc_1}$$

and taking account of the support of $v_{0,2q}$ we have from Proposition 4.5 we have

(4.12) $$|I(x, t)| \leqslant Ck\, e^{-(c_0+c_1)t}$$

and from Proposition 4.6

(4.13) $$|II(x, t)| \leqslant Ce^{-(c_0+c_1)t}.$$

Let $\operatorname{Re}\mu > 0$. We have

$$\begin{aligned}\hat{E}_0(x, \mu; k) &= \int_{-\infty}^{\infty} e^{-\mu t} E_0(x, t; k)dt \\ &= \int e^{-\mu t} \sum_{q=0}^{\infty} (\lambda\tilde{\lambda})^q F(x, t-2dq; k)dt \\ &\quad + \int e^{-\mu t}(I(x, t)+II(x, t))dt \\ &= \sum_{q=0}^{\infty} (\lambda\tilde{\lambda})^q e^{-2d\mu q} \hat{F}(x, \mu; k) + \hat{I}(x, \mu) + \hat{II}(x, \mu).\end{aligned}$$

From (4.12) and (4.13) we see that $\hat{I}$, $\hat{II}$ can be prolonged analytically into $\operatorname{Re}\mu < -(c_0+c_1)$. Since supp $F(x, \cdot\,; k)$ is compact $\hat{F}(x, \mu; k)$ is an entire function. Thus we have in $\operatorname{Re}\mu > 0$

$$\hat{E}_0(x, \mu; k) = \frac{1}{1-\lambda\tilde{\lambda}e^{-2\mu d}} \hat{F}(x, \mu; k) + \hat{I}(x, \mu) + \hat{II}(x, \mu),$$

and the right hand side can be prolonged analytically into $\operatorname{Re}\mu > -(c_0+c_1)$ beside μ's such that $1-\lambda\tilde{\lambda}e^{-2d\mu q}=0$, that is, μ_j's. Similarly we have the same representation for $\tilde{E}_0^{\wedge}(x, \mu; k)$.

Repeating this argument we have for

$$E_j(x, t; k) = \sum_{q=0}^{\infty} e^{ik(\varphi_q(x)-t)} v_{j,q}(x, t) \qquad (j \geqslant 1)$$

a representation of its Laplace transformation

$$\hat{E}_j(x, \mu; k) = \left(\frac{1}{1-\lambda\tilde{\lambda}e^{-2\mu d}}\right)^{j+1} F_j(x, \mu; k)$$

where F_j is holomorphic in $\operatorname{Re}\mu > -c_0-c_1$.

Making estimations for $u^{(2)}$ and the difference of the exact solution for an oscillatory boundary data (3.5) from $u(x, t; k)$ we can show that $\hat{u}(x, \mu; k)$ can be prolonged analytically into $\mathscr{D}_\varepsilon$. By applying this result to $w(x, t; \eta, \tau)$ in (3.4) we have a representation $\hat{u}(x, \mu)$ of (3.3). Then using (4.2) we obtain the desired analytic continuation of $U(\mu)$.

In order to show Theorem 4.3 we estimate directly the variation of argument of $\hat{u}(x, \mu; k)$ along a contour C_j containing μ_j in its interior, and by using the exact form of $\hat{u}$ derived in the above we see that the variation is not zero for some point on segment a_1a_2.

4.2. *Case of the principal curvatures vanish at a_j.*[3)]

In this case, if we make a formal calculus in (4.11), λ, $\tilde{\lambda}$ in Proposition 4.6 are equal to 1. Then c_0 equals 0. This suggests the existence of poles of $\mathscr{S}(z)$ converging to the real axis. But the argument used for strictly convex objects is no more valid for this case.

Here we shall use the argument of Bados, Guillot and Ralston [1] which is based on the trace formula

$$\text{(4.14)} \qquad \operatorname{tr}_{L^2(\boldsymbol{R}^3)} \int \rho(t)(\cos t\sqrt{-\Delta} \oplus 0 - \cos t\sqrt{-\Delta_0})dt = \sum_{z_j:\text{poles}} \hat{\rho}(z_j),$$

$$\hat{\rho}(z) = \int e^{izt}\rho(t)dt$$

for $\rho \in C_0^\infty(\boldsymbol{R})$, where Δ denotes the selfadjoint realization in $L^2(\Omega)$ of the Laplacian with the Dirichlet boundary condition, and Δ_0 the selfadjoint realization in $L^2(\boldsymbol{R}^3)$ of the Laplacian in $\boldsymbol{R}^3$ and

$$((\cos t\sqrt{-\Delta} \oplus 0)f)(x) = \begin{cases} (\cos t\sqrt{-\Delta}\, f_1)(x) & \text{for } x \in \Omega \\ 0 & \text{for } x \in \mathcal{O} \end{cases}$$

for $f = f_1 + f_2$, $f_1 \in L^2(\Omega)$, $f_2 \in L^2(\mathcal{O})$.

To show Theorem 2 we derive a contradiction from the trace formula by a substitution of a sequence of $\rho(t)$. Let $\rho_0(t) \in C_0^\infty(-1, 1)$ verifying $\rho_0(t) \geqslant 0$, $\hat{\rho}_0(-k) = \hat{\rho}_0(k) \geqslant 0$ for all $k \in \boldsymbol{R}$ and $\hat{\rho}_0(0) = 1$. We set

$$\rho_q(t) = \rho_0((q+1)^l(t - 2dq)) \qquad \text{for } q = 1, 2, \cdots.$$

Lemma 4.7. *Suppose that all the poles $\{z_j\}_{j=1}^\infty$ of $\mathscr{S}(z)$ verify*

$$\operatorname{Im} z_j \geqslant \alpha$$

for some $\alpha > 0$. Then we have

$$\text{(4.15)} \qquad \sum_{j=1}^\infty |\hat{\rho}_q(z_j)| \leqslant C(q+1)^{4l} e^{-2d\alpha q} \qquad \textit{for all } q.$$

---

3) A detailed proof is given in [7].

Concerning the left hand side of (4.14) we have

Proposition 4.8. *Suppose that $\mathcal{O}$ satisfies the condition in Theorem 2. Then we have*

$$\text{(4.16)} \qquad \operatorname{tr}_{L^2(\mathbf{R}^3)} \int \rho_q(t)(\cos t\sqrt{-\Delta} \oplus 0 - \cos t\sqrt{-\Delta_0})dt$$
$$\geqslant cq^{(1-2/e_0)(l-1)} - C_l q^{(1-3/e_0)l}$$

for all $q \geqslant q_0$ if $l \geqslant l_0$, where $e_0 = e+2$ and l_0 is a positive integer, c is a positive constant independent of l.

Evidently Lemma 4.7 and Proposition 4.8 imply the existence of a sequence of poles which converges to the real axis. Indeed, if there exists $\alpha > 0$ such that

$$\operatorname{Im} z_j \geqslant \alpha \qquad \text{for all } j,$$

we have from (4.15) and (4.16)

$$C(q+1)^{4l}e^{-2d\alpha q} \geqslant cq^{(1-2/e_0)(l-1)}(1+o(1)),$$

which shows a contradiction.

We shall give a brief explanation on the plan of the proof of Proposition 4.8.

Let $\psi(x) \in C_0^\infty(\Omega)$. Then for $f \in C^\infty(\bar{\Omega})$ we have

$$\psi(x)f(x) = w(x)\int dy \int_0^\infty k^2 dk \int_{S^2} d\omega\, e^{ik(x-y)\cdot\omega}\psi(y)f(y)$$

where $w \in C_0^\infty(\Omega)$ such that $w=1$ on supp ψ. Let $w(x, t; k, \omega)$ be the solution of

$$\text{(4.17)} \qquad \begin{cases} \Box u = 0 & \text{in } \Omega\times(0, \infty) \\ u(x, t) = 0 & \text{on } \Gamma\times(0, \infty) \\ u(x, 0) = w(x)e^{ikx\cdot\omega} & \\ \dfrac{\partial u}{\partial t}(x, 0) = 0. & \end{cases}$$

Then the kernel $E(t, x, y)\psi(y)$ of $\cos t\sqrt{-\Delta}\,\psi$ can be represented formally as

$$\text{(4.18)} \qquad E(t, x, y)\psi(y) = \int_0^\infty k^2 dk \int_{S^2} d\omega\, w(x, t: k, \omega)e^{-iky\cdot\omega}\psi(y).$$

Then we have

$$\text{(4.19)} \qquad \operatorname{tr}\int \rho_q(t)\cos t\sqrt{-\Delta}\,\psi\,dt$$
$$=\int_\Omega dx\int_0^\infty k^2dk\int_{S^2}d\omega\left(\int \rho_q(t)w(x,t;k,\omega)dt\right)e^{-ikx\cdot\omega}\psi(x).$$

In order to estimate the right hand side for large q we have to know the behavior of $w(x, t; k, \omega)$ for large t. To this end we shall use asymptotic solutions of (4.17). An asymptotic solution of Cauchy problem

$$\text{(4.20)} \qquad \begin{cases} \Box u=0 & \text{in } \boldsymbol{R}^3\times(0,\infty) \\ u(x,0)=w(x)e^{ikx\cdot\omega} & \\ \dfrac{\partial u}{\partial t}(x,0)=0, & \end{cases}$$

can be obtained in a form

$$u_c(x,t;k,\omega)=e^{ik(x\cdot\omega-t)}\sum_{j=0}^{N}v_j^+(x,t)(ik)^{-j}$$
$$+e^{ik(x\cdot\omega+t)}\sum_{j=0}^{N}v_j^-(x,t)(ik)^{-j}=u^+ + u^-.$$

If we set $m^\pm=u^\pm|_{\Gamma\times(0,\infty)}$ and denote by $u_b^\pm$ an asymptotic solution for

$$\text{(4.21)} \qquad \begin{cases} \Box u=0 & \text{in } \Omega\times\boldsymbol{R} \\ u=m^\pm & \text{on } \Gamma\times\boldsymbol{R} \\ \operatorname{supp} u\subseteq\Omega\times\{t>0\}. & \end{cases}$$

Then

$$u(x,t;k,\omega)=u_c(x,t;k,\omega)-u_b^+(x,t;k,\omega)-u_b^-(x,t;k,\omega)$$

is an asymptotic solution of (4.17). Following the process in Section 3 the main part of $u_b^\pm$ can be represented as

$$\text{(4.22)} \qquad \begin{cases} u_b^\pm(x,t;k,\omega)=\displaystyle\sum_{q=0}^{\infty}(-1)^q u_q^\pm(x,t;k,\omega), \\ \quad u_q^\pm(x,t;k,\omega)=e^{ik(\varphi_q(x,\omega)\pm t)}\displaystyle\sum_{j=0}^{N}v_{j,q}^\pm(x,t)(ik)^{-j}. \end{cases}$$

Remark that for $\omega_0=(0,0,\pm 1)$ and for $x_1=x_2=0$, $v_{0,q}(x,t)=1$, which is a consequence of $\lambda=\tilde{\lambda}=1$. Note that only $u_{2q-1}^\pm\neq 0$ on $\operatorname{supp}\rho_q(t)$. The following lemma is a key point for the proof of Proposition 4.8.

Lemma 4.9. *For each x_3 fixed we have*

$$\int dx_1 dx_2 \int_{S^2} d\omega\, e^{ik(\varphi_{2q-1}(x,\omega)-x\cdot\omega)} v^{\pm}_{0,2q-1}(x,t)\psi(x)$$
$$=e^{i2dqk}\{c\psi(0,0,x_3)w(0,0,x_3+2dq-t)k^{-1-2/e_0}+O(k^{-1-3/e_0})\}.$$

The critical point in variables x_1, x_2 and $\omega \in S^2$ is $x_1=x_2=0$ and $\omega=(0,0,\pm 1)$. But these critical points are degenerate. Therefore we use Varčenko's theorem for an estimation of the above oscillating integral.

By using Lemma 4.9 we see that the main part of the right hand side of (4.19) becomes

$$c\int_0^\infty dk \int dt\, \rho_q(t)e^{\pm ikt}k^2k^{-1-2/e_0}\int \psi(0,0,x_3)dx_3$$
$$=cq^{(1-2/e_0)(l-1)}\int \psi(0,0,x_3)dx_3.$$

By making estimations of the other terms and the difference of the asymptotic solution from the actual solution of (4.17), we have Proposition 4.8.

4.3. *On the asymptotic behavior of the scattering phase.*[4)]

We shall use a trace formula in Petkov and Popov [19]

(4.23)
$$2\mathrm{tr}\int_{-\infty}^{\infty}\rho(t)(\cos t\sqrt{-\Delta}\oplus 0-\cos t\sqrt{-\Delta_0})dt$$
$$=\int_{-\infty}^{\infty}\frac{d\hat{\rho}}{d\lambda}(\lambda)s(\lambda)d\lambda \qquad \text{for } \rho\in C_0^\infty(\boldsymbol{R}),$$

where $\hat{\rho}(\lambda)=\int \rho(t)e^{-i\lambda t}dt$. Following [19] denote by $\sigma(t)$ the Fourier transformation of $s(\lambda)$. Let $\chi_1(t)\in C_0^\infty(\boldsymbol{R})$ verifying

$$\chi_1(t)=\begin{cases}1 & t\leqslant |d|/4\\ 0 & t\geqslant |d|/2,\end{cases}$$

and let us set $\chi_2(t)=1-\chi_1(t)$.

By using the respresentation (4.18) of the distribution kernel of $\cos t\sqrt{-\Delta}\,\psi$ we have $t\sigma(t)\in C^\infty([-d/2,d/2]-\{0\})$. Then $t\sigma(t)\chi_1(t)\in C^\infty(R-\{0\})$. From the consideration in [19] on the singularity of $t\sigma(t)$ at $t=0$ we have

4) A detailed proof will be given in [8].

$$\mathscr{F}(t\sigma(t)\chi_1(t))(\lambda)=\frac{2}{\pi^2}|\mathcal{O}|\lambda^2+\frac{1}{8\pi}|\Gamma|\lambda+O(1). \tag{4.24}$$

Next consider

$$I_2(\rho)=2\,\mathrm{tr}\int\rho(t)(\cos t\sqrt{-\Delta}\oplus 0-\cos t\sqrt{-\Delta_0})\chi_2(t)dt.$$

In a representation of an asymptotic solution $u(x, t; k, \omega)$ of (4.17) $u_b^{\pm}(x, t; k, \omega)$ of the form (4.22) satisfies

$$|v_{j,q}^{\pm}(x, t)|\leqslant Cq^j\gamma^q, \qquad 0<\gamma<1$$

because of the assumption on the principal curvatures in Theorem 3. By modifying the argument in the previous subsection we have for $t>0$

$$\begin{aligned}\int_{S^2}d\omega\int e^{ik(\varphi_{2q-1}(x,\omega)-x\cdot\omega-t)}\,v_{0,2q-1}^{\pm}(x, t)\psi(x)dx\\ =\gamma^{2q}\,e^{ik(2dq-t)}c\int\psi(0, 0, x_3)dx_3k^{-3/2-1/e_0}(1+O(k^{-1/e_0}))\end{aligned}$$

where c is a non-zero constant. Then for $\rho\in C_0^{\infty}(0, \infty)$

$$I_2(\rho)=\iint e^{-ikt}\rho(t)k^{-3/2-1/e_0}\{g(k)+a(t, k)\}k^2dkdt$$

where

$$g(k)=\sum_{q=0}^{\infty}\gamma^{2q}e^{i2dqk}$$

and $a(t, k)$ satisfies

$$|\partial_t^{\beta}a(t, k)|\leqslant C_{\beta}k^{-3/2-2/e_0-(1/2+\varepsilon)\beta}e^{-\alpha|t|} \qquad (\alpha>0)$$

for any $\varepsilon>0$. Similar estimate holds for $t<0$.

Putting $\rho_T(t)=\chi_1(t/T)$ we have

$$I_2(\rho_Te^{-i\lambda t})\longrightarrow\mathscr{F}(t\sigma(t)\chi_2(t))(\lambda) \qquad \text{as } T\longrightarrow\infty.$$

On the other hand for large

$$\begin{aligned}I_2(\rho_Te^{-i\lambda t})\longrightarrow c\lambda^{1/2-1/e_0}\{g(\lambda)+g(-\lambda)+O(\lambda^{-1/e_0+\varepsilon})\}\\ \text{as } T\longrightarrow\infty.\end{aligned}$$

Then we have

(4.25) $$\mathscr{F}(t\sigma(t)\chi_2(t))(\lambda)=2c\lambda^{1/2-1/e_0}\{\mathrm{Re}\, g(\lambda)+O(\lambda^{-1/2e_0})\}.$$

Combining (4.24) and (4.25) we have the desired expansion.

References

[1] C. Bardos, J. C. Guillot and J. Ralston, La relation de Poisson pour l'équation des ondes dans un ouvert non borné. Application à la théorie de la diffusion, Comm. Partial Differential Equations, **7** (1982), 905–958.

[2] M. Ikawa, Decay of solutions of the wave equation in the exterior of two convex obstacles, Osaka J. Math., **19** (1982), 459–509.

[3] ——, Mixed problems for the wave equation, Proc. NATO Advanced Study Inst., Singularities in Boundary Value Problems, Reidel Publ., Dordrecht, 1981, 97–119.

[4] ——, On the poles of the scattering matrix for two strictly convex obstacles, J. Math. Kyoto Univ., **23** (1983), 127–194.

[5] ——, On the poles of the scattering matrix for two strictly convex obstacles: An addendum, J. Math. Kyoto Univ., **23** (1983), 795–802.

[6] ——, On the distribution of the poles of the scattering matrix for two strictly convex obstacles, Hokkaido Math. J., **12** (1983), 343–359.

[7] ——, Trapping obstacles with a sequence of poles of the scattering matrix converging to the real axis, Osaka J. Math., **22** (1985), 657–689.

[8] ——, On asymptotic behavior of the scattering phase, in preparation.

[9] P. D. Lax and R. S. Phillips, Scattering Theory, Academic Press, New York, 1967.

[10] P. D. Lax and R. S. Phillips, A logarithmic bound on the location of the poles of the scattering matrix, Arch. Rational Mech. Anal., **40** (1971), 268–280.

[11] A. Jensen and T. Kato, Asymptotic behavior of the scattering phase for exterior domains, Comm. Partial Differential Equations, **3** (1978), 1165–1195.

[12] R. B. Melrose, Singularities and energy decay in acoustical scattering, Duke Math. J., **46** (1979), 43–59.

[13] ——, Polynomial bound on the number of scattering poles, J. Funct. Anal., **53** (1983), 287–303.

[14] R. B. Melrose and J. Sjöstrand, Singularities of boundary value problems, I, II, Comm. Pure Appl. Math., **31** (1978), 593–617, **35** (1982), 129–168.

[15] A. Majda and J. Ralston, An analogue of Weyl's theorem for unbounded domains, I, II, III, Duke Math. J., **45** (1978), 183–196, 513–536, **46** (1979), 725–731.

[16] ——, Geometry in the scattering phase, Proc. Symp. Pure Math., AMS, **36**, 1980, 253–255.

[17] S. Mizohata, Sur l'analyticité de la fonction spectrale de l'opérateur △ relatif au problème extérieur, Proc. Japan Acad., **39** (1964), 352–357.

[18] C. S. Morawetz, J. Ralston and W. A. Strauss, Decay of solutions of the wave equation outside nontrapping obstacles, Comm. Pure Appl. Math., **30** (1977), 447–508.

[19] V. Petkov and G. Popov, Asymptotic behavior of the scattering phase for non-trapping obstacles, Ann. Inst. Fourier, **32** (1982), 111–149.

[20] V . Petkov, La distribution des poles de la matrice de diffusion, Séminaire Goulaouic-Meyer-Schwartz, 1982–1983, Exposé N° VII.

[21] J. Ralston, Trapped rays in spherical symmetric media and poles of the scattering matrix, Comm. Pure Appl. Math., **24** (1971), 571–582.

[22] ——, The first variation of the scattering matrix; An addendum, J. Differen-

tial Equations, **28** (1978), 155–162.
[23] J. Ralston, Propagation of singularities and the scattering matrix, Proc. NATO Advanced Study Inst., Singularities in Boundary Value Problems, Reidel Publ., Dordrecht, 1981, 161–181.
[24] A. N. Varčenko, Newton polyhedra and estimation of oscillating integrals, Functional Anal. Appl., **10** (1976), 175–196.

Department of Mathematics
Osaka University
Toyanaka, Osaka 560
Japan

Taniguchi Symp. HERT
Katata 1984, pp. 85–88

Three Spectral Problems Revised

Victor IVRIĬ

In this paper we give a more general and refined form of the results known.

1. Let X be a compact closed d-dimensional C^∞-manifold, dx a C^∞-density on X, E a Hermitian C^∞-fibering over X. Let A: $L_2(X, E)\to L_2(X, E)$ be a pseudo-differential operator, $A=A_0+A_1$ where $A_0 \in L^1_{\mathrm{cl}}(X, E)$ is elliptic and $A_1 \in L^1_{1,0}(X, E)$ such that in every local card its symbol a_1 satisfies the following condition:

$$D^\beta_\xi a_1(x, \xi)=o(|\xi|^{1-|\beta|}) \quad \text{as} \quad |\xi|\longrightarrow\infty, |\beta|\leqslant 1. \tag{1}$$

Let A be a symmetric and hence a selfadjoint operator in $L_2(X, E)$ and let $E(s)$ be its spectral projectors, $e^\pm(x, y, k)$ the Schwartz kernels of $\pm(E(\pm k)-E(0))$ and $N^\pm(k)$ a number of eigenvalues of A lying between 0 and $\pm k$, $k>0$. It should be noted that $\pm A$ is semibounded from above if and only if $\pm a$ is negative definite where a is the principal symbol of A_0.

Theorem 1. *If $\pm a$ is not negative definite then the following asymptotics holds as $k\to\infty$:*

$$N^\pm(k)=M^\pm(k)+O(k^{d-1}), \tag{2}$$

$$M^\pm(k)=(2\pi)^{-d}\iint n^\pm(x, \xi, k)dx\, d\xi, \tag{3}$$

$$e^\pm(x, x, k)=(2\pi)^{-d}\int n^\pm(x, \xi, k)d\xi+O(k^{d-1}) \tag{4}$$

uniformly with respect to x where $n^\pm(x, \xi, k)$ is a number of eigenvalues of a^W lying between 0 and $\pm k$, a^W is the Weyl symbol of A.

Moreover, if the measure of the set of all periodic generalized rays corresponding to a [2] equals 0 and

$$D_x a_1(x, \xi)=o(|\xi|) \quad as \quad |\xi|\longrightarrow\infty \tag{5}$$

then

(6) $$N^{\pm}(k)=M^{\pm}(k)+c_1^{\pm}k^{d-1}+o(k^{d-1}) \quad as \quad k\longrightarrow\infty.$$

2. Let $A_h=a(x, hD, h)$ be an h-pseudo-differential operator in $L_2(X, \boldsymbol{C})$ with the symbol

$$a(x, p, h)\sim\sum_{n=0}^{\infty} a_n(x, p)h^n$$

where

$$|D_p^\alpha D_x^\beta a_n(x, p)|\leqslant C_{\alpha\beta n}(1+|p|)^{m-|\alpha|},$$
$$a_0(x, p)\geqslant c|p|^m-C, \qquad C>0,$$

$m>0$ (if $d=1$ then the latter inequality can be replaced by $|a_0(x, p)|\geqslant c|p|^m-C$ and all the following considerations must be changed in the obvious way).

Let A_h: $L_2(X, \boldsymbol{C})\rightarrow L_2(X, \boldsymbol{C})$ be symmetric and hence selfadjoint for every $h\in(0, h_0]$; here $h_0>0$ is not fixed and can be decreased if necessary. Let $N(k, h)$ be a number of eigenvalues of A_h not exceeding k.

It is well-known [3] that if k_0 is not a critical value of a_0 then the following asymptotics holds as $h\rightarrow+0$:

(7) $$N(k, h)=M(k, h)+O(h^{1-d}),$$

(8) $$M(k, h)=(2\pi)^{-d}\iint\limits_{a_0(x, p)+a^s(x, p)h<k} dx\, dp$$

uniformly with respect to k in the neighbourhood of k_0; here

$$a^s=a_1+\frac{i}{2}\sum_j\frac{\partial^2}{\partial x_j\partial p_j}a_0$$

and

$$M(k, h)\sim\sum_{n=0}^{\infty} c_n(k)h^{-d+n},$$

$c_n\in C^\infty$ in the neighbourhood of k_0.

Moreover, if the measure of the set of all points lying on the surface $\{a_0(x, p)=k_0\}$*) and periodic with respect to Hamiltonian flow generated by a_0 equals 0 then

(9) $$N(k_0, h)=M(k_0, h)+o(h^{1-d}) \quad \text{as} \quad h\longrightarrow+0.$$

Theorem 2. *Let k_0 be non-degenerate critical value of a_0 (i.e. surface*

*) There is a natural measure $dxdp$: da_0 on this surface.

$\{a_0(x, p)=k_0\}$ contains no degenerate critical point of a_0); then asymptotics (7)–(8) holds. Moreover, if $d\geqslant 2$ and the measure of the set of the points lying on the surface $\{a_0(x, p)=k_0\}$ and periodic with respect to Hamiltonian flow generated by a_0 equals 0 then asymptotics (9) holds.

Moreover, if $d\geqslant 2$ then

$$M(k, h)=c_0(k)h^{-d}+c_1(k)h^{1-d}+o(h^{1-d});$$

if $d=1$ then

$$M(k, h)=c_0(k)h^{-1}+O(\ln h);$$

we can replace $O(\ln h)$ by $O(1)$ provided $\{a_0(x, p)=k_0\}$ contains no minimax point.

Similar results can be obtained for matrix operators with constant multiplicities of eigenvalues of their principal symbols.

3. Let us consider the application of these results (using the Birman-Schwinger principle) to the degenerate spectral problem

$$(10) \qquad Au=\mu Bu \qquad u\in L_2(X, \mathbf{C})\setminus 0$$

where A is a strongly elliptic positive definite selfadjoint differential operator of order p and B is a symmetric differential operator of order q, $m=p-q>0$.

Let $N^{\pm}(k)$ be a number of eigenvalues of problem (10) lying between 0 and $\pm k$, $k>0$. It should be noted that $N^{\pm}(k)=O(1)$ as $k\to\infty$ if and only if $\pm B$ is negative definite (we assume that $q=0, 1$).

Theorem 3. *Let $\pm B$ be not negative definite and let one of the following two conditions be fulfilled:*

(i) *$p=2$, $q=0$, $B=b(x)$ and 0 is not a degenerate critical value of $b(x)$;*

(ii) *$q=1$, $B=\langle b(x), D\rangle+\beta(x)$ where $b(x)$ is a vector field and if $b(x^0)=0$ then* $\det((\partial b_j/\partial x_k)(x^0))\neq 0$.

Then the following asymptotics holds as $k\to\infty$:

$$(11) \qquad N^{\pm}(k)=c_0^{\pm}k^{d/m}+O(k^{(d-1)/m}) \qquad \textit{provided } d\geqslant 2 \textit{ or } q=0,$$

$$(12) \qquad N^{\pm}(k)=c_0^{\pm}k^{1/m}+c_1^{\pm}\ln k+O(1) \qquad \textit{provided } d=1,\ q=1;$$

here $c_0^{\pm}>0$.

Moreover, if $d\geqslant 2$ and the measure of the set of the points lying on the *surface* $\{a \mp b = 0\}$ *and periodic with respect to Hamiltonian flow generated by* $a \mp b$ *equals* 0 *where* a, b *are the principal symbols of* A, B *then*

$$N^{\pm}(k) = c_0^{\pm} k^{d/m} + c_1^{\pm} k^{(d-1)/m} + o(k^{(d-1)/m}) \quad \text{as } k \longrightarrow \infty. \tag{13}$$

References

[1] V. Ivriĭ, Accurate spectral asymptotics for elliptic operators that act on vector bundles, Functional Anal. Appl., **16** (1982), 101–108.
[2] R. Melrose, The trace of the wave group, Contemp. Math., **27**, 1984, 127–168.
[3] J. Chazarain, Spectre d'un Hamiltonien quantique et mechanique classique, Comm. Partial Differential Equations., **5** (1980), 695–744.
[4] V. Ivriĭ, On exact quasiclassical spectral asymptoties for h-pseudodifferential operators acting in bundles (in Russian), Trudy Sem. S. L. Sobolev, 1983, No. 1, 30–54.

Mining-Metallurgical Institute
Magnitogorsk 455000
Lenina 38
U.S.S.R.

Taniguchi Symp. HERT
Katata 1984, pp. 89–100

The Cauchy Problem for Effectively Hyperbolic Equations (Remarks)

Nobuhisa IWASAKI

§ 0. Introduction

We mention the results [6, 7, 8] of the Cauchy problem for effectively hyperbolic equations and give three remarks about them. The first remark is concerned with the method of proof which will be more simplified than our old one [6]. The second one is about the bicharacteristic curves. And the last one will give an application to the hyperbolic Monge-Ampère equation.

We consider the following system P of partial differential operators of order m which has the diagonal principal part with the same single symbol p_m.

$$P=p_m I+L, \tag{0.1}$$

where L is a system of partial differential operators of order not greater than $m-1$. It is well known that any Volevič type of system is reduced to this type by multiplication of its formal cofactor operator. We assume that the principal symbol p_m is effectively hyperbolic with respect to the direction dx_0. Naturally we consider these systems on a suitable open set of $\boldsymbol{R}^{n+1}$ with the variable $(x_0, x)=(x_0, x_1, \cdots, x_n)$. The effective hyperbolicity for a real valued single symbol p_m is exactly defined as follows.

Definition 0.1. 1) *p_m is called effectively hyperbolic with respect to the direction $\theta \neq 0$ if p_m is hyperbolic with respect to the direction $\theta \neq 0$ and if the fundamental matrix F_p at any critical point of $p_m=0$ has non zero real eigenvalues. (F_p is defined as $\sigma(u, F_p v)=H_p(u, v)$ by means of the Hessian H_p of p_m and the canonical two form σ on $T^*\boldsymbol{R}^{n+1}$.)*

2) *p_m is called strongly hyperbolic with respect to the direction $\theta \neq 0$ if the Cauchy problem for P with any arbitrary lower order term is well posed with respect to the direction θ. Here "well posed" means the solvability on the space of infinitely differentiable functions.*

The proposition for single equations that p_m should be effectively hyperbolic if p_m were strongly hyperbolic is one of the results by V. Ya.

Ivriĭ and V. Petkov [3]. We will mention here the results on the converse proposition. The main result is as follows.

Theorem 0.1. *Let P be a system of partial differential operators of order m as* (0.1). *If the principal symbol p_m is effectively hyperbolic with respect to dx_0 at a neighborhood of the origin, then P is well posed at the origin with respect to dx_0.*

Moreover there exists a lens neighborhood Ω of the origin contained in a neighborhood given beforehand such that for any infinitely differentiable function u on Ω supported on $x_0 \geq 0$, it holds the estimates

$$\|u\|_s \leq C_s(\|Pu\|_{s+l} + \|a\|_{s+l}\|Pu\|_l)$$

for any real positive s and for a fixed l, where a is coefficients of P and where $\|\cdot\|_s$ stand for the Sobolev norms on the lens domain

$$\Omega = \{(x_0, x) : -\varepsilon < x_0 < \varepsilon - |x|^2\}.$$

Ω, the constants C_s and l are uniform as far as the principal symbol p_m being hyperbolic and coefficients a of P belong to a small neighborhood of a fixed effectively hyperbolic one and a bounded set in the space of infinitely differentiable functions, respectively.

A simple example. On $\boldsymbol{R}^3$,

$$Q = \begin{pmatrix} \xi_0 - \xi_1, & \xi_2 \\ x_0^2\xi_2, & \xi_0 + \xi_1 \end{pmatrix} + \text{lower terms}.$$

Then,

$$P = Q^{co}Q \ (QQ^{co}, \text{resp.}) = (\xi_0^2 - \xi_1^2 - x_0^2\xi_2^2)I + \text{lower terms}.$$

§ 1. How to prove

Recently R. Melrose [10] got the regularity estimate for effectively hyperbolic equations, and T. Nishitani [11] improved the method by O. A. Oleinik [13] by refering to Melrose's idea to prove the well-posedness. The method, we used and show here, is somehow different. It is descended from the factorization method on the Hermite differential equation. And also it is one tried by Ivriĭ [4] after Oleinik.

The characteristics of the effectively hyperbolic p_m are at most double. So it is essential to prove it for a system of second order. We consider the operator P, which is a classical pseudodifferential operator in the variable x and a differential operator in the variable x_0 of second order. The principal symbol of P is p_2I such that

$$p_2 = -\xi_0^2 + a_1^{\sim}(x_0, x, \xi)\xi_0 + b_2^{\sim}(x_0, x, \xi), \tag{1.1}$$

and L is a system of order 1. Moreover we assume that P depends smoothly on another parameter t varying on the interval [0, 1].

At first we standardize the principal symbol p_2.

Theorem 1.1. *Any effectively hyperbolic p_2 with infinitely differentiable coefficients is written locally as*

$$p_2 = -(\xi_0 - \Lambda_1)(\xi_0 - \Lambda_0) + b_2, \tag{1.2}$$

where Λ_0, Λ_1 and b_2 are infinitely differentiable functions in (x_0, x, ξ, t) of homogeneous order 1, 1 *and* 2 *in ξ, respectively, such that*

$$b_2 \geq 0,$$

$$\{\xi_0 - \Lambda_1, \xi_0 - \Lambda_0\} > 0 \qquad \text{at } \Lambda_0 - \Lambda_1 = b_2 = 0 \tag{1.3}$$

and

$$\{\xi_0 - \Lambda_0, b_2\} + cb_2 = 0$$

with an infinitely differentiable function c in (x_0, x, ξ, t). (The negative case of the left hand side of (1.3) *is also possible.)*

Next we estimate the single equation with the symbol p_2 and get the result to be able to deal with the lower terms L as one of the perturbed terms. This step in our original proof was complicated because we would remove exactly the effect of first order terms and had to use an intricate intertwining operator.

We may assume that $\Lambda_0 \equiv 0$ in (1.2) so that $b_2 = c(x_0, x, \xi)b_2'(x, \xi)$ with $c \neq 0$ and $b_2' \geq 0$ (strictly positive outside a bounded set in x). And we add a parameter λ such that

$$p = a_1 a_0 + b_2, \tag{1.4}$$

$$a_1 = i\xi_0 - i\Lambda_1 + \lambda$$

and

$$a_0 = i\xi_0 + \lambda.$$

The intertwining operator F is defined as

$$F = (i\xi_0 + \lambda)^{\nu(\omega(\xi_0, \lambda) - i\theta)},$$

where ω is defined by

$$\omega(\xi_0, \lambda)=\omega_0(\xi_0\langle\xi_0, \lambda\rangle^{-1})$$

with an infinitely differentiable function $\omega_0(s)$ on $[-1, 1]$ such that $\omega_0(s)=\omega_0(-s)$, $(d/ds)\omega_0(s)=\omega_1(s)^2: 0<s$, $\omega_0(s)=0: 0\leq s\leq 1/3$ and $\omega_0(s)=1: 2/3\leq s\leq 1$. F is a pseudodifferential operator in x_0. It is clear that F is invertible.

Let us consider $F^{-1}pF$.

$$\begin{aligned}p_b&=F^{-1}pF\\&=p+i\nu(\omega-i\theta)(\partial/\partial x_0)\Lambda_1\\&\quad+\nu(i\xi_0+\lambda)\log(i\xi_0+\lambda)(\partial/\partial x_0)\Lambda_1(\partial/\partial\xi_0)\omega\\&\quad+(i\xi_0+\lambda)^{-1}[(\log\lambda)^2b_1+(\log\lambda)c_2b_2]\\&\quad+d_0,\end{aligned}$$

where b_1 (c_2, d_0) is a pseudodifferential operator of order 0 (0, resp.) in x_0 and of order 1 (0, resp.) in x.

Remark. Here the calculus are one of pseudodifferential operators in x_0 valued in pseudodifferential operators in x. In other words, it is one of pseudodifferential operators with separate weights such that

$$\begin{aligned}&|(\partial/\partial\xi_0)^{\alpha_0}(\partial/\partial x_0)^{\beta_0}(\partial/\partial\xi)^{\alpha}(\partial/\partial x)^{\beta}g(x_0, x, \xi_0, \xi)|\\&\quad\leq C[\log\lambda]^{|\beta_0|}\langle\xi_0, \lambda\rangle^{(m_0-|\beta_0|)}\langle\xi\rangle^{m-|\beta|}.\end{aligned}$$

There is no difference from usual ones except for taking care of decreasing orders.

$(\partial/\partial x_0)\Lambda_1$ is positive on the double characteristics. So $(\partial/\partial x_0)\Lambda_1$ is elliptic modulo Λ_1 and b_2. Therefore there exists h a uniformly positive pseudodifferential operator of order 1 in x such that

$$\begin{aligned}p_b=p&+ih\times(\omega-i\theta)\\&+h\nu(\partial/\partial\xi_0)\omega[\log(i\xi_0+\lambda)](i\xi_0+\lambda)\\&+d_1(\log\lambda)\Lambda_1+d_2(\log\lambda)b_2\\&+(i\xi_0+\lambda)^{-1}[(\log\lambda)^2b_1+(\log\lambda)c_2b_2],\end{aligned}$$

where d_1, d_2, b_1 and c_2 are pseudodifferential operators of order 0 in x_0 and of order 0, -1, 1 and 0 in x, respectively.

Let us consider $\mathrm{Re}\,(p_bu, a_0u)$. If θ is chosen as large as it covers the lack of estimate by $\omega\xi_0$ at the region that $\omega=0$, and also if ν is large, then

$$\begin{aligned}\mathrm{Re}\,(p_bu, a_0u)&\geq(\lambda-|\log\lambda|C)(\|a_0u\|_0^2+\|u\|_{(b)}^2)\\&\quad+C_0\nu(\|u\|_{(h)}^2+\|u\|_{(h\log)}^2)\\&\quad-C(\nu)|\log\lambda|^2(\|u\|_{1/2}^2+\|u\|_{(b)}^2),\end{aligned}$$

where

$$\|u\|_s^2 = \|\langle\xi\rangle^s u\|^2,$$
$$\|u\|_{(h)}^2 = \|h^{1/2}\langle\xi_0, \lambda\rangle^{1/2}u\|_0^2 + C_1\|\langle\xi_0, \lambda\rangle^{1/2}u\|_0^2$$
$$\geq C_2\|\langle\xi_0, \lambda\rangle^{1/2}\|_{1/2}^2 \geq C_2\lambda\|u\|_{1/2}^2,$$
$$\|u\|_{(h\log)}^2 = \|h^{1/2}\langle\xi_0, \lambda\rangle^{1/2}\omega_1^{\sim}(\mathrm{Log}\,\langle\xi_0, \lambda\rangle)^{1/2}u\|_0^2$$
$$: \omega_1^{\sim -2} = |(d/ds)\omega_0(\xi_0\langle\xi_0, \lambda\rangle^{-1})|,$$
$$\|u\|_{(b)}^2 = \mathrm{Re}\,(b_2 u, u) + C_3\|u\|_{1/2}^2$$

and C_j ($j=0, \cdots, 3$) are positive constants. Thus we conclude a lemma as follows.

Lemma 1.1. *If θ and ν_0 are chosen sufficiently large, then there exists $\lambda_0(\nu)$ for any $\nu \geq \nu_0$ such that for any $\lambda \geq \lambda_0(\nu)$,*

$$\mathrm{Re}\,(p_b u, a_0 u) \geq C_0(\lambda\|a_0 u\|_0^2 + \lambda\|u\|_{(b)}^2 + \nu\|\langle\xi_0, \lambda\rangle^{1/2}u\|_{1/2}^2),$$

where C_0 is a positive constant independent of ν and λ.

It is clear that this estimate admits the perturbation $F^{-1}LF$ by any lower term L. In fact we may take the parameter ν large in proportion to the size of the terms of order 1. And the size of parameter ν is not only independent of the perturbation of lower terms L of order 0 but also of scalar lower terms expanded by $i\xi_0 + \lambda$, Λ_1, b_2 and ∇b_2, though the parameter λ has to be large depending on them.

Theorem 1.2. *Let P be a type (0.1) of system, whose principal symbol p_2 is (1.4) such that $(\partial/\partial x_0)\Lambda_1 \neq 0$ at $\Lambda_1 = b_2 = 0$ and the lower terms L is $L = L_0 a_0 + L_1$ with L_j pseudodifferential operators of order j in x. Then, there exists a constant l such that, for any f, $\langle\xi_0, \lambda\rangle^l f$ belonging to $\boldsymbol{H}^s$, there exists a unique solution u, $\langle\xi_0, \lambda\rangle^{-l}u$ belongs to $\boldsymbol{H}^s$, of $Pu = f$ if the parameter λ is large. It satisfies the estimate that*

$$\|\langle\xi_0, \lambda\rangle^{-l}u\|_s \leq C_s\lambda^{-2}\|\langle\xi_0, \lambda\rangle^l Pu\|_s,$$

where $\|v\|_s = \|\langle\xi, \xi_0, \lambda\rangle^s v\|$.

The extension of Theorem 1.2 to the general case Theorem 0.1 is easily executed by means of two following lemmas. The dependence of estimates on the coefficients of equation will be obtained by differentiation of both sides of the equation after obtaining an estimate whose dependence on the coefficients is not clear.

Lemma 1.2 (Recombination of characteristics). *Let P be a type* (0.1) *of system of order $m \geq 2$ with the principal symbol p_m of double character-*

istics and be pseudodifferential operators in the variable x and differential operators in x_0. If the domain in x_0 is narrowed, then there exists a same type of operator Q of order $m-2$ with a strictly hyperbolic principal symbol such that

$$QP-R_{(1)}R_{(2)}\cdots R_{(m-1)}=\sum_{j=0}^{2m-3} a_j(\partial_0)^j,$$

where the principal symbol of QP has the double characteristics, $R_{(j)}$ ($j=1,\cdots,m-1$) are also same type of operators of order 2 with hyperbolic principal symbol $r_{(j)2}$ and a_j are systems of pseudodifferential operators of order $-l-j$ in x ($0\leq l\leq+\infty$). The double characteristic points of $r_{(j)2}$ coincide exactly with ones of p_m near there so that, if p_m is effectively hyperbolic, then all $r_{(j)2}$ are also effectively hyperbolic.

Let P and Q be two systems of order m such that

$$\partial_0^m+\sum_{j=1}^m a_j(x_0, x, \partial_x)\partial_0^{m-j}, \tag{1.5}$$

where the order of systems a_j of pseudodifferential operators in x do not exceed j.

$$L=P-Q=\sum_{j=0}^l b_j(x_0, x, \partial_x)\partial_0^{l-j} \tag{1.6}$$

is called an operator of order k and of differential order l when $k=\max_j\{l-j+\text{order } b_j\}$ and b_0 is not absolutely 0.

Let us consider the Cauchy problem on the space $\boldsymbol{H}^s(t, T)$ ($t<T$), which stands for the subspaces of the Sobolev space $\boldsymbol{H}^s$ on $\{(x_0, x)\colon x_0<T\}$ consisting of all elements supported on $\{(x_0, x)\colon x_0\geq t\}$. Then, P is an operator from $\boldsymbol{H}^s(t, T)$ to $\boldsymbol{H}^{s-m}(t, T)$.

We say here the Cauchy problem for P strongly well posed on the level l if there exists the operator R from $\boldsymbol{H}^s(0, 1)$ to $\boldsymbol{H}^{s+l}(0, 1)$ with a constant l such that R is the inverse of P for sufficiently many s with respect to l and also satisfies the estimate

$$\|Rf\|_{s+l,t}\leq C_s\int_0^t \|f\|_{s,\tau}d\tau$$

with respect to the norm $\|\cdot\|_{s,T}$ of $\boldsymbol{H}^s(0, T)$.

Lemma 1.3 (Stability of smooth perturbations). *Let P, Q and L be defined as* (1.5) *and* (1.6). *If P is strongly well posed on the level l and if the order of L is less than or equal to l, then $Q=P+L$ is strongly well posed on the level l.*

§2. Bicharacteristic curves

The theme coming next to the solvability of the Cauchy problem is the propagation of singularity of solutions. R. Melrose's result micro-localizing the Oleinik's method and T. Nishitani's one [12] showing the validity of V. Ya Ivriĭ's method [5 and related], which S. Wakabayashi [14 and related] attempts to apply to general hyperbolic equations with analytic coefficients, claim that the wave front set of solutions never propagate outside bicharacteristic curves starting from data's ones in the effectively hyperoblic cases as well as in the strictly hyperbolic cases. We note here that Theorem 1.1 makes the behavior of bicharacteristic curves clear.

Theorem 2.1. *Let p_m be effectively hyperbolic. The bicharacteristic curves through a critical point of the characteristics $\{p_m=0\}$ (double characteristic point) consist of union of two smooth curves which cross transversally each other and a hypersurface, including critical points, at the critical point, accordingly, which never pass through other critical points at a neighborhood of the critical point.*

We denote $(\xi_0-\Lambda_0)$ at Theorem 1.1 by $(\xi_0-\Lambda_+)$ and ones satisfying (1.3) with the negative sign by $(\xi_0-\Lambda_-)$, and also b_2 by b_+ and b_-, respectively. The curves in Theorem 2.1 are given by the bicharacteristic curves of $(\xi_0-\Lambda_\pm)$ through the points $b_\pm=0$. These curves cross transversally each other and also the surface $\Lambda_+-\Lambda_-=0$ since it is checked that $\{\xi_0-\Lambda_-, \xi_0-\Lambda_+\}>0$ on the critical points. The surface $\Lambda_+-\Lambda_-=0$ includes all critical points near there so that they leave immediately the critical points after coming there. Theorem 1.1 means microlocally that p_2 is written as $p_2=a(y_0\eta_0+cb)$, where $a=a(y_0, y, \eta_0, \eta)\neq 0$, $c=c(y_0, y, \eta_0, \eta)\neq 0$

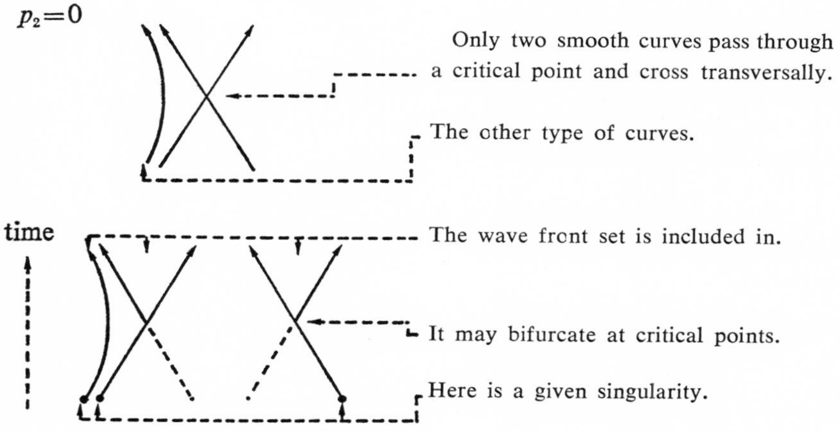
$p_2=0$
Only two smooth curves pass through a critical point and cross transversally.
The other type of curves.
time
The wave front set is included in.
It may bifurcate at critical points.
Here is a given singularity.

and $b=b(y, \eta_0, \eta)\geq 0$ (independent of y_0) with a canonical coordinate (y_0, y, η_0, η). There are two cases that c is positive or negative. By this expression, we are able easily to see that other bicharacteristic curves of p_2 never come to the critical points. Therefore we can conclude that the bicharacteristic curves through the critical points behave as well as ones in the case that $b_2\equiv 0$ do, namely, they bifurcate at the critical points. We suppose but don't know whether the propagation of singularity also bifurcate at the critical points in general.

§ 3. An application

The Monge-Ampère equation is one of typical nonlinear equations with application to the geometry. It includes many types of partial differential equations, elliptic, hyperbolic, ultrahyperbolic, etc.. Recent works are for the most parts ones concerned with the elliptic Monge-Ampère equation. Here we try to deal with the hyperbolic Monge-Ampère equation as an application of the previously mentioned results. (Refer to D. Gilbarg and N. S. Trudinger [1].)

The $n+1$ dimensional Monge-Ampère equation is one as follows because we consider local problems.

$$(3.1)\qquad \Phi(u)=\det(\nabla^2 u+C(x_0, x, u, \nabla u))-f(x_0, x, u, \nabla u)=0,$$

where u is an unknown real valued function, $\nabla u=((\partial/\partial x_i)u)_{i=0,\dots,n}$, $\nabla^2 u=((\partial^2/\partial x_i\partial x_j)u)_{i,j=0,\dots,n}$, $C(x_0, x, v, w)$ is an $(n+1)\times(n+1)$ real symmetric matrix valued infinitely differentiable function on $\boldsymbol{R}\times\boldsymbol{R}^n\times\boldsymbol{R}\times\boldsymbol{R}^{n+1}$, and $f(x_0, x, v, w)$ is a real valued infinitely differentiable function on $\boldsymbol{R}^{2n+3}$.

A typical example is the Gauss curvature of hypersurfaces of $\boldsymbol{R}^{n+1}$, which we denote by K. Then

$$\det(\nabla^2 u)=K(x)(1+|\nabla u|^2)^{(n+2)/2}$$

if a hypersurface of $\boldsymbol{R}^{n+1}$ is given by a function $y=u(x)$ on $\boldsymbol{R}^n$.

We consider the noncharacteristic Cauchy problem with the initial surface $\{x_0=0\}$. Let us put as

$$A(u)=(a_{ij})_{i,j=0,\dots,n}=\nabla^2 u+C,$$

and denote minor matrices of $n\times n$ with respect to a_{ij} and the cofactor matrix by

$$A_{ij}=(a_{kl})_{k,l\neq i,j}$$

and

$$A^{co}=(a_{ij}^{co}),\qquad a_{ij}^{co}=(-1)^{i+j}\det A_{ji}.$$

We assume that the equation (3.1) is kowalevskian so that the cofactor $\det A_{00}$ of a_{00} should not vanish. Since we consider the hyperbolic cases, the initial condition

$$(u, (\partial/\partial x_0)u)|_{x_0=0} = (g_0, g_1) \tag{3.2}$$

are necessarily restricted. We assume that the matrix $A_{00}(u)|_{x_0=0}$, which is determined by the initial datum (g_0, g_1), is strictly positive,

$$A_{00}(u)|_{x_0=0} > 0, \tag{3.3}$$

and that the function f is non positive,

$$f \leq 0. \tag{3.4}$$

Lemma 3.1. *The Fréchet derivative* $\Phi'(u:\phi)$ *of* $\Phi(u)$ *in* u *is a linear hyperbolic partial differential operator of order* 2 *in* ϕ *with respect to the direction* dx_0 *at a neighborhood of the origin under the condition* (3.4) *if* u *is a solution of* (3.1) *and* (3.2) *at a neighborhood of the origin and if the initial datum* (g_0, g_1) *satisfies the condition* (3.3). *Moreover if we assume that*

$$(Ł)^2[f(x_0, x, u, \nabla u)] < 0 \qquad \text{at the origin,} \tag{3.5}$$

when $f(x_0, x, u, \nabla u)$ *vanishes there, then* $\Phi'(u:\phi)$ *is effectively hyperbolic, where* Ł *is a vector field defined by*

$$Ł = \sum_{j=0}^{n} a_{0j}^{\mathrm{co}}(\partial/\partial x_j).$$

(If f *is negative, then* Φ' *is strictly hyperbolic.)*

The following Theorem 3.1 is a corollary of the general results described continuously.

Theorem 3.1. *The Cauchy problem for the Monge-Ampère equation* (3.1) *and* (3.2) *has a unique infinitely differentiable solution* u *on a neighborhood of the origin if the initial datum* (g_0, g_1) *and the function* f *satisfy the conditions* (3.3) *and* (3.4), *respectively, and if for a formal solution* $u^{\sim}$ *at* $x_0 = 0$, *which exists under* (3.4), *the condition* (3.5) *holds when* $f(x_0, x, g_0, g_1, \nabla_x g_0) = 0$ *at the origin.*

A simple example. On $\boldsymbol{R}^2$,

$$\Phi(u) = u_{xx}u_{yy} - u_{xy}^2 = -x^2,$$
$$u = 2^{-1}y^2 - 12^{-1}x^4$$

and

$$\Phi'(u\colon \phi)=(\partial_x^2-x^2\partial_y^2)\phi.$$

Let us consider a single nonlinear partial differential equation of order m

$$pu=p(x_0, x, \partial^\alpha u)=0,$$

where $(\partial^\alpha u)_{|\alpha|=l}$ are l-th derivatives of a real valued function u and p is a function in $(x_0, x, \eta_\alpha)_{|\alpha|<m}$.

Remark. All functions are infinitely differentiable and a neighborhood of a function is one in such a function space.

We define the principal symbol p_m of p as

$$p_m(x_0, x, \partial^\alpha u, \xi)=\sum_{|\beta|=m}(\partial/\partial\eta_\beta)p(x_0, x, \partial^\alpha u)\xi^\beta.$$

Definition 3.1. *We call effectively hyperbolic with respect to the direction dx_0 at $(x_0^\sim, x^\sim, u^\sim, h^\sim)$ if there exist linear partial differential operators $q\phi=\sum_{|\alpha|\le m} q_\alpha(x_0, x, \partial^\beta u)\partial^\alpha\phi$ and $r\phi=\sum_{|\alpha|\le m} r_\alpha(x_0, x, \partial^\beta u, \partial^\beta h)\partial^\alpha\phi$ of order m such that the following* 1), $\cdots$, 4) *hold for all (x_0, x, u) belonging to a neighborhood of $(x_0^\sim, x^\sim, u^\sim)$.*

1) $p_m=q+r$ *with* $h=pu-h^\sim$.

2) *q is effectively hyperbolic with respect to the direction dx_0.*

3) *q_α and r_α are functions in $(x_0, x, \eta_\beta)_{|\beta|\le m'}$ and in $(x_0, x, \eta_\beta, \zeta_\beta)_{|\beta|\le m''}$, respectively.*

4) $r_\alpha=0$ *at* $(\zeta_\beta)_{|\beta|\le m''}=0$.

Theorem 3.2 (Unique extension). *Let p be effectively hyperbolic with respect to dx_0 at $(x_0^\sim, x^\sim, u^\sim, h^\sim)$. If there exists a neighborhood Ω_0 of $(x_0^\sim, x^\sim)$ such that $u^\sim$ is a solution of $pu^\sim=h^\sim$ at $\{x_0\le x_0^\sim\}\wedge\Omega_0$, then there exists a unique solution u of $pu=h^\sim$ on a neighborhood Ω of $(x_0^\sim, x^\sim)$ such that $u=u^\sim$ on $\{x_0\le x_0^\sim\}\wedge\Omega$.*

This theorem is proved directly from the result of linear cases by means of the Nash-Moser implicit function theorem.

Corollary (Unique existence). *Let p be effectively hyperbolic with respect to dx_0 at (x_0, x, u, pu) for all (x_0, x, u) of a neighborhood of $(x_0^\sim, x^\sim, u^\sim)$. If $u^\sim$ is a formal solution of $pu^\sim=0$ at $x_0=x_0^\sim$, then there exists a unique solution u of $pu=0$ at a neighborhood of $(x_0^\sim, x^\sim)$ such that $u-u^\sim$ is flat at $x_0=x_0^\sim$.*

Another example. We consider the non-linear wave equation.

$$\Phi(u)=\partial_0^2 u-|\partial u|\sum_{i=1}^n \partial_i(|\partial u|^{-1}\partial_i u),$$
$$\Phi(u)=f \qquad (\partial u \neq 0).$$

Since $\Phi(u)$ is developed as

$$\Phi(u)=\partial_0^2 u-\Delta u+\langle z, \partial\rangle^2 u-z_0\langle z, \partial\rangle\partial_0 u+\text{lower terms},$$
$$z=\partial u/|\partial u|,$$

so the linear part is

$$\Phi'(u: \phi)=\partial_0^2\phi-\Delta\phi+\langle z, \partial\rangle^2\phi-z_0\langle z, \partial\rangle\partial_0\phi+\text{lower terms},$$

which is always hyperbolic. If $\partial_0 u\neq 0$, then $\Phi'(u: \phi)$ is strictly hyperbolic. If $\partial\partial_0 u$ and ∂u are linearly independent at $\partial_0 u=0$, then $\Phi'(u: \phi)$ is effectively hyperbolic. Therefore there exists a unique local smooth solution.

1) Put

$$u=2^{-1}x_0^2+x_1.$$

Then

$$\Phi'(u: \phi)=\partial_0^2\phi+(1+x_0^2)^{-1}x_0\partial_0\partial_1\phi-(1+x_0^2)^{-1}x_0^2\partial_1^2\phi-\Delta'\phi$$
$$+\text{lower terms}$$
$$\Phi(u)=1.$$

2) Let the initial data be

$$u|_{x_0=0}=x_1+2^{-1}x_2^2 \quad \text{and} \quad \partial_0 u|_{x_0=0}=0.$$

Then for a formal solution $u_0^{\sim}$ of $\Phi(u_0^{\sim})=0$ at $x_0=0$, it holds that

$$\partial_0^2 u_0^{\sim}|_{x_0=0}=(1+x_2^2)^{-1}\neq 0.$$

Therefore $\Phi(u)$ is effectively hyperbolic near $u_0^{\sim}$ so that a solution exists for $\Phi(u)=0$.

References

[1] D. Gilbarg and N. S. Trudinger, Elliptic Partial Differential Equations of Second Order (second ed.), Springer-Verlag, 1983.

[2] L. Hörmander, The Cauchy problem for differential equations with double characteristics, J. Analyse Math., **32** (1977), 118–196.

[3] V. Ya. Ivriĭ and V. M. Petkov, Necessary conditions for the Cauchy problem for non-strictly hyperbolic equations to be well-posed, Uspehi Mat. Nauk, **29** (1974), 3–70. (Russian Math. Surveys, **29** (1974), 1–70.)

[4] Y. Ya. Ivriĭ, Sufficient conditions for regular and completely regular hyper-

bolicity, Trudy Moskov. Mat. Obšč., **33** (1976), 3–66. (Trans. Moscow Math. Soc. **33** (1978), 1–65.)

[5] ——, Wave fronts of solutions of certain pseudodifferential equations, Trudy Moskov. Mat. Obšč., **39** (1979), 49–82.

[6] N. Iwasaki, The Cauchy problem for effectively hyperbolic equations (a special case), J. Math. Kyoto Univ., **23** (1983), 503–562.

[7] ——, The Cauchy problem for effectively hyperbolic equations (a standard type), Publ. RIMS, Kyoto Univ., **20** (1984), 551–592.

[8] ——, The Cauchy problem for effectively hyperbolic equations (general cases), RIMS preprint 468, RIMS Kyoto Univ., 1984.

[9] ——, The Cauchy problem for hyperbolic equations with double characteristics, Publ. RIMS, Kyoto Univ., **19** (1983), 927–942.

[10] R. Melrose, The Cauchy problem for effectively hyperbolic operators, Hokkaido Math. J., **12** (1983), 371–391.

[11] T. Nishitani, Local energy integrals for effectively hyperbolic operators, (preprint).

[12] ——, On Wave front sets of solutions for effectively hyperbolic operators, (preprint).

[13] O. A. Olejnik, On the Cauchy problem for weakly hyperbolic equations, Comm. Pure Appl. Math., **23** (1970), 569–586.

[14] S. Wakabayashi, Singularities of solutions of the Cauchy problem for symmetric hyperbolic systems, (preprint).

RESEARCH INSTITUTE FOR
MATHEMATICAL SCIENCES
KYOTO UNIVERSITY
KYOTO, 606
JAPAN

Taniguchi Symp. HERT
Katata 1984, pp. 101–123

The Cauchy Problem for Uniformly Diagonalizable Hyperbolic Systems in Gevrey Classes

Dedicated to Professor Sigeru Mizohata on his sixtieth birthday

Kunihiko KAJITANI

Introduction

We consider the Cauchy problem for uniformly diagonalizable hyperbolic systems of linear partial differential equations in Gevrey classes. Here we do not assume the smoothness of the diagonalizer for the system. If we do so, we have many results. For example, Friedrichs in [3] and [4] treated the symmetric or more generally, symmetrizable hyperbolic systems, and Mizohata in [13] investigated the diagonalizable hyperbolic systems.

We shall prove that the Cauchy problem for uniformly diagonalizable hyperbolic systems whose coefficients are in Gevrey class $\gamma^{(s_0)}$ is well posed in $\gamma^{(s)}$ for any s $(s_0 \leq s \leq 2)$.

Historically Petrovski in [14] has given the definition of the uniformly diagonalizable hyperbolic system and an interesting example. Following Petrovski, Kasahara and Yamaguti in [10] proved that if the coefficients of the system are constant, the uniformly diagonalizable hyperbolic system is equivalent to strongly hyperbolic one, namely, this system is stable under the perturbation of lower order term of the system. Moreover Kreiss [11], Strang [16] and Vaillant [17] have given the various characterizations of uniformly diagonalizable hyperbolic systems.

We consider the Cauchy problem for the following first order system in $[\tau, \bar{T}] \times \boldsymbol{R}^n$, $(\tau < \bar{T})$

$$
(0.1)\qquad \begin{aligned} &L[u] = \Big(D_0 + \sum_{j=1}^{n} A_j(x) D_j + B(x)\Big) u(x) = f(x), \ x \in (\tau, \bar{T}) \times \boldsymbol{R}^n, \\ &u|_{x_0=\tau} = u_0(x'), \quad x' \in \boldsymbol{R}^n, \end{aligned}
$$

where $A_j(x)$ and $B(x)$ are $d \times d$ matrices and $u(x)$ an unknown vector-valued function of d-components, and $D_j = i^{-1}\partial/\partial x_j$. For V an open set in $\boldsymbol{R}^n$, we denote by $\gamma_A^{(s)}(V)$ the Gevrey class consisting of functions $f(x)$ such that

$$|f|_{\gamma_A^{(s)}(V)}=\sup_{\substack{x\in V\\ \alpha\in N^n}}\frac{|D_x^\alpha f(x)|}{A^{|\alpha|}|\alpha|!^s}<\infty,$$

and $\gamma_A^{(s)}=\gamma_A^{(s)}(\boldsymbol{R}^n)$, and by $C^k(\tau, T;\gamma_A^{(s)})$, ($k$ a non negative integer) the class of functions which are k times continuously differentiable in $\gamma_A^{(s)}$ with respect to $x_0\in[\tau, T]$, and define

$$C^k(\tau, T, \gamma^{(s)})=\bigcup_{A>0} C^k(\tau, T;\gamma_A^{(s)}),$$

$$C^\infty(\tau, T;\gamma^{(s)})=\bigcap_{k=0}^{\infty} C^k(\tau, T;\gamma^{(s)}),$$

$$C^\infty(R^1;\gamma^{(s)})=\bigcap_{\tau<T} C^\infty(\tau, T;\gamma^{(s)}).$$

We assume the following conditions

[A.I] *The coefficients of* $A_j(x)$ *and* $B(x)$ *are in* $C^\infty(\boldsymbol{R}^1;\gamma^{(s_0)})$ *where* $1<s_0\leq 2$.

[A.II] $L_0(x,\xi)=\xi_0+\sum_{j=1}^n A_j(x)\xi_j$ *is hyperbolic with respect to* x_0, *that is,*

$$\det L_0(x,\xi)=\prod_{j=1}^{d}(\xi_0-\lambda_j(x,\xi')),$$

where $\lambda_j(x,\xi')$, $j=1,\cdots,d$, *are real valued functions in* $\boldsymbol{R}^{n+1}\times\boldsymbol{R}^n$.

We note that the characteristic roots $\lambda_j(x,\xi')$ are continuous but not necessarily differentiable in $\boldsymbol{R}^{n+1}\times\boldsymbol{R}^n$

[A.III] $L_0(x,\xi)$ *is uniformly diagonalizable, that is, there exists a non singular matrix* $N(x,\xi')$ *such that,*

$$N(x,\xi')L_0(x,\xi)N^{-1}(x,\xi')=\begin{bmatrix}\xi_0-\lambda_1(x,\xi') & & 0\\ & \ddots & \\ 0 & & \xi_0-\lambda_d(x,\xi')\end{bmatrix}, \tag{0.2}$$

$$\sup_{\substack{x\in[\tau,T]\times R^n\\ |\xi'|=1}}\{|N(x,\xi')|+|N^{-1}(x,\xi')|\}\leq C, \tag{0.3}$$

where the norm of $N(x,\xi')$ *is given by the maximum of the absolute value of elements of matrix.*

We note that the diagonalizer $N(x,\xi')$ is not necessarily continuous in $\boldsymbol{R}^{n+1}\times\boldsymbol{R}^n\backslash 0$. As mentioned above, we do assume not here the smoothness of $N(x,\xi')$.

Then we shall prove,

Theorem 0.1. *Assume that* [A.I], [A.II] *and* [A.III] *are valid and* $s_0^{-1} \geq \kappa > 1/2$. *Then for any* $u_0 \in \gamma^{(s)}$ *and* $f \in C^\infty(\tau, \bar{T}; \gamma^{(s)})$ $(s^{-1} = \kappa)$ *there exists a solution* $u(x) \in C^\infty(\tau, T; \gamma^{(s)})$ *of the Cauchy problem* (0.1).

Next we shall determine the dependence domain of the Cauchy problem (0.1). We define

$$(0.4) \qquad \begin{aligned} &\Gamma_+ = \{\theta \in \boldsymbol{R}^{n+1};\ \inf_{\substack{x \in [\tau, \bar{T}] \times \boldsymbol{R}^n \\ j=1,\cdots,d}} (\theta_0 - \lambda_j(x, \theta')) > 0\}, \\ &\Gamma_- = -\Gamma_+, \end{aligned}$$

and for $\hat{x} \in [\tau, \bar{T}] \times \boldsymbol{R}^n$, we define the dual cone of $\Gamma_\pm$

$$\Gamma^0_\pm(\hat{x}) = \{x \in \boldsymbol{R}^{n+1};\ \langle x - \hat{x}, \theta \rangle > 0 \text{ for any } \theta \in \Gamma_\pm\},$$

where $\langle\ ,\ \rangle$ stands for the inner product of R^{n+1}. Then we have

Theorem 0.2. *Assume that* [A.I], [A.II] *and* [A.III] *are valid. Let* $\hat{x}$ *be in* $(\tau, \bar{T}] \times \boldsymbol{R}^n$. *If* $u(x) \in C^1(\tau, \bar{T}; \gamma^{(s)})$ $(s_0 \leq s \leq 2)$ *satisfies*

$$\begin{aligned} &L(x, D)u(x) = 0 && \textit{in } \Gamma^0_-(\hat{x}) \cap \{x_0 > \tau\}, \\ &u(\tau, x') = 0 && \textit{on } \Gamma^0_-(\hat{x}) \cap \{x_0 = \tau\}, \end{aligned}$$

then $u(x) = 0$ *in* $\Gamma^0_-(\hat{x}) \cap \{x_0 > \tau\}$.

The case of $s = s_0 = 2$ is treated in §5.

In order to prove Theorem 0.1 and Theorem 0.2, we transform the operator $L(x, D)$ by e^Λ as follows,

$$(0.5) \qquad L_\Lambda(x, D) = e^\Lambda L(x, D) e^{-\Lambda},$$

where $\Lambda = (T - x_0)(h + \langle D' \rangle^\kappa)$, $s_0^{-1} \geq \kappa > 1/2$ and $h > 0$. Then the operator $L_\Lambda(x, D)$ becomes a pseudo-differential operator of which symbol is in $S^{(1)}_{\kappa, 1-\kappa}$ and if we assume that $L(x, D)$ is hyperbolic then $L_\Lambda(x, D)$ is nearly elliptic. Therefore we can construct a parametrix $Q(x, D)$ of the operator $L_\Lambda(x, D)$ in $L^2(\boldsymbol{R}^{n+1})$. We note that when we are concerned with the operator $L_\Lambda(x, D)$, we need not consider it in the Gevrey class and it is enough to treat it in the usual L^2 space. The parametrix $Q(x, D)$ is a pseudo-differential operator of which symbol is in $S^{(-\kappa)}_{\kappa, 1-\kappa}$. If we assume that $1 - \kappa \leq \kappa$, we can construct the parametrix $Q(x, D)$ for $L_\Lambda(x, D)$ similarly to the elliptic operator. It should be remarked that Bronshtein [1] has constructed a parametrix of $L(x, D)$ directly in Gevrey class, which is a pseudo-differential operator in Gevrey class $\gamma^{(s)}$ of type (ρ, δ), where $\rho = \kappa$, $\delta = 1 - \kappa$ and $\kappa^{-1} = s$, and also he has proved that its parametrix is bounded in $\gamma^{(s)}$. His analysis is very delicate. Because we do not know as a general theory if the pseudo-differential operators in Gevrey class $\gamma^{(s)}$ of type (ρ, δ) $(\delta \neq 0)$ are bounded in $\gamma^{(s)}$.

As mentioned above, in the case of constant coefficients Kasahara-Yamaguti proved that the Cauchy problem (0.1) is C^∞-well posed for any lower order term B, if and only if $L_0(\xi)=\xi_0+\sum A_j\xi_j$ is uniformly diagonalizable. In the case of variable coefficients, Kasahara-Yamaguti theorem does not hold generally, but the multiplicity of the characteristic roots of $L_0(x, \xi)$ is constant, this theorem is true ([8]). In general, the characterization of the strong hyperbolicity in the C^∞-sense for the systems of variable coefficients is an open problem. We note that Ivrii-Petkov [6] and Iwasaki [7] have proved that for a single higher order hyperbolic equation the strong hyperbolicity in the C^∞-sense is equivalent to the effective one.

In this paper we shall prove that the uniform diagonalizability for the hyperbolic systems is a sufficient condition in order that the Cauchy problem is well posed in the sense of Gevrey class $\gamma^{(s)}$ $(s \leq 2)$.

§ 1. Pseudo-differential operators in Gevrey class

Denote by $S^{(m)}_{\rho,\delta}$ the class of symbols of pseudo-differential operators such that

$$|a|^{(m)}_l = \sup_{\substack{(x,\xi)\in \boldsymbol{R}^{2n} \\ |\alpha+\beta|\leq l}} \frac{|a^{(\alpha)}_{(\beta)}(x, \xi)|}{\langle\xi\rangle^{m-\rho|\alpha|+\delta|\beta|}} < \infty$$

where $\langle\xi\rangle=(1+|\xi|^2)^{1/2}$. We define $a(x, D)$ as usual

$$a(x, D)u(x)=\int e^{ix\xi}a(x, \xi)u(\xi)d\xi,$$

where $\hat{u}(\xi)$ is a Fourier transform of $u(x)$ and $d\xi=(2\pi)^n d\xi$.

Lemma 1.1 ([5] and [2]). *Let be $a(x, \xi)$ in $S^{(0)}_{\rho,\delta}$ where $\rho \geq \delta$. Then there eixst l_0 and C_0 dependent only on n, m, ρ and δ such that*

$$\|a(x, D)u\|_{L^2(\boldsymbol{R}^n)} \leq C_0|a|^{(0)}_{l_0}\|u\|_{L^2(\boldsymbol{R}^n)}, \tag{1.1}$$

for u in $L^2(\boldsymbol{R}^n)$.

Denote by $A^{(m)}_{s,A,l}$ all functions satisfying

$$|a|_{A^{(m)}_{s,A,l}} = \sup_{\substack{(x,\xi)\in \boldsymbol{R}^{2n} \\ \alpha\in \boldsymbol{N}^n \\ |\beta|\leq l}} \frac{|a^{(\alpha)}_{(\beta)}(x, \xi)|}{A^{|\alpha|}|\alpha|!^s\langle\xi\rangle^{m+|\beta|}} < \infty. \tag{1.2}$$

We put

$$\Lambda(\varepsilon, D)=\varepsilon(h+\langle D\rangle^\kappa)$$

where $\varepsilon \in \boldsymbol{R}^1$ and $h>0$, and define e^{Λ} as follows

$$e^{\Lambda(\varepsilon, D)}u(x)=\int e^{ix\xi+\Lambda(\varepsilon,\xi)}\hat{u}(\xi)d\xi.$$

We define also $a_{\Lambda}(x, D)$ as

$$a_{\Lambda}(x, D)=e^{\Lambda}a(x, D)e^{-\Lambda}.$$

Lemma 1.2. *Let $a(x, \xi)$ be in $A^{(m)}_{s, A_0 l_0}$, $\Lambda=\varepsilon(h+\langle D\rangle^{\kappa})$, $\kappa=s^{-1}$. If $|\varepsilon|\leq (24n^k A_0^{\kappa})^{-1}$, then $a_{\Lambda}(x, D)$ is a pseudo-differential operator and its symbol $a_{\Lambda}(x, \xi)$ is given as a form,*

$$a_{\Lambda}(x, \xi)=a(x, \xi)+\varepsilon R(a)(x, \xi) \tag{1.3}$$

where $R(a)(x, \xi)$ is in $S^{(m-1+\kappa)}_{1,0}$ and

$$|R(a)|^{(m-1+\kappa)}_{l}\leq C_l|a|_{[l(1-\kappa)-1]+l+1, A^{(m)}_{s,A_0}}, \tag{1.4}$$

here $[\ \cdot\ ]$ stands for Gauss symbol and C_l is independent of h, ε.

Proof. It follows from Kumano-go [12] that the symbol $a_{\Lambda}(x, \xi)$ is given by

$$a_{\Lambda}(x, \xi)=\text{os-}\iint e^{-iy\eta+\Lambda(\xi+\eta)-\Lambda(\xi)}a(x+y, \xi)dyd\eta,$$

where os-$\iint$ means an oscillatory integral. Hence we have by Taylor expansion,

$$a_{\Lambda}(x, \xi)=a(x, \xi)+\sum_{0<|\gamma|<N+1}\gamma!^{-1}a_{(\gamma)}(x, \xi)\mu_{\gamma}(\xi)$$
$$+\text{os-}\iiint_0^1(1-\theta)^N e^{-iy\eta+\Lambda(\xi+\eta)-\Lambda(\xi)}\mu_{\gamma}(\xi+\eta)a_{(\gamma)}(x+\theta y, \xi)dyd\eta d\theta$$

where $\mu_{\gamma}(\xi)=e^{-\Lambda(\xi)}D^{\gamma}e^{\Lambda(\xi)}$. Since $\Lambda(\xi)=\varepsilon(h+\langle\xi\rangle^{\kappa})$, we can write

$$\mu_{\gamma}(\xi)=\varepsilon\lambda_{\gamma}(\xi), \qquad |\lambda^{(\beta)}_{\gamma}(\xi)|\leq C_{\gamma\beta}\langle\xi\rangle^{-(1-\kappa)|\gamma|-|\beta|},$$

where $C_{\gamma\beta}$ is independent of h and ε. We can prove similarly to Proposition 2.3 in [9] that

$$R(a)(x, \xi)=\sum_{0<|\gamma|<N-1}\gamma!^{-1}a_{(\gamma)}(x, \xi)\lambda_{\gamma}(\xi)$$
$$+\sum_{|\gamma|=N}\gamma!^{-1}\,\text{os-}\iiint_0^1(1-\theta)^N e^{-iy\eta+\Lambda(\xi+\eta)-\Lambda(\xi)}\lambda_{\gamma}(\xi+\eta)a_{(\gamma)})x+\theta y, \xi)dyd\eta d\theta$$

which is in $S^{(m-1+\kappa)}_{1,0}$ and satisfies (1.4).

§ 2. The construction of a parametrix

For $L_0(x, \xi)=\xi_0+\sum_{j=1}^{n} A_j(x)\xi_j$, we put

$$Q_0(x, \xi)=L^{-1}(x, \xi_A),$$

where $x=(x_0, x') \in [\tau, \overline{T}]\times \boldsymbol{R}^n$, $\xi=(\xi_0, \xi') \in \boldsymbol{R}^{n+1}$,

$$\xi_A=(\xi_{0A}, \xi'), \qquad \xi_{0A}=\xi_0+i\delta(h+\langle\xi'\rangle^{\kappa}),$$

$h>0$, $0<\delta<1$, and $\langle\xi'\rangle=(1+|\xi'|^2)^{1/2}$. Then we have,

Lemma 2.1. *Assume that* [A.II] *and* [A.III] *are valid. Then* $Q_0(x, \xi)$ *satisfies the following estimate,* $x \in [\tau, \overline{T}]\times \boldsymbol{R}^n$,

$$(2.1) \qquad |Q_0(x, \xi)|\leq \begin{cases} C(\delta h+\langle\xi\rangle)^{-1}, & |\xi_0|\geq M_0|\xi'|, \\ C\delta^{-1}(h+\langle\xi'\rangle^{\kappa})^{-1}, & |\xi_0|\leq M_0|\xi'|, \end{cases}$$

where M_0 *and* C *are positive constants independent of* $h>0$, $\xi \in \boldsymbol{R}^{n+1}$ *and* δ $(0<\delta<1)$.

Proof. Since $L_0(x, \xi)$ is uniformly diagonalizable, we have from (0.2),

$$Q_0(x, \xi)=N(x, \xi')\begin{bmatrix} (\xi_{0A}-\lambda_1(x, \xi'))^{-1} & & 0 \\ & \ddots & \\ 0 & & (\xi_{0A}-\lambda_d(x, \xi'))^{-1} \end{bmatrix} N^{-1}(x, \xi').$$

Hence we can estimate from (0.3),

$$|Q_0(x, \xi)|\leq C \max_{1\leq j\leq d} |\xi_0-i\delta(h+\langle\xi'\rangle^{\kappa})-\lambda_j(x, \xi')|^{-1}.$$

We put

$$M_0=\max_{\substack{x\in[\tau, \overline{T}]\times \boldsymbol{R}^n \\ |\xi'|=1 \\ 1\leq j\leq d}} 2|\lambda_j(x, \xi')|.$$

For $|\xi_0|\geq M_0|\xi'|$, we have

$$\begin{aligned} |Q_0(x, \xi)| &\leq C(\delta h+2^{-1}|\xi_0|)^{-1}\leq C(\delta h+4^{-1}|\xi_0|+2^{-1}M_0|\xi'|)^{-1} \\ &\leq C(\delta h+\langle\xi\rangle)^{-1}. \end{aligned}$$

For $|\xi_0|\leq M_0|\xi'|$

$$\begin{aligned} |Q_0(x, \xi)| &\leq C\delta^{-1}(h+\langle\xi'\rangle^{\kappa})^{-1}\leq C\delta^{-1}(h+2^{-1}\langle\xi'\rangle^{\kappa}+2^{-1}M_0^{-\kappa}|\xi_0|^{\kappa})^{-1} \\ &\leq C\delta^{-1}(h+\langle\xi\rangle^{\kappa}), \end{aligned}$$

where all constants are independent of h and δ.

Proposition 2.2. *Assume that* [A.II] *and* [A.III] *are valid. Then* $Q_0(x, \xi)=L_0(x, \xi_\Lambda)^{-1}$ *satisfies for* $\alpha=(\alpha_0, \alpha')$, $\beta \in N^{n+1}$, $x \in [\tau, \bar{T}]\times R^n$

$$(2.2)\qquad |Q_{0(\beta)}^{(\alpha)}(x, \xi)|\leq\begin{cases} C_{\alpha\beta\delta}(h+\langle\xi\rangle)^{-1-|\alpha|}, & |\xi_0|\geq M_0|\xi'|,\\ C_{\alpha\beta\delta}\dfrac{\langle\xi'\rangle^{-|\alpha'|\kappa+(1-\kappa)|\beta|}}{(h+\langle\xi\rangle^\kappa)^{1+\alpha_0}}, & |\xi_0|\leq M_0|\xi'|.\end{cases}$$

Proof. It follows from (2.1) that (2.2) is true for $|\alpha+\beta|=0$. Assume that (2.2) is valid for $|\alpha+\beta|\leq k-1$. Since $Q_0(x, \xi)L_0(x, \xi_\Lambda)=I$, we have for $|\alpha+\beta|=k$,

$$D_x^\alpha D_\xi^\beta(Q_0L_0)(x, \xi)=\sum_{\substack{\hat\alpha\leq\alpha\\ \hat\beta\leq\beta}}\binom{\alpha}{\hat\alpha}\binom{\beta}{\hat\beta}Q_{0(\beta-\hat\beta)}^{(\alpha_0-\hat\alpha)}L_{0(\hat\beta)}^{(\hat\alpha)}=0.$$

Hence we have

$$Q_{0(\beta)}^{(\alpha)}(x, \xi)=-\sum_{\hat\alpha+\hat\beta\neq 0}\binom{\alpha}{\hat\alpha}\binom{\beta}{\hat\beta}Q_{0(\beta-\hat\beta)}^{(\alpha-\hat\alpha)}L_{0(\hat\beta)}^{(\hat\alpha)}L_0^{-1}.$$

Since $L_0(x, \xi_\Lambda)$ is elliptic for $|\xi_0|\geq M_0|\xi'|$, it is evident that (2.2) is valid for $|\xi_0|\geq M_0|\xi'|$. Noting that

$$L_0(x, \xi_\Lambda)=\xi_0+i\delta(h+\langle\xi'\rangle^\kappa)+\sum A_j(x)\xi_j,$$

we have

$$(2.3)\qquad \begin{aligned} &|L_0^{(\alpha)}|\leq\begin{cases} C, & |\alpha|=1,\\ C\langle\xi'\rangle^{\kappa-|\alpha|}, & |\alpha|\geq 2,\quad \alpha_0=0,\end{cases}\\ &L_{0(\beta)}^{(\alpha)}=0,\quad |\alpha+\beta|\geq 2,\quad \alpha_0\neq 0,\\ &|L_{0(\beta)}^{(\alpha)}|\leq C\langle\xi'\rangle^{1-|\alpha|},\quad |\alpha'|\leq 1,\quad \alpha_0=0,\quad |\beta|\neq 0,\\ &L_{0(\beta)}^{(\alpha)}=0,\quad |\alpha|\geq 2,\quad |\beta|\neq 0.\end{aligned}$$

Therefore we obtain in all cases,

$$|Q_{0(\beta-\beta)}^{(\alpha-\alpha)}L_{0(\beta)}^{(\alpha)}L_0^{-1}|\leq C_{\alpha\beta\delta}\frac{\langle\xi'\rangle^{-|\alpha'|\kappa+(1-\kappa)|\beta|}}{(h+\langle\xi\rangle^\kappa)^{1+\alpha_0}},$$

which implies (2.2).

Let $\phi^{(1)}(\xi)$ be in $S_{1,0}^{(0)}$ such that

$$\phi^{(1)}(\xi)=\begin{cases}1, & |\xi_0|\geq 2M_0|\xi'|,\\ 0, & |\xi_0|\leq M_0|\xi'|,\end{cases}$$

and

$$\phi^{(2)}(\xi)=1-\phi^{(1)}(\xi). \tag{2.4}$$

Then noting that $L_0(x, \xi_A)$ is elliptic for $\xi \in \operatorname{supp} \phi^{(1)}$, we can prove easily,

Proposition 2.3. *Assume that* [A.I], [A.II] *and* [A.III] *are valid. Then*

$$R^{(1)}(x, D)=L_A(x, D)Q_0(x, D)\phi^{(1)}(D)-\phi^{(1)}(D)$$

is a pseudo-differential operator of which symbol is in $S_{1,0}^{(0)}$ *and satisfies*

$$|R_{(\beta)}^{(1)(\alpha)}(x, \xi)|\leq C_{\alpha\beta\delta}(h+\langle\xi\rangle)^{-1-\alpha_0}\langle\xi'\rangle^{\kappa-|\alpha'|} \tag{2.5}$$

for $\alpha=(\alpha_0, \alpha') \in \boldsymbol{N}^{n+1}$, $x \in [\tau, \overline{T}]\times\boldsymbol{R}^n$ *and* $\xi \in \boldsymbol{R}^{n+1}$, *where* $C_{\alpha\beta\delta}$ *is independent of* h.

Proof. Noting that it follows from Lemma 1.2

$$L_A(x, \xi)=L_0(x, \xi_A)+\delta(T-x_0)R(L_0)(x, \xi)+B_A(x, \xi')$$

where $R(L_0)(x, \xi)$ satisfies (1.4) with $m=1$ and $B_A(x, \xi')$ in $S_{1,0}^{(0)}$, we have

$$\begin{aligned} R^{(1)}(x, \xi)&=\text{os-}\iint e^{-iy\eta}L_A(x, \xi+\eta)Q_0(x+y, \xi)\phi^{(1)}(\xi)dyd\eta-\phi^{(1)}(\xi)\\ &=\{\delta(T-x_0)R(L_0)(x, \xi)+B_A(x, \xi')\}Q_0(x, \xi)\phi^{(1)}(\xi)\\ &\quad+\sum_{|\gamma|=1}\text{os-}\iiint_0^1 e^{-iy\eta}L_A^{(\gamma)}(x, \xi+\eta)Q_{0(\gamma)}(x+\theta y, \xi)\phi^{(1)}(\xi)dyd\eta d\theta \end{aligned}$$

which satisfies (2.5) from (2.2) for $|\xi_0|\geq M_0|\xi'|$.

Therefore we can estimate by virtue of (2.5),

$$\begin{aligned} \|\sigma R^{(1)}(x, D)u\|_{W_l(\boldsymbol{R}^{n+1})}&\leq C_{l\delta}h^{-(1-\kappa)}\|u\|_{W_l(\boldsymbol{R}^{n+1})}\\ &\leq(1/4)\|u\|_{W_l(\boldsymbol{R}^{n+1})}, \end{aligned} \tag{2.6}$$

for any $h\geq h_0(l, \delta)$ and u in $W_l(\boldsymbol{R}^{n+1})$ with $\operatorname{supp} u\subset\{x_0\geq\tau\}$, where $W_l(\boldsymbol{R}^{n+1})$ is Sobolev space and $\sigma(x_0) \in C_0^\infty(R^1)$ with $\operatorname{supp} \sigma\subset[\tau, \overline{T}]$.

Proposition 2.4. *Assume that* [A.I], [A.II] *and* [A.III] *are valid. Let be* $\Lambda=\delta(T-x_0)(h+D'\rangle^\kappa)$. *Then there are positive constants* ε_0 *independent of* δ *and* $h_0(\delta)$ *depending on* δ *such that for* $|T-x_0|\leq\varepsilon_0$ *and* $h\geq h_0(\delta)$ *the symbol* $L_A(x, \xi)$ *has an inverse* $M(x, \xi)=L_A(x, \xi)^{-1}$ *satisfying*

$$|M_{(\beta)}^{(\alpha)}(x, \xi)|\leq C_{\alpha\beta\delta}\frac{\langle\xi'\rangle^{-|\alpha'|\kappa+(1-\kappa)|\beta|}}{(h+\langle\xi\rangle^\kappa)^{1+\alpha_0}} \tag{2.7}$$

for $\alpha=(\alpha_0, \alpha')$ *and* $\beta \in N^{n+1}$, $x \in [\tau, \bar{T}]\times R^n$ *and* $\xi \in R^{n+1}$, *where* $C_{\alpha\beta\delta}$ *is independent of* h.

Proof. It follows from Lemma 1.2 that

$$L_\Lambda(x, \xi)=L_0(x, \xi_\Lambda)+\delta(T-x_0)R(L_0)(x, \xi)+B_\Lambda(x, \xi').$$

Hence we have

$$L_\Lambda(x, \xi)=L_0(x, \xi_\Lambda)\{I+\delta(T-x_0)L_0^{-1}R(L_0)+L_0^{-1}B\}.$$

If we choose ε_0 and $h_0(\delta)$ such that from (2.1)

$$|\delta(T-x_0)L_0^{-1}R(L_0)(x, \xi)|\leq C\frac{\langle\xi\rangle^\kappa|T-x_0|}{h+\langle\xi\rangle^\kappa}\leq C\varepsilon_0\leq 1/4,$$

$$|L_0^{-1}B_\Lambda(x, \xi')|\leq C\delta^{-1}(h+\langle\xi\rangle^\kappa)^{-1}\leq C(\delta h_0)^{-1}\leq 1/4,$$

$L_\Lambda(x, \xi)^{-1}$ exists and from (2.2) we obtain (2.7) directly.

Proposition 2.5. *Assume that* [A.I], [A.II] *and* [A.III] *are valid. Let be* $\Lambda=\delta(T-x_0)(h+\langle D'\rangle^\kappa)$, *where* $s_0^{-1}\geq\kappa>1/2$. *Then*

$$R^{(2)}(x, D)=L_\Lambda(x, D)M(x, D)\phi^{(2)}(D)-\phi^{(2)}(D)$$

is a pseudo-differential operator of which symbol is in $S^{(1-2\kappa)}_{\kappa,1-\kappa}$ *and satisfies for* $|T-x_0|\leq\varepsilon_0$ *and* $h>h_0(\delta)$,

$$|R^{(2)(\alpha)}_{(\beta)}(x, \xi)|\leq C_{\alpha\beta\delta}\frac{\langle\xi\rangle^{1-\kappa(1+|\alpha|)+(1-\kappa)|\beta|}}{h+\langle\xi\rangle^\kappa} \tag{2.8}$$

where $C_{\alpha\beta\delta}$ *is independent of* h *and* T.

Proof. We have

$$R^{(2)}(x, \xi)=\text{os-}\iint e^{-iy\eta}L_\Lambda(x, \xi+\eta)M(x+y, \xi)\phi^{(2)}(\xi)dyd\eta-\phi^{(2)}(\xi).$$

Let $\chi_1(\xi', \eta')$ be a function in R^{2n} such that

$$\chi_1(\xi', \eta')=\begin{cases}1 & \text{for } |\eta'|\geq 2^{-1}|\xi'|,\\ 0 & \text{for } |\eta'|\leq 4^{-1}|\xi'|,\end{cases}$$

$$|D^\alpha_{\xi'}D^\beta_{\eta'}\chi_1(\xi', \eta')|\leq C_{\alpha\beta}|\xi'+\eta'|^{-|\alpha+\beta|}. \tag{2.9}$$

Since $L_\Lambda(x, \xi)M(x, \xi)=I$, we have by Taylor expansion,

$$R^{(2)}(x, \xi)=\text{os-}\iint \sum_{\substack{k=1\\ -\phi^{(2)}(\xi)}}^{2} e^{-iy\eta}\chi_k(\xi', \eta')L_A(x, \xi+\eta)M(x+y, \xi)\phi^{(2)}(\xi)dyd\eta$$

$$=\phi^{(2)}(\xi)\sum_{|\gamma|=1}\text{os-}\iiint_0^1 e^{-iy\eta}D_\eta^\gamma(\chi_1 L_A)M_{(\gamma)}(x+\theta y, \xi)dyd\eta d\theta$$

$$+\sum_{1\le|\gamma|<l+1}\gamma!^{-1}D_\eta^\gamma(\chi_2(\xi', \eta')L_A(x, \xi+\eta))|_{\eta=0}M_{(\gamma)}(x, \xi)\phi^{(2)}(\xi)$$

$$+\phi^{(2)}(\xi)\sum_{|\gamma|=l+1}\gamma!^{-1}\text{os-}\iiint_0^1 e^{-iy\eta}D_\eta^\gamma(\chi_2 L_A)M_{(\gamma)}(x+\theta y, \xi)dyd\eta d\theta$$

$$\equiv I_1+I_2+I_3.$$

For $I_1(x, \xi)$, we have for any positive integers p and q,

$$I_{1(\beta)}^{(\alpha)}(x, \xi)=\sum_{|\gamma|=1}\sum\binom{\alpha}{\hat{\alpha}}\binom{\beta}{\hat{\beta}}\text{os-}\iiint_0^1\frac{e^{-iy\eta}(\langle D_\eta\rangle^{2p}D_\eta^\gamma\chi_1 L_{A(\hat{\beta})}^{(\hat{\alpha})})}{\langle y\rangle^{2p}\langle\eta\rangle^{2q}}$$
$$\times\langle D_y\rangle^{2q}D_x^{\beta-\hat{\beta}}M_{(\gamma)}^{(\alpha-\hat{\alpha})}(x+\theta y, \xi)\phi^{(2)}dyd\eta d\theta.$$

Noting that for $|\eta|\ge 4^{-1}|\xi|$ and $|\gamma|=1$

$$|\langle D_\eta\rangle^{2p}D_\eta^\gamma\chi_1(\xi', \eta')L_{A(\hat{\beta})}^{(\hat{\alpha})}(x, \xi+\eta)|\le C,$$

and taking account of (2.7), we obtain

$$|I_{2(\beta)}^{(\alpha)}|\le C\frac{\langle\xi\rangle^{1-\kappa(1+|\alpha|)+(1-\kappa)|\beta|}}{h+\langle\xi\rangle^\kappa}\iint_{|\eta|\ge 4^{-1}|\xi|}\frac{\langle\xi\rangle^{2q(1-\kappa)+\kappa(1+|\alpha|)}}{\langle y\rangle^{2p}\langle\xi\rangle^{2q}}dyd\eta$$
$$\le C\frac{\langle\xi\rangle^{1-\kappa(1+|\alpha|)+(1-\kappa)|\beta|}}{h+\langle\xi\rangle^\kappa}$$

if we choose p and q as $2p\ge n+2$ and $2p\ge n+2+2q(1-\kappa)+\kappa(1+|\alpha|)$. It follows from (2.7) and (2.9) that I_2 satisfies (2.10). For I_3, we have

$$I_{3(\beta)}^{(\alpha)}=\sum_{|\gamma|=l+1}\gamma!^{-1}\binom{\alpha}{\hat{\alpha}}\binom{\beta}{\hat{\beta}}\text{os-}\iiint_0^1\frac{e^{-iy\eta}(1-\theta)^l(\langle D_\eta\rangle^{2p}D_\eta^\gamma D_\xi^{\hat{\alpha}}\chi_2 L_{A(\hat{\beta})})}{\langle y\rangle^{2p}\langle\eta\rangle^{2q}}$$
$$\times(\langle D_y\rangle^{2q}D_\xi^{\alpha-\hat{\alpha}}M_{(\gamma+\beta-\hat{\beta})}(x+\theta y, \xi)\phi^{(2)})dyd\eta d\theta.$$

Noting that for $\xi\in\text{supp}\,\phi^{(2)}$ and $(\xi, \eta)\in\text{supp}\,\chi_2$,

$$|\langle D_\eta\rangle^{2p}D_\eta^\gamma D_\xi^{\hat{\alpha}}\chi_2(\xi, \eta)L_{A(\hat{\beta})}(x, \xi+\eta)|\le C\langle\xi\rangle^{1-|\hat{\alpha}+\hat{\beta}|},$$

and choosing p and q as $2p\ge n+2$ and $2q\ge n+2$, we obtain

$$|I_{3(\beta)}^{(\alpha)}|\le C\frac{\langle\xi\rangle^{1-\kappa(1+|\alpha|)+(1-\kappa)|\beta|+2q(1-\kappa)-\kappa l}}{h+\langle\xi\rangle^\kappa}$$
$$\le C\frac{\langle\xi\rangle^{1-\kappa(1+|\alpha|)+(1-\kappa)|\beta|}}{h+\langle\xi\rangle^\kappa}$$

if we take l as $\kappa l \geq 2q(1-\kappa)$. This completes the proof of our proposition.

Proposition 2.6. *Assume that* [A.I] [A.II] *and* [A.III] *are valid. Let be* $\Lambda = \delta(T-x_0)(h+\langle D'\rangle^\kappa)$, *and* $s_0^{-1} \geq \kappa > 2^{-1}$. *Then there is a right inverse* $Q(x, D)$ *of* $L_\Lambda(x, D)$, *that is, for* $|T-x_0| \leq \varepsilon_0/2$ *and for* $h \geq h_0(\delta)$,

$$L_\Lambda(x, D)Q(x, D) = I, \tag{2.10}$$

and for any $l \geq 0$ *there is a positive constant* $h_0(\delta, l)$ *such that for* $h \geq h_0(\delta, l)$,

$$\|Qu\|_{W_l(\boldsymbol{R}^{n+1})} \leq C_{\delta l} \|u\|_{W_{l-\kappa}(\boldsymbol{R}^{n+1})}, \tag{2.11}$$

where $C_{\delta l}$ *is independent of* h.

Proof. Let $\sigma(x_0)$ be a C^∞-function in R^1 as follows

$$\sigma(x_0) = \begin{cases} 1 & \text{for } |T-x_0| \leq \varepsilon_0/2, \\ 0 & \text{for } |T-x_0| \geq \varepsilon_0. \end{cases}$$

We define

$$Q'(x, D) = \sigma(x_0)\{Q^{(1)}(x, D) + Q^{(2)}(x, D)\},$$
$$R(x, D) = \sigma(x_0)\{R^{(1)}(x, D) + R^{(2)}(x, D)\}.$$

Then from Propositions 2.3 and 2.5, and from (2.4) it follows that we have

$$L_\Lambda(x, D)Q'(x, D) = I + R(x, D),$$

for $|T-x_0| \leq \varepsilon_0/2$ and for $h \geq h_0(\delta)$. Moreover it follows from (2.5), (2.8) and Lemma 1.1 that for any l,

$$\|Ru\|_{W_l(\boldsymbol{R}^{n+1})} \leq C_{\delta l} h^{1-2\kappa} \|u\|_{W_l(\boldsymbol{R}^{n+1})}, \tag{2.12}$$

for u in $W_l(\boldsymbol{R}^{n+1})$. Hence if we choose $h_0(\delta, l)$ such that

$$C_{\delta l} h_0^{1-2\kappa} \leq 1/2,$$

we have the inverse $(I + R(x, D))^{-1}$ in $W_l(\boldsymbol{R}^{n+1})$. Therefore we define

$$Q(x, D) = Q'(x, D)(I + R(x, D))^{-1},$$

which satisfies (2.10) and (2.11).

We can obtain the left inverse of $L_\Lambda(x, D)$ similarly,

Proposition 2.7. *Assume that the conditions* [A.I], [A.II] *and* [A.III] *are satisfied. Let be* $\Lambda = \delta(T-x_0)(h+\langle D'\rangle^\kappa)$ *and* $s_0^{-1} \geq \kappa > 1/2$. *Then there*

exists the left inverse of $L_\Lambda(x, D)$, $P(x, D)$, *such that for* $|T-x_0|\leq\varepsilon_0/2$ *and* $h\geq h_0(\delta)$

$$P(x, D)L_\Lambda(x, D)=I,$$

and for any l *in* R^1, *there is a positive constant* $h_0(\delta, l)$ *such that*

$$\|Pu\|_{W_l(\boldsymbol{R}^{n+1})}\leq C_{\delta l}\|u\|_{W_{l-\kappa}},$$

for u *in* $W_l(\boldsymbol{R}^{n+1})$, *where* $C_{\delta l}$ *is independent of* h *and* T.

§ 3. The existence theorem of Cauchy problem

We begin to consider the following problem,

$$(3.1) \qquad L(x, D)u(x)=f(x), \qquad \text{for } |T-x_0|\leq\varepsilon_0/2,$$

where $f(x)$ in $W^\Lambda(\boldsymbol{R}^{n+1})$, $\Lambda=\delta(T-x_0)(h+\langle D'\rangle^\kappa)$ and $s_0^{-1}\geq\kappa>1/2$. Then it follows from Proposition 2.6 that we have,

Theorem 3.1. *Assume that* [A.I], [A.II] *and* [A.III] *are valid. Let be* $\Lambda=\delta(T-x_0)(h+\langle D'\rangle^\kappa)$ *and* $s_0^{-1}\geq\kappa>1/2$. *Then for any* f *in* $W_{l-\kappa}^\Lambda(\boldsymbol{R}^{n+1})$ *there exists a solution* u *in* $W_l^\Lambda(R^{n+1})$ *of* (3.1) *for* $|T-x_0|\leq\varepsilon_0/2$, *which satisfies for any* $h\geq h_0(\delta, l)$,

$$(3.2) \qquad \|u\|_{W_l^{(\Lambda)}(\boldsymbol{R}^{n+1})}\leq C_{\delta l}\|f\|_{W_{l-\kappa}^{(\Lambda)}(\boldsymbol{R}^{n+1})},$$

where ε_0 *and* $h_0(\delta, l)$ *are given in Proposition* 2.6.

Proof. We define

$$u(x)=e^{-\Lambda}Q(x, D)e^\Lambda f(x).$$

It follows from (2.10) and (2.11) that $u(x)$ satisfies (3.1) and (3.2) evidently.

Theorem 3.2. *Assume that* [A.I], [A.II] *and* [A.III] *are valid. Then if for* $\Lambda=\delta(T_0-x_0)(h+\langle D'\rangle^\kappa)$, *where* $T_0\geq\hat{x}_0$, $\delta>0$ *and* $s_0^{-1}\geq\kappa>1/2$, $u(x)$ *is in* $W_1^{(\Lambda)}(\boldsymbol{R}^{n+1})$ *and* $\operatorname{supp} Lu$ *is in* $[\hat{x}_0, \infty)\times\boldsymbol{R}^n$, *we have*

$$(3.3) \qquad \operatorname{supp} u\subset[\hat{x}_0, \infty)\times\boldsymbol{R}^n.$$

Proof. We put $Lu=f$. Since u is in $W_1^{(\Lambda_0)}(\boldsymbol{R}^{n+1})$, f is in $W_0^{(\Lambda_0)}(\boldsymbol{R}^{n+1})$. Moreover since $\operatorname{supp} f$ is in $[\hat{x}_0, \infty)\times\boldsymbol{R}^n$, f is in $W_0^{(\Lambda)}(\boldsymbol{R}^{n+1})$ for $\Lambda=\delta(T-x_0)(h+\langle D'\rangle^\kappa)$, $T\leq\hat{x}_0$ and $h\geq h_0$. Hence it follows from (2.12) and (2.13) that

$$u(x)=e^{-\Lambda}Pe^\Lambda Lu=e^{-\Lambda}P(x, D)e^\Lambda f(x),$$

for $|T-x_0|\leq\varepsilon_0/2\equiv\varepsilon_1$ and

$$\|u\|_{W_0^{(\Lambda)}([T-\varepsilon_1, T+\varepsilon_1]\times R^n)}\leq C\|f\|_{W_0^{(\Lambda)}(R^{n+1})}$$

where $\Lambda=\delta(T-x_0)(h+\langle D'\rangle^\kappa)$ and C is independent of h. On the other hand we have

$$\|f\|^2_{W_0^{(\Lambda)}}=\int_{\hat{x}_0}^{\infty}dx_0\int|e^{\delta(T-x_0)(h+\langle D'\rangle^\kappa)}f(x)|^2dx'$$

is bounded for $h\to\infty$, because of $\delta(T-x_0)\leq 0$ for $x_0\geq\hat{x}_0$. Therefore $\|u\|_{W_0^{(\Lambda)}}$ is also bounded for $h\to\infty$. This implies that $u(x)=0$ for $x_0\in[(T-\varepsilon_0)/2, T]$. Since T is arbitrary in $(-\infty, x_0]$, we obtain (3.3).

Lemma 3.3. *Let be* $\Lambda=\delta(T-x_0)(h+\langle D'\rangle^\kappa)$ *and* $f(x)$ *in* $W_l^{(\Lambda)}((-\infty, T_1]\times R^n)$, *where l is a positive integer. Then there is an extension* $\tilde{f}(x)$ *of* $f(x)$ *such that* $\tilde{f}(x)=f(x)$ *for* $x_0\leq T_1$ *and*

$$\|\tilde{f}\|_{W_l^{(\Lambda)}(R^{n+1})}\leq C_l\|f\|_{W_l^{(\Lambda)}((-\infty, T_1]\times R^n)}, \tag{3.4}$$

where C_l is independent of h.

Proof. We put

$$\tilde{f}(x)=\begin{cases} f(x), & x_0\leq T_1,\\ \sum_{i=1}^{l} a_i f(T_1-i(T_1-x_0)), & x_0>T_1,\end{cases}$$

where

$$\sum_{i=1}^{l} a_i i^k=1,\quad k=0,\cdots,l-1.$$

Then $\tilde{f}(x)$ is in $W_l^{(\Lambda)}(R^{n+1})$ and satisfies

$$\begin{aligned}\|\tilde{f}\|_{W_l^{(\Lambda)}(x_0\geq T_1)}&=\left\|e^{\Lambda(\delta(T-x_0))}\sum_{i=1}^{l}a_i f(T_1-i(T_1-x_0))\right\|_{W_l(x_0\geq T_1)}\\ &\leq\sum_{i=1}^{l}\|e^{\Lambda(\delta(T-T_1+i^{-1}(x_0-T_1)))}a_i f\|_{W_l((-\infty, T_1)\times R^n)}\\ &\leq C\|e^{\Lambda(\delta(T-x_0))}f\|_{W_l((-\infty, T_1]\times R^n)},\end{aligned}$$

because of $\delta\{T-T_1+i^{-1}(x_0-T_1)\}(h+\langle\xi'\rangle^\kappa)\leq\delta(T-x_0)(h+\langle\xi'\rangle^\kappa)$ for $x_0\leq T_1$ and $i=1,\cdots,l$.

Theorem 3.4. *Assume that* [A.I], [A.II] *and* [A.III] *are valid. Let be* $\Lambda=\delta(T-x_0)(h+\langle D'\rangle^\kappa)$, $s_0^{-1}\geq\kappa>1/2$, $\delta>0$, *l an integer* (≥ 0) *and* $\hat{x}_0$ *in* R^1.

Then there are positive numbers ε_1 *(independent of* δ *and* l*) and* $h_0(\delta, l)$ *such that for any* $f(x)$ *in* $W_l^{(\Lambda)}((-\infty, \hat{x}_0+\varepsilon_1]\times \boldsymbol{R}^n)$ *with* $\operatorname{supp} f$ *in* $\{x_0 \geq \hat{x}_0\}$, *where* $|T-\hat{x}_0|\leq \varepsilon_1$, *there exists a solution* $u(x)$ *in* $W_l^{(\Lambda)}((-\infty, \hat{x}_0+\varepsilon_1]\times \boldsymbol{R}^n)$ *which satisfies* (3.1) *for* $|x_0-\hat{x}_0|\leq \varepsilon_1$ *and*

$$\operatorname{supp} u \subset \{x_0 \geq \hat{x}_0\}, \tag{3.5}$$

$$\|u\|_{W_l^{(\Lambda)}([\hat{x}_0, \hat{x}_0+\varepsilon_1]\times \boldsymbol{R}^n)} \leq C\|f\|_{W_l^{(\Lambda)}([\hat{x}_0, \hat{x}_0+\varepsilon_1]\times \boldsymbol{R}^n)}. \tag{3.6}$$

Proof. We put $\varepsilon_1=\varepsilon_0/2$. It follows from Lemma 3.3 that we have an extension $\tilde{f}(x)$ of $f(x)$ satisfying (3.4) with $T_1=x_0+\varepsilon_1$. Then we obtain a solution $\tilde{u}(x)$ of (3.1) for $\tilde{f}(x)$ by virtue of Theorem 3.1, which satisfies from (3.2)

$$\|\tilde{u}\|_{W_l^{(\Lambda)}(\boldsymbol{R}^{n+1})} \leq C_l \|\tilde{f}\|_{W_l^{(\Lambda)}(\boldsymbol{R}^{n+1})}.$$

We define $u(x)=\tilde{u}(x)|_{x_0\leq T_1}$. It follows from Theorem 3.2 and (3.4) that $u(x)$ satisfies (3.5) and (3.6), and Theorem 3.2 also assures that the solution u does not depend on the extension of f.

Next we consider the Cauchy problem for $L(x, D)$ with non-zero initial data,

$$\begin{cases} L(x, D)u(x)=f(x), & \hat{x}_0<x_0<\hat{x}_0+\varepsilon_2, \\ u(\hat{x}_0, x')=u_0(x'). \end{cases} \tag{3.7}$$

We give the initial data $u_0(x')$ in $W^{(\Lambda_0)}(\boldsymbol{R}^n)=\{v \in L^2(\boldsymbol{R}^n): e^{\Lambda_0}v \in L^2(\boldsymbol{R}^n)\}$, where $\Lambda_0=c_0(h+\langle D'\rangle^\kappa)$, $c_0>0$ and $f(x) \in W_l^{(\Lambda_0)}([x_0, x_0+\varepsilon_1]\times \boldsymbol{R}^n)$, l an integer ≥ 0. Then we have,

Theorem 3.5. *Assume that* [A.I], [A.II] *and* [A.III] *are valid. Let be* $s_0^{-1}\geq \kappa > 1/2$ *and* $\varepsilon_2=\varepsilon_1/2$, *where* ε_1 *is given in Theorem* 3.4. *Then for any* $u_0(x')$ *in* $W^{(\Lambda_0)}(\boldsymbol{R}^n)$ *and* $f(x)$ *in* $W_l^{(\Lambda_0)}([x_0, x_0+\varepsilon_2]\times \boldsymbol{R}^n)$, $\Lambda_0=c_0(h+\langle D'\rangle^\kappa)$, $c_0>0$, *there exists a solution* $u(x)$ *in* $W_l^{(\Lambda_1)}([\hat{x}_0, \hat{x}_0+\varepsilon_2]\times \boldsymbol{R}^n)$ *of* (3.7), *where* $\Lambda_1=(c_0/2)(h+\langle D'\rangle^\kappa)$.

Proof. We take $\Lambda=\delta(T-x_0)(h+\langle D'\rangle^\kappa)$, where $T=\hat{x}_0+\varepsilon_1$ and $\delta\varepsilon_1=c_0$. Then $f(x)$ is also in $W_l^{(\Lambda)}([\hat{x}_0, \hat{x}_0+\varepsilon_2]\times \boldsymbol{R}^n)$ evidently. We can find $w(x) \in W_l^{(\Lambda)}([\hat{x}_0, \hat{x}_0+\varepsilon_2]\times \boldsymbol{R}^n)$ such that

$$L(x, D)w(x)-f(x)=g(x) \qquad \text{for } x_0<\hat{x}_0+\varepsilon_2 \tag{3.8}$$

where $g(x)$ is in $W_l^{(\Lambda)}((-\infty, x_0+\varepsilon_2]\times \boldsymbol{R}^n)$ and $\operatorname{supp} g \subset [\hat{x}_0, \hat{x}_0+\varepsilon_2]\times \boldsymbol{R}^n$. Therefore from Theorem 3.4 we have a solution $v(x)$ in $W_l^{(\Lambda)}((-\infty, \hat{x}_0+\varepsilon_1]\times \boldsymbol{R}^n)$ of the equation $L(x, D)v(x)=g(x)$ for $x_0<\hat{x}_0+\varepsilon_2$. Then supp $v\subset[\hat{x}_0, \hat{x}_0+\varepsilon_1]\times \boldsymbol{R}^n$ and the solution $v(x)$ is in $W_l^{(\Lambda)}([\hat{x}_0, \hat{x}_0+\varepsilon_2]\times\boldsymbol{R}^n)$. Hence we have the solution $u(x)=w(x)+v(x)$ of (3.7) which is in $W_l^{(\Lambda)}([\hat{x}_0, \hat{x}_0+\varepsilon_2]\times\boldsymbol{R}^n)$.

Thus we have obtained the local existence theorem of the Cauchy problem for $L(x, D)$. Noting that the interval $[\hat{x}_0, \hat{x}_0+\varepsilon_2]$ of the existence of the solution does not depend on the data $\{u(x', 0), f(x)\}$, we can get the global solution of the Cauchy problem for $L(x, D)$. We consider the following Cauchy problem,

$$(3.9)\qquad \begin{cases} L(x, D)u(x)=f(x), & x\in[\tau, \bar{T}]\times\boldsymbol{R}^n,\\ u(0, x')=u_0(x'), & x'\in\boldsymbol{R}^n,\end{cases}$$

where $\bar{T}>\tau$.

Theorem 3.6. *Assume that* [A.I], [A.II] *and* [A.III] *are valid and* $s_0^{-1}\geq\kappa>1/2$. *Then for any* $f(x)$ *in* $W_l^{(\Lambda_0)}([\tau, \bar{T}]\times\boldsymbol{R}^n)$, ($l$ *an imteger* ≥ 0), *and for any* $u_0(x')$ *in* $W^{(\Lambda_0)}(\boldsymbol{R}^n)$, *where* $\Lambda_0=c_0(h+\langle D'\rangle^\kappa)$, $(c_0>0)$, *there exists a solution* $u(x)$ *of* (3.9) *which is in* $W_l^{(\Lambda_1)}([\tau, \bar{T}]\times\boldsymbol{R}^n)$, *where* $\Lambda_1=c_1(h+\langle D'\rangle^\kappa)$, $c_1=c_0/2^N$, $N=[(\bar{T}-\tau)/\varepsilon_2]+1$, *and* ε_2 *given in Theorem* 3.5.

Proof. Let ε_2 be a positive number in Theorem 3.5. We take N as $N\varepsilon_2\geq\bar{T}-\tau$, that is, $N=[(\bar{T}-\tau)/\varepsilon_2]+1$. We put $t_0=\tau$, $t_j=j\varepsilon_2+\tau$ $(j=1, \cdots, N-1)$ and $t_N=\bar{T}$. We denote by $u^{(0)}(x)$ the solution of the Cauchy problem,

$$\begin{cases} L(x, D)u^{(0)}(x)=f(x), & t_0<x_0<t_1,\\ u^{(0)}(t_0, x')=u_0(x'). \end{cases}$$

Since $u_0(x')$ is in $W^{(\Lambda_0)}(\boldsymbol{R}^n)$, it follows from Theorem 3.5 that $u^{(0)}(x)$ is in $W^{(\Lambda^{(1)})}([t_0, t_1]\times\boldsymbol{R}^n)$, where $\Lambda^{(1)}=(c_0/2)(h+\langle D'\rangle^\kappa)$. Next we define $u^{(j)}(x)$ $(j=1, \cdots, N-1)$ inductively as the solution of the following problem,

$$(3.10)\qquad \begin{cases} L(x, D)u^{(j)}(x)=f(x), & t_j<x_0<t_{j+1},\\ u^{(j)}(t_j, x')=u^{(j-1)}(t_j, x'). \end{cases}$$

We assume that $u^{(j-1)}(t_j, x')$ is in $W^{(\Lambda^{(j)})}(\boldsymbol{R}^n)$, where $\Lambda^{(j)}=(c_0/2^j)(h+\langle D'\rangle^\kappa)$. Since $f(x)$ is in $W^{(\Lambda^{(j)})}([t_j, t_{j+1}]\times\boldsymbol{R}^n)$ it follows from Theorem 3.5 that we have the solution $u^{(j)}$ of (3.10) in $W^{(\Lambda^{(j+1)})}([t_j, t_{j+1}]\times\boldsymbol{R}^n)$. We define $u(x)$ as $u(x)=u^{(j)}(x)$, $x_0\in[t_j, t_{j+1})$. Then it follows from (3.10) that $u(x)$ is in $W^{(\Lambda_1)}([0, \bar{T}]\times\boldsymbol{R}^n)$, where $\Lambda_1=(c_0/2^N)(h+\langle D'\rangle^\kappa)$.

§ 4. The dependence domain

We shall determine the dependence domain for the Cauchy problem (0.1). Since our assumption [A.II] is not invariant under the Holmgren transform, we shall apply the Paley-Wiener method, following Sakamoto [15]. We define

$$(4.1)\quad \begin{cases} \Gamma_+=\{\theta \in \boldsymbol{R}^{n+1};\ \theta_0-\lambda_j(x,\theta')>0, \text{ for } x\in[\tau,\bar{T}]\times \boldsymbol{R}^n, j=1,\cdots,d\}, \\ \Gamma_-=-\Gamma_+, \end{cases}$$

where $\lambda_j(x,\theta')$ $(j=1,\cdots,d)$ are the eigen values of $\sum A_j(x)\theta_j$ and define also the dual cone of $\Gamma_\pm$ as follows,

$$\Gamma^0_\pm(\hat{x})=\{x\in \boldsymbol{R}^{n+1};\ (x-\hat{x})\theta\geq 0 \text{ for all } \theta\in\Gamma_\pm\}.$$

Theorem 4.1. *Assume that* [A.I], [A.II] *and* [A.III] *are valid and* $s_0^{-1}\geq \kappa>1/2$. *Let be* $\Lambda_0=c_0\langle D'\rangle^\kappa$ $(c_0>0)$. *If* $u(x)$ *is in* $W_1^{(\Lambda_0)}((-\infty,\bar{T}]\times\boldsymbol{R}^n)$, $\operatorname{supp} u\subset\{x_0\geq\tau\}$ *and* $\operatorname{supp} Lu\subset\Gamma^0_+(\hat{x})$, *where* $\hat{x}\in[\tau,\bar{T}]\times\boldsymbol{R}^n$, *then* $\operatorname{supp} u\subset\Gamma^0_+(\hat{x})$.

This theorem is a direct result of the following theorem.

Theorem 4.2. *Assume that* [A.I], [A.II] *and* [A.III] *are valid and* $s_0^{-1}\geq\kappa>1/2$. *Let be* $\Lambda=\delta(h+\langle D'\rangle^\kappa)$ *and* $T\in[\tau,\bar{T}]$. *Then for* $\theta\in\Gamma_+$ *there are positive constants* ε_1, $h_0(\delta)$ *and* C *independent of* $|\theta|$ *such that*

$$(4.2)\quad \|e^{-x\theta}u\|_{W_0^{(\Lambda)}([T-\varepsilon_1,T])}\leq C\|e^{-x\theta}Lu\|_{W_0^{(\Lambda)}([T-\varepsilon_1,T])}$$

for any $u\in W_1^{(\Lambda)}((-\infty,\bar{T}]\times\boldsymbol{R}^n)$.

Proof of Theorem 4.1. We choose $T\in(\tau,\bar{T}]$ arbitrarily and $\delta>0$ such that $\delta(T-x_0)<c_0$ for $x_0\in[T-\varepsilon_1,T]$, where ε_1 is given in Theorem 4.2. We note that for $\Lambda=\delta(T-x_0)(h+\langle D'\rangle^\kappa)$,

$$\begin{aligned}\|e^{-(x-\hat{x})\theta}f\|_{W_0^{(\Lambda)}[T-\varepsilon_1,T]}&=\|e^{\Lambda-(x-\hat{x})\theta}f\|_{L^2([T-\varepsilon_1,T]\times\boldsymbol{R}^n)}\\ &\leq\sum_{j=0}^\infty j!^{-1}(\delta\varepsilon_1)^j\|\langle D'\rangle^{\kappa j}e^{-(x-\hat{x})\theta+T\delta h}f\|_{L^2[T-\varepsilon_1,T]}\\ &\leq\sum_{j=0}^\infty j!^{-1}(\delta\varepsilon_1)^j\sum_{|\alpha|=[j\kappa]+1}\|e^{-(x-\hat{x})\theta+T\delta h}(D'+i\theta')^\alpha f\|_{L^2}.\end{aligned}$$

Since $W_0^{(\Lambda)}((-\infty,\bar{T}]\times\boldsymbol{R}^n)\cap C_0^\infty(\Gamma^0_+(\hat{x}))$ is dense in

$$W_0^{(\Lambda)}((-\infty,\bar{T}]\times\boldsymbol{R}^n)\cap\{\operatorname{supp} f\subset\Gamma^0_+(\hat{x})\}$$

we may assume that f is in $W_0^{(\Lambda_0)}((-\infty,\bar{T}]\times\boldsymbol{R}^n)\cap C_0^\infty(\Gamma_+(\hat{x}))$. Hence we have $\eta_0>0$ such that $-(x-\hat{x})\theta\leq-\eta_0|\theta|$ for $x\in\operatorname{supp} f$ and $\vartheta\in\Gamma_+$.

Therefore, noting that $\|D'^{\alpha}f\|^2_{L^2[T-\varepsilon_1,T]}\leq C_1^{|\alpha|}|\alpha|!^s\|f\|_{W_0^{(A_0)}(-\infty,T]}$ we have

$$\|e^{-(x-\hat{x})\theta+\delta T h}(D'+i\theta')^{\alpha}f\|_{L^2[T-\varepsilon_1,T]}\leq e^{-\eta_0|\theta|}\|(D'+\theta')^{\alpha}f\|_{L^2}$$
$$\leq C(C_1(|\theta|\eta_0)^{-1})^{|\alpha|}|\alpha|!^s\|f\|_{W_0^{(A_0)}[T-\varepsilon_1,T]}$$

where C and C_1 are independent of $|\theta|$. Hence if we take $\delta>0$ such that $\delta\varepsilon_1C_1\eta_0^{-1}<1$, we have

$$\|e^{-(x-\hat{x})\theta}u\|_{L^2[T-\varepsilon_1,T]}\leq\|e^{-(x-\hat{x})\theta}u\|_{W^{(A)}[T-\varepsilon_1,T]}$$
$$\leq C\|f\|_{W_0^{(A_0)}[T-\varepsilon_1,T]},$$

which is bounded for $|\theta|\to\infty$. This implies that $\operatorname{supp} u\cap[T-\varepsilon_1, T]\subset\Gamma_+^0(x)$. Since T is arbitrary, we have $\operatorname{supp} u\subset\Gamma_+^0(x)$.

Noting that Theorem 4.1 holds for the adjoint operator $L^{(*)}(x, D)=D_0+\sum D_jA_j(x)^{(*)}+B(x)^{(*)}$, we have,

Theorem 4.3. *Assume that* [A.I], [A.II] *and* [A.III] *are valid and* $s_0^{-1}\geq\kappa>1/2$. *Let* $u(x)$ *be in* $W_1^{(A_0)}([\tau, \bar{T}]\times\boldsymbol{R}^n)$, *where* $A_0=c_0\langle D'\rangle^{\kappa}$ $(c_0>0)$ *and* $\hat{x}\in(\tau, \bar{T}]\times\boldsymbol{R}^n$. *If* $u(x)$ *satisfies*

$$u(\tau, x')=0 \qquad \text{for } x'\in\Gamma_-(\hat{x})\cap\{x_0=\tau\},$$
$$L(x, D)u(x)=0 \qquad \text{for } x\in\Gamma_-(\hat{x})\cap\{x_0>\tau\},$$

then we have

$$u(x)=0 \qquad \text{for } x\in\Gamma_-(\hat{x})\cap\{x_0>\tau\}.$$

To prove Theorem 4.2, we need the following lemmas,

Lemma 4.4 (Strang [13]). *Let* $L_0(x, \xi)=\xi_0+\sum_{j=1}^n A_j(x)\xi_j$, *where* $A_j(x)$ *are* $d\times d$*-matrices. Then the following conditions are equivalent.*

(i) $L_0(x, \xi)$ *satisfies* [A.II] *and* [A.III], *that is, hyperbolic and uniformly diagonalizable.*

(ii) *There is a positive constant* C *independent of* $(x, \xi)\in\boldsymbol{R}^{2(n+1)}$ *such that for* $\lambda\in\boldsymbol{C}^1$ $(\operatorname{Im}\lambda\neq0)$ *and* $(x, \xi)\in\boldsymbol{R}^{2(n+1)}$,

$$|L_0(x, \xi_0+\lambda, \xi')^{-1}|\leq C|\operatorname{Im}\lambda|^{-1}. \tag{4.3}$$

Definition 4.5. *We say that* $L_0(x, \xi)=\xi_0+\sum_{j=1}^n A_j(x)\xi_j$, *where* $A_j(x)$ *are* $d\times d$*-matrices, is hyperbolic and uniformly diagonalizable with respect to* $\omega\in\boldsymbol{R}^{n+1}$ $(|\omega|=1)$, *if there is a positive constant* $C(\omega)$ *such that*

$$|L_0(x, \xi+\omega\mu)^{-1}|\leq C(\omega)|\operatorname{Im}\mu|^{-1}, \tag{4.4}$$

for $(x, \xi)\in\boldsymbol{R}^{2(n+1)}$ *and* $\mu\in\boldsymbol{C}^1$ $(\operatorname{Im}\mu\neq0)$.

We note that if $L_0(x, \xi)$ satisfies [A.II] and [A.III], $L_0(x, \xi)$ is hyperbolic and uniformly diagonalizable with respect to $\omega=(1, 0, \cdots, 0) \in \boldsymbol{R}^{n+1}$.

Lemma 4.6. *Assume that* $L_0(x, \xi)=\xi_0+\sum A_j(x)\xi_j$ *satisfies* [A.II] *and* [A.III]. *Let* Γ_+ *be a cone defined in* (4.1). *Then for any* $\omega \in \Gamma_+$ $(|\omega|=1)$, $L_0(x, \xi)$ *is hyperbolic and uniformly diagonalizable with resepct to* ω.

The proof of this lemma will be given in the appendix.

Proof of Theorem 4.2. We put $v(x)=e^{-x\theta}u(x)$. Then it suffices to prove that the following estimate, instead of (4.2) is derived,

$$\|v\|_{W_0^{(A)}[T-\varepsilon_1,T]}\leq C\|L(\theta)v\|_{W_0^{(A)}[T-\varepsilon_1,T]} \tag{4.5}$$

where $L(\theta)=e^{-x\theta}L(x, D)e^{x\theta}=L(x, D+i\theta)=L_0(x, D+i\theta)+B(x)$. Therefore putting $w(x)=e^{\Lambda}v(x)$, it suffices to derive the following estimate instead of (4.5),

$$\|w\|_{L^2([T-\varepsilon_1,T]\times\boldsymbol{R}^n)}\leq C\|L_\Lambda(\theta)w\|_{L^2([T-\varepsilon_1,T]\times\boldsymbol{R}^n)} \tag{4.6}$$

where $L_\Lambda(\theta)=e^{\Lambda}L(\theta)e^{-\Lambda}$ and $\Lambda=\delta(T-x_0)(h+\langle D'\rangle^\kappa)$. By use of Lemma 4.6 we can construct a left parametrix of $L_\Lambda(\theta)$ similarly to Proposition 2.5. In fact the symbol $L_\Lambda(\theta, x, \xi)$ of $L_\Lambda(\theta)$ is given by

$$L_\Lambda(\theta, x, \xi)=L_0(x, \xi_\Lambda+i\theta)+\delta(T-x_0)R(L_0)(x, \xi)+B_\Lambda(x, \xi),$$

where $R(L_0)(x, \xi)$ in $S_{1,0}^{(\kappa)}$ satisfies from Lemma 1.2,

$$|R(L_0)_{(\beta)}^{(\alpha)}(x, \xi)|\leq C_{\alpha\beta}(|\theta|+\langle\xi\rangle^\kappa)\langle\xi'\rangle^{-|\alpha|}.$$

Moreover, since $(\delta(h+\langle\xi'\rangle^\kappa)+\theta_0, \theta')$ is in Γ_+ for $\theta \in \Gamma_+$, it follows from Lemma 4.6 and (4.3) that we have

$$|L_0(x, \xi_\Lambda+i\theta)^{-1}|\leq C\delta^{-1}(h+\langle\xi\rangle^\kappa+|\theta|)^{-1},$$

where C is independent of h, ξ and $|\theta|$. Hence we have

$$|L_\Lambda(\theta, x, \xi)^{-1}|\leq C_\delta(h+\langle\xi\rangle^\kappa)^{-1},$$

for $|T-x_0|\leq\varepsilon_0$ and $h\geq h_0(\delta)$, where C_δ is independent of $|\theta|$, and Proposition 2.4 holds for $M(\theta, x, \xi)=L_\Lambda(\theta, x, \xi)^{-1}$ and $C_{\alpha\beta}$ in (2.3) is independent of $|\theta|$. Therefore we can construct a left parametrix of $L_\Lambda(\theta, x, D)$ similarly to Proposition 2.5 and derive the estimate (4.6).

§ 5. The case of $s_0^{-1}=\kappa=1/2$

In the case of $s_0^{-1}=\kappa=1/2$, we can obtain only a local solution of Cauchy problem of (0.1). In this case, in general the interval of existence of the solution of (0.1) depends on the data $\{f, u_0\}$. In fact, in the proof of Proposition 2.6 in order to estimate the remainder term $R^{(2)}(x, D)$ as follows,

$$\|R^{(2)}u\|_{W_l([T-\varepsilon_1,T])}\leq(1/2)\|u\|_{W_l[T-\varepsilon_1,T]}$$

we must take a parameter δ as a large number. For, from Lemma 2.1 it follows that we have $C_{\alpha\beta\delta}$ in (2.8) of Proposition 2.5, as $C_{\alpha\beta\delta}=C_{\alpha\beta}(\delta^{-1}+\varepsilon_0)$. Therefore we obtain instead of (2.12)

$$\|R^{(2)}u\|_{W_l}\leq C_l(\delta^{-1}+\varepsilon_0)\|u\|_{W_l}\leq(1/2)\|u\|_{W_l},$$

if we take δ and ε_0 such that $\delta^{-1}+\varepsilon_0\leq(2C_l)^{-1}$.

Thus we have obtained,

Theorem 5.1. *Assume that* [A.I], [A.II] *and* [A.III] *are valid and* $s_0^{-1}=\kappa=1/2$. *Then for any* $f\in W_l^{(\Lambda_0)}([\tau, \bar{T}]\times \boldsymbol{R}^n)$ *and for any* $u_0(x')\in W^{(\Lambda_0)}(\boldsymbol{R}^n)$, *where* $\Lambda_0=c_0(h+\langle D'\rangle^\kappa)$ $(c_0>0)$, *there is* $T_0\in(\tau, \bar{T}]$ *depending on* c_0 *and* l *such that we have a solution* $u(x)\in W^{(\Lambda_1)}([\tau, T_0]\times\boldsymbol{R}^n)$ *of* (0. 1), *where* $\Lambda_1=c_1(h+\langle D'\rangle^\kappa)$ *and* c_1 *is a positive number depending on* c_0 *and* l.

Remark. In this case we can also determine the dependence domain of (0.1) analogously to Theorem 4.3.

Appendix

To prove Lemma 4.6, we need the following well known lemma. We put

$$L(x, \xi)=\xi_0+\sum_{j=1}^{n} A_j(x)\xi_j, \tag{A.1}$$

$$p(x, \xi)=\det L(x, \xi). \tag{A.2}$$

Assume that $p(x, \xi)$ is hyperbolic with respect to $(1, 0, \cdots, 0)$ that is, $p(x, \xi)=\prod_{j=1}^{d}(\xi_0-\lambda_j(x, \xi'))$, where $\lambda_j(x, \xi')$ are real valued in $[0, T]\times\boldsymbol{R}^n$ and satisfies

$$\sup_{\substack{|\xi'|=1\\ x\in[\tau,T]\times\boldsymbol{R}^n\\ j=1,\cdots,d}} |\lambda_j(x, \xi')|\leq M<\infty. \tag{A.3}$$

We define Γ_+ by (0.4).

Lemma A. *Let $p(x, \xi)$ be a polynomial of degree d defined by* (A.2). *Assume that $p(x, \xi)$ is hyperbolic with respect to* $(1, \cdots, 0)$ *and satisfies* (A.3). *Then for any $\omega \in \Gamma_+$ with $|\omega|=1$ $p(x, \xi)$ is hyperbolic with respect to ω, that is,*

$$p(x, \xi+\omega\mu)=p(x, \omega)\prod_{j=1}^{d}(\mu-\mu_j(x, \xi, \omega)),$$

where $\mu_j(x, \xi, \omega)$ are real valued in $[\tau, T]\times \boldsymbol{R}^n \times\{\Gamma_+\cap S^n\}$ and satisfies

$$\text{(A.4)} \qquad \sup_{\substack{|\xi|=1 \\ x\in[\tau,T]\times \boldsymbol{R}^n \\ j=1,\cdots,d}} |\mu_j(x, \xi, \omega)|\leq M(\omega)<\infty.$$

Proof. We have

$$p(x, \xi+\omega\mu)=\det L(x, \xi+\omega\mu)=\prod_{j=1}^{d}(\xi_0+\omega_0\mu-\lambda_j(x, \xi'+\omega'\mu)).$$

We shall prove that the following equation in μ,

$$\text{(A.5)} \qquad f_j(x, \xi+\omega\mu)=\xi_0+\omega_0\mu-\lambda_j(x, \xi'+\omega'\mu)=0$$

has at least one real root. In fact,

$$f_j(x, \xi+\omega\mu)=\mu\{\omega_0-\lambda_j(x, \omega')+o(\mu)\}\to\pm\infty \qquad \text{for } \mu\to\pm\infty,$$

because of $\omega \in \Gamma_+$. Denote by $\mu_j(x, \xi, \omega)$ a root of the equation $f_j(x, \xi+\omega\mu)=0$, $j=1, \cdots, d$. Then we have d number of roots of the equation $p(x, \xi+\omega\mu)=0$. Therefore the equation $f_j(x, \xi+\omega\mu)=0$ has only one real root μ_j and f_j can be factorized

$$\text{(A.6)} \qquad f_j(x, \xi+\omega\mu)=g_j(x, \xi, \omega)(\mu-\mu_j(x, \xi, \omega)),$$

where g_j satisfies from the fact that $|\det L(x, \omega)|\geq\delta_1>0$,

$$\text{(A.7)} \qquad |g_j(x, \xi, \omega)|\geq\delta>0.$$

It is evident that μ_j and g_j are homogeneous degree one in ξ and zero respectively. Next we shall prove that μ_j satisfies (A.4). Assume that there are j_0 and a sequence $\{x^{(q)}, \xi^{(q)}\}$ in $[\tau, T]\times S^n$ and $\omega \in \Gamma_+$ such that $\mu^{(q)}=\mu_{j_0}(x^{(q)}, \xi^{(q)}, \omega)\to\infty$, for $q\to\infty$. Then since

$$\xi_0^{(q)}+\omega_0\mu^{(q)}-\mu_{j_0}(x^{(q)}, \xi^{(q)}+\omega'\mu^{(q)})=0,$$

we have

$$\begin{aligned} 0&=\lim_{q\to\infty}\{\xi_0^{(q)}\mu^{(q)-1}+\omega_0-\lambda_{j_0}(x^{(q)}, \omega'+\xi'^{(q)}\mu^{(q)-1})\} \\ &=\lim_{q\to\infty}(\omega_0-\lambda_{j_0}(x^{(q)}, \omega')) \end{aligned}$$

which contradicts to the fact that $\omega \in \Gamma_+$.

Proof of Lemma 4.5. Since $L(x, \xi)$ is uniformly diagonalizable, it follows from Lemm 4.3 that we can factorize

$$L(x, \xi)^{-1} = \sum_{j=1}^{d} \frac{a_j(x, \xi')}{\xi_0 - \lambda_j(x, \xi')}$$

where $a_j(x, \xi')$ are homogeneous degree zero in ξ' and satisfies

$$\sup_{\substack{x \in [\tau, T] \times R^n \\ |\xi'|=1 \\ j=1,\cdots,d}} |a_j(x, \xi')| \leq M_0 < \infty. \tag{A.8}$$

Hence we have

$$L(x, \xi + \omega\mu)^{-1} = \sum_{j=1}^{d} \frac{a_j(x, \xi' + \omega'\mu)}{\xi_0 + \omega_0\mu - \lambda_j(x, \xi' + \omega'\mu)}.$$

On the other hand we expand into partial fraction,

$$L(x, \xi + \omega\mu)^{-1} = \sum_{j=1}^{r} \frac{b_j(x, \xi, \mu)}{(\mu - \tilde{\mu}_j(x, \xi))^{m_j}}$$

where $\tilde{\mu}_j$ $(j=1, \cdots, r)$ are rearranged $\mu_1, \cdots, \mu_d$ such that $\tilde{\mu}_1 < \tilde{\mu}_2 < \cdots < \tilde{\mu}_r$ and r and m_j may depend on (x, ξ), and $b_j(x, \xi, \mu)$ is homogeneous degree $m_j - 1$ in (ξ, μ) which satisfies

$$b_j(x, \xi, \tilde{\mu}_j(x, \xi)) \neq 0. \tag{A.9}$$

We shall prove that $m_j = 1$. If $m_j > 1$ for some (j, x, ξ), then

$$\begin{aligned} b_j(x, \xi, \tilde{\mu}_j(x, \xi, \omega)) &= L(x, \xi + \omega\mu)^{-1}(\mu - \tilde{\mu}_j)^{m_j}|_{\mu=\mu_j} \\ &= \lim_{\mu \to \tilde{\mu}_j} \sum_{l=1}^{d} \frac{a_l(x, \xi' + \omega'\mu)(\mu - \tilde{\mu}_j)^{m_j}}{\xi_0 + \omega_0\mu - \lambda_l(x, \xi' + \omega'\mu)} \\ &= 0, \end{aligned}$$

which contradicts to (A.9). For, the equation (A.5) $f_j = 0$ has only a simple real root μ_j. Next we shall prove

$$\sup_{\substack{x \in [\tau, T] \times R^n \\ |\xi'|=1 \\ j=1,\cdots,d}} |b_j(x, \xi')| \leq M < \infty. \tag{A.10}$$

Assume that there is a sequence $(x^{(q)}, \xi^{(q)}) \in [\tau, T] \times S^n$ such that for j_0,

$$|b_{j_0}(x^{(q)}, \xi^{(q)})| \to \infty, \quad (q \to \infty). \tag{A.11}$$

If it is necessary, by taking the subsequence of $(x^{(q)}, \xi^{(q)})$, we may assume that $\det L(x, \xi+\omega\mu)=\det L(x, \omega) \prod_{j=1}^{r}(\mu-\tilde{\mu}_j(x^{(q)}, \xi^{(q)})^{m_j}$ where r and m_j do not depend on q, that is,

(A.12) $$|\tilde{\mu}_j(x^{(q)}, \xi^{(q)})-\tilde{\mu}_{j_0}(x^{(q)}, \xi^{(q)})|>0 \quad (j\neq j_0).$$

We put

$$\mu^{(q)}=\tilde{\mu}_{j_0}(x^{(q)}, \xi^{(q)})+\varepsilon^{(q)},$$

$$\varepsilon^{(q)}=\frac{\prod_{j\neq j_0} |\tilde{\mu}_j(x^{(q)}, \xi^{(q)})-\tilde{\mu}_{j_0}(x^{(q)}, \xi^{(q)})|}{\sum_{j=1}^{r}|b_j(x^{(q)}, \xi^{(q)})|^2}.$$

Then we have

(A.13)
$$\begin{aligned} L(x^{(q)}, \xi^{(q)}+\omega\mu^{(q)})^{-1} &= \sum_{j=1}^{r} \frac{b_j(x^{(q)}, \xi^{(q)})}{\mu^{(q)}-\tilde{\mu}_j(x^{(q)}, \xi^{(q)})} \\ &= \sum_{j=1}^{r} \frac{\tilde{a}_j(x^{(q)}, \xi'^{(q)}+\omega'\mu^{(q)})}{\xi_0+\omega_0\mu^{(q)}-\tilde{\lambda}_j(x^{(q)}, \xi'^{(q)}+\omega'\mu^{(q)})}, \end{aligned}$$

where $\tilde{\lambda}_j(x^{(q)}, \xi'^{(q)}+\omega'\mu^{(q)})$ $(j=1, \cdots, r)$ is a rearrangement of $\tilde{\lambda}_j$ $(j=1, \cdots, d)$ as $\tilde{\lambda}_1<\cdots<\tilde{\lambda}_r$, that is, we can choose a subsequence of $(x^{(q)}, \xi^{(q)})$ such that the multiplicity of $\lambda_j(x^{(q)}, \xi'^{(q)}+\omega'\mu^{(q)})$ is independent of q. Then we can factorize

$$\xi_0+\omega_0\mu^{(q)}-\lambda_j(x^{(q)}, \xi'^{(q)}+\omega'\mu)=g_j(\mu^{(q)})(\mu^{(q)}-\tilde{\mu}_j(x^{(q)}, \xi^{(q)})).$$

Here we note from (A.7),

(A.14) $$\inf_{q\in N} |g_j(\mu^{(q)})|\neq 0.$$

From (A.11) and (A.4) we have

$$\varepsilon^{(q)}\to 0, \quad (q\to\infty)$$
$$\mu^{(q)}-\tilde{\mu}_{j_0}(x^{(q)}, \xi^{(q)})\to 0, \quad (q\to\infty).$$

Therefore we have from (A.13) and (A.14),

$$\begin{aligned} b_{j_0}(x^{(q)}, \xi^{(q)}) = &-\varepsilon^{(q)} \sum_{j\neq j_0} \frac{b_j(x^{(q)}, \xi^{(q)})}{\mu^{(q)}-\tilde{\mu}_j(x^{(q)}, \xi^{(q)})} \\ &+\varepsilon^{(q)} \sum_{j\neq j_0} \frac{\tilde{a}_j(x^{(q)}, \xi'^{(q)})}{g_j(\mu^{(q)})(\mu^{(q)}-\tilde{\mu}_j(x^{(q)}, \xi'^{(q)})} \\ &+\tilde{a}_{j_0}(x^{(q)}, \xi'^{(q)})g_j(\mu^{(q)})^{-1}, \end{aligned}$$

where $\tilde{a}_j(x^{(q)}, \xi'^{(q)})$ is a rearrangement of $a_j(x^{(q)}, \xi'^{(q)})$ corresponding to $\tilde{\lambda}_j(x^{(q)}, \xi'^{(q)})$. Hence we obtain

$$\lim_{q\to\infty} \{b_{j_0}(x^{(q)}, \xi'^{(q)}) - \tilde{a}_{j_0}(x^{(q)}, \xi'^{(q)}) g_{j_0}(\mu^{(q)})^{-1}\} = 0$$

which contradicts to (A.8), (A.14) and (A.11). Thus we have proved (A.10). It is evident that (A.10) implies (4.4).

References

[1] M. Bronshtein, The parametrix of the Cauchy problem for hyperbolic operators with characteristics of variable multiplicity, Trudy Moscov. Mat. Obšč., **41** (1980), 83–99.

[2] A. P. Caldéron and R. Vaillancourt, A class of bounded pseudo-differential operators, Proc. Nat. Acad. Sci. USA, **69** (1972), 1185 1187.

[3] K. O. Friedrichs, Symmetric hyperbolic systems of linear differential equations, Comm. Pure Appl. Math. **7** (1954), 354–392.

[4] ——, Pseudo-differential operators, Lecture Notes in Courant Inst., 1968.

[5] L. Hörmander, Pseudo-differential operators and hypoelliptic equations, Proc. Symposium on Singular Integral, Amer. Math. Soc., 1967, 138–183.

[6] Y. Ya. Ivriĭ and V. M. Petkov, Necessary conditions for the Cauchy problem for non-strictly hyperbolic equations to be well-posed, Russian Math. Survey **29** (1974), 1–70.

[7] N. Iwasaki, The Cauchy problem for effectively hyperbolic equations, (a special case), J. Math. Kyoto Univ. **23** (1983), 503–562.

[8] K. Kajitani, Strongly hyperbolic systems with variable coefficients, Publ. RIMS Kyoto Univ. **9** (1974), 597–612.

[9] ——, Cauchy problem for nonstrictly hyperbolic systems in Gevrey class, J. Math. Kyoto Univ. **23** (1983), 599–616.

[10] K. Kasahara and M. Yamaguti, Strongly hyperbolic systems of linear partial differential equations with constant coefficients, Mem. College Sci. Kyoto Univ. Ser. A **33** (1960), 1–23.

[11] H. O. Kreiss, Über die Stabilitatsdefinition für Differenzengleichungen die partialle Differentialgleichungen approximieren, B.I.T. **2** (1962), 153–181.

[12] H. Kumano-go, Pseudo-differential operators, MIT Press, Cambridge, 1982.

[13] S. Mizohata, Le problème de Cauchy pour les systèmes hyperboliques et paraboliques, Mem. College Sci., Kyoto Univ. **32** (1959), 181–212.

[14] I. G. Petrovski, Über das Cauchysche Problem für ein System linearen partialler Differentialgleichungen im Gebiete der nicht-analytischen Function, Bull. Univ. Etat Moscow (1938), 1–74.

[15] R. Sakamoto, Hyperbolic Boundary Value Problems, Cambridge Univ. Press, 1978.

[16] G. Strang, On strongly hyperbolicity, J. Math. Kyoto Univ. **6** (1967), 397–417.

[17] J. Vaillant, Remarques sur les systèmes fortement hyperboliques, J. Math. Pures Appl. **50** (1971), 24–51.

Institute of Mathematics
University of Tsukuba
Ibaraki 305
Japan

Taniguchi Symp. HERT
Katata 1984, pp. 125–141

Quasi-Positivity for Pseudodifferential Operators and Microlocal Energy Methods

Kiyômi KATAOKA

Abstract

It is well-known that analyticity and positivity for Hermite kernels are deeply connected. For example, the Hermite positivity is preserved under analytic continuations. Recently, in [10], we combined these facts with the theory of microfunctions and pseudodifferential operators. In particular, the quasi-positivity for Hermite pseudodifferential operators is the key notion. In this paper, we survey the positivity and the quasi-positivity in the microlocal analysis with some applications.

Introduction

Let X be an abstract set. Then, a $\boldsymbol{C}$-valued function $K(x, u)$ defined on $X\times X$ is said to be an Hermite kernel if $K(x, u)=\overline{K(u, x)}$ for every x, $u \in X$. Moreover $K(x, u)$ is said to be positive if $(K(x_i, x_j))_{i,j}$ is positive semi-definite for every $N=1, 2, \cdots$ and every $x_1, \cdots, x_N \in X$. We employ the notation "$K\gg 0$" for a positive Hermite kernel K. The following is a simple, but basic fact concerning positive Hermite kernels.

Proposition 1. *If $K_1\gg 0$ and $K_2\gg 0$, then $K_1+K_2\gg 0$ and $K_1K_2\gg 0$. Here, the product K_1K_2 is defined by the product as scalar functions on $X\times X$.*

Hence, the Hermite positivity induces an order structure in the real algebra of Hermite kernels; tha is,

$$K_1\ll K_2 \underset{\text{definition}}{\Longleftrightarrow} K_2-K_1\gg 0.$$

Let X be a domain in $\boldsymbol{C}^n$ and X^c be the complex conjugate of X. Then, a holomorphic function $G(z, w)$ on $X\times X^c$ is said to be an analytic Hermite kernel on $X\times X$ if $G(z, \overline{w})$ is an Hermite kernel on $X\times X$. The following is a non-trivial example of analytic positive Hermite kernels.

Example. Set $X=\{z \in \boldsymbol{C};\ \mathrm{Im}\, z>0\}$. Then for any $\alpha>-1$,

$$\{(z-\overline{w})/i\}^{-1-\alpha}=\Gamma(\alpha+1)^{-1}\int_0^\infty e^{itz}\cdot\overline{e^{itw}}\, t^\alpha\, dt$$

is positive.

The most important examples of analytic positive Hermite kernels are the Bergman kernels $K_X(z, w)$ associated with X. We denote by $A^2(X)$ the Hilbert space of all the square-integrable holomorphic functions on X. Then the Bergman kernel is written as

$$K_X(z, w)=\sum_{j=1}^{\infty}\varphi_j(z)\cdot\overline{\varphi_j(w)}, \tag{1}$$

where $\{\varphi_j(z)\}_j$ is any complete orthonormal system for $A^2(X)$ ([3], [11], [12]). Therefore $K_X(z, w)$ is the reproducing kernel for $A^2(X)$. In our theory, the following fact on Bergman kernels is the most important. This is easily obtained from the Hilbert-Schmidt expansions of analytic Hermite kernels.

Proposition 2. *Let $G(z, w)$ be an analytic Hermite kernel on $X\times X$. Suppose that G is square-integrable on $X\times X$. Then, we have*

$$-\|G\|_{X\times X}\cdot K_X\ll G\ll\|G\|_{X\times X}\cdot K_X, \tag{2}$$

where $\|G\|_{X\times X}$ is the $L^2(X\times X)$-norm of G.

Another important fact on the Bergman kernels is the monotonicity with respect to domain-increasing. Hereafter, we denote by X, X' a pair of domains in $\boldsymbol{C}^n$ such that $X'\subset X$.

Theorem 3 ([3], [12]). *The Bergman kernels K_X, $K_{X'}$ satisfy the following inequality:*

$$K_{X'}(z, w)\gg K_X(z, w) \qquad \text{on } X'\times X'.$$

As a generalization of Theorem 3, we know the following beautiful result ([6], [11], [12], [7]).

Theorem 4. *Let K_1, K_2 be analytic Hermite kernels on $X\times X$. Then we have*

i) *If $K_1\ll K_2$ on $X'\times X'$, then $K_1\ll K_2$ on $X\times X$.*

ii) *Let K_3 be another Hermite kernel (not necessarily analytic). If $K_1\ll K_3\ll K_2$ on $X'\times X'$, then there exists a unique analytic Hermite kernel $\tilde{K}_3$ such that $\tilde{K}_3=K_3$ on $X'\times X'$ and $K_1\ll\tilde{K}_3\ll K_2$ on $X\times X$.*

As a direct corollary of this theorem, we know the following: Let $F_1(z), \cdots, F_N(z)$ be holomorphic functions on X'. Suppose that the "energy form"

(3) $$E(z, w)=\sum_{j=1}^{N} F_j(z)\overline{F_j(w)}$$

is analytically extensible to $X\times X$. Then each $F_j(z)$ is analytically extensible to X. Such arguments are very like the so-called energy methods. Further, this fact is generalized to the case that $E(z, w)$ is defined by integration (see [10] § 1). We state here only the C^∞- (or C^ω-) version.

Theorem 5. *Let T be an open set in $\boldsymbol{R}^m$, and let $K(z, w; t)$ be a C^∞- (or C^ω-) function on $X'\times X'\times T$. We assume the following:*

i) *For every fixed $t \in T$, $K(z, w; t)$ is an analytic positive Hermite kernel on $X'\times X'$.*

ii) *For every compact subset L of X', $K(z, w; t)$ is integrable on $L\times L\times T$ by the usual Lebesgue measure.*

Further we suppose that the energy form

(4) $$E(z, w)=\int_T K(z, w; t)dt$$

is analytically extensible to $X\times X$. Then, there exists a unique C^∞- (respectively, C^ω-) function $\tilde{K}(z, w; t)$ on $X\times X\times T$ such that $\tilde{K}$ satisfies the modified version of i) *and* ii) *(replace X' by X), and that $\tilde{K}=K$ on $X'\times X'\times T$.*

This theorem is proved by using a vector-valued version of the Radon-Nikodym theorem.

§ 1. The microlocalization of Hermite positivity

Let $f_1(x), \cdots, f_N(x)$ be hyperfunctions defined on a neighborhood of $0 \in \boldsymbol{R}^n$. Then we can consider $E(x, u)=\sum_{j=1}^{N} f_j(x)\overline{f_j(u)}$ as a positive Hermite hyperfunction-kernel, though the restriction of E to the diagonal $\{x=u\}$ is non-sense in general. However, the relationship between $E(x, u)$ and $f_j(x)$ is obscure in the category of "functions". On the other hand it becomes clear once we consider them as microfunctions at an anti-diagonal point p_0 of

$$\Delta^a(iS^*\boldsymbol{R}^{n+n})=i\{(x, u; i\xi dx+i\eta du); x=u, \xi=-\eta\neq 0\}.$$

In fact, let p_0 be $(x_0, x_0; i\xi_0(dx-du))$ with $|\xi_0|=1$. Then each $f_j(x)$ is decomposed as follows:

$$f_j(x)=F_j(x+i0\Gamma)+g_j(x),$$

where $g_j(x)$ is a hyperfunction satisfying $\mathrm{SS}(g(x, u)) \not\ni (x_0; i\xi_0)$, and $F_j(z)$ is a holomorphic function defined in

$$(5) \qquad V=\{z \in \boldsymbol{C}^n;\ |z-z_0|<r,\ \operatorname{Im} z \in \Gamma\}$$

with some small $r>0$ and

$$(6) \qquad \Gamma=\{y \in \boldsymbol{R}^n;\ \langle y, \xi_0\rangle > r(|y|^2-\langle y, \xi_0\rangle^2)^{1/2}\}.$$

Therefore we have

$$E(x, u)=\sum_{j=1}^{N} F_j(x+i0\Gamma)F_j^*(u-i0\Gamma)+R(x, u),$$

where $SS(R(x, u)) \not\ni p_0$ and $F_j^*(z)=\overline{F_j(\bar{z})}$ for every $j=1, \cdots, N$. Hence we know that $E(x, u)$ coincides microlocally at p_0 with the boundary value of one analytic Hermite kernel $K(z, w)=\sum_{j=1}^{N} F_j(z)\overline{F_j(w)}$. From this observation, we can formulate the microlocalization of Hermite positivity as follows:

Definition 6. Let p_0 be a point of $\Delta^a(iS^*\boldsymbol{R}^{n+n})$, and let $k(x, u)$ be a microfunction defined in a neighborhood of p_0. Then $k(x, u)$ is said to be an Hermite microkernel at p_0 if $k(x, u)=\overline{k(u, x)}$ holds as a microfunction at p_0. Further, $k(x, u)$ is said to be positive at p_0 if

$$k(x, u)=[K(x+i0\Gamma, \overline{u-i0\Gamma})]$$

holds at p_0 for some analytic positive Hermite kernel $K(z, w)$ on $V\times V$ (as for V, Γ, see (5) and (6)). For two Hermite microkernels $k_1(x, u)$, $k_2(x, u)$ defined at p_0, we write "$k_1 \ll k_2$ at p_0" when k_2-k_1 is positive at p_0.

Theorem 7 ([10] § 2). *The relation "$k_1 \ll k_2$" for Hermite microkernels is an order relation. That is, if $k \gg 0$ and $-k \gg 0$ at p_0, then $k=0$ at p_0. Further, if $k_1 \gg 0$ and $k_2 \gg 0$ at p_0, then $k_1+k_2 \gg 0$ at p_0.*

The n-dimensional delta-kernel $\delta(x-u)$ is a typical example of positive Hermite microkernels because it is expressed as follows:

$$\delta(x-u)=\frac{(n-1)!}{(-2\pi i)^n}\int_{S^{n-1}}\frac{d\sigma(\xi)}{(\langle x-u, \xi\rangle+i0)^n}$$

(the plane wave expansion formula), where $d\sigma(\xi)$ is the volume element of the unit sphere S^{n-1}. In fact, as seen in the example in Introduction, the defining function

$$(n-1)!\ \{i/2\pi(\langle z, \xi\rangle-\langle \overline{w}, \xi\rangle)\}^n$$

is positive on $\{\langle \operatorname{Im} z, \xi\rangle>0\}\times\{\langle \operatorname{Im} w, \xi\rangle>0\}$ for every $\xi \in S^{n-1}$.

In order to treat energy forms defined by integration, we must introduce positive Hermite microkernels with real analytic parameter. Let $f(t, x)$ be a hyperfunction depending real analytically on t, and let $P(t, x, u, D_x, D_u)$ be a pseudodifferential operator of positive Hermite type with real analytic parameter t (see § 2). Then,

$$(7) \qquad P(t, x, u, D_x, D_u)(f(t, x)\overline{f(t, u)})$$

is a typical example of positive Hermite microkernel with real analytic parameter t. Further, for the application to boundary value problems, we must consider real analyticity on t up to the boundary; such a notion can be formulated by using mild hyperfunctions and their canonical extensions (see [8]). However, we are not going to the detailed definitions. At any rate, the order relation "$\ll$" and the related theorems are generalizable to the case with real analytic parameter. Further, the following is fundamental for our theory.

Theorem 8. *Let $f(t, x)$ be a hyperfunction defined in $\{(t, x) \in \boldsymbol{R}^{m+n}; |t-t_0|<r, |x-x_0|<r\}$ depending real analytically on t. Then, $f(t, x)\overline{f(t, u)}$ is a positive Hermite microkernel at $p_0=(t_0, x_0, x_0; i\xi_0(dx-du))$ for every $\xi_0 \in S^{n-1}$ depending real analytically on t. Moreover, if $f(t, x)\overline{f(t, u)}$ is microanalytic on*

$$\{(t, x, u; i\tau dt+i\xi dx+i\eta du) \in iS^*\boldsymbol{R}^{m+2n};$$
$$t=t_0, x=u=x_0, \xi=-\eta=\xi_0, \tau \in \boldsymbol{R}^m\},$$

then $f(t, x)$ is microanalytic on

$$\{(t, x; i\tau, i\xi) \in iS^*\boldsymbol{R}^{m+n}; t=t_0, x=x_0, \xi=\xi_0, \tau \in \boldsymbol{R}^m\}.$$

Remark. This theorem is also generalized to the case that t_0 is on the boundary of the parameter space.

As the last of this section, we state the main theorem of § 2 of [10].

Theorem 9. *Let T be a bounded open subset of $\boldsymbol{R}^m$ with real analytic boundary ∂T, and let $k(t, x, u)$ be a hyperfunction defined in $\{(t, x, u) \in T\times \boldsymbol{R}^{n+n}; |x-x_0|<r, |u-x_0|<r\}$ such that k is mild on $\partial T\times\{|x-x_0|<r, |u-x_0|<r\}$. We denote by* ext $(k(t, x, u))$ *the canonical extension of k across $\{t \in \partial T\}$. Suppose that the microfunction* [ext $(k(t, x, u))$] *is a positive Hermite microkernel depending real analytically on $t \in T$ everywhere in $\overline{T}\times \{(x_0, x_0; i\xi_0(dx-du))\}$. Let $\rho(t)$ be an analytic function defined in a neighborhood of $\overline{T}$ such that $\rho(t)\geqq 0$ and $\rho(t)\not\equiv 0$ on every connected component of T. Then, the microfunction induced by*

(8) $$E(x, u)=\int_T k(t, x, u)\rho(t)dt$$

is a positive Hermite microkernel at $p_0=(x_0, x_0, ; i\xi_0(dx-du))$. *Further, if* $\mathrm{SS}(E(x, u)) \not\ni p_0$, *then* $\mathrm{ext}\,(k(t, x, u))$ *is microanalytic on a neighborhood of*

$$\{(t, x, u; i\tau dt+i\xi dx+i\eta du);\ t \in \overline{T},\ x=u=x_0,\ \xi=-\eta=\xi_0,\ \tau \in \boldsymbol{R}^m\}.$$

§ 2. The positivity for pseudodifferential operators

In this section, we introduce the pseudodifferential operators which operate on positive Hermite microkernels. Firstly we recall the symbol theory for pseudodifferential operators (see [1], [2], [5]). For a point $p_0=(z_0; \xi_0 dz) \in T^*\boldsymbol{C}^n$ with $|\xi_0|=1$, we call an open subset of $T^*\boldsymbol{C}^n$ of the following type "a conic neighborhood of p_0".

(9) $$\{(z; \xi dz) \in \boldsymbol{C}^{n+n};\ |z-z_0|<r,\ |(\xi/|\xi|)-\xi_0|<r,\ |\xi|>r^{-1}\}$$

with a small $r>0$. Then a holomorphic function $P(z, \xi)$ defined in a conic neighborhood V of p_0 is said to be a (simple) symbol at p_0 if

$$|P(z, \xi)|\leqq C_\varepsilon e^{\varepsilon|\xi|} \qquad \text{on } V$$

holds for any $\varepsilon>0$ with some constant $C_\varepsilon>0$. Then, a pseudodifferential operator $P(z, D_z)$ (here, $D_z=\partial/\partial z$) is defined as the equivalence class of symbol $P(z, \xi)$. That is, two symbols $P_j(z, \xi)$ $(j=1, 2)$ at p_0 are said to be equivalent if there exist some positive constants C, δ such that

$$|P_1(z, \xi)-P_2(z, \xi)|\leqq Ce^{-\delta|\xi|}$$

holds in a small conic neighborhood of p_0.

Definition 10. Let $p_0=(z_0, \overline{z}_0; \xi_0 dz+\overline{\xi}_0 dw)$ be a point of $T^*\boldsymbol{C}^{n+n}$ with $|\xi_0|=1$. Then a pseudodifferential operator $P(z, w, D_z, D_w)$ at p_0 is said to be of "Hermite type at p_0" if any symbol of P satisfies

$$P(z, w, \xi, \eta)\sim\overline{P(\overline{w}, \overline{z}, \overline{\eta}, \overline{\xi})} \qquad \text{at } p_0.$$

In particular, P has an Hermite symbol $(P(z, w, \xi, \eta)+\overline{P(\overline{w}, \overline{z}, \overline{\eta}, \overline{\xi})})/2$. It is clear that, when $p_0 \in \Delta^a(iS^*\boldsymbol{R}^{n+n})$, such a pseudodifferential operator operates on the stalk of Hermite microkernels at p_0.

On the other hand, as for positive Hermite symbols, we meet with some difficulty, which is caused by the unboundedness of conic neighborhoods. In fact, if $P(z, w, \xi, \eta)$ is a symbol of positive Hermite type defined in a conic neighborhood W of p_0, then P is necessarily extensible analytically to a conic neighborhood $V \times V^c$ of product type. Hence, we must consider symbols defined on some conic neighborhoods of product type.

Definition 11. Let p_0 be a point of $T^*\boldsymbol{C}^{n+n}$. Then, a simple symbol $P(z, w, \xi, \eta)$ at p_0 is said to be a symbol at p_0 of "product Hermite type" if the following conditions are satisfied:

i) P is holomorphic in $V \times V^c$, where

$$V = \{(z; \xi) \in \boldsymbol{C}^{2n}; |z - z_0| < r, |(\xi/|\xi|) - \xi_0| < r, |\xi| > r^{-1}\}. \tag{10}$$

ii) For any $\varepsilon > 0$, P has the following estimate

$$|P(z, w, \xi, \eta)| \leq C_\varepsilon e^{\varepsilon(|\xi| + |\eta|)} \quad \text{on } V \times V^c \tag{11}$$

with some constant $C_\varepsilon > 0$.

iii) $P(z, w, \xi, \eta) = \overline{P(\overline{w}, \overline{z}, \overline{\eta}, \overline{\xi})}$ on $V \times V^c$.

Further, P is said to be a symbol at p_0 of "positive Hermite type" if P satisfies i)$\sim$iii), and $P(z, \overline{w}, \xi, \overline{\eta})$ is of positive Hermite type on $V \times V$. Moreover, a symbol P at p_0 of product Hermite type is said to be of 0-class if there exist a positive number δ and a small conic neighboorhood V' like (10) such that

$$|P(z, w, \xi, \eta)| = |P(\overline{w}, \overline{z}, \overline{\eta}, \overline{\xi})| \leq C_\varepsilon e^{\varepsilon|\xi| - \delta|\eta|} \text{ on } V' \times V'^c \cap \{|\xi| \geq |\eta|\} \tag{12}$$

holds for every $\varepsilon > 0$ with some constant $C_\varepsilon > 0$.

The above notion of symbols of product Hermite type or positive Hermite type is generalized to formal symbols (cf. [1], [2]); that is, a formal symbol of product Hermite type is a double series $\sum_{j,k=0}^{\infty} P_{jk}(z, w, \xi, \eta)$ of symbols P_{jk} of product Hermite type satisfying some growth conditions. Then, we can define the symbol calculus for formal symbols of product Hermite type in the usual way ([1], [2], [5]). In particular, the operator product of two formal symbols of positive Hermite type is also positive. Further, when $p_0 \in \Delta^a(iS^*\boldsymbol{R}^{n+n})$, the pseudodifferential operators of positive Hermite type operate on the stalk of Hermite microkernels at p_0 preserving the order structure.

Example. Let $P(z, \xi)$ be a symbol defined in a conic neighborhood V of $(z_0; \xi_0 dz)$ in $T^*\boldsymbol{C}^n$. Put $P^*(w, \eta) = \overline{P(\overline{w}, \overline{\eta})}$. Then,

$$P(z, \xi) + P^*(w, \eta) \quad \text{and} \quad P(z, \xi)P^*(w, \eta) \tag{13}$$

are symbols at $(z_0, \overline{z_0}; \xi_0 dz + \overline{\xi_0} dw)$ of product Hermite type and of positive Hermite type respectively. Further if $P(z, \xi)$ satisfies

(14) $$C|\xi|^m \leqq \operatorname{Re} P \leqq |P| \leqq C^{-1}|\xi|^m \qquad \text{on } V$$

for some constants $C>0$, $m\geqq 0$, then the operator inverse

(15) $$(P(z, D_z)+P^*(w, D_w))^{-1}$$

is of positive Hermite type at $(z_0, \bar{z}_0; \xi_0 dz+\bar{\xi}_0 dw)$ (recall the example in Introduction).

In many problems, we often meet the pseudodifferential operators of type $P(z, D_z)+P^*(w, D_w)$. Such operators are not positive, but "quasi-positive" in the following sense: That is, there exists an elliptic pseudodifferential operator $Q(z, w, D_z, D_w)$ at p_0 of positive Hermite type such that

(16) $$Q(z, w, D_z, D_w)(P(z, D_z)+P^*(w, D_w))$$

is of positive Hermite type (under the assumption (14)). In fact, we can take $Q=(P(z, D_z)+P^*(w, D_w))^{-1}$ in this case. Before stating the general result, we introduce symbols of "restricted Hermite type".

Definition 12. Let $p_0=(z_0, \bar{z}_0; \xi_0 dz+\bar{\xi}_0 dw)$ be a point of $T^*\boldsymbol{C}^{n+n}$ with $|\xi_0|=1$. Then, a symbol $\exp(P(z, w, \xi, \eta))$ is said to be of "restricted Hermite type" at p_0 if P is a holomorphic function defined in

(17) $$V(r)=\{(z, w; \xi, \eta) \in \boldsymbol{C}^{4n}; |z-z_0|<r, |w-\bar{z}_0|<r, |(\xi/|\xi|)-\xi_0|<r, \\ |(\eta/|\eta|)-\bar{\xi}_0|<r, |\xi|>r^{-1}, |\eta|>r^{-1}\}$$

with some $r>0$ satisfying the following: For some constants σ, $C>0$ $(0<\sigma<1/2)$,

(18) $$\begin{cases} |\operatorname{grad}_{(z,w)} P| \leqq C \min\{|\xi|^\sigma, |\eta|^\sigma\} \\ |\operatorname{grad}_{(\xi,\eta)} P| \leqq C(|\xi|+|\eta|)^{\sigma-1} \end{cases} \qquad \text{on } V(r),$$

and $P(z, w, \xi, \eta)=\overline{P(\bar{w}, \bar{z}, \bar{\eta}, \bar{\xi})}$ on $V(r)$. In particular e^P is a symbol of product Hermite type at p_0 because

(19) $$|P(z, w, \xi, \eta)| \leqq M(|\xi|+|\eta|)^\sigma \qquad \text{on } V(r)$$

for some constant $M>0$.

The operators as above are of infinite order in general, but they have very useful properties concerning the exponential calculus (cf. [1], [2], [5]). Hereafter we often employ the notation $:P(z, w, \xi, \eta):$ instead of $P(z, w, D_z, D_w)$ for a symbol $P(z, w, \xi, \eta)$.

Theorem 13. *We inherit the notation p_0, $V(r)$, σ and C from above. Let $\exp(P(z, w, \xi, \eta))$ and $\exp(Q(z, w, \xi, \eta))$ be any symbols at p_0 of restricted Hermite type such that P and Q satisfy (18). Then, for any positive r' ($<r$), there exist a constant B ($>r^{-1}$) depending only on r, r', σ, and C, and holomorphic functions $S_l(z, w, \xi, \eta)$ ($l=1, 2$) defined on $V(r') \cap \{|\xi|>B, |\eta|>B\} = V'$ satisfying the following conditions:*

i) $S_l(z, w, \xi, \eta)=\overline{S_l(\overline{w}, \overline{z}, \overline{\eta}, \overline{\xi})}$ *on V' for $l=1, 2$.*

ii) $|S_l(z, w, \xi, \eta)| \leqq B(|\xi|+|\eta|)^{2\sigma-1}$ *on V' for $l=1, 2$.*

iii) $:\exp(P): :\exp(Q):=:\exp(P+Q+S_1):$
and $:\exp(P):^{-1}=:\exp(-P+S_2):$

hold as equivalence relations for formal symbols of product Hermite type.

A typical and important example of symbols of positive and restricted Hermite type is the following:

$$\exp(P)=\exp\left\{K(z, \overline{w})\frac{(\xi\eta)^{N+\sigma/2}}{(\xi^N+\eta^N)^2}\right\}, \tag{20}$$

which is defined in

$$\{(z, w; \xi, \eta) \in \boldsymbol{C}^4; |z|<r, |w|<r, |\arg(\xi-r^{-1})|<\pi/4N, \\ |\arg(\eta-r^{-1})|<\xi/4N\}. \tag{21}$$

Here, $r>0$, $0<\sigma<1/2$, $N=2, 3, 4, \cdots$, and $K(z, w)$ is any analytic positive Hermite kernel in z, w. By using symbols of this type and Proposition 2, we can get the following main theorem.

Theorem 14. *We inherit the notation p_0 and $V(r)$. Let $\{Q_\lambda(z, w, \xi, \eta)\}_{\lambda\in\Lambda}$ be a family of simple symbols at p_0 of product Hermite type satisfying the following* i) ~ iii):

i) *every Q_λ is holomorphic in $V(r)$ with some constant $r>0$ independent of $\lambda \in \Lambda$.*

ii) *Quasi-positivity*: $Q_\lambda(z, w, \xi, \eta) \notin \{x \in \boldsymbol{R}; x \leqq 0\}$ *on $V(r)$ for every $\lambda \in \Lambda$.*

iii) *Ellipticity*: *There exist some constants A, $\sigma>0$ ($0<\sigma<1$) independent of $\lambda \in \Lambda$, and constants $l_\lambda \in \boldsymbol{R}$ depending on $\lambda \in \Lambda$ such that*

$$\begin{cases} |l_\lambda| \leqq A \\ A^{-1} \leqq |Q_\lambda(z, w, \xi, \eta)|/(|\xi|+|\eta|)^{l_\lambda}| \leqq A \\ |\mathrm{grad}_{(\xi,\eta)} Q_\lambda| \leqq A(|\xi|+|\eta|)^{l_\lambda+\sigma-1} \end{cases} \tag{22}$$

hold in $V(r)$ for every $\lambda \in \Lambda$.

Then, there exist a positive number r' ($<r$) independent of $\lambda \in \Lambda$ and a

symbol $\exp(P(z, w\ \xi, \eta))$ *at* p_0 *of restricted Hermite type such that* $:\exp(P):$ *and* $:\exp(P)::Q_\lambda:$ *are positive in a conic neighborhood* $V(r')$ *for every* $\lambda \in \Lambda$.

Let $p_0=(x_0, x_0; i\xi_0(dx-du))$ be a point of $\Delta^a(iS^*\boldsymbol{R}^{n+n})$ and T_j be a bounded open set in $\boldsymbol{R}^{m_j}$ with real analytic boundary for every $j=1, \cdots, N$ $(m_j=0, 1, 2, \cdots)$. Further, let $Q_j(t_j, z, \xi)$ be a simple symbol with real analytic parameter t_j for every $j=1, \cdots, N$ such that $Q_j(t_j, z, \xi)$ is holomorphic in

$$X_j=\{(t_j, z, \xi) \in \boldsymbol{C}^{m_j+n+n};\ t_j \in U_j, |z-x_0|<r,\ |(\xi/|\xi|)-i\xi_0|<r, |\xi|>r^{-1}\}$$

$(|\xi_0|=1)$ and satisfies

$$C^{-1}|\xi|^{l_j} \leqq \operatorname{Re} Q_j \leqq |Q_j| \leqq C|\xi|^{l_j} \quad \text{on } X_j \tag{23}$$

for every $j=1, \cdots, N$. Here C, r and the l_j's are positive constants, and U_j is a complex neighborhood of $\overline{T_j}$. Setting

$$Q_j^*(t_j, z, \xi)=\overline{Q_j(,\ \overline{t_j}\bar{z}, \bar{\xi})},$$

we consider the pseudodifferential operators $Q_j(t_j, z, D_z)+Q_j^*(t_j, w, D_w)$ for $j=1, \cdots, N$. Then, as a direct corollary of Theorems 14, 9 and 8, we obtain the following:

Theorem 15. *Let* $f_j(t_j, x)$ *be a hyperfunction defined on* $T_j \times \{x \in \boldsymbol{R}^n; |x-x_0|<r\}$ *with some* $r>0$ *for every* $j=1, \cdots, N$ *such that* f_j *is depending real analytically on* $t_j \in T_j$ *at every point of* $\overline{T_j}\times\{x_0\}$ *(up to the boundary* $\partial T_j\times\{x_0\}$*). Let* $\rho_j(t_j)$ *be a real analytic function defined on a neighborhood of* $\overline{T_j}$ *for every* $j=1, \cdots, N$ *such that* $\rho_j(t_j)\geqq 0$ *on* T_j *and* $\rho_j \not\equiv 0$ *on any connected component of* T_j. *Put the energy form*

$$\begin{aligned} &E(x, u) \\ &=\sum_{j=1}^{N}\int_{T_j}(Q_j(t_j, x, D_x)+Q_j^*(t_j, u, D_u))\{f_j(t_j, x)\overline{f_j(t_j, u)}\}\rho_j(t_j)dt_j. \end{aligned} \tag{24}$$

Then, $E(x, u)$ *is well-defined as a "quasi-positive Hermite microkernel" at* p_0. *That is, there exists a symbol* $\exp(P(z, w, \xi, \eta))$ *at* p_0 *of restricted Hermite type such that* $:e^P:$ *is positive and* $:e^P: E(x, u)$ *is a positive Hermite microkernel at* p_0. *In particular, if* $E(x, u)\ll 0$ *at* p_0, *then we have*

$$\mathrm{SS}(\operatorname{ext}(f_j(t_j, x)))\cap\{(t_j, x; i\tau_j dt_j+i\xi dx);\ t_j \in \overline{T}_j, \tau_j \in \boldsymbol{R}^{m_j},\ x=x_0, \xi=\xi_0\}=\varnothing$$

for every $j=1, \cdots, N$.

§ 3. Some applications

In this section, we give some easy applications. The first one is concerned with real analyticity of solutions of heat equations with the Dirichlet conditions. This is well-known in the case of distribution-solutions. In the case of hyperfunction-solutions, however, the situation is very different. In fact, the space of hyperfunctions defined on an open set has no locally convex and Hausdorff topology, and so we cannot apply the semi-group theory (cf. [14]).

Theorem 16. *Let W be a bounded domain in $\boldsymbol{R}^n$ with real analytic boundary ∂W. Let $f(x, t)$ be an arbitrary hyperfunction-solution of the following*:

$$(25)\qquad \begin{cases} \partial_t f = \Delta_x f & \text{in } W\times I, \\ f=0 & \text{on } \partial W\times I. \end{cases}$$

Then, $f(x, t)$ is analytic in a neighborhood of $\overline{W}\times I$.

Proof. Since $\partial W\times I$ is non-characteristic for $\partial_t - \Delta_x$, $f(x, t)$ is mild on $\partial W\times I$. In addition, by the general theory of microdifferential equations and boundary value problems, we easily know that $f(x, t)$ depends real analytically on $x\in W$ at every point of $\overline{W}\times I$ (cf. [15], [9]). Hence we can consider the energy form

$$E(t, s)=\int_W f(x, t)\overline{f(x, s)}dx.$$

Then, using the Dirichlet condition, we have

$$\begin{aligned} D_t E(t, s) &= \int_W \Delta_x f(x, t)\cdot\overline{f(x, s)}dx = \int_W f(x, t)\cdot\overline{\Delta_x f(x, s)}dx \\ &= D_s E(t, s). \end{aligned}$$

Therefore $(D_t - D_s)E(t, s)=0$, and so $E(t, s)$ is microanalytic on the anti-diagonal of $iS^*(I\times I)$. Hence the proof is completed by Theorem 9 and Remark in Theorem 8.

The second application is concerned with analytic hypoellipticity. This is not contained in the recent results of [20], [19], [13] because the principal symbol vanishes not exactly to the second order on the symplectic submanifold.

Theorem 17. *Let $A_1(t, x)$, $A_2(t, x)$ be real analytic functions defined near $(t, x)=(t_1, t_2, x)=0\in \boldsymbol{R}^3$ such that* $\mathrm{Re}\, A_1(0)>0$ *and* $\mathrm{Re}\, A_2(0)>0$. *Let m_1, m_2 be positive integers. Then,*

$$P(t, x, D_t, D_x) = D_{t_1}^2 + D_{t_2}^2 + (A_1(t, x)t_1^{2m_1} + A_2(t, x)t_2^{2m_2})D_x^2$$

is analytically hypoelliptic at the origin.

Proof. Let $f(t, x)$ be an arbitrary hyperfunction solution of $Pf=0$ in a neighborhood of the origin. Easily to see, we have

$$\text{(26)} \qquad \mathrm{SS}(f(t, x)) \cap iS^*U \subset \{t_1 = t_2 = 0, \ \xi_{t_1} = \xi_{t_2} = 0\}$$

for some neighborhood U of $0 \in \boldsymbol{R}^3$. In particular, $f(t, x)$ depends real analytically on $t=(t_1, t_2)$, and so we can consider the following positive Hermite microkernel at $(0; \pm idx)$:

$$E(x, u) = \int_{\{|t|<r\}} \mathrm{grad}_t f(t, x) \cdot \overline{\mathrm{grad}_t f(t, u)} dt_1 dt_2.$$

Then, since $f(t, x)$ is analytic on $\{|t|=r\}$ for any small $r>0$ (see (26)), by integration by parts, we get the following microlocal equality at $(0; \pm idx)$:

$$E(x, u) = \frac{1}{2} \sum_{j=1}^{2} \int_{\{|t|<r\}} t_j^{2m_j}(A_j(t, x)D_x^2 + A_j^*(t, u)D_u^2)[f(t, x)\overline{f(t, u)}]dt.$$

Here, we note that the each term of the right-hand side is quasi-negative at $(0; \pm idx)$. Therefore, $[f(t, x)]$ must vanish at $(0; \pm idx)$ by Theorem 15, that is, $f(t, x)$ is analytic at 0.

The last application is concerned with propagation of analytic singularities. This problem has a deep connection with the diffraction problem (Theorem 0.3 of [17]). That is, only for the operator $P = D_1^2 - (x_1 - x_2)D_3^2$ with the Dirichlet condition on $x_1 = +0$, we will give another proof of Theorem 0.3 of [17] after this proposition.

Proposition 18. *Let $g(s, x_1, x_2)$ be a hyperfunction defined on $\{0<s<1, |x_1|+|x_2|<r\}$ for some $r>0$ such that g is mild on $\{s=+0\}$ and on $\{s=1-0\}$. Suppose that g satisfies the equations*

$$\text{(27)} \qquad (D_s^2 + (1-s)x_1^3 D_{x_2}^2)[g] = 0 \ \textit{in} \ 0<s<1 \ (\textit{up to} \ s=+0 \ \textit{and} \ s=1-0),$$

$$\text{(28)} \qquad [g(+0, x_1, x_2)] = 0,$$

$$\text{(29)} \qquad [D_s g|_{s=1-0}] = -\lambda e^{\pi i/3} x_1 (D_{x_2}/i)^{2/3} [g|_{s=1-0}]$$

microlocally along the regular involutory submanifold

$$\text{(30)} \qquad V = \{(s, x_1, x_2; i\xi_s, i\xi_1, i\xi_2) \in iS^*\boldsymbol{R}^3; \ \xi_s = 0, x_1 = 0, |x_2|<r, \\ \xi_2 > 0, |\xi_1/\xi_2 - \theta_0| < r\}.$$

Here, $\lambda>0$ and θ_0 are real constants. Then, $SS(D_s g(+0, x_1, x_2)) \not\ni (0, 0; i\theta_0 dx_1 + idx_2)$ *provided*

(31) $$SS(D_s g(+0, x_1, x_2)) \cap \{(0, 0; i\theta dx_1 + idx_2);\ \theta_0 < \theta < \theta_0 + r\} = \emptyset.$$

Proof. Since $SS(g) \cap \{x_1 = 0, 0 < s < 1\} \cap W \subset \{\xi_s = 0\}$ for some neighborhood W of V, the following positive Hermite microkernel is well-defined at $p_0 = (0, 0, 0, 0; i\theta_0 dx_1 + idx_2 - i\theta_0 du_1 - idu_2)$:

$$E(x_1, x_2, u_1, u_2) = \int_0^1 D_s g(s, x) \cdot \overline{D_s g(s, u)}\, ds \gg 0.$$

By integration by parts and equations (27) ~ (29), we have

(32) $$\begin{aligned} E(x, u) &= x_1^3 D_{x_2}^2 H(x, u) - \lambda e^{\pi i/3} x_1 (D_{x_2}/i)^{2/3} h(x, u) \\ &= u_1^3 D_{u_2}^2 H(x, u) - \lambda e^{-\pi i/3} u_1 (D_{u_2}/(-i))^{2/3} h(x, u) \end{aligned}$$

at p_0, where

(33) $$H(x, u) = \int_0^1 (1-s) g(s, x) \overline{g(s, u)}\, ds \gg 0,$$

(34) $$h(x, u) = g(1-0, x) \overline{g(1-0, u)} \gg 0.$$

Put

$$Q(x_1, D_{x_2}) = x_1 (D_{x_2}/i)^{2/3}.$$

Hence, from (32), we obtain

(35) $$\begin{aligned} &(Q(x_1, D_{x_2})^3 - Q^*(u_1, D_{u_2})^3) H \\ &\quad + \lambda (e^{\pi i/3} Q(x_1, D_{x_2}) - e^{-\pi i/3} Q^*(u_1, D_{u_2})) h = 0, \end{aligned}$$

(36) $$\begin{aligned} &(Q(x_1, D_{x_2})^3 + Q^*(u_1, D_{u_2})^3) H \\ &\quad + \lambda (e^{\pi i/3} Q(x_1, D_{x_2}) + e^{-\pi i/3} Q^*(u_1, D_{u_2})) h \ll 0. \end{aligned}$$

Here we claim the following:

(37) $$\begin{aligned} &SS(H(x, u)) \cap \{(x, u; i\theta dx_1 + idx_2 - i\tau du_1 - idu_2); \\ &\quad x = u = 0,\ |\theta - \theta_0| < r,\ |\tau - \theta_0| < r\} \subset \{\theta \leqq \theta_0,\ \tau \leqq \theta_0\}, \end{aligned}$$

(38) $$\begin{aligned} &SS(h(x, u)) \cap \{(x, u: i\theta dx_1 + idx_2 - i\tau du_1 - idu_2); \\ &\quad x = u = 0,\ |\theta - \theta_0| < r,\ |\tau - \theta_0| < r\} \subset \{\theta \leqq \theta_0,\ \tau \leqq \theta_0\}. \end{aligned}$$

In fact, since the micro-principal symbol of (27) along regular involutory submanifold V is just ξ_s^2, the micro-analyticity of g on V propagates along integral curves of $\partial/\partial s$ (see [4], [16] and the latter part of the proof of Theorem 2.4 of [9]). Hence we get

$$SS(Y(s)Y(1-s)g(s, x))\cap\{\xi_s=0, x_1=x_2=0, \xi_2>0, \theta_0<\xi_1/\xi_2<\theta_0+r\}=\emptyset,$$

where $Y(s)$ is Heaviside's function. From this we directly obtain (37) and (38). Note that, in (35), $Q(x_1, D_{x_2})^3-Q^*(u_1, D_{u_2})^3$ is divisible by $e^{\pi i/3}Q(x_1, D_{x_2})-e^{-\pi i/3}Q^*(u_1. D_{u_2})$. As a result, we have the following reduced form of (35) at p_0:

$$(39)\qquad (e^{2\pi i/3}Q(x_1, D_{x_2})^2+e^{-2\pi i/3}Q^*(u_1, D_{u_2})^2+Q(x_1, D_{x_2})Q^*(u_1, D_{u_2}))H =\lambda h\gg 0.$$

Indeed, the difference of both sides is a homogeneous solution at p_0 of $L=e^{\pi i/3}Q(x_1, D_{x_2})-e^{-\pi i/3}Q^*(u_1, D_{u_2})$ with singular spectrum like (37). Hence the difference must be zero at p_0 because L is microlocally of partial Cauchy-Riemann type; that is, the micro-analyticity propagates along integral manifolds of $\{\partial/\partial\theta, \partial/\partial\tau\}$. Then, the combination of (36) and (39) yields

$$(40)\qquad 2Q(x_1, D_{x_2})Q^*(u_1, D_{u_2})(e^{\pi i/3}Q(x_1, D_{x_1})+e^{-\pi i/3}Q^*(u_1, D_{u_2}))H\ll 0$$

at p_0. Here, apply an Hermite positive operator $2^{-1}Q(x_1-i0, D_{x_2})^{-1}\times Q^*(u_1+i0, D_{u_2})^{-1}$ to (40); this operator operates on microfunctions at p_0 with support like (37) as an autohomomorphism. Thus we obtain

$$(41)\qquad (e^{\pi i/3}Q(x_1, D_{x_2})+e^{-\pi i/3}Q^*(u_1, D_{u_2}))H(x, u)\ll 0 \qquad \text{at } p_0.$$

Moreover, apply an Hermite positive operator

$$(iQ(x_1-i0, D_{x_2})-iQ^*(u_1+i0, D_{u_2}))^{-1}$$

to (39) (see (15)); this operator also operates on microfunctions at p_0 with support like (37) as an autohomomorphism. Then we have

$$(42)\qquad (e^{\pi i/6}Q(x_1, D_{x_2})+e^{-\pi i/6}Q^*(u_1, D_{u_2}))H\gg 0 \qquad \text{at } p_0.$$

Hence, considering $\sqrt{3}\times(41)-(42)$, we get

$$(iQ(x_1, D_{x_2})-iQ^*(u_1, D_{u_2}))H(x, u)\ll 0 \qquad \text{at } p_0.$$

Then, by the same reason as above, we conclude $H(x, u)\ll 0$ at p_0, and so $H(x, u)=0$ at p_0 (recall (33)). Consequently the proof is completed by Theorem 9 and Remark in Theorem 8.

Let $f(x_0, x_1, x_2)$ be any microlocal solution at $p_0=(0; i\theta_0 dx_1+idx_2)\in iS^*\boldsymbol{R}^2$ of the following Dirichlet problem in the sense of [9]:

$$(43)\qquad \begin{cases} Pf=(D_0^2-(x_0-x_1)D_2^2)f(x)=0 & \text{in } x_0>0\,, \\ f(+0, x_1, x_2)=0. \end{cases}$$

Here, P is Sato's operator which is a typical example concerning diffraction effect. γ denotes the null bicharacteristic passing through $\rho(0)=(0, 0, 0; i\theta_0 dx_1+idx_2) \in iS^*\boldsymbol{R}^3$:

$$(44)\qquad \gamma=\{\rho(t)=(t^2, 0, 2t^3/3; -itdx_0+i(\theta_0+t)dx_1+idx_2);\ t \in \boldsymbol{R}\}.$$

Then, Theorem (0.3) of J. Sjöstrand [17] claims the following:

"If $SS(f)\cap\{\rho(t)\in\gamma;\ -r<t<0\}=\emptyset$ and $SS(D_{x_0}f(+0, x_1, x_2))\cap\{(x_1, x_2; i\theta dx_1+idx_2);\ x_1=x_2=0,\ \theta_0<\theta<\theta_0+r\}=\emptyset$

for some $r>0$, then $SS(D_{x_0}f(+0, x_1, x_2)) \not\ni p_0=(0, 0; i\theta_0 dx_1+idx_2)$, that is, $f(x)=0$ at p_0."

In order to prove this, we employ the trick in § 2 of [9]. Firstly, by the hyperbolicity of P in $x_0-x_1>0$, we know that $f(x)$ extends to a microlocal solution on $\{x \in U;\ x_0>0 \text{ or } x_0-x_1>0\}$ for a small neighborhood U of $x=0$ (as for the rigorous version, see Proposition 2.3 of [9]). Therefore we can identify $f(x)$ with the following $\tilde{f}(s, x_0, x_1, x_2)$:

$$(45)\qquad \begin{cases} D_s\tilde{f}(s, x)=0 & \\ (D_{x_0}^2-(x_0-x_1)D_{x_2}^2)\tilde{f}(s, x)=0 & \text{in } \{0<s<1,\ x_0-sx_1>0\}, \\ \tilde{f}(+0, x)=f(x) & \text{in } \{x_0>0\}, \end{cases}$$

where $\tilde{f}$ satisfies the above equations only microlocally with respect to (x_1, x_2)-variables; that is, "at p_0". In particular, $\tilde{f}(1-0, x)$ satisfies the Tricomi equation

$$(D_0^2-(x_0-x_1)D_2^2)\tilde{f}(1-0, x)=0 \qquad \text{in } x_0-x_1>0$$

and $SS(\tilde{f}(1-0, x))\cap\{\rho(t)\in\gamma;\ -r<t<0\}=\emptyset$. Hence, by employing the microlocal Green formula and a suitable fundamental solution of the boundary value problem for P in $x_0-x_1>0$, we can show the following relationship between boundary values on $x_0-x_1=+0$:

$$(46)\qquad D_{x_0}\tilde{f}(1-0, x)|_{x_0-x_1=+0}=-\lambda e^{\pi i/3}(D_{x_2}/i)^{2/3}\tilde{f}(1-0, x)|_{x_0-x_1=+0}$$

at p_0, where $\lambda=3^{1/3}\Gamma(2/3)/\Gamma(1/3)>0$ (the proof goes in the similar way like Example 1.5 of [9]). Change the coordinates s, x_0, x_1, x_2 into $s,\ t=x_0-sx_1,\ x_1, x_2$, and put

(47) $$h(s, t, x_1, x_2) \equiv \tilde{f}(s, t+sx_1, x_1, x_2).$$

Then, h satisfies the following equations:

$$\begin{cases} (D_s - x_1 D_t)h = 0 \\ (D_t^2 + ((1-s)x_1 - t)D_{x_2}^2)h = 0 \end{cases} \quad \text{in } \{t>0,\ 1>s>0\}$$

at p_0 (in the sense as above). In particular, $g(s, x_1, x_2) \equiv h(s, +0, x_1, x_2)$ satisfies

(48) $$\begin{cases} (D_s^2 + (1-s)x_1^3 D_{x_2}^2)g = 0 \qquad \text{in } 0<s<1, \\ g(+0, x_1, x_2) = 0, \\ D_s g(1-0, x_1, x_2) = -\lambda e^{\pi i/3} x_1 (D_{x_2}/i)^{2/3} g(1-0, x_1, x_2), \\ D_s g(+0, x_1, x_2) = x_1 D_{x_0} f(+0, x_1, x_2) \end{cases}$$

at p_0. Therefore, all the assumptions for $g(s, x_1, x_2)$ in Proposition 18 are fulfilled, and so we conclude $x_1 D_{x_0} f(+0, x_1, x_2) = 0$ at p_0. Hence $D_{x_0} f(+0, x_1, x_2) = 0$ at p_0 because the microanalyticity of any solution $u(x)$ of $x_1 u(x) = 0$ propagates along integral curves of $\partial/\partial\theta$.

References

[1] T. Aoki, Invertibility for microdifferential operators of infinite order, Publ. Res. Inst. Math. Sci., Kyoto Univ., **18** (1982), 1–29.

[2] ——, The exponential calculus of microdifferential operators of infinite order, II, Proc. Japan Acad., **58** (1982), 154–157.

[3] S. Bergman, The Kernel Function and Conformal Mapping, Mathematical Surveys Number V, Amer. Math. Soc., New York, 1950/Second Edition, 1970.

[4] J.-M. Bony, Extension du théorème de Holmgren, Sém. Goulaouic-Schwartz 1975–1976, exposé 17.

[5] L. Boutet de Monvel, Opérateurs pseudo-différentiels analytiques et opérateurs d'ordre infini. Ann Inst. Fourier, Grenoble, **22** (1972), 229–268.

[6] H. Bremermann, Holomorphic continuation of the kernel function and the Bergman metric in several complex variables, Univ. Michigan Press, Ann Arbor, 1955, 349–383.

[7] W. F. Donoghue, Jr., Reproducing kernel spaces and analytic continuation, Rocky Mountain J. Math., **10–1** (1980), 85–97.

[8] K. Kataoka, Micro-local theory of boundary value problems. I, J. Fac. Sci. Univ. Tokyo, Sect. IA Math., **27** (1980), 355–399.

[9] ——, Micro-local theory of boundary value problems. II, J. Fac. Sci. Univ. Tokyo Sect. IA Math., **28** (1981), 31–56.

[10] ——, Microlocal energy methods and pseudo-differential operators, Invent. Math., **81** (1985), 305–340.

[11] M. G. Kreĭn: Hermitian-positive kernels on homogeneous spaces. I, Ukrain. Mat. Ž. 1, **4** (1949), 64–98; Amer. Math. Soc. Transl. Ser. 2, **34** (1963), 69–108.

[12] H. Meschkowski, Hilbertsche Räume mit Kernfunktion, Grundl. math. Wiss. Einzeldarstellungen, Band 113, Springer, Berlin-Göttingen-Heidelberg, 1962.

[13] G. Métivier, Analytic hypoellipticity for operators with multiple characteristics, Comm. Partial Differential Equations, **6** (1981), 1–90.
[14] Ôuchi, On abstract Cauchy problems in the sense of hyperfunction, Lecture Notes in Math. Vol. 287, Springer Berlin-Heidelberg-New York, 1973, 135–152.
[15] M. Sato, T. Kawai and M. Kashiwara, Microfunctions and Pseudo-differential Equations. Lecture Notes in Math. Vol. 287, Springer, Berlin-Heidelberg-New York, 1973, 265–529.
[16] P. Schapira, Propagation au bord et réflexion des singularités analytiques des solutions des équations aux dérivées partielles, II. Sém. Goulaouic-Schwartz 1976–77, exposé 9.
[17] J. Sjöstrand, Propagation of analytic singularities for second order Dirichlet problems III. Comm. Partial Differential Equations, **6** (1981), 499–567.
[18] F. Sommer and J. Mehring, Kernfunktion und Hüllenbildung in der Funktionentheorie mehrer Veränderlichen. Math. Ann., **131** (1956), 1–16.
[19] D. S. Tartakoff, The local real analyticity of solutions to $\square_b$ and the $\bar{\partial}$-Neumann problem. Acta Math., **145** (1980), 177–204.
[20] F. Trèves, Analytic hypoellipticity of a class of pseudo-differential operators with double characteristics. Comm. Partial Differential Equations, **3** (1978), 475–642.

DEPARTMENT OF MATHEMATICS
FACULTY OF SCIENCE
TOKYO METROPOLITAN UNIVERSTY
TOKYO, 158 JAPAN

Taniguchi Symp. HERT
Kyoto 1984, pp. 143–154

Systems of Microdifferential Equations of Infinite Order

Takahiro KAWAI

Although it is now commonly accepted that microdifferential operators of infinite order are crucially important in studying the structure of microdifferential equations of finite order ([17], [8], [18], $\cdots$), the importance of microdifferential *equations* of infinite order has just been being recognized. In view of this situation we want to show in this report

(i) how (micro)differential equations of infinite order arises in analysis,
(ii) how we can analyse them,
and
(iii) what kind of novelties we have so far observed.

To illustrate (i), the best way will be to analyze the classical elliptic theta function $\vartheta_3(z|t)=\sum_{\nu\in Z}\exp(\pi i\nu^2 t+2\pi i\nu z)$ from our viewpoint, which was first advocated by Professor M. Sato. Here, and in what follows, we denote by i the imaginary unit $\sqrt{-1}$.

For the convenience of the explanation, let us modify $\vartheta_3(z|t)$ and introduce two functions $\vartheta(x, y|t)$ and $\vartheta^*(x, y|t)$ ($x, y\in \boldsymbol{C}$, $t\in \boldsymbol{C}^+ \underset{\text{def}}{=} \{t\in \boldsymbol{C};\ \operatorname{Im} t>0\}$, or according to the situations $x, y, t\in \boldsymbol{R}$) as follows:

$$\vartheta(x, y|t)=\exp(\pi ix(xt+y))\vartheta_3(xt+y|t), \tag{1}$$

$$\vartheta^*(x, y|t)=\left(\frac{\partial}{\partial y}+\pi ix\right)\vartheta(x, y|t). \tag{2}$$

More explicitly, they can be rewritten as follows:

$$\vartheta(x, y|t)=\sum_{\nu\in \boldsymbol{Z}}\exp(\pi i(\nu+x)^2t+\pi i(2\nu+x)y), \tag{3}$$

$$\vartheta^*(x, y|t)=\sum_{\nu\in \boldsymbol{Z}}2\pi i(\nu+x)\exp(\pi it(\nu+x)^2+\pi i(2\nu+x)y). \tag{4}$$

Denoting ${}^t(\vartheta, \vartheta^*)$ by $\vec{\vartheta}$, we immediately find

$$\left(\frac{\partial}{\partial x}-\pi iy\right)\vec{\vartheta}=\begin{bmatrix} 0 & t \\ 4\pi i\left(t\dfrac{\partial}{\partial t}+\dfrac{1}{2}\right) & 0 \end{bmatrix}\vec{\vartheta} \tag{5}$$

and

$$(6) \qquad \left(\frac{\partial}{\partial y}+\pi ix\right)\vec{\vartheta}=\begin{bmatrix} 0 & 1 \\ 4\pi i\dfrac{\partial}{\partial t} & 0 \end{bmatrix}\vec{\vartheta}.$$

In what follows, we want to regard $\vec{\vartheta}(x, y|t)$ to solve the hyperbolic system of equations (5) and (6) having (x, y) as time variables. This viewpoint makes a clear contrast to the usual standpoint of grasping elliptic theta functions as solutions of heat equation $(\partial^2/\partial z^2-4\pi(\partial/\partial t))\vec{\vartheta}(z|it)=0$ having t as the time variable.

Let P_1 (resp., P_2) denote the matrix of differential operators appearing in the right-hand side of (5) (resp., (6)). Then they satisfy the following commutation relation:

$$(7) \qquad [P_1, P_2](\underset{\text{def}}{=}P_1P_2-P_2P_1)=-2\pi iI_2.$$

Here and in what follows, I_r denotes the $r\times r$ identity matrix. Thanks to this relation, we find that

$$(8) \qquad f(x, y|t)\underset{\text{def}}{=}\exp(xP_1+yP_2)\vec{\vartheta}(0, 0|t)$$

solves equations (5) and (6) (in the framework of holomorphic functions or hyperfunctions). To verify this, it suffices to note that (7) combined with the Campbell-Hausdorff formula entails

$$(9) \qquad \begin{aligned}\exp(xP_1+yP_2)&=\exp(\pi ixy)\exp(xP_1)\exp(yP_2)\\ &=\exp(-\pi ixy)\exp(yP_2)\exp(xP_1).\end{aligned}$$

Since $f(0, 0|t)=\vec{\vartheta}(0, 0|t)$ holds, the uniqueness theorem for the Cauchy problem implies $f(x, y|t)=\vec{\vartheta}(x, y|t)$. To put it in another way, we otain

$$(10) \qquad \vec{\vartheta}(x, y|t)=\exp(xP_1+yP_2)\vec{\vartheta}(0, 0|t).$$

Now, let us recall that the elliptic theta function satisfies the so-called quasi-periodicity conditions, that is,

$$(11) \qquad \vartheta_3(z+1|t)=\vartheta_3(z|t)$$

and

$$(12) \qquad \vartheta_3(z+t|t)=\exp(-2\pi iz-\pi it)\vartheta_3(z|t).$$

Inheriting this property, $\vec{\vartheta}(x, y|t)$ also satisfies the following:

(13) $$\vec{\vartheta}(x+1, y\,|\,t)=\exp(-\pi i y)\vec{\vartheta}(x, y\,|\,t)$$

and

(14) $$\vec{\vartheta}(x, y+1\,|\,t)=\exp(\pi i x)\vec{\vartheta}(x, y\,|\,t).$$

Or, more generally, we obtain

(15) $$\vec{\vartheta}(x+l, y+m\,|\,t)=(-1)^{lm}\exp(\pi i(mx-ly))\vec{\vartheta}(x, y\,|\,t)$$

for each (l, m) in $\boldsymbol{Z}\times\boldsymbol{Z}$. In particular, we have

(16) $$\vec{\vartheta}(l, m\,|\,t)=(-1)^{lm}\vec{\vartheta}(0, 0\,|\,t).$$

Combining (10) and (16), we find

(17) $$\exp(lP_1+mP_2)\vec{\vartheta}(0, 0\,|\,t)=(-1)^{lm}\vec{\vartheta}(0, 0\,|\,t).$$

In particular,

(18) $$(\exp P_1)\vec{\vartheta}(0, 0\,|\,t)=\vec{\vartheta}(0, 0\,|\,t)$$

and

(19) $$(\exp P_2)\vec{\vartheta}(0, 0\,|\,t)=\vec{\vartheta}(0, 0\,|\,t).$$

Note that (9) enables us to deduce (17) from the special cases (18) and (19). Hence we usually discuss only (18) and (19).

Here we observe two important facts:

First, $\exp P_1$ and $\exp P_2$ *do* commute, although P_1 and P_2 do not. This is an immediate consequence of (9) with $x=y=1$.

Still more important is the fact that $\exp P_1$ and $\exp P_2$, or, more generally, $\exp(xP_1+yP_2)$ $(x, y \in \boldsymbol{C})$, are linear differential operators of infinite order (in the sense of [17], Chap. I, §2.1), and hence preserve local properties. This follows from the fact that each entry of P_1^2 and P_2^2 is a linear differential operator of order at most 1. That is, ord P_1 and ord P_2 are 1/2 (in the sense of [15], Definition 1.1). Thus $\vec{\vartheta}(0, 0\,|\,t)$ satisfies a system of linear differential equations of infinite order. At first this might sound mysterious. But, we can see through the background of this if we recall how the operator $\exp(xP_1+yP_2)$ comes into our discussion. It gives a solution to the Cauchy problem for the system of equations (5) and (6), while their characteristic variety is $\{(x, y, t; \xi, \eta, \tau) \in T^*(\boldsymbol{C}^3);\ \xi=\eta=0\}$. (Here we have used the fact that P_1 and P_2 are of lower order compared with $\partial/\partial x$ and $\partial/\partial y$, respectively.) Since the singularities propagate only along bicharacteristics, $\exp(xP_1+yP_2)$ does *not* change the location of the singularities of $\vec{\vartheta}(0, 0|t)$ in view of the above form of the characteristic variety.

Incidentally, one can easily imagine that, if we pick up a solution $h(t)$ of the equations (18) and (19), then we can construct a solution $f(x, y|t)$ of (5) and (6) satisfying the quasi-periodicity condition (15) by defining

$$f(x, y|t) = \exp(xP_1 + yP_2)h(t).$$

This is indeed the case. (See [15], Theorem 2.10 for the proof.) It means that the global condition on $f(x, y|t)$ (i.e., the quasi-periodicity condition) is equivalent to the purely local conditions on $f(0, 0|t)$ (i.e., the system of equations (18) and (19)).

The discussions given so far have explained how linear differential equations of infinite order come into the study of such an interesting object as elliptic theta functions. Then the following question naturally arises:

Can the system of equations (18) and (19) characterize the theta-zero-value $\vec{\vartheta}(0, 0|t)$?

The answer to this question is affirmative.

Since $\exp P_1$ and $\exp P_2$ commute, we can construct the following complex $c^{\bullet}(t)$:

$$(20) \qquad 0 \to \mathcal{O}^2_{C,t} \xrightarrow{(\exp P_1 - I_2,\, \exp P_2 - I_2)} \mathcal{O}^2_{C,t} \oplus \mathcal{O}^2_{C,t} \xrightarrow{\begin{bmatrix} -\exp P_2 + I_2 \\ \exp P_1 - I_2 \end{bmatrix}} \mathcal{O}^2_{C,t} \to 0.$$

Here $\mathcal{O}_{C,t}$ denotes the germ of the sheaf $\mathcal{O}_C$ at t. Then we can calculate the cohomology groups of the complex explicitly ([11]):

$$(21) \qquad H^0(c^{\bullet}(t)) \cong \begin{cases} \boldsymbol{C}, & \text{if } \operatorname{Im} t > 0 \\ 0, & \text{if } \operatorname{Im} t \leq 0 \end{cases}$$

$$(22) \qquad H^1(c^{\bullet}(t)) \cong \begin{cases} 0, & \text{if } \operatorname{Im} t > 0 \\ \boldsymbol{C}, & \text{if } \operatorname{Im} t \leq 0 \end{cases}$$

and

$$(23) \qquad H^2(c^{\bullet}(t)) = 0 \qquad \text{for every } t.$$

Since we know that $\vec{\vartheta}(0, 0|t)$ solves (18) and (19), (21) implies that it is the unique solution of (18) and (19) up to a constant multiple.

An important feature of this fact is that this characterization is done purely locally; the hitherto known way of characterizing the theta-zero-values is based upon their automorphic properties such as Jacobi's imaginary transformation, whereas our approach is designed to deduce such a global property of the theta-zerovalue from its local property, i.e., from the fact that it solves a system of linear differential equations.

In order to illustrate this, let us discuss how Jacobi's imaginary transformation is related to the system of equations (18) and (19). We first note that P_1 (resp., P_2) assumes the following form $\tilde{P}_1$ (resp., $\tilde{P}_2$) if we apply the coordinate transformation $t \mapsto s = -1/t$:

$$\tilde{P}_1 = \begin{bmatrix} 0 & -s^{-1} \\ 4\pi i\left(-s\dfrac{\partial}{\partial s} + \dfrac{1}{2}\right) & 0 \end{bmatrix}, \tag{24}$$

$$\tilde{P}_2 = \begin{bmatrix} 0 & 1 \\ 4\pi i s^2 \dfrac{\partial}{\partial s} & 0 \end{bmatrix}. \tag{25}$$

Let S denote the following matrix:

$$S = \begin{bmatrix} s^{1/2} & 0 \\ 0 & s^{3/2} \end{bmatrix}. \tag{26}$$

Clearly S is well-defined and invertible on $\{s \in \boldsymbol{C};\ \mathrm{Im}\, s > 0\}$, and it satisfies the following:

$$S^{-1}\tilde{P}_1 S = \begin{bmatrix} 0 & -1 \\ -4\pi i \dfrac{\partial}{\partial s} & 0 \end{bmatrix}, \tag{27}$$

$$S^{-1}\tilde{P}_2 S = \begin{bmatrix} 0 & s \\ 4\pi i\left(s\dfrac{\partial}{\partial s} + \dfrac{1}{2}\right) & 0 \end{bmatrix}. \tag{28}$$

Now, it follows from (18) and (19) that

$$(\exp \tilde{P}_1)\vec{\vartheta}(0, 0| -1/s) = \vec{\vartheta}(0, 0| -1/s) \tag{29}$$

and

$$(\exp \tilde{P}_2)\vec{\vartheta}(0, 0| -1/s) = \vec{\vartheta}(0, 0| -1/s) \tag{30}$$

hold on $\{s \in \boldsymbol{C};\ \mathrm{Im}\, s > 0\}$. In view of (27) and (28), we find

$$(\exp(-P_2))S^{-1}(0\vec{\vartheta}, 0| -1/s) = S^{-1}\vec{\vartheta}(0, 0| -1/s) \tag{31}$$

and

(32) $$(\exp P_1)S^{-1}\vec{\vartheta}(0, 0|-1/s)=S^{-1}\vec{\vartheta}(0, 0|-1/s).$$

Since the equation $(\exp P_2)h=h$ is equivalent to the equation $\exp(-P_2)h=h$, the relation (21) combined with (31) and (32) entails that there exists a constant C for which

$$S^{-1}\vec{\vartheta}(0, 0|-1/s)=C\vec{\vartheta}(0, 0|s)$$

holds on $\{s \in \boldsymbol{C}; \operatorname{Im} s>0\}$. In particular, we obtain

(33) $$s^{-1/2}\vartheta(0, 0|-1/s)=C\vartheta(0, 0|s).$$

Since $\vartheta(0, 0|i)=\sum_{\nu\in \boldsymbol{Z}}\exp(-\pi\nu^2)\neq 0$, we can fix C by evaluating the both sides of (33) at $s=i$; C is equal to $\exp(-\pi i/4)$.

Summing up, we have thus deduced the following global relation (34) from purely local properties of the theta-zerovalue;

(34) $$\vartheta(0, 0|s)=\exp(\pi i/4)s^{-1/2}\vartheta(0, 0|-1/s).$$

This is exactly what Jacobi's imaginary transform of the theta-zerovalue should satisfy.

We also note that we can interpret the transformation $t\mapsto -1/t$ used in Jacobi's imaginary transform as $\exp((\pi/2)X)$, where X denotes a vector field $-(1+t^2)\partial/\partial t$. This vector field is determined by the operator

$$\frac{1}{4\pi i}(P_1^2+P_2^2)=\begin{bmatrix}(1+t^2)\dfrac{\partial}{\partial t}+\dfrac{t}{2} & 0\\ 0 & (1+t^2)\dfrac{\partial}{\partial t}+\dfrac{3}{2}t\end{bmatrix},$$

and the reason for the appearance of this operator is tied up with the general automorphic property under the action of $Sp(n; \boldsymbol{Z})$ on some special functions (called Jacobi functions in [14], [15] and [16]) which are characterized by some particular systems of microdifferential equations of infinite order. (See [16], § 2 for the precise formulation of the automorphic property of Jacobi functions.)

Now, in view of the success of the study of the theta-zerovalue by linear differential equations, which we have so far explained, we may expect that we can manage such wild functions as theta-zerovalues by their local properties without resorting to their global informations. Needless to say, too straightforward approach should fail; linear differential equations of finite order obviously cannot work for this purpose. We do need opeartors of infinite order, while the treatment of equations determined by such wild operators has been considered to be terribly difficult.

However, the recent progress in microlocal analysis has enabled us to cope with the trouble. In fact, we have recently succeeded in realizing such an expectation in a pretty satisfactory manner ([16]).

Before proceeding further, we note that the relations (21), (22) and (23) resemble the structure of holomorphic solutions of holonomic systems; the cohomology groups in question are of finite-dimension, and, if we suitably stratify the underlying space, then the dimension is constant on each stratum. However, there is one very important difference; the stratification we need here is not complex analytic. This is reasonable, as the system in question arises from the analysis of the theta-zerovalue $\vartheta(t) = \sum_{\nu\in \boldsymbol{Z}} \exp(\pi i\nu^2 t)$, for which $\{t \in \boldsymbol{C};\ \mathrm{Im}\, t=0\}$ is a natural boundary.

These observations lead us to the expectation that we can find a suitable class of systems of (micro)differential equations of infinite order, which includes the usual holonomic systems and enjoys some finiteness theorems. At first sight, such an expectation might sound too optimistic. But, luckily enough, we have recently found such a class, namely, the class of $\boldsymbol{R}$-holonomic complexes ([16]).

To state the definition of $\boldsymbol{R}$-holonomic complexes, we prepare some notions and notations.

First of all, to incorporate into the theory the above observation to the effect that non-complex-analytic stratification is necessary, we have to use the symplectic structure of T^*X regarded as a real manifold for a complex manifold X. To emphasize this, we use the notation $(T^*X)^{\boldsymbol{R}}$ to denote the real homogeneous symplectic manifold endowed with the real canonical 1-form $\omega^{\boldsymbol{R}}=\omega+\bar{\omega}$, where ω denotes the holomorphic canonical 1-form on T^*X and $\bar{\omega}$ denotes its complex conjugate. An $\boldsymbol{R}$-holonomic complex, whose precise definition will be given below (Definition 1), is, roughly speaking, a system whose "characteristic variety" is Lagrangian with respect to this symplectic structure. To formulate it, however, we need some more preparations.

Let W be an open subset of $T^*X - T^*_X X$ and let $\mathscr{K}^{\cdot}$ be a complex of $\mathscr{E}^{\boldsymbol{R}}_X$-Modules defined on W. We say $\mathscr{K}^{\cdot}$ is a good complex (of $\mathscr{E}^{\boldsymbol{R}}$-Modules) if $\mathscr{K}^{\cdot}$ is (locally quasi-isomorphic to) a bounded complex of free $\mathscr{E}^{\boldsymbol{R}}$-Modules of finite rank. Here, and in what follows, $\mathscr{E}^{\boldsymbol{R}}_X$ denotes the sheaf of holomorphic microlocal operators. A holomorphic microlocal opeator is, by definition, an integral operator whose kernel function is a section of the sheaf $\mathscr{C}^{\boldsymbol{R}}_{X|X\times X}$ ([17], Chap. II, § 1,1. Definition 1.1.4). [The terminology "holomorphic microlocal operator" is not used in [17]. It was coined by Professor K. Kataoka.]

For a good complex $\mathscr{K}^{\cdot}$, its characteristic set $\mathrm{Ch}(\mathscr{K}^{\cdot})$ is, by definition, the union of the closure of $\mathrm{Supp}\,\mathscr{H}_l(\mathscr{K}^{\cdot})$ for all l's.

Example 1. Let $\mathcal{D}_C^\infty$ denote the sheaf of linear differential operators of infinite order defined on $\boldsymbol{C}$. Let $\mathcal{M}.$ denote the following complex of $\mathcal{D}_C^\infty$-Modules:

$$0 \longleftarrow (\mathcal{D}_C^\infty)^2 \xleftarrow{\begin{bmatrix}\exp P_1 - I_2 \\ \exp P_2 - I_2\end{bmatrix}} (\mathcal{D}_C^\infty)^2 \oplus (\mathcal{D}_C^\infty)^2 \xleftarrow{(-\exp P_2 + I_2,\ \exp P_1 - I_2)} (\mathcal{D}_C^\infty)^2 \longleftarrow 0.$$

Let $\mathcal{K}.$ denote $\mathcal{E}_C^R \otimes \mathcal{M}..$ Then it is clear that $\mathcal{K}.$ is a good complex. Further we can verify (cf. Theorem 1 below)

$$\text{(35)} \qquad \operatorname{Ch}(\mathcal{K}.) \subset \{(r, s; \rho, \sigma) \in (T^*\boldsymbol{C} - T_C^*\boldsymbol{C})^R \cong \boldsymbol{R}^2 \times (\boldsymbol{R}^2 - \{0\});\ s=0,\ \rho=0,\ \sigma>0\}.$$

Here r (resp., s) denotes the real (resp., imaginary) part of t, and (ρ, σ) denotes the corresponding cotangent vector. Hence (35) implies that $\operatorname{Ch}(\mathcal{K}.)$ is contained in a Lagrangian subset of $(T^*\boldsymbol{C})^R$.

Note that the holomorphic solution complex determined by $\mathcal{M}.$, i.e., $\mathcal{H}om_{\mathcal{D}_C^\infty}(\mathcal{M}., \mathcal{O}_C)_t$, is the complex $c^\cdot(t)$ given by (20).

Example 2. Let Q_j ($j=1, 2$) denote the following matrices of linear differential operators:

$$\text{(36)} \qquad Q_1 = \begin{bmatrix} 0 & -2it \\ -\left(2t\dfrac{d}{dt}+1\right) & 0 \end{bmatrix},$$

$$\text{(37)} \qquad Q_2 = \begin{bmatrix} 0 & 1 \\ -i\dfrac{d}{dt} & 0 \end{bmatrix}.$$

Let $J(z)$ denote the entire function $1/\Gamma(z)$, where $\Gamma(z)$ denotes the gamma function. Then, using the commutation relation $[Q_1, Q_2]=1$, we can verify

$$\text{(38)} \qquad J(Q_2) \exp Q_1 = \exp Q_1 J(Q_2 - 1).$$

This relation enables us to construct the following complex $\mathcal{N}.$ of $\mathcal{D}_C^\infty$-Modules, although $\exp Q_1$ and $J(Q_2)$ are not commutative.

$$\text{(39)} \qquad \mathcal{N}.: 0 \longleftarrow (\mathcal{D}_C^\infty)^2 \xleftarrow{\begin{bmatrix} Q_2 \exp Q_1 - I_2 \\ J(Q_2) \end{bmatrix}} (\mathcal{D}_C^\infty)^2 \oplus (\mathcal{D}_C^\infty)^2 \xleftarrow{(\Phi,\ \Psi)} (\mathcal{D}_C^\infty)^2 \longleftarrow 0,$$

where $\Phi \underset{\text{def}}{=} -Q_2^{-1} J(Q_2)$ and $\Psi \underset{\text{def}}{=} (\exp Q_1)(Q_2 - I_2) - I_2 = Q_2 \exp Q_1 - I_2$.

(Note that, since $J(z)$ has a simple zero at $z=0$, $Q_2^{-1}J(Q_2)$ is a well-defined linear differential operator.)

An interesting feature of this complex is that the structure of its holomorphic solutions is the most elementary one of the sort: letting $\tilde{c}^{\cdot}(t)$ denote the holomorphic solution complex at t which is determined by $\mathcal{N}.$, we can prove the following relations (40) and (41) by a similar reasoning used in [11]:

(40) $$H^0(\tilde{c}^{\cdot}(t)) \cong \begin{cases} \boldsymbol{C}, & \text{if } \operatorname{Im} t > 0 \\ 0, & \text{if } \operatorname{Im} t \leqq 0 \end{cases},$$

and

(41) $$H^1(\tilde{c}^{\cdot}(t)) = H^2(\tilde{c}^{\cdot}(t)) = 0 \qquad \text{for every } t.$$

It is clear that $\mathscr{E}_{\boldsymbol{C}}^{\boldsymbol{R}} \otimes \mathcal{N}.$ is also a good complex.

Example 3. Let $a(n)$ ($n \in \boldsymbol{N} = \{0, 1, 2, \cdots\}$) be a sequence of complex numbers which satisfies the following conditions (42) and (43):

(42) $$a(n) \neq 0 \quad \text{and} \quad \lim_{n \to \infty} n/a(n) = 0,$$

(43) there exists a strictly positive constant c such that

$$|a(n) - a(n')| \geqq c|n - n'|$$

holds for every (n, n') in $\boldsymbol{N} \times \boldsymbol{N}$.

Then the infinite product $P(d/dt)$ of linear differential operators given by

$$\prod_{n=0}^{\infty} \left(1 + \frac{(d/dt)^2}{a(n)^2}\right)$$

determines a linear differential operator of infinite order. Hence the following complex $\mathscr{P}.$ is trivially well-difined:

(44) $$\mathscr{P}.:\ 0 \longleftarrow \mathscr{D}_{\boldsymbol{C}}^{\infty} \xleftarrow{P(d/dt)} \mathscr{D}_{\boldsymbol{C}}^{\infty} \longleftarrow 0.$$

It is then clear that we obtain another good complex $\mathscr{E}_{\boldsymbol{C}}^{\boldsymbol{R}} \otimes \mathscr{P}..$

Unlike good complexes given in Examples 1 and 2, this good complex is *not* $\boldsymbol{R}$-holonomic in the sense to be defined below. This makes a clear contrast to (micro)differential equations of finite order; a non-zero scalar linear ordinary differential operator of finite order always determines a holonomic system. Of course, it is not surprising; the solution space of the equation

(45) $$P(d/dt)f(t)=0$$

is clearly of infinite dimension, as any function of the form

$$\sum_{n=1}^{\infty} c(n) \exp(ia(n)t) \qquad (c(n) \in \boldsymbol{C})$$

satisfies (45) if the series is suitably convergent.

We mention that the equation (45) plays an essential role when we discuss the Fabry-type gap theorems by a differential equation theoretic method. (See [12] and [13] for the details.)

Once the notion of the characteristic set is given, it is natural to introduce the following

Definition 1. *A good complex* $\mathscr{K}.$ *of* $\mathscr{E}^{\boldsymbol{R}}$*-Modules is called* ***R****-holonomic if* $\mathrm{Ch}(\mathscr{K}.)$ *is contained in a subanalytic subset of* $(T^*X-T^*_XX)^{\boldsymbol{R}}$ *that is Lagrangian (with respect to* $\omega^{\boldsymbol{R}}$*).*

Of course, if we were unable to find a way to calculate $\mathrm{Ch}(\mathscr{K}.)$, this definition might not be practical. Such a concern is not groundless, as a system of linear differential equations of infinite order has been thought to be a terrible object even when it is with constant coefficients. (See [6], p. 319–p. 320, for example. Note, however, here we have bypassed some of the troubles mentioned there by restricting our consideration to good complexes.) But, several results of Aoki ([1], [2], [3] and references cited there) are now available. The results of Aoki give us enough information on the invertibility of holomorphic microlocal operators (and, in particular, microdifferential operators of infinite order). We refer the reader to [4], [12], [13] and [16] for the way how to use his results to calculate the characteristic set of a good complex, and here we content ourselves with quoting just the following simple and useful result from [4] in order to illustrate the nature of the invertibility theorem.

Theorem 1 ([4], Theorem 1). *Let* (z_0, ζ_0) $(\zeta_0 \neq 0)$ *denote a point in* $T^*\boldsymbol{C}^n$, *and let U denote a conical neighborhood of* (z_0, ζ_0), *that is, U is a neighborhood of* (z_0, ζ_0) *which is stable under the dilation map* $(z, \zeta) \mapsto (z, c\zeta)$ $(c>0)$. *Let* $P(z, D_z)$ *be a linear differential operator defined on* $\pi(U)$, *where* π *denotes the projection from* $T^*\boldsymbol{C}^n$ *onto the base manifold* $\boldsymbol{C}^n$. *Let* $P(z, \zeta)$ *denote its symbol. Suppose that* $P(z, \zeta)$ *never vanishes on* $\{(z, \zeta) \in U; |\zeta|>C\}$ *for some constant C. Then* $P(z, D_z)$ *is invertible in* $\mathscr{E}^{\boldsymbol{R}}_{\boldsymbol{C}^n, (z_0, \zeta_0)}$, *the germ at* (z_0, ζ_0) *of the sheaf of rings of holomorphic microlocal operators on* $\boldsymbol{C}^n$.

Now, the importance of the notion of ***R***-holonomic complexes will be most clearly visible through the follownig result:

Theorem 2. *Let X be a complexification of a real analytic manifold M. Let W be an open subset of* $T^*X - T^*_X X$ *and let* $\mathcal{K}^{\cdot}$ *be an* **R**-*holonomic complex of* $\mathcal{E}^{\mathbf{R}}$-*Modules defined on W. Then, for each j and each* $p \in W \cap T^*_M X$,

$$\dim_{C} \mathcal{E}xt^{j}_{\mathcal{E}^{\mathbf{R}}}(\mathcal{K}^{\cdot}, \mathcal{C}_M)_p < \infty .$$

Here $\mathcal{C}_M$ *denotes the sheaf of microfunctions.*

See [16], Theorem 1.4 for the proof.

Although we do not go into the details here, we can let a holomorphic microlocal operator act upon (relative cohomology groups of) holomorphic functions as an integral operator. ([10] [8], [5], [7], $\cdots$.) Using this technique, we can discuss "holomorphic solutions" of $\boldsymbol{R}$-holonomic complexes, and then we can generalize the reconstruction theorem for holonomic $\mathcal{D}_X$-Modules ([8], Chap. I, § 4, Theorem 1.4.9) to $\boldsymbol{R}$-holonomic complexes, that is, we can recover in a canonical manner the original $\boldsymbol{R}$-holonomic complex from its "holomorphic solutions". Since we need several preliminaries to state this generalized reconstruction theorem in a precise form, we content ourselves with refering to [16], § 1 (in particular, Theorem 1.5) and giving some of its consequences.

Theorem 3 ([16], p. 288). *Let* $\mathcal{K}^{\cdot}$ *be the complex given in Example* 1. *Then the sheaf* $\mathcal{C}_{\boldsymbol{R}}$ *of microfunctions regarded as an* $\mathcal{E}^{\mathbf{R}}$-*Module coincides with* $\mathcal{K}^{\cdot}$.

This is an immediate consequence of the definition of $\mathcal{C}_{\boldsymbol{R}}$ and the structure of $H^j(c^{\cdot}(t))$ ((21), (22) and (23)). Note also that the same result holds for $\mathcal{C}_{\boldsymbol{R}^n}$ if we replace $\mathcal{K}^{\cdot}$ by a suitable Koszul complex $\tilde{\mathcal{K}}^{\cdot}$ that is determined by the analysis of n-dimensional theta-zerovalue

$$\sum_{\nu \in \boldsymbol{Z}^n} \exp\left(\pi i \left(\sum_{l=1}^{n} \nu_l^2 t_l\right)\right).$$

The following intriguing Theorem 4 also follows from the generalized reconstruction theorem and the fact that $H^j(c^{\cdot}(t))$ is invariant under the real translation $t \mapsto t+a$ ($a \in \boldsymbol{R}$). A rather explicit construction of the operator G_a is given in [9].

Theorem 4. *Let* $\vartheta(t)$ *denote the theta-zerovalue* $\sum_{\nu \in \boldsymbol{Z}} \exp(\pi i \nu^2 t)$ *considered on* $\{t \in \boldsymbol{C}; \operatorname{Im} t > 0\}$. *Let a be a real number. Then we can find a linear differential operator* G_a *for which the following holds:*

$$\vartheta(t+a)=G_a\vartheta(t). \tag{46}$$

These theorems make it now manifest that (micro)differential equations of infinite order, in particular, $\boldsymbol{R}$-holonomic complexes, are deeply related to several branches of analysis. We end this report by expressing our hope that they will put a new complexion upon microlocal analysis.

References

[1] T. Aoki, Invertibility for microdifferential operators of infinite order, Publ. RIMS, Kyoto Univ., **18** (1982), 421–449.

[2] ——, Calcul exponentiel des opérateurs microdifférentiels d'ordre infini, I, Ann. Inst. Fourier, Grenoble, **33** (1983), 227–250.

[3] ——, The exponential calculus of microdifferential operators of infinite order. V, Proc. Japan Acad., Ser. A, **60** (1984), 8–9.

[4] T. Aoki, M. Kashiwara and T. Kawai, On a class of linear differential operators of infinite order with finite index, to appear in Adv. in Math.

[5] J. M. Bony and P. Schapira, Propagation des singularités analytiques pour les équations aux dérivées partielles, Ann. Inst. Fourier, Grenoble, **26** (1976), 81–140.

[6] L. Ehrenpreis, Fourier Analysis in Several Complex Variables, Wiley-Interscience, New York-London-Sydney-Toronto, 1970.

[7] M. Kashiwara and T. Kawai, Micro-hyperbolic pseudo-differential operators I., J. Math. Soc. Japan, **27** (1975), 359–404.

[8] ——, On holonomic systems of micro-differential equations. III—systems with regular singularities—, Publ. RIMS, Kyoto Univ., **17** (1981), 813–979.

[9] ——, A differential relation between $\vartheta(t+a)$ and $\vartheta(t)$, RIMS Technical Report No. 485 (1984).

[10] M. Kashiwara and P. Schapira, Micro-hyperbolic systems, Acta Math., **142** (1979), 1–55.

[11] T. Kawai, An example of a complex of linear differential operators of infinite order, Proc. Japan Acad. Ser. A, **59** (1983), 113–115.

[12] ——, The Fabry-Ehrenpreis gap theorem and linear differential equations of infinite order, to appear in Amer. J. Math.

[13] ——, The Fabry-Ehrenpreis gap theorem for hyperfunctions, Proc. Japan Acad., Ser. A. **60** (1984), 276–278.

[14] M. Sato, Pseudo-differential equations and theta functions, Astérisque, **2** et **3** (1973), 286–291.

[15] M. Sato, M. Kashiwara and T. Kawai, Linear differential equations of infinite order and theta functions, Adv. in Math., **47** (1983), 300–325.

[16] ——, Microlocal analysis of theta functions, Advanced Studies in Pure Math., **4**, 1984, 267–289.

[17] M. Sato, T. Kawai and M. Kashiwara, Microfunctions and pseudo-differential equations, Lecture Notes in Math., No. 287, Springer, Berlin-Heidelberg-New York, 1973, pp. 265–529.

[18] K. Uchikoshi, Microlocal analysis of partial differential operators with irregular singularities, J. Fac. Sci. Univ. Tokyo, Sect. IA, **30** (1983), 299–332.

RESEARCH INSTITUTE FOR
MATHEMATICAL SCIENCES
KYOTO UNIVERSITY
KYOTO 606, JAPAN

Taniguchi Symp. HERT
Katata 1984, pp. 155–179

Irregularity of Hyperbolic Operators

Dedicated to Professor Sigeru Mizohata on his sixtieth birthday

Hikosaburo KOMATSU

The irregularity condition the author introduced earlier in [15] and [17] is shown to be necessary in general in order that a formally hyperbolic equation with real analytic coefficients be well posed in a corresponding Gevrey class of functions and ultradistributions.

§ 1. Irregularity

Let $P(x, \partial)$ be either a partial differential operator with real analytic coefficients [15] or a formally hyperbolic partial differential operator of constant multiplicity [17] defined on a domain Ω and let $(\mathring{x}, \mathring{\xi})$ be a *non-singular characteristic element* of $P(x, \partial)$, that is, a point in the non-singular part of the characteristic variety $\mathrm{Ch}(P)=\{(x, \xi) \in T^*\Omega; \sigma(P)(x, \xi)=0\}$, where $\sigma(P)$ is the characteristic polynomial. Then there is a partial differential operator $K(x, \partial)$ which is simple characteristic at $(\mathring{x}, \mathring{\xi})$ and such that $\sigma(K)$ is a factor of $\sigma(P)$ ([15], Matsuura [24]). Moreover, we can find partial differential operators $Q_i(x, \partial)$ defined near $\mathring{x}$, such that

$$P(x, \partial)=\sum_{i=0}^{m} Q_i(x, \partial)K(x, \partial)^{d_i}, \tag{1.1}$$

where either $Q_i\equiv 0$ and $d_i=\infty$ or else the characteristic polynomial $\sigma(Q_i)(x, \xi)$ does not vanish identically on the characteristic variety $\mathrm{Ch}(K)$ near $(\mathring{x}, \mathring{\xi})$ in $\boldsymbol{R}^n\times\boldsymbol{C}^n$ and the order of $Q_iK^{d_i}$ is equal to i. m is the order of P and $d=d_m$ is the *multiplicity* of the characteristic element $(\mathring{x}, \mathring{\xi})$. The assumption of non-singularity implies $\sigma(Q_m)(\mathring{x}, \mathring{\xi})\neq 0$.

Then we define the *irregularity* σ of $P(x, \partial)$ at $(\mathring{x}, \mathring{\xi})$ by

$$\sigma=\max\left\{1, \max_{0\leq i<m} \frac{d_m-d_i}{m-i}\right\}. \tag{1.2}$$

Clearly we have $1\leqq\sigma\leqq d$. When $\sigma=1$, $P(x, \partial)$ is said to satisfy *Levi's condition* at $(\mathring{x}, \mathring{\xi})$. We call (1.1) the *De Paris decomposition* after De Paris [5]. The De Paris decomposition is not unique and depends on the coordinate system but the irregularity is uniquely determined by P and $(\mathring{x}, \mathring{\xi})$.

Actually the irregularity is a microlocal invariant. Aoki [1] defined the irregularity of a microdifferential operator (=analytic pseudodifferential operator) $P(x, \partial)$ relative to a microdifferential operator $K(x, \partial)$ of simple characteristic at $(\mathring{x}, \mathring{\xi})$ as follows. The operator $P(x, \partial)$ can be written in the form

$$P(x, \partial)=\sum_{i\in I} Q_i(x, \partial)K(x, \partial)^i, \tag{1.3}$$

where (i) I is a subset of $\{0, 1, 2, \cdots, d\}$, (ii) $Q_i(x, \partial)$ is a microdifferential operator whose principal symbol $\sigma(Q_i)(x, \xi)$ does not vanish identically on the characteristic variety $\mathrm{Ch}(K)$ of K, (iii) $\omega(i)=\text{order }(Q_iK^i)$ increases strictly with i on I. Then the irregularity σ of P is defined by

$$\sigma=\max\left\{1, \max_{i\in I\setminus\{d\}} \frac{d-i}{\omega(d)-\omega(i)}\right\}. \tag{1.4}$$

Aoki has also proved the compatibility of two definitions of irregularity when P and K are differential operators.

§ 2. Gevrey classes and the irregularity condition

Let $s>1$ and let Ω be an open set in $\boldsymbol{R}^n$. An infinitely differentiable function f on Ω is said to belong to the *Gevrey class* $\mathscr{E}^{\{s\}}(\Omega)$ (resp. $\mathscr{E}^{(s)}(\Omega)$) if for any compact set K in Ω there are constants h and C (resp. and any $h>0$ there is a constant C) such that

$$\sup_{x\in K} |\partial^\alpha f(x)|\leqq Ch^{|\alpha|}|\alpha|!^s. \tag{2.1}$$

We denote by $*$ either $\{s\}$ or (s). $\mathscr{D}^*(\Omega)$ stands for the function space $\mathscr{E}^*(\Omega)\cap\mathscr{D}(\Omega)$ equipped with a natural locally convex topology. The dual $\mathscr{D}^{*\prime}(\Omega)$ of $\mathscr{D}^*(\Omega)$ is defined to be the space of *ultradistributions of class* $*$ on Ω (cf. [18, 19, 20]). We also admit $\{1\}$ and (∞), and understand that $\mathscr{E}^{\{1\}}=\mathscr{A}$ the space of real analytic functions, $\mathscr{D}^{\{1\}\prime}=\mathscr{B}$ the space of hyperfunctions, $\mathscr{E}^{(\infty)}=\mathscr{E}$ the space of infinitely differentiable functions and $\mathscr{D}^{(\infty)\prime}=\mathscr{D}'$ the spcae of distributions.

Let $P(x, \partial)$ be a *formally hyperbolic operator of constant multiplicity* on a domain Ω in $\boldsymbol{R}^{n+1}$, that is, a linear partial differential operator on Ω satisfying the conditions that the hypersurfaces $\{x_0=\text{constants}\}$ are non-characteristic and that the characteristic polynomial $\sigma(P)\,(x; \xi_0, \xi')$ has only real zeros ξ_0 of locally constant multiplicity for any $x\in\Omega$ and $\xi'\in \boldsymbol{R}^n\setminus 0$.

Let σ_0 be the maximum of the irregularities σ at all characteristic elements of P and set $s_0=\sigma_0/(\sigma_0-1)$. We mean by the *irregularity condition* the following inequality:

(2.2) $$1\leqq s<s_0 \quad \text{if } *=\{s\}, \qquad 1<s\leqq s_0 \quad \text{if } *=(s).$$

When the irregularity condition is satisfied, the associated Cauchy problem is well posed in $\mathscr{E}^*$ and in $\mathscr{D}'^*$. Namely we have the following.

Theorem 1. *Let $P(x,\partial)$ be a formally hyperbolic operator of constant multiplicity, of order m and defined on $\Omega_T=(-T,T)\times \boldsymbol{R}^n$. Suppose that the irregularity condition (2.2) is satisfied for the Gevrey class $*$, the coefficients of $P(x,\partial)$ are in $\mathscr{E}^*(\Omega_T)$ and the characteristic roots ξ_0 are uniformly bounded on $\Omega_T\times S^{n-1}$. Then for any data*

(2.3) $$f\in\mathscr{E}((-T,T),\mathscr{D}^{*\prime}(\boldsymbol{R}^n)),$$

(2.4) $$g_j\in\mathscr{D}^{*\prime}(\boldsymbol{R}^n),\quad j=0,1,\cdots,m-1,$$

the Cauchy porblem

(2.5) $$\begin{cases} P(x,\partial)u(x)=f(x), \\ \partial_0^j u(0,x')=g_j(x'), \quad j=0,1,\cdots,m-1, \end{cases}$$

has a unique solution

(2.6) $$u\in\mathscr{E}((-T,T),\mathscr{D}^{*\prime}(\boldsymbol{R}^n)).$$

If

(2.7) $$f\in\mathscr{E}((-T,T),\mathscr{E}^*(\boldsymbol{R}^n)) \qquad (\text{resp. } \mathscr{E}^*(\Omega_T)),$$

(2.8) $$g_j\in\mathscr{E}^*(\boldsymbol{R}^n), \qquad j=0,1,\cdots,m-1,$$

then

(2.9) $$u\in\mathscr{E}((-T,T),\mathscr{E}^*(\boldsymbol{R}^n)) \qquad (\textit{resp. } \mathscr{E}^*(\Omega_T)).$$

If

(2.10) $$f\in\mathscr{D}^{*\prime}(\Omega_T) \quad \textit{and} \quad \operatorname{supp} f\subset\{x_0\geqq 0\}$$

then there is a unique solution

(2.11) $$u\in\mathscr{D}^{*\prime}(\Omega_T) \quad \textit{with} \quad \operatorname{supp} u\subset\{x_0\geqq 0\}$$

of $P(x,\partial)u(x)=f(x)$.

In case $*=\{1\}$ Bony-Schapira [2] have proved that if $P(x, \partial)$ is a formally hyperbolic operator with real analytic coefficients, then the associated Cauchy problem is well posed in $\mathcal{E}^*=\mathcal{A}$ and in $\mathcal{D}^{*\prime}=\mathcal{B}$ without the assumption of constant multiplicity.

In case $*=(\infty)$ the theorem is due to Mizohata-Ohya [27] and Chazarain [4]. A more elementary proof is in Komatsu [17].

The case $*=\{s\}$ has been discussed by Ohya [29], Leray-Ohya [22, 23], Hamada-Leray-Wagschal [11], Ivriĭ [12] and De Paris-Wagschal [6]. The theorem as it is is in Komatsu [17] including the case $*=(s)$.

We are interested in the necessity of the irregularity condition (2.2) in order that the Cauchy problem be well posed in the class $*$. In case $*=(\infty)$ the necessity of Levi's condition $\sigma_0=1$ has been proved by Mizohata-Ohya [28], Flaschka-Strang [8] and Ivriĭ-Petkov [14]. Our main result is the following.

Theorem 2. *Let $P(x, \partial)$ be a formally hyperbolic operator with real analytic coefficients in a neighborhood of $\mathring{x} \in \boldsymbol{R}^{n+1}$. Suppose that $(\mathring{x}, \mathring{\xi})$ is a non-singular characteristic element at which P has irregularity $\sigma>1$ and that the equation*

$$\sum_{d-d_i=\sigma(m-i)} \sigma(Q_i)(\mathring{x}, \mathring{\xi})\tau^{d_i}=0 \tag{2.12}$$

has distinct roots $\tau_1, \cdots, \tau_d$ all different from zero.

Then there is a neighborhood Ω_0 of $\mathring{x}$ such that for any neighborhood Ω of $\mathring{x}$ in Ω_0 we can find a solution

$$u \in \mathcal{D}^{(s)\prime}(\Omega) \setminus \mathcal{D}^{\{s\}\prime}(\Omega) \tag{2.13}$$

of $P(x, \partial)u(x)=0$ whose Cauchy data

$$\partial_0^j u(\mathring{x}_0, x') \in \mathcal{E}^{\{s\}}(\Omega'), \quad j=0, 1, \cdots, \tag{2.14}$$

where $s=\sigma/(\sigma-1)$ and $\Omega'=\{x' \in \boldsymbol{R}^n \mid (\mathring{x}_0, x') \in \Omega\}$.

We note that the condition on the roots of (2.12) is a generic one under the assumption that P has irregularity σ at $(\mathring{x}, \mathring{\xi})$. We also remark that we assume the constant multiplicity of characteristics only microlocally. Therefore the irregularity condition (2.2) is necessary in general even for a formally hyperbolic equation of variable multiplicity in order that it be well posed in class $*$. In particular, the results of Trépreau [33] and Bronsteĭn [3] cannot be improved in general. For other necessary conditions, which should be equivalent to our irregularity condition (2.2), see Ivriĭ [12] and Mizohata [26]. In [16] we claimed a stronger proposition than Theorem 2 but the proof sketched there did not work except for the case $\sigma=1$ because the kernel theorem [19, 20] gave us estimate (3.12) below only on the real domain.

§ 3. Formal solutions

We employ the methods of Hamada [9, 10] and Ōuchi [30, 31] when they discussed the Cauchy problem in the complex domain for meromorphic data.

Let $P(z, \partial)$ be an analytic extension of $P(x, \partial)$. We construct a *holomorphic phase function* $\varphi(z)$ as a solution of

$$\begin{cases} \sigma(K)(z, \operatorname{grad} \varphi(z))=0 \\ \varphi(\mathring{x})=0 \\ \operatorname{grad} \varphi(\mathring{x})=\mathring{\xi}, \end{cases} \tag{3.1}$$

where $K(z, \partial)$ is an analytic extension of the operator $K(x, \partial)$ in the De Paris decomposition (1.1). We choose $\varphi(z)$ so that it takes real values on the real domain.

Let $w(\chi)$ be an arbitrary *wave form* which is a (generalized) function of one variable χ. By Leibniz' rule we can find partial differential operators $P^i_\varphi(z, \partial)$ of order at most i such that

$$P(z, \partial)\,(w(\varphi(z))u(z))=\sum_{i=0}^{m} w^{(i)}(\varphi)P^{m-i}_\varphi(z, \partial)u(z). \tag{3.2}$$

The operators P^i_φ depend only on φ and do not on w, so that, if we take $w(\chi)=e^{\lambda\chi}$ with a parameter λ, we have

$$P(z, \partial)(e^{\lambda\varphi(z)}u(z))=e^{\lambda\varphi(z)}\sum_{i=0}^{m} \lambda^i P^{m-i}_\varphi(z, \partial)u(z). \tag{3.3}$$

We set

$$P_\varphi(z, \partial, \lambda)=\sum_{i=0}^{m} P^{m-i}_\varphi(z, \partial)\lambda^i. \tag{3.4}$$

A formal power series

$$u(z, \lambda)=\sum_{j=-\infty}^{\infty} u_j(z)\lambda^j \tag{3.5}$$

is called a *formal operator solution* associated with the phase function φ if it satisfies

$$P_\varphi(z, \partial, \lambda)u(z, \lambda)=0. \tag{3.6}$$

Then for any sequence $w^{(j)}(\chi)$ of (generalized) functions such that

(3.7) $$dw^{(j)}(\chi)/d\chi = w^{(j+1)}(\chi)$$

the (generalized) function

(3.8) $$u(z)=u(z, \partial_\varphi)w^{(0)}(\varphi(z))=\sum_{j=-\infty}^{\infty} u_j(z)w^{(j)}(\varphi(z))$$

satisfies the equation

(3.9) $$P(z, \partial)u(z)=0$$

formally in the sense that all the coefficients of $w^{(j)}(\varphi(z))$ vanish. If we take $w^{(j)}(\chi)=\lambda^j e^{\lambda\chi}$, we obtain an asymptotic solution in the sense of Lax [21]. In this case (3.8) does not converge. But for a suitable choice of sequence $w^{(j)}(\chi)$ it may converge in a suitable topology and represent a genuine solution of (3.9).

Improving the estimates by Hamada [10], we have proved in [15] the following estimates for the coefficients of formal operator solutions.

Theorem 3. *Let $P(z, \partial)$ be a linear partial differential operator with holomorphic coefficients and of irregularity σ at the non-singular characteristic element $(\mathring{x}, \mathring{\xi})$ of multiplicity d. If the hypersurface $z_0=\mathring{x}_0$ is transversal to the bicharacteristic curve of $K(z, \partial)$ through $(\mathring{x}, \mathring{\xi})$, then for any holomorphic functions $h_0(z'), \cdots, h_{d-1}(z')$ defined in a neighborhood of $\mathring{x}'$ there is a formal operator solution* (3.5) *on a complex neighborhood Ω_0 of $\mathring{x}$ satisfying the initial condition*

(3.10) $$\partial_0^k u_j(\mathring{x}_0, z')=\delta_{j,0}h_k(z'), \qquad 0\leqq k<d,$$

and the estimates

(3.11) $$|u_j(z)|\leqq C^{-j+1}(-j)!, \qquad j\leqq 0,$$

(3.12) $$|u_j(z)|\leqq\begin{cases} C^{j+1}\left(\dfrac{|z_0-\mathring{x}_0|^j}{j!}\right)^{\sigma/(\sigma-1)}, & \sigma>1 \text{ and } j>0,\\ 0, & \sigma=1 \text{ and } j>0,\end{cases}$$

with a constant C.

Moreover, we can prove that there are no other solutions satisfying (3.10), (3.11) and

(3.13) $$|u_j(z)|\leqq C_\varepsilon \varepsilon^j/j!, \quad j>0,$$

for any $\varepsilon>0$ with a constant C_ε.

With these estimates we can prove the convergence of (3.8) for many wave forms $w^{(j)}(\chi)$. If we take $w^{(j)}(\chi)=f^{(j+k)}(\chi)$, where

$$(3.14)\qquad f^{(j)}(\chi)=\begin{cases}\dfrac{(-1)^j j!}{\chi^{j+1}}, & j\geqq 0,\\[2mm] \dfrac{\chi^{-j-1}}{(-j-1)!}\left(\log\chi-1-\dfrac{1}{2}-\cdots-\dfrac{1}{-j-1}\right), & j<0,\end{cases}$$

then we obtain Hamada's solution [10]. Ōuchi [30, 31] investigated its asymptotic behavior as z tends to the characteristic surface $\{\varphi(z)=0\}$. Earlier Mizohata [25] constructed null solutions and non-analytic solutions by this method when the characteristic element $(\mathring{x}, \mathring{\xi})$ is simple.

The function $w_1(\chi)$ defined by

$$(3.15)\qquad w_1(\chi)=\begin{cases}\exp(-\chi^{-1/(s-1)}), & \chi>0,\\ 0, & \chi\leqq 0,\end{cases}$$

belongs to the Gevrey class $\mathscr{E}^{\{s\}}(\boldsymbol{R})$ but does not to $\mathscr{E}^{(s)}(\boldsymbol{R})$. If $1<s\leqq\sigma/(\sigma-1)$, then it follows from estimates (3.11) and (3.12) that (3.8) converges in $\mathscr{E}^{\{s\}}(\Omega)$ for $w^{(0)}(\chi)=w_1(\chi)$ and its derivatives and primitives with support in $\{\chi\geqq 0\}$. Hence it represents a *null solution* in a neighborhood Ω of $\mathring{x}$ ([15]).

To prove Theorem 2 we consider the case $s=\sigma/(\sigma-1)$ and take $w^{(0)}(\chi)=w_1(c\chi)$ for a $c>0$. Then, as in [15], it is easily proved that (3.8) converges in $\mathscr{D}^{(s)\prime}(\Omega_0)$ and represents a solution u of (3.9). Moreover, it converges in $\mathscr{E}^{\{s\}}(\Omega_1)$ if Ω_1 is a sufficiently small neighborhood of Ω'. In particular, the initial values $\partial_0^j u(\mathring{x}_0, x')$ are all in $\mathscr{E}^{\{s\}}(\Omega')$. In the following sections we show that there are initial data $h_j(z')$ such that for any neighborhood Ω of $\mathring{x}$ the solution u is not in $\mathscr{D}^{\{s\}\prime}(\Omega)$ if c is sufficiently large.

§ 4. The Heaviside calculus

The expression (3.8) is ambiguous because of the constants of integration in $w^{(j)}(\chi)$ for $j<0$. When $w^{(0)}(\chi)$ has support in the half line $[a, \infty)$ (or $(-\infty, b]$), the constants of integration may be fixed so that $w^{(j)}(\chi)$ have support in the same half line. In this case an effective way of computing $w^{(j)}(\chi)$ and $v(\chi, \partial_\chi)w^{(0)}(\chi)$ for the formal power series

$$(4.1)\qquad v(\chi, \lambda)=\sum_{j=-\infty}^{\infty} v_j(\chi)\lambda^j$$

is the Heaviside calculus based on the Laplace transform (see e.g. [34]). Let $w^{(0)}(\chi)$ be a hyperfunction of exponential type and with support in

$[a, \infty)$. Then its Laplace transform

(4.2) $$\hat{w}^{(0)}(\lambda)=\int_{-\infty}^{\infty} e^{-\lambda\chi}w^{(0)}(\chi)d\chi$$

is a holomorphic function on a half plane Re $\lambda>\Lambda_0$. Let $\Lambda>\Lambda_0$ and define

(4.3) $$W^{(j)}(\zeta)=\frac{1}{2\pi i}\int_{\Lambda}^{\infty} e^{\lambda\zeta}\lambda^j\hat{w}^{(0)}(\lambda)d\lambda.$$

Then $W^{(j)}(\zeta)$ is a holomorphic function on $\boldsymbol{C}\diagdown[a, \infty)$ and we have

(4.4) $$w^{(j)}(\chi)=W^{(j)}(\chi+i0)-W^{(j)}(\chi-i0)$$

in the sense of hyperfunction. Hence if (4.1) converges in a half plane Re $\lambda>\Lambda_1$ and is bounded by $C_\varepsilon e^{\varepsilon|\lambda|}$ for any $\varepsilon>0$, then $v(\chi)=v(\chi, \partial_\chi)w^{(0)}(\chi)$ is represented as the boundary value

(4.5) $$v(\chi)=V(\chi+i0)-V(\chi-i0),$$

of the holomorphic function

(4.6) $$V(\zeta)=\frac{1}{2\pi i}\int_{\Lambda}^{\infty} e^{\lambda\zeta}v(\zeta, \lambda)\hat{w}^{(0)}(\lambda)d\lambda,$$

where $\Lambda>\max\{\Lambda_0, \Lambda_1\}$.

Unfortunately Hamada's formal operator solution $u(z, \lambda)$ does not converge as a function because of terms of negative powers. However, if we use integral kernels for iterated integrals,

(4.7) $$u_-(z, \partial_\varphi)=\sum_{j=-\infty}^{-1} u_j(z)\partial_\varphi^j$$

can be realized as an integral operator with bounded kernel for sufficiently small Ω_0 and hence turns out to be a continuous operator in each $\mathscr{E}^*(\Omega)$ (see [15]). On the other hand,

(4.8) $$u_+(z, \lambda)=\sum_{j=0}^{\infty} u_j(z)\lambda^j$$

is an entire function of order $O(\exp B|z_0-\mathring{x}_0||\lambda|^{(\sigma-1)/\sigma})$. Therefore its operation is defined by the integral (4.6). However, as Ōuchi [30, 31] pointed out, it is more convenient to decompose $u(z, \lambda)$ into another way. We will prove the following theorem in §§ 5–7.

Theorem 4. *In addition to the assumptions of Theorem 3 assume that $\sigma > 1$ and that the roots τ_i of* (2.12) *are distinct. Then the formal operator solution $u(z, \lambda)$ obtained in Theorem 3 is written*

$$u(z, \lambda) = u_{\mathrm{I+II}}(z, \lambda) + u_{\mathrm{III}}(z, \lambda), \tag{4.9}$$

where

$$u_{\mathrm{III}}(z, \lambda) = \sum_{j=-\infty}^{-1} u_{\mathrm{III},j}(z)\lambda^{j/q} \tag{4.10}$$

is a formal power series in $\lambda^{-1/q}$ for an integer $q > 0$ whose coefficients have the estimates

$$|u_{\mathrm{III},j}(z)| \leqq M^{1-j}\Gamma(1-j/q) \tag{4.11}$$

with a constant M. The first term $u_{\mathrm{I+II}}(z, \lambda)$ is a power series in $\lambda^{1/q}$ and $\lambda^{-1/q}$ which converges for $|\lambda| > \Lambda_1$ for a Λ_1. On each sector Σ of opening less than $s\pi$ it is decomposed into the sum $u_{\mathrm{I}}(z, \lambda) + u_{\mathrm{II}}(z, \lambda)$ which have the asymptotic expansions

$$u_{\mathrm{I}}(z, \lambda) \sim \sum_{j=1}^{d} e^{\lambda^{\alpha}\psi_i(z,\lambda)} \sum_{j=0}^{\infty} a_{i,j}(z, \lambda)\lambda^{-\alpha j}, \tag{4.12}$$

$$u_{\mathrm{II}}(z, \lambda) \sim \sum_{j=0}^{\infty} b_j(z, \lambda)\lambda^{-\alpha j}, \tag{4.13}$$

as $|\lambda| \to \infty$ in Σ provided that $|z_0 - \mathring{x}_0|$ is sufficiently small, where

$$\alpha = (\sigma - 1)/\sigma = 1/s, \tag{4.14}$$

and $\psi_i(z, \lambda)$, $a_{i,j}(z, \lambda)$, and $b_j(z, \lambda)$ are holomorphic functions of z and $\lambda^{-1/q}$ on $\Omega_0 \times \{\lambda; |\lambda| > \Lambda_1\}$ with the asymptotic behavior

$$\psi_i(z, \lambda) = \tilde{\tau}_i(z_0 - \mathring{x}_0) + O((|z - \mathring{x}| + |\lambda|^{-1/q})^2), \tag{4.15}$$

$$a_{i,j}(z, \lambda) = a_{i,j} + O(|z - \mathring{x}| + |\lambda|^{-1/q}), \tag{4.16}$$

$$b_j(z, \lambda) = b_j + O(|z - \mathring{x}| + |\lambda|^{-1/q}) \tag{4.17}$$

as (z, λ) tends to $(\mathring{x}, \infty)$. Here $\tilde{\tau}_i$, $a_{i,j}$ and b_j are constants. If the roots τ_i of (2.12) *are all non-zero, and $h_0(\mathring{x}') \neq 0$, then we have for any i*

$$\tilde{\tau}_i = \tau_i / b \neq 0, \tag{4.18}$$

and

$$a_{i,0} \neq 0. \tag{4.19}$$

We note that α is a rational number satisfying

(4.20) $$0<\alpha<1.$$

The natural number q in the theorem is the denominator of α:

(4.21) $$\alpha=l/q,\ l \text{ and } q \in \boldsymbol{N} \text{ with } (l, q)=1.$$

Ōuchi considered the case where $\hat{w}^{(0)}(\lambda)=\lambda^k$ and discussed the asymptotic behavior of integral (4.6). In our case we have the following estimates for the Laplace transform of $w^{(0)}(\chi)=w_c(\chi)=w_1(c\chi)$.

Theorem 5. *Let $c>0$. Then the Laplace transform $\hat{w}_c(\lambda)$ is holomorphic in the sector $|\arg \lambda|<\pi s/2$ and in each subsector $|\arg \lambda|\leqq\pi s'/2$, $s'<s$, it has the uniform asymptotic expansion*

(4.22) $$\hat{w}_c(\lambda)=(c_0/c)(\lambda/c)^{-1+1/(2s)}\, e^{-c_1(\lambda/c)^{1/s}}(1+O((\lambda/c)^{-1/s}))$$

as λ tends to ∞, where c_0 and c_1 are positive constants depending only on s.

Proof. We assume that $c=1$. The general case is reduced to this case by a change of variable. In (4.2) we change the variable into $x=\lambda^{(s-1)/s}\chi$. If $|\arg \lambda|<\pi s/2$, we can rotate the path of integration and obtain

(4.23) $$\hat{w}_1(\lambda)=\lambda^{-(s-1)/s}\int_0^\infty \exp(-\lambda^{1/s}(x^{-1/(s-1)}+x))dx.$$

The function $x^{-1/(s-1)}+x$ takes a unique critical value $s/(s-1)^{(s-1)/s}$ at $x=(s-1)^{-(s-1)/s}$ and its second derivative is equal to $s/(s-1)^{1/s}$ at that point. Hence the standard method of steepest descent gives (4.20) with $c_0=(2\pi s^{-1}(s-1)^{1/s})^{1/2}$ and $c_1=s/(s-1)^{(s-1)/s}$ (cf. the proof of Theorem 6 below).

Now the proof of Theorem 2 is not difficult. Firstly estimates (4.11) imply that

$$u_{\mathrm{III}}(z, \partial_\varphi)=\sum_{j=-\infty}^{-1} u_{\mathrm{III},j}(z)\partial_\varphi^{j/q}$$

is an integral operator with bounded holomorphic kernel if Ω_0 is a sufficiently small complex neighborhood of $\mathring{x}$ because $\partial_\varphi^{j/q}$ for $j<0$ are represented by the Riemann-Liouville integrals. Hence $u_{\mathrm{III}}(x, \partial_\varphi)w_c(\varphi(x))$ is a function in $\mathscr{E}^{\{s\}}(\Omega)$ for any real neighborhood Ω of $\mathring{x}$ in Ω_0.

Next in order to prove that $u_{\mathrm{I}+\mathrm{II}}(x, \partial_\varphi)w_c(\varphi(x))$ is not in $\mathscr{D}^{\{s\}\prime}(\Omega)$ we consider the inverse Laplace transforms

$$U_{\mathrm{I}}(z, \zeta)=\frac{1}{2\pi i}\int_{\Lambda}^{\infty} e^{\lambda\zeta}u_{\mathrm{I}}(z, \zeta, \lambda)\hat{w}_c(\lambda)d\lambda, \tag{4.24}$$

$$U_{\mathrm{II}}(z, \zeta)=\frac{1}{2\pi i}\int_{\Lambda}^{\infty} e^{\lambda\zeta}u_{\mathrm{II}}(z, \zeta, \lambda)\hat{w}_c(\lambda)d\lambda, \tag{4.25}$$

where $u_{\mathrm{I}}(z, \lambda)$ and $u_{\mathrm{II}}(z, \lambda)$ have uniform asymptotic expansions (4.12) and (4.13) on the sector $\Sigma=\{\lambda \in C; |\lambda|>\Lambda_1, |\arg\lambda|<\pi s'/2\}$ for any $s'<s$. Clearly U_{I} and U_{II} are holomorphic on $\Omega_0\times\{\zeta \in \boldsymbol{C}; |\arg(-\zeta)|<(1+s')\pi/2\}$ if Ω_0 is a sufficiently small complex neighborhood of $\mathring{x}$.

Let z be a point in Ω_0. Then the exponent

$$\lambda^{\alpha}\psi_i(z, \lambda)-c_1(\lambda/c)^{1/s}=\lambda^{\alpha}(\psi_i(z, \lambda)-c_1c^{-\alpha})$$

of the product $u_{\mathrm{I}}\hat{w}_c$ is expanded into the convergent series

$$t_l(z)\lambda^{l/q}+t_{l-1}(z)\lambda^{(l-1)/q}+\cdots+t_0(z)+t_{-1}(z)\lambda^{-1/q}+\cdots$$

and the leading coefficient $t_l(z)$ has the asymptotic expansion

$$t_l(z)\sim\tilde{\tau}_i(z_0-\mathring{x}_0)-c_1c^{-\alpha}+O(|z-\mathring{x}|^2) \tag{4.26}$$

as z tends to $\mathring{x}$.

We consider the integral

$$U(\zeta)=\frac{1}{2\pi i}\int_{\Lambda}^{\infty} e^{\lambda\zeta}e^{t_l\lambda^{l/q}+\cdots+t_1\lambda^{1/q}}k(\lambda)d\lambda \tag{4.27}$$

assuming that $t=t_l\neq0$ and that $k(\lambda)$ is a holomorphic function of order $O(\lambda^{\beta})$, $\beta \in \boldsymbol{R}$, on the domain $\Sigma=\{\lambda \in C; |\lambda|>\Lambda_1, |\arg\lambda|<\pi s'/2\}$.

If there is a ϑ with $|\vartheta|<\pi s'/2$ such that $|\arg t+\alpha\vartheta|>\pi/2$ $(\alpha=l/q)$, then the second factor of the integrand becomes bounded on the ray from Λ to $e^{i\vartheta}\infty$ and hence $|\zeta|^{\gamma}U(\zeta)$ is bounded on the sector $|\arg(-\zeta)+\vartheta|<\pi/2-\varepsilon$, $\varepsilon>0$, where $\gamma=\max\{\beta+1, 0\}$. Thus it follows that if

$$\arg t+\pi\alpha s'/2>\pi/2 \tag{4.28}$$

(resp.

$$\arg t-\pi\alpha s'/2<-\pi/2\,), \tag{4.29}$$

then $|\zeta|^{\gamma}U(\zeta)$ is bounded on each proper subsector of the sector

$$-\frac{\pi}{2}-\frac{\pi s'}{2}<\arg(-\zeta)<\frac{\pi}{2}-\frac{\pi}{2\alpha}+\frac{\arg t}{\alpha} \tag{4.30}$$

(resp.

$$-\frac{\pi}{2}+\frac{\pi}{2\alpha}-\frac{\arg t}{\alpha}<\arg(-\zeta)<\frac{\pi}{2}+\frac{\pi s'}{2}\Big). \tag{4.31}$$

Theorem 6. *If*

$$|\arg t|<(1+\alpha)\pi/2 \tag{4.32}$$

or more generally if $|\arg t|<(1+\alpha s')\pi/2$, *then integral* (4.27) *has the asymptotic behavior*

$$\begin{aligned} U(\zeta)=c_2\exp\{t^{1/(1-\alpha)}(-\zeta)^{-\alpha/(1-\alpha)}\bar{\omega}(\zeta)\}t^{1/(2-2\alpha)}(-\zeta)^{-(2-\alpha)/(2-2\alpha)} \\ \times(k((-\zeta/t)^{-1/(1-\alpha)}\bar{\mu}(\zeta))+O((-\zeta)^{-\beta/(1-\alpha)+1/(q-l)})) \end{aligned} \tag{4.33}$$

as ζ *tends to zero in a proper subsector of the sector defined by the inequalities*

$$|\arg(-\zeta)-\arg t|<(1-\alpha)s'\pi/2, \tag{4.34}$$

$$\left|\arg(-\zeta)-\frac{\arg t}{\alpha}\right|<\frac{\pi}{2\alpha}-\frac{\pi}{2}. \tag{4.35}$$

Here c_2 *is a constant, and*

$$\bar{\omega}(\zeta)=c_3+\sum_{j=1}^{\infty}\omega_j(-\zeta)^{j/(q-1)}, \tag{4.36}$$

$$\bar{\mu}(\zeta)=c_4+\sum_{j=1}^{\infty}\mu_j(-\zeta)^{j/(q-1)} \tag{4.37}$$

are convergent power series with positive constants c_3 *and* c_4, *and coefficients* ω_j *and* μ_j *depending on* $t_1, \cdots, t_l$.

Proof. Changing the variable λ in (4.27) into

$$\mu=(-\zeta/t)^{1/(1-\alpha)}\lambda=(-\zeta/t)^{q/(q-l)}\lambda, \tag{4.38}$$

we have

$$\begin{aligned} U(\zeta)=\frac{1}{2\pi i}\int_{(-\zeta/t)^{1/(1-\alpha)}\Lambda}^{\infty}\exp\{t^{1/(1-\alpha)}(-\zeta)^{-\alpha/(1-\alpha)}\omega(\mu)\} \\ \times k((-\zeta/t)^{-1/(1-\alpha)}\mu)(-\zeta/t)^{-1/(1-\alpha)}d\mu, \end{aligned}$$

where

$$\omega(\mu)=-\mu+\mu^{l/q}+\tilde{t}_{l-1}(-\zeta)^{1/(q-l)}\mu^{(l-1)/q}+\cdots+\tilde{t}_1(-\zeta)^{(l-1)/(q-l)}\mu^{1/q}$$

with constants $\bar{t}_j$ depending on t_j.

When $|\zeta|$ is sufficiently small, $\omega(\mu)$ has the critical point

$$\bar{\mu}(\zeta)=(l/q)^{q/(q-l)}+\sum_{j=1}^{\infty}\mu_j(-\zeta)^{j/(q-l)},$$

and it takes there the critical value

$$\bar{\omega}(\zeta)=\omega(\bar{\mu}(\zeta))=(l/q)^{l/(q-l)}(q-l)/q+\sum_{j=1}^{\infty}\omega_j(-\zeta)^{j/(q-l)}.$$

Condition (4.34) assures that the corresponding value of λ is in the sector $|\arg\lambda|<s'\pi/2$ if $|\zeta|$ is small. Condition (4.35) implies

$$\operatorname{Re}(t^{1/(1-\alpha)}(-\zeta)^{\alpha/(1-\alpha)}\bar{\omega}(\zeta))>0$$

for sufficiently small $|\zeta|$. Since

$$\frac{\partial^2\omega(\bar{\mu}(\zeta),\zeta)}{\partial\mu^2}=-(l/q)^{-q/(q-l)}(q-l)/q+O((-\zeta)^{1/(q-l)}) \tag{4.39}$$

does not vanish for small $|\zeta|$, the critical point is of the second order. If we take a path of integral which goes through the critical point along the steepest arc, the asymptotic behavior of the integral is determined by the part of the integral near the critical point. Changing the variable into y defined by

$$-y^2=\omega(\mu,\zeta)-\omega(\bar{\mu}(\zeta),\zeta), \tag{4.40}$$

we have

$$\begin{aligned} U(\zeta)&\sim\exp\{t^{1/(1-\alpha)}(-\zeta)^{-\alpha/(1-\alpha)}\bar{\omega}(\zeta)\}(-\zeta/t)^{-1/(1-\alpha)}\\ &\quad\times\frac{1}{2\pi i}\int_{-\varepsilon}^{\varepsilon}\exp\{-t^{1/(1-\alpha)}(-\zeta)^{-\alpha/(1-\alpha)}y^2\}\tilde{k}(y,\zeta)dy, \end{aligned} \tag{4.41}$$

where

$$\tilde{k}(y,\zeta)=k((-\zeta/t)^{-1/(1-\alpha)}\mu(y,\zeta))\partial\mu/\partial y. \tag{4.42}$$

It is easily proved that $(-\zeta)^{\beta/(1-\alpha)}\partial^j\tilde{k}(y,\zeta)/\partial y^j$, $j=0$, 1 and 2, are uniformly bounded in the ε neighborhood of 0 if ζ is in a proper subsector of the sector defined by (4.34) and (4.35) and if $|\zeta|$ is sufficiently small, because $\partial^j\mu/\partial y^j$, $j=1$, 2 and 3, are bounded and $\lambda^{j-\beta}k^{(j)}(\lambda)$, $j=0$, 1 and 2, are uniformly bounded by the Cauchy inequality. Moreover, we have

$$\begin{aligned} \tilde{k}(0,\zeta)&=k((-\zeta/t)^{-1/(1-\alpha)}\bar{\mu}(\zeta))(-\omega^{(2)}(\bar{\mu}(\zeta),\zeta)/2)^{-1/2}\\ &\sim k((-\zeta/t)^{-1/(1-\alpha)}\bar{\mu}(\zeta))(((l/q)^{q/(q-l)}2q/(q-l))^{1/2}+O((-\zeta)^{1/(q-l)})). \end{aligned}$$

Expand $\tilde{k}(y, \zeta)$ in (4.41) into the Taylor series up to order 2 and evaluate the integral. Then we obtain (4.33).

If condition (4.32) is satisfied, then the sector defined by (4.34) and (4.35) intersects the sector $|\arg(-\zeta)|<\pi$.

Suppose that at least one $\tilde{\tau}_i$ in Theorem 4 has non-negative (resp. negative) real part. We take such a $\tilde{\tau}_i$ with the smallest (resp. largest) $|\arg \tilde{\tau}_i|$. Then for any point $x \in \Omega$ sufficiently close to $\mathring{x}$ and satisfying $\varphi(x)=0$ and $x_0-\mathring{x}_0>0$ (resp. <0) we can find a large $c>0$ such that $t_l(z)$ of (4.26) is away from 0 and satisfies (4.32) uniformly for all z in a complex neighborhood Ω_1 of x in Ω_0.

Theorems 5 and 6 enable us to compute the asymptotic expansion of the integrals

$$U_{\mathrm{I},i,j}(z, \zeta)=\frac{1}{2\pi i}\int_{\Lambda}^{\infty} e^{\lambda\zeta} e^{\lambda^{\alpha}\psi_i(z,\lambda)} a_{i,j}(z, \lambda)\lambda^{-\alpha j}\hat{w}_c(\lambda)d\lambda$$

as ζ tends to 0 in a sector Z satisfying (4.34) and (4.35) for $t=t_l(z)$. If we take a sufficiently small Ω_1 and then subdivide Z if necessary, we may assume that all ζ in Z satisfy for each i' either (4.34) and (4.35) or else (4.30) or else (4.31) for all $z \in \Omega_1$. Then only one $U_{\mathrm{I},i',0}(z, \zeta)$ becomes dominant in Z. Hence it follows that there are positive constants C_1 and C_2 such that

$$|U_{\mathrm{I}}(z, \zeta)| \geq C_1 \exp(C_2|\zeta|^{-1/(s-1)}) \tag{4.43}$$

for $z \in \Omega_1$ and $\zeta \in Z$ with sufficienlty small $|\zeta|$.

On the other hand, it is easily proved that $U_{\mathrm{II}}(z, \zeta)$ is bounded on $\Omega_0 \times \{\zeta; |\arg(-\zeta)|<\pi\}$. Therefore setting $\omega=\Omega_1 \cap \boldsymbol{R}^{n+1}$, we have the estimate

$$\sup_{x\in\omega} |U_{\mathrm{I+II}}(x+iy, \varphi(x+iy))| \geq C_3 \exp(C_4|y|^{-1/(s-1)}) \tag{4.44}$$

with positive constants C_3 and C_4 for the defining function $U_{\mathrm{I+II}}(z, \varphi(z))$ of the hyperfunction $u_{\mathrm{I+II}}(x, \partial_\varphi)w_c(\varphi(x))$. This proves that the hyperfunction is not in $\mathscr{D}^{\{s\}\prime}(\Omega)$ by the characterization theorem of ultradistributions of class $\{s\}$ ([18], Petzsche [32] and de Roever [7]).

§ 5. Ōuchi's first solution

We denote the characteristic polynomial $\sigma(A)(z, \xi)$ of a differential operator $A(z, \partial)$ by the corresponding small letter $a(z, \xi)$. A simple calculation yields

$$A_\varphi^0(z, \partial)=a(z, \operatorname{grad}\varphi(z)), \tag{5.1}$$

(5.2) $$A^1_\varphi(z,\partial)=\sum_{k=0}^{n}\frac{\partial a(z,\operatorname{grad}\varphi(z))}{\partial\xi_k}\partial_k+\text{a function}.$$

If

(5.3) $$A(z,\partial)=B(z,\partial)C(z,\partial),$$

then

(5.4) $$A^i_\varphi(z,\partial)=\sum_{j+k=i}B^j_\varphi(z,\partial)C^k_\varphi(z,\partial).$$

Hence we have for the operator $P(z,\partial)$ with the De Paris decomposition (1.1)

(5.5) $$P_\varphi(z,\partial,\lambda)=\sum_{i=0}^{m}\lambda^{i-d_i}\Big\{q_i(z,\operatorname{grad}\varphi)K^1_\varphi(z,\partial)^{d_i}+\sum_{j=1}^{i-d_i}\lambda^{-j}P_i^{d_i+j}(z,\partial)\Big\},$$

where $P_i^{d_i+j}=(Q_iK^{d_i})_\varphi^{d_i+j}$.

We have to solve the equation

(5.6) $$P_\varphi(z,\partial,\lambda)u(z,\lambda)=0,$$

(5.7) $$u(z,\lambda)=\sum_{j=-\infty}^{\infty}u_j(z)\lambda^j,$$

under the initial condition

(5.8) $$\partial_0^k u_j(\mathring{x}_0,z')=\delta_{j,0}h(z'),\quad 0\leqq k<d.$$

Ōuchi [31] transforms $u(z,\lambda)$ as

(5.9) $$u(z,\lambda)=\frac{1}{2\pi i}\int_\Gamma e^{-\lambda^\alpha r}v(z,\lambda,r)dr,$$

where α is the number defined by (4.14) and Γ is the path consisting of the ray from $\infty e^{i\vartheta}$ to $Re^{i\vartheta}$, the circle $\{r=Re^{i\omega};\ \vartheta\leqq\omega<\vartheta+2\pi\}$ and the ray from $Re^{i\vartheta}$ to $\infty e^{i\vartheta}$ with suitable ϑ and R. Since $\lambda^\alpha e^{-\lambda\alpha r}=(-\partial_r)e^{-\lambda^\alpha r}$, the multiplication of u by λ^α is transformed into the differentiation ∂_r of v by integration by parts. Let

(5.10) $$I_0=\{i\in\{0,1,\cdots,m\};\ d-d_i=\sigma(m-i)\},$$

(5.11) $$I_1=\{i\in\{0,1,\cdots,m\};\ i\notin I_0,\ d_i<\infty\}.$$

If $i\in I_1$ and $d_i>d$, then we change $Q_iK^{d_i-d+1}$ into new Q_i so that we always have $d_i<d$. The exponent of λ of each term of (5.5) is written

(5.12) $$i-d_i-j=(1-\alpha)(m-d-j)+\alpha(m-d_i-j)+\beta_i,$$

where $\beta_i=0$ for $i\in I_0$ and $\beta_i>0$ for $i\in I_1$. Transforming $\lambda^{\alpha(m-d_i-j)}$ into $\partial_r^{m-d_i-j}$, we obtain

(5.13) $$P_\varphi(z,\partial,\lambda)u(z,\lambda)=\lambda^{(1-\alpha)(m-d)}\frac{1}{2\pi i}\int_\Gamma e^{-\lambda^\alpha r}L(z,\lambda,\partial_z,\partial_r)v(z,\lambda,r)dr,$$

(5.14) $$L(z,\lambda,\partial_z,\partial_r)=(P_0^d+P_1^d(\lambda))\partial_r^{m-d}+\sum_{k=d+1}^{m}P^k(\lambda)\partial_r^{m-k},$$

where

(5.15) $$P_0^d(z,\partial_z,\partial_r)=\sum_{i\in I_0}q_i(z,\operatorname{grad}\varphi)K_\varphi^1(z,\partial_z)^{d_i}\partial_r^{d-d_i},$$

(5.16) $$P_1^d(z,\lambda,\partial_z,\partial_r)=\sum_{i\in I_1}\lambda^{-\beta_i}q_i(z,\operatorname{grad}\varphi)K_\varphi^1(z,\partial_z)^{d_i}\partial_r^{d-d_i},$$

(5.17) $$P^k(z,\lambda,\partial_z,\partial_r)=\sum_{i=0}^{m}\lambda^{-(1-\alpha)(k-d)-\beta_i}P_i^{k+d_i-d}\partial_r^{d-d_i}.$$

The coefficients of these operators are polynomials in $\lambda^{-1/q}$ and converge to zero as $|\lambda|\to\infty$ except for P_0^d.

Firstly we look for a formal solution $v(z,\lambda,r)$ of

(5.18) $$L(z,\lambda,\partial_z,\partial_r)v(z,\lambda,r)=0$$

of the form

(5.19) $$v(z,\lambda,r)=\sum_{k=0}^{\infty}v_k(z,\lambda,r),$$

(5.20) $$v_k(z,\lambda,r)=\sum_{j=-k}^{\infty}v_{k,j}(z,\lambda)f^{(j)}(r),$$

where $f^{(j)}(r)$ are functions defined by (3.14). If $j\geqq 0$, we have clearly

(5.21) $$\frac{1}{2\pi i}\int_\Gamma e^{-\lambda^\alpha r}f^{(j)}(r)dr=\lambda^{\alpha j}.$$

The same is true for $j<0$ if we choose the argument ϑ of Γ suitably according to the argument of λ. Hence we assign the initial conditions

(5.22) $$\partial_0^l v_{k,j}(\mathring{x}_0,z',\lambda)=\delta_{k,0}\delta_{j,0}h_l(z'),\qquad 0\leqq l<d.$$

We determine v_0 as a solution of

(5.23) $$(P_0^d+P_1^d(\lambda))\partial_r^{m-d}v_0=0,$$

and, when $v_0, \cdots, v_{k-1}$ are determined, v_k as a solution of

$$(P_0^d + P_1^d(\lambda))\partial_r^{m-d} v_k = -\sum_{l=d+1}^{m} P^l(\lambda)\partial_r^{m-l} v_{k+d-l}. \tag{5.24}$$

For $i=0, 1, \cdots, m$ let

$$L_i(z, \partial, \lambda) = \lambda^{-\beta_i} q_i(z, \operatorname{grad}\varphi) K_\varphi^1(z, \partial)^{d_i}. \tag{5.25}$$

Then the coefficients $v_{k,j}$ of v_k are obtained as solutions of

$$L_m(z, \partial) v_{0,j} = -\sum_{i=0}^{m-1} L_i(z, \partial, \lambda) v_{0, j+d_i-d}, \tag{5.26}$$

for $k=0$ and

$$\begin{aligned} L_m(z, \partial) v_{k,j} &= -\sum_{i=0}^{m-1} L_i(z, \partial, \lambda) v_{k, j+d_i-d} \\ &\quad - \sum_{l=d+1}^{m} \lambda^{-(1-\alpha)(l-d)} \sum_{i=0}^{m} \lambda^{-\beta_i} P_i^{l+d_i-d}(z, \partial) V_{k+d-l, j-d+l+d_i-d} \end{aligned} \tag{5.27}$$

for $k>0$ under initial conditions (5.22). Here $v_{k,j}$ for $k<0$ and for $j<-k$ are defined to be zero.

We note that $v_{0,0}$ is independent of λ. The other coefficients $v_{k,j}$ are polynomials in $\lambda^{-1/q}$ whose coefficients are holomorphic functions in z.

Let β be the maximum of β_i for $i \in I_1$. Employing the method of Wagschal [35] and Hamada [10] (cf. [15] and Ōuchi [31] § 3), we obtain the following.

Theorem 7. *The solutions* $v_{k,j}$ *of* (5.22), (5.26) *and* (5.27) *are holomorphic functions in* z *on a neighborhood* Ω_0 *of* $\mathring{x}$ *and there are constants* A, B *and* M *such that the following inequalities hold for* $z \in \Omega_0$ *and* $0<|\lambda|<\infty$:

(i) *If* $j \geqq 0$,

$$|v_{k,j}(z, \lambda)| \leqq M A^k B^{k+j} (1+|\lambda|^{-\beta})^{k+j} |\lambda|^{-(1-\alpha)k} \frac{|z_0 - \mathring{x}_0|^j}{j!}, \tag{5.28}$$

(ii) *If* $-k \leqq j \leqq 0$,

$$|v_{k,j}(z, \lambda)| \leqq M A^k B^{k+j} (1+|\lambda|^{-\beta})^{k+j} |\lambda|^{-(1-\alpha)k} |j|!. \tag{5.29}$$

These estimates prove that

$$v^+(z, \lambda, r) = \sum_{k=0}^{\infty} \sum_{j=0}^{\infty} v_{k,j}(z, \lambda) f^{(j)}(r) \tag{5.30}$$

and

$$v^-(z,\lambda,r)=\sum_{k=0}^{\infty}\sum_{j=-k}^{-1} v_{k,j}(z,\lambda)f^{(j)}(r) \tag{5.31}$$

converge. In fact, we have

$$\begin{aligned}\sum_{k=0}^{\infty}\sum_{j=0}^{\infty}&|v_{k,j}(z,\lambda)||f^{(j)}(r)|\\ &\leqq\sum_{k=0}^{\infty} MA^kB^k(1+|\lambda|^{-\beta})^k|\lambda|^{-(1-\alpha)k}\\ &\quad\times\sum_{j=0}^{\infty} B^j(1+|\lambda|^{-\beta})^j|z_0-\mathring{x}_0|^j/|r|^{j+1},\end{aligned} \tag{5.32}$$

$$\begin{aligned}\sum_{k=0}^{\infty}\sum_{j=-k}^{-1}&|v_{k,j}(z,\lambda)||f^{(j)}(r)|\\ &\leqq\sum_{j=1}^{\infty} MA^j|\lambda|^{-(1-\alpha)j}j|r|^{j-1}\left|\log r-1-\cdots-\frac{1}{j-1}\right|\\ &\quad\times\sum_{k=0}^{\infty} A^kB^k(1+|\lambda|^{-\beta})^k|\lambda|^{-(1-\alpha)k}.\end{aligned} \tag{5.33}$$

Hence there are constants Λ_1, R_1 and a such that if $|\lambda|>\Lambda_1$, then $v^+(z,\lambda,r)$ converges and is holomorphic on the domain $|r|>a|z_0-\mathring{x}_0|$ and $v^-(z,\lambda,r)$ converges and is a multivalued analytic function on the domain $|r|<R_0$ with a logarithmic singularity along the hypersurface $r=0$.

The formal power series $u_{\mathrm{I+II}}(z,\lambda)$ in Theorem 4 is given by

$$u_{\mathrm{I+II}}(z,\lambda)=\frac{1}{2\pi i}\int_\Gamma e^{-\lambda^\alpha r}v^+(z,\lambda,r)dr, \tag{5.34}$$

where Γ is the path of integral (5.9). However, $v^+(z,\lambda,r)$ being single-valued, Γ may be replaced by the circle $\Gamma_0=\{|r|=R\}$ for a sufficiently large R. Since the series (5.30) converges uniformly on Γ_0, the integral is evaluated as

$$u_{\mathrm{I+II}}(z,\lambda)=\sum_{k=0}^{\infty}\sum_{j=0}^{\infty} v_{k,j}(z,\lambda)\lambda^{\alpha j}. \tag{5.35}$$

This is a holomorphic function of z and $\lambda^{1/q}$ on the domain $\Omega_0\times\{\lambda;|\lambda|>\Lambda_1\}$ and we have the estimate

$$|u_{\mathrm{I+II}}(z,\lambda)|\leqq M_{\mathrm{I+II}}\exp\{B(1+|\lambda|^{-\beta})|z_0-\mathring{x}_0||\lambda|^\alpha\} \tag{5.36}$$

with a constant $M_{\mathrm{I+II}}$.

The other term $u_{\mathrm{III}}(z,\lambda)$ is also defined by

$$u_{\mathrm{III}}(z,\lambda)=\frac{1}{2\pi i}\int_\Gamma e^{-\lambda^\alpha r}v^-(z,\lambda,r)dr \tag{5.37}$$

at least formally. In order to have this make sense, we write

$$v^{-}(z, \lambda, r)=\sum_{j=-\infty}^{-1} v_j^{-}(z, \lambda)f^{(j)}(r), \tag{5.38}$$

putting

$$v_j^{-}(z, \lambda)=\sum_{k=-j}^{\infty} v_{k,j}(z, \lambda). \tag{5.39}$$

Since

$$\sum_{k=-j}^{\infty} |v_{k,j}(z, \lambda)| \leqq MA^{-j}|\lambda|^{(1-\alpha)j}(-j)!\sum_{k=0}^{\infty} A^k B^k(1+|\lambda|^{-\beta})^k|\lambda|^{-(1-\alpha)k}, \tag{5.40}$$

$$v_{\mathrm{III},j}(z, \lambda)=\frac{1}{2\pi i}\int_{\Gamma} e^{-\lambda^{\alpha}r}v_j^{-}(z, \lambda)dr \tag{5.41}$$

converges and is equal to $v_j^{-}(z, \lambda)\lambda^{\alpha j}$ for sufficiently large $|\lambda|$.

Now, we define $u_{\mathrm{III}}(z, \lambda)$ by

$$u_{\mathrm{III}}(z, \lambda)=\sum_{j=-\infty}^{-1} v_{\mathrm{III},j}(z, \lambda)=\sum_{j=-\infty}^{-1} v_j^{-}(z, \lambda)\lambda^{\alpha j} \tag{5.42}$$

as a formal power series in $\lambda^{-1/q}$. Let $t=\lambda^{-1/q}$. Then the inequalities (5.40) imply the majorization

$$u_{\mathrm{III}}(z, \lambda) \ll M\sum_{j=-\infty}^{-1} A^{-j}(-j)!\,t^{-qj}\sum_{k=0}^{\infty} A^k B^k(t(1-t)^{-1})^k. \tag{5.43}$$

Since the second factor is majorized by $(1-(1+AB)t)^{-1}$, we have

$$u_{\mathrm{III}}(z, \lambda) \ll M_{\mathrm{III}}\sum_{j=1}^{\infty} C^j\Gamma(1+j/q)t^j \tag{5.44}$$

for some constants M_{III} and C uniformly in $z \in \Omega_0$.

From the above construction it is easily proved that the formal power series $u(z, \lambda)=u_{\mathrm{I+II}}(z, \lambda)+u_{\mathrm{III}}(z, \lambda)$ in $\lambda^{1/q}$ and $\lambda^{-1/q}$ satisfies equation (3.6) and initial condition (3.10). If we write

$$u(z, \lambda)=\sum_{k=0}^{q-1} \lambda^{k/q}u^k(z, \lambda) \tag{5.45}$$

with formal power series $u^k(z, \lambda)$ in λ and λ^{-1}, then each term $u^k(z, \lambda)$ also satisfies the equation and the initial condition (3.10) for $k=0$ and the zero initial condition for $k\neq 0$. Hence it follows from the uniqueness remarked after Theorem 3 that $u(z, \lambda)$ is the solution of Theorem 3. Thus we have (4.11).

§ 6. Ōuchi's second solution

To investigate the asymptotic behavior or $u_{\mathrm{I}+\mathrm{II}}(z, \lambda)$ as $|\lambda|\to\infty$, we construct another solution $\tilde{v}(z, \lambda, r)$ of (5.18) according to Ōuchi [31] § 5. Let ζ and ρ be the dual variables of z and r respectively. In view of (5.14) the characteristic polynomial $l(z, \lambda, \zeta, \rho)$ of $L(z, \lambda, \partial_z, \partial_r)$ is written

$$\begin{aligned} l(z, \lambda, \zeta, \rho) &= (p_0^d(z, \zeta, \rho)+p_1^d(z, \lambda, \zeta, \rho))\rho^{m-d} \\ &\qquad + \sum_{k=d+1}^{m} p^k(z, \lambda, \zeta, \rho)\rho^{m-k} \\ &= \sum_{i\in I_0} q_i(z, \operatorname{grad}\varphi(z))k_\varphi^1(z, \zeta)^{d_i}\rho^{m-d_i}+l_1(z, \lambda, \zeta, \rho), \end{aligned} \tag{6.1}$$

where $l_1(z, \lambda, \zeta, \rho)$ is a polynomial of degree m in ζ and ρ whose coefficients are of order $O(\lambda^{-1/q})$ as $|\lambda|\to\infty$. By (5.2) $k_\varphi^1(z, \zeta)$ is the Hamilton vector field of $K(z, \partial)$ evaluated at $\zeta d\zeta$. Since the hypersurface $z_0=\mathring{x}_0$ is non-characteristic with respect to $K(z, \partial)$, we have

$$b=\frac{\partial k(\mathring{x}, \mathring{\xi})}{\partial \xi_0}\neq 0. \tag{6.2}$$

Hence if we look at the characteristic equation

$$l(z, \lambda, \zeta, \rho)=0 \tag{6.3}$$

as an equation of the ratio $\zeta_0: \rho$, then in a neighborhood of $(z, \lambda, \zeta')=(\mathring{x}, \infty, 0)$ there are d simple roots $\zeta_0^{(i)}(z, \lambda, \zeta')$ close to τ_i/b and $m-d$ roots close to ∞, where τ_i are the roots of (2.12). For each $i=1, \cdots, d$ let $\psi_i(z, \lambda)$ be the solution of

$$\begin{cases} \partial_0\psi_i(z, \lambda)=\zeta_0^{(i)}(z, \lambda, \partial'\psi_i), \\ \psi_i(\mathring{x}_0, z'; \lambda)=0. \end{cases} \tag{6.4}$$

Then $\psi_i(z, \lambda)+r$ are characteristic phase functions of the operator $L(z, \lambda, \partial_z, \partial_r)$.

We mean by *Ōuchi's second solution* the solution $\tilde{v}(z, \lambda, r)$ of (5.18) of the form

$$\tilde{v}(z, \lambda, r)=\sum_{i=1}^{d}\sum_{j=-\infty}^{0} \tilde{v}_{i,j}(z, \lambda)f^{(j)}(\psi_i(z, \lambda)+r) \tag{6.5}$$

which satisfies the initial condition

$$\partial_0^k\tilde{v}(\mathring{x}_0, z', \lambda, r)=\partial_0^k v(\mathring{x}_0, z', \lambda, r), \qquad 0\leqq k<d. \tag{6.6}$$

In view of (5.22) this amounts to the condition

$$\partial_0^k \tilde{v}(\mathring{x}_0, z', \lambda, r) = h_k(z') f^{(0)}(r), \qquad 0 \leqq k < d. \tag{6.7}$$

The initial surface $z_0 = \mathring{x}_0$ is not necessarily non-characteristic with respect to the operator $L(z, \lambda, \partial_z, \partial_r)$ but the method of Hamada [9] and Wagschal [35] in the non-characteristic case works without modification and gives the following result.

Theorem 8. *Let $\Lambda_1 > 0$ be a sufficiently large number and Ω_0 be a sufficiently small complex neighborhood of $\mathring{x}$. Then there are holomorphic functions $\tilde{v}_{i,j}(z, \lambda)$ of z and $\lambda^{-1/q}$ on $\Omega_0 \times \{\lambda; |\lambda| > \Lambda_1\}$ having the estimates*

$$|\tilde{v}_{i,j}(z, \lambda)| \leqq M^{1-j}(-j)! \tag{6.8}$$

with a constant M and such that $\tilde{v}(z, \lambda, r)$ defined by (6.5) *is a solution of* (5.18) *satisfying the initial condition* (6.7).

Moreover, the first terms $\tilde{v}_{i,0}(\mathring{x}_0, z', \lambda)$ on the initial surface are solutions of the linear equations

$$\sum_{i=1}^{d} \tilde{\tau}_i(z')^k \tilde{v}_{i,0}(\mathring{x}_0, z') = \delta_{k,0} h_0(z'), \qquad k = 0, \cdots, d-1, \tag{6.9}$$

where

$$\tilde{\tau}_i(z') = \zeta_0^{(i)}(\mathring{x}_0, z', 0). \tag{6.10}$$

In particular, they are independent of λ and all different from 0 on a neighborhood of $\mathring{z}'$ if $h_0(\mathring{z}') \neq 0$.

The last statement follows from the fact that $\tilde{\tau}(\mathring{x}') = \tau_i / b$ are distinct and different from 0 by the assumption.

The difference

$$v^0(z, \lambda, r) = v^+(z, \lambda, r) + v^-(z, \lambda, r) - \tilde{v}(z, \lambda, r) \tag{6.11}$$

of two solutions $v = v^+ + v^-$ and $\tilde{v}$ of (5.18) satisfies the zero initial condition

$$\partial_0^k v^0(\mathring{x}_0, z', \lambda, r) = 0, \qquad 0 \leqq k < d, \tag{6.12}$$

and is defined on the universal covering space of the domain

$$\begin{aligned}\{(z, \lambda, r);\ z \in \Omega_0,\ |\lambda| > \Lambda_1,\ a|z_0 - \mathring{x}_0| < |r| < R_0, \\ \psi_i(z, \lambda) + r \neq 0\ (i = 1, \cdots, d)\}\end{aligned} \tag{6.13}$$

as was proved by Theorems 7 and 8.

Employing the existence and the uniqueness of solutions of the Goursat problem

$$(6.14)\quad \begin{cases} L(z, \lambda, \partial_z, \partial_r)v(z, \lambda, r)=0, \\ K_\varphi^1(z, \partial_z)^k v(\mathring{x}_0, z', \lambda, r)=0, & 0\leqq k<d, \\ \partial_r^k v(z, \lambda, \mathring{r})=\partial_r^k v^0(z, \lambda, \mathring{r}), & 0\leqq k<m-d, \end{cases}$$

Ōuchi [31] has proved the following.

Theorem 9. *Suppose that* Λ_1 *is sufficiently large. If* $v(z, \lambda, r)$ *is a solution of* (5.18) *on the universal covering space of the domain* (6.13) *and if it satisfies the initial condition* (6.12), *then it is continued to a holomorphic function on*

$$(6.15)\quad \{(z, \lambda, r);\ z\in\Omega_1,\ |\lambda|>\Lambda_1,\ |r|<R_1\},$$

where Ω_1 *is a complex neighborhood of* $\mathring{x}$ *in* Ω_0 *and* $R_1>0$.

§ 7. Proof of Theorem 4

In the definition (5.34) of $u_{\mathrm{I+II}}(z, \lambda)$ we may take for the path of integral the circle $\Gamma=\{r;\ |r|=R\}$ with radius R satisfying $a|z_0-\mathring{x}_0|<R<R_1$. When a direction ϑ is fixed, we regard Γ as the path $\{Re^{i\omega};\ \vartheta\leqq\omega<\vartheta+2\pi\}$ and write Γ_ϑ. Substituting $v^+=\tilde{v}-v^-+v^0$ into (5.34), we have

$$(7.1)\quad u_{\mathrm{I+II}}(z, \lambda)=u_{\mathrm{I}}(z, \lambda)+u_{\mathrm{II}}(z, \lambda)$$

where

$$(7.2)\quad u_{\mathrm{I}}(z, \lambda)=\frac{1}{2\pi i}\int_{\Gamma_\vartheta} e^{-\lambda^\alpha r}\tilde{v}(z, \lambda, r)dr,$$

$$(7.3)\quad u_{\mathrm{II}}(z, \lambda)=\frac{-1}{2\pi i}\int_{\Gamma_\vartheta} e^{-\lambda^\alpha r}v^-(z, \lambda, r)dr.$$

The integral of v^0 disappears by Cauchy's theorem.

In view of the convergent expansions (6.5) and (5.31) we have only to consider the integral

$$(7.4)\quad u_j(z, \lambda)=\frac{1}{2\pi i}\int_{\Gamma_\vartheta} e^{-\lambda^\alpha r}f^{(j)}(\psi(z, \lambda)+r)dr,$$

where

$$(7.5)\quad \psi(z, \lambda)=\sum_{k=0}^\infty t_k(z)\lambda^{-k/q}\quad (\text{or } =0)$$

is convergent on $\Omega_1 \times \{\lambda; |\lambda| > \Lambda_1\}$.

We know that

(7.6)
$$\sup_{|\lambda| \geqq \Lambda_1} |\psi(z, \lambda)| \leqq a|z_0 - \mathring{x}_0|$$

for $\psi(z, \lambda) = \psi_i(z, \lambda)$ with a constant a.

If $j \geqq 0$, then

$$u_j(z, \lambda) = \frac{1}{2\pi i} \int_{\Gamma_\vartheta} e^{-\lambda^\alpha r} \frac{(-1)^j j!}{(\psi(z, \lambda) + r)^{j+1}} dr$$
$$= e^{\lambda^\alpha \psi(z, \lambda)} \lambda^{\alpha j}.$$

If $j < 0$, then we have

$$u_j(z, \lambda) = \int_{-\psi(z,\lambda)}^{Re^{i\vartheta}} e^{-\lambda^\alpha r} \frac{(\psi(z, \lambda) + r)^{-j-1}}{(-j-1)!} dr$$
$$= e^{\lambda^\alpha \psi(z,\lambda)} \int_0^{Re^{i\vartheta} + \psi(z,\lambda)} e^{-\lambda^\alpha \rho} \rho^{-j-1} / (-j-1)! \, d\rho.$$

Suppose that Σ is the sector

(7.7)
$$\Sigma = \{\lambda; |\alpha \arg \lambda + \vartheta| \leqq \pi/2 - \sin^{-1}(a|z_0 - \mathring{x}_0|/R) - \varepsilon\}$$

for an $\varepsilon > 0$. Then we have the uniform estimates

$$-\mathrm{Re}\,(\lambda^\alpha(Re^{i\vartheta} + \psi(z, \lambda)) \leqq -c_5 |\lambda|^\alpha$$

for $\lambda \in \Sigma$ with a positive constant c_5. Hence it follows that

(7.8)
$$u_j(z, \lambda) = e^{\lambda^\alpha \psi(z,\lambda)} (\lambda^{\alpha j} + O(e^{-c_6 |\lambda|^\alpha}).$$

This completes the proof of Theorem 4 and hence Theorem 2.

References

[1] T. Aoki, An invariant measuring the irregularity of a differential operator and a microdifferential operator, J. Math. Pures Appl., **61** (1982), 131–148.

[2] J.-M. Bony et P. Schapira, Solutions hyperfonctions du problème de Cauchy, Lecture Notes in Math., **287**, 1973, pp. 82–98.

[3] M. D. Bronsteĭn, The Cauchy problem for hyperbolic operators with characteristics of variable multipilicity, Trans. Moscow Math. Soc., **41** (1980), 87–103 (Original Russian: Trudy Moskov. Mat. Obšč., **41** (1980), 83–99).

[4] J. Chazarain, Opérateurs hyperboliques à caractéristiques de multiplicité constante, Ann. Inst. Fourier, Grenoble, **24** (1974), 173–202.

[5] J.-C. De Paris, Problème de Cauchy oscillatoire pour un opérateur différentiel à caractéristiques multiples; lien avec l'hyperbolicité, J. Math. Pures Appl., **51** (1972), 231–256.

[6] J.-C. De Paris et C. Wagschal, Problème de Cauchy non caractéristique à données Gevrey pour un opérateur analytique à caractéristiques multiples, J. Math. Pures Appl., **57** (1978), 157–172.

[7] J. W. de Roever, Hyperfunctional singular support of ultradistributions, J. Fac. Sci., Univ. Tokyo, Sec. IA, **31** (1985), 585–631.

[8] H. Flaschka and G. Strang, The correctness of the Cauchy problem, Advances in Math., **6** (1971), 347–379.

[9] Y. Hamada, The singularities of the solutions of the Cauchy problem, Publ. RIMS, Kyoto Univ., **5** (1969), 21–40.

[10] ——, Problème analytique de Cauchy à caractéristiques multiples dont les données de Cauchy ont des singularités polaires, C. R. Acad. Sci. Paris, Sér. A, **276** (1973), 1681–1684.

[11] Y. Hamada, J. Leray et C. Wagschal, Système d'équations aux dérivées partielles à caractéristiques multiples: problème de Cauchy ramifié; hyperbolicité partielle, J. Math. Pures Appl., **55** (1976), 297–352.

[12] V. Ya. Ivriĭ, Conditions for correctness in Gevrey classes of the Cauchy problem for weakly hyperbolic equations, Siberian Math. J., **17** (1976), 422–435 (Original Russian: Sibirsk. Mat. Ž., **17** (1976), 547–563).

[13] ——, Cauchy problem conditions for hyperbolic operators with characteristics of variable multiplicity for Gevrey classes, Siberian Math. J., **17** (1976), 921–931 (Original Russian: Sibirsk. Mat. Ž., **17** (1976), 1256–1270).

[14] V. Ya. Ivriĭ and V. M. Petkov, Necessary conditions for the Cauchy problem for non-strictly hyperbolic equations to be well-posed, Russian Math. Surveys, **29** (1974), no. 5, 1–70 (Original Russian: Uspehi Mat. Nauk, **29** (1974), no. 5, 3–70).

[15] H. Komatsu, Irregularity of characteristic elements and construction of null-solutions, J. Fac. Sci., Univ. Tokyo, Sec. IA, **23** (1976), 297–342.

[16] ——, Irregularity of characteristic elements and hyperbolicity, Publ. RIMS., Kyoto Univ., **12** Suppl. (1977), 233–245.

[17] ——, Linear hyperbolic equations with Gevrey coefficients, J. Math. Pures Appl., **59** (1980). 145–185.

[18] ——, Ultradistributions, I, Structure theorems and a characterizations, J. Fac. Sci., Univ. Tokyo, Sec. IA, **20** (1973), 25–105.

[19] ——, Ultradistributions, II, The kernel theorem and ultradistributions with support in a submanifold, J. Fac. Sci., Univ. Tokyo, Sec. IA, **24** (1977), 607–628.

[20] ——, Ultradistributions, III, Vector valued ultradistributions and the theory of kernels, J. Fac. Sci., Univ. Tokyo, Sec. IA, **29** (1982), 653–718.

[21] P. D. Lax, Asymptotic solutions of oscillatory initial value problems, Duke Math. J., **24** (1957), 627–646.

[22] J. Leray et Y. Ohya, Systèmes linéaires, hyperboliques non stricts, Colloque C. B. R. M., 1964, pp. 105–144.

[23] ——, Equations et systèmes non-linéaires, hyperboliques non-stricts, Math. Ann., **170** (1967), 167–205.

[24] S. Matsuura, On non-strict hyperbolicity, Proc. Internat. Conf. on Functional Analysis and Related Topics, 1969, Univ. of Tokyo Press, Tokyo, 1970, pp. 171–176.

[25] S. Mizohata, Solutions nulles et solutions non analytiques, J. Math. Kyoto Univ., **1** (1962), 272–302.

[26] ——, Sur l'indice de Gevrey, Propagation des Singularités et Opérateurs Différentiels, Séminaire de Vaillant 1984–1985, Hermann, Paris, 1985, pp. 106–120.

[27] S. Mizohata et Y. Ohya, Sur la condition de E. E. Levi concernante des équations hyperboliques, Publ. RIMS, Kyoto Univ., **4** (1968), 511–526.
[28] ——, Sur la condition d'hyperbolicité pour les équations à caractéristiques multiples, II, Japan. J. Math., **40** (1971), 63–104.
[29] Y. Ohya, Le problème de Cauchy pour les équations hyperboliques à caractéristique multiple, J. Math. Soc. Japan, **16** (1964), 268–286.
[30] S. Ōuchi, Asymptotic behaviour of singular solutions of linear partial differential equations in the complex domain, J. Fac. Sci., Univ. Tokyo, Sec. IA, **27** (1980), 1–36.
[31] ——, An integral representation of singular solutions of linear partial differential equations in the complex domain, J. Fac. Sci., Univ. Tokyo, Sec. IA. **27** (1980), 37–85.
[32] H.-J. Petzsche, Generalized functions and the boundary values of holomorphic functions, J. Fac. Sci., Univ. Tokyo, Sec. IA, **31** (1984), 391–431.
[33] J.-M. Trépreau, Le problème de Cauchy hyperbolique dans les classes d'ultrafonctions et d'ultradistributions, Comm. Partial Differential Equations, **4** (1979), 339–387.
[34] B. Van der Pol and H. Bremmer, Operational Calculus Based on the Two-sided Laplace Integral, 2nd ed., Cambridge Univ. Press, 1959.
[35] C. Wagschal, Problème de Cauchy analytique, à données méromorphes, J. Math. Pures Appl., **51** (1972), 375–397.

DEPARTMENT OF MATHEMATICS
FACULTY OF SCIENCE
UNIVERSITY OF TOKYO
TOKYO 113
JAPAN

Taniguchi Symp. HERT
Katata 1984, pp. 181–192

Propagation for the Wave Group of a Positive Subelliptic Second-Order Differential Operator

Richard Melrose

§ 1. Introduction

In this note two propagation results, one local the other microlocal, for the wave group of a subelliptic non-negative self-adjoint second order differential operator on a compact manifold are discussed. Let A be such a second-order non-negative differential operator on the compact manifold X:

$$\langle Au, u\rangle = \langle u, Au\rangle \geq 0, \qquad \forall\, u \in C^\infty(X),$$

where:

$$\langle u, v\rangle = \int u(x)\bar{v}(x)\nu$$

is the inner product with respect to some positive C^∞ density ν. For the most part we shall assume that A is (microlocally) subelliptic:

(1.1) There exists $s>0$ such that $Au \in H^k$ microlocally near any point $\rho \in T^*X$ implies $u \in H^{k+2s}$ near ρ.

This of course implies subellipticity:

$$\langle Au, u\rangle \geq \langle \Delta^s u, u\rangle - c\|u\|^2 \quad s>0, \quad u \in C^\infty(X), \tag{1.2}$$

where Δ is the Laplacian of any metric. The assumption of formal self adjointness implies that A has a self-adjoint extension with the domain:

$$\mathscr{D}(A) = \{u \in C^{-\infty}(X);\ Au \in L^2(X)\}.$$

By the spectral theorem the self-adjoint operator

$$G(t) = A^{-1/2} \sin(tA^{1/2}) \tag{1.3}$$

is a well-defined bounded operator on $L^2(X)$, in fact it maps $L^2(X)$ into $\mathscr{D}(A^{1/2})$. Together with the self-adjoint operator:

$$G'(t) = \cos(tA^{1/2}),$$

this allows the Cauchy problem for the wave opeator:

$$(1.4) \qquad \begin{aligned} (D_t^2 - A)u = Pu = 0 \qquad &\text{in } \boldsymbol{R}\times X, \\ u = u_0, \quad D_t u = u_1 \quad &\text{at} \quad t = 0, \end{aligned}$$

to be solved by:

$$(1.5) \qquad u(t, x) = G'(t)u_0 + G(t)u_1.$$

The unitary wave group is defined by:

$$U(t)(u_0, u_1) = (u(t, \cdot), D_t u(t, \cdot)).$$

In § 3 it is shown that P has finite propagation speed, using the notion of A-distance $\delta_A(x, y)$, $x, y \in X$, derived from a discussion of rays (A-geodesics) in § 2. This is most simply expressed in terms of the support of the Schwartz' kernel, $G \in C^{-\infty}(\boldsymbol{R}\times X\times X)$, of G:

$$(1.6) \qquad \operatorname{supp}(G) \subset \{(t, x, y) \in \boldsymbol{R}\times X\times X;\ \delta_A(x, y) \le |t|\}.$$

In fact subellipticity is not needed here. This result is used in § 4 to derive a bound for the heat kernel, when A is subelliptic. Namely the Schwartz' kernel of

$$H(t) = \exp(-tA) \qquad t \ge 0,$$

satisfies, near $t = 0$:

$$(1.7) \qquad H(t, x, y) \le Ct^{-k}\exp(-\delta_A(x, y)/4t) \qquad \text{as } t \downarrow 0.$$

For subelliptic sums of squares of vector fields a similar result has recently been proved by D. Jerison and A. Sanchez (private communication).

In § 6, using results from §§ 2,5 the following theorem on the 'classical' propagation of singularities for P is discussed.

(1.8) **Theorem.** *If A is a self-adjoint non-negative real subelliptic differential operator of second order then*:

$$WF(G) \subset \{(t, x, y, \tau, \xi, -\eta) \in \boldsymbol{T}^*(\boldsymbol{R}\times X\times X);$$
$$\text{there is a null ray of length } |t| \text{ from } (y, \eta) \text{ to } (x, \xi)\}.$$

The same result is true if A is non-negative (not necessarily real) and hypoelliptic with loss of one derivative, or is a subelliptic Hermitian form in vector fields.

The subelliptic real second-order differential operators have been characterized by Fefferman and Phong.

It is important to note that this ray relation is quite computable in examples and seems to correspond to the best conic propagation result. It has been previously obtained in various special cases. In particular in the elliptic case it is Hörmander's theorem originating from [2]. In case A has only double characteristics on a symplectic submanifold it was obtained in [5] (in codimension 2) and by B. and R. Lascar [3], [4] in the general case. Other cases such as degenerate constant coefficient operators and operators with involutive double characteristics are not included in the theorem stated here, because of the assumption of subellipticity. However, it is only at the last step in the argument that difficulties arise and it seems likely that similar methods can be applied.

§ 2. Rays

Let X be a compact C^∞ manifold with $a \in C^\infty(T^*X\diagdown 0)$ a non-negative function homogeneous of degree two:

$$a(x, \xi)\geq 0, \quad a(x, r\xi)=r^2a(x, \xi), \quad r>0,$$

in any canonical coordinates (x, ξ). We will consider both forward-pointing and forward-null rays for the Hamiltonian

$$p=\tau^2-a \in C^\infty(M), \qquad M=[T^*\boldsymbol{R}\times T^*X]\diagdown 0,$$

where (t, τ) are the natural canonical coordinates on $T^*\boldsymbol{R}$. These rays will be defined in terms of the microlocal propagation cones, since we need to allow rays which do not have the natural bicharacteristic parametrization.

In the cotangent bundle the forward and forward-null sets of p are defined by:

$$M_+=\{m \in M; p(m)\geq 0, \tau\geq 0\},$$
$$\Sigma=\{m \in M; p(m)=0, \tau\geq 0\}.$$

For $m \in M_+$ consider the set:

$$\mathscr{H}_m=\boldsymbol{R}^+\cdot H_p(m)\subset T_mM,$$

where H_p is the Hamilton vector field of p, defined in terms of ω, the symplectic 2-form on M. Of course if m is a double characteristic for p then

$\mathscr{H}_m=\{0\}$. We shall therefore extend the definition of the 'bicharacteristic directions' at m.

Let $\Sigma_{(2)}$ be the subset of Σ on which $a=0$. If $m \in \Sigma_{(2)}$ the Hessian, a_m, of a is well-defined. Then $p_m=(d\tau)^2-a_m$, $d\tau$ being a linear function on $T_m M$. More precisely, let $\Sigma_2 \subset \Sigma_{(2)}$ be the subset where a_m is not identically zero and Σ_3 the complementary subset, of points m such that a_m vanishes identically. For $m \in \Sigma_{(2)}$ set

$$\Lambda_m=\{v \in T_m M;\ d\tau(v)\geq 0,\ p_m(v)\geq 0\}, \tag{2.1}$$

the forward-pointing microlocal normal cone at m. Thus, Λ_m is a half-space if $m \in \Sigma_3$. In any case define:

$$\Gamma_m=\{v \in T_m M;\ \omega(v, w)\leq 0 \quad \forall\, w \in \Lambda_m\}, \tag{2.2}$$

the microlocal propagation cone at m, obtained from Λ_m by duality with respect to the symplectic form. If $m \in \Sigma_0=M_+\backslash\Sigma$ or $m \in \Sigma_1=\Sigma\backslash\Sigma_{(2)}$ then set:

$$\Gamma_m=\mathscr{H}_m.$$

Notice that Γ_m is continuous in its variation on $\Sigma_{(2)}$ and on $\Sigma_0 \cup \Sigma_1$.

(2.3) **Definition.** A forward-pointing ray for p is a Lipschitz curve $\gamma: I \rightarrow M_+$ defined on some interval $I \subset \boldsymbol{R}$ with set-valued derivative $\gamma'(s) \subset \Gamma_{\gamma(s)}$, $\forall\, s \in I$. Such a ray is forward-null if $\gamma(s) \in \Sigma\ \forall\, s \in I$.

To anlayse this definition we need to examine the defining cones Γ_m.

(2.4) **Lemma.** *The assignment $(m, a) \mapsto \Gamma_m$ is inner semi-continuous in the C^∞ topology on a in its set of definition.*

Proof. The homogeneity of a is not used in the definition of the propagation cones so in examining their behavior along a sequence $\{m_j\}$ of points and $\{a_j\}$ of non-negative functions the variation of m can be reduced to $m_j=(0, \tau_j, 0)$ by a translation, at the expense of introducing C^∞ parameters in a. The translation is symplectic, so only the inner semi-continuity under simultaneous change of a and τ needs to be considered. Thus consider a sequence $\{a_j\}$ of positive functions in a neighborhood of 0 in $\boldsymbol{R}^k$ converging uniformly with all derivatives to a. Passing to subsequences it is enough to consider separately the two cases $a_j(0)=0$ and $a_j(0)>0$ for all j. In the first case $\tau_j=0$ and the microlocal propagation cones are defined by (2.2) for all j. Since the Hessians converge to the Hessian of a there is clearly continuity of both normal and propagation cones.

Thus it can be supposed that $a_j(0)>0$ for all j. Of course if $a(0)>0$ then the propagation cone consists of just the positive multiples of the Hamilton vector field and convergence is again obvious, so it can be assumed that $a(0)=0$, so $\tau=0$. By assumption $m \in M_+$ with respect to each element of the sequence, so:

$$\tau_j \geq (a_j(0))^{1/2}. \tag{2.5}$$

The inner continuity now consists in the statement that if $\{v_j\}$ is a convergent sequence in TM with v_j a positive multiple of the Hamilton vector field of a_j at $m(j)=(0, \tau_j, 0)$ then the limit $v \in \Gamma_m$, computed with respect to a. Consider the form of v_j:

$$v_j = r_j[\partial_t + \tfrac{1}{2}(a_j(0)^{1/2}/\tau_j)a_j(0)^{-1/2}\Xi_j]$$

where Ξ_j is the Hamilton vector field of a_j at 0. In view of (2.5) we can always pass to a subsequence on which $a_j(0)^{1/2}/\tau_j$ converges in [0, 1], so it is enough to show that every limit point of $\partial_t + \frac{1}{2}a_j(0)^{-1/2}\Xi_j$ is in the propagation cone of a at m.

Unraveling the definition (2.2) it is easily seen that, computed with respect to a,

$$\begin{aligned} &\Gamma_m = \boldsymbol{R}^+(\partial_t + B), \\ &B = \operatorname{cxhl}(\lim \tfrac{1}{2}a_m^{-1/2}\Xi) \end{aligned} \tag{2.6}$$

where the limit in the definition of B is over sequences of tangent vectors converging to 0 and along which a_m does not vanish and with Ξ the Hamilton vector field of a_m at v. Thus it is enough to show that all limits of

$$a_j(0)^{-1/2}da_j(0) \tag{2.7}$$

are in the convex hull of the same limit set for a_m. In the directions of the kernel of a_m, all limits of (2.7) must vanish since in such variables the a_j form a sequence of positive functions, tending to 0 at 0 with second derivatives also tending to 0 at 0. Thus it can be assumed that a_m is positive definite; the cubic terms in a_j can not then affect the limits of (2.7) so it suffices to assume that a_j is a quadratic polynomial, positive near 0. The result is then easily proved.

This inner continuity of the cone fields leads to stability results for the rays they define. From the definition of the cones Γ_m there is a uniform constant c, such that:

$$v \in \Gamma_m \Rightarrow v = T\partial_t + v', \qquad |v'| \leq cT,$$

where v' is tangent to T^*X. Thus, along any forward-pointing ray t is a Lipschitz function of the parameter with strictly positive dreivative. So t can always be introduced as a Lipschitz parameter for γ. From now on by a *ray* we shall mean a forward-pointing ray which is parametrized by t, such a ray being null if it lies in Σ.

(2.8) **Lemma.** *Let $\{a_j\}$ be a sequence of non-negative functions converging in C^∞ to a and suppose $\{\gamma(j)\}$ is a sequence of rays for the corresponding p_j. If the end points of the $\gamma(j)$ converge in M and the length:*

$$\mathscr{L}(\gamma(j))=\int_{\gamma(j)} dt$$

is bounded above then any subsequence of $\{\gamma(j)\}$ has a subsequence which converges uniformly to a ray for p, the Hamiltonian defined by a.

Proof. From the definition of a ray there is a uniform Lipschitz estimate on the sequence of rays so any subsequence does indeed have a uniformly convergent subsequence. It remains only to show that any such limit is a ray of p, where we can assume that $\gamma_j \to \gamma$ uniformly in I. For $t' \in I$ consider $\bar{m}=\gamma(t')$. We work in some canonical local coordinates near $\bar{m}$. Let Γ be a closed conic neighborhood of $\Gamma_{\bar{m}}$. By the inner semicontinuity proved in Lemma 2.4 for $j>J(\Gamma)$ and some $\varepsilon>0$ each curve γ_j satisfies:

$$\gamma_j(t) \in \gamma_j(t')+\Gamma, \qquad t \in [t'-\varepsilon,\ t'+\varepsilon].$$

Thus for $t \in [t'-\varepsilon,\ t'+\varepsilon]$, $\gamma(t) \in \bar{m}+\Gamma$, hence any limit of the difference quotient for γ at t' is in the cone Γ, hence in $\Gamma_{\bar{m}}$. This proves that γ is a ray of p as desired.

Of course it follows immediately that a limit of null rays of the p_j is a null ray of p.

§ 3. Propagation speed

Let A be a non-negative self-adjoint differential operator of second order on a compact manifold, X. If Δ is the Laplacian with respect to some fixed Riemannian metric on X then the regularized operator:

$$A_\varepsilon=A+\varepsilon\Delta, \qquad \varepsilon>0,$$

is non-negative and self-adjoint with domain the Sobolev space $H^2(X)$, the same as Δ; this follows simply from the ellipticity of A_ε. Moreover,

(3.1) $$\lim_{\varepsilon \downarrow 0} A_\varepsilon=A \qquad \text{on } H^2(X).$$

It follows easily from this that the wave kernels converge:

(3.2) $$G_\varepsilon(t, x, y) \longrightarrow G(t, x, y) \quad \text{as} \quad \varepsilon \downarrow 0 \quad \text{in} \quad C^{-\infty}(\boldsymbol{R}\times X \times X; \Omega).$$

Indeed it follows from (3.1) that the groups converge weakly on a dense subset, and hence weakly by unitarity.

(3.3) **Definition.** The A-distance on X is:

$$\delta_A(x, y) = \inf \mathscr{L}(\gamma)$$

with the infimum taken over rays with initial point in M projecting to x and terminal point projecting to y. If there is no such linking ray then:

$$\delta_A(x, y) = \infty.$$

(3.4) **Proposition.** *If A is a positive self-adjoint differential operator fo second order on X then*:

$$\operatorname{supp} G \subset \{(t, x, y) \in \boldsymbol{R}\times X \times X; \delta_A(x, y) \leq |t|\}.$$

Proof. For an elliptic operator A this is a standard result. Suppose that (t, x, y) is in the support of G. From the distributional convergence (3.2) there must be a sequence of points (t_j, x_j, y_j) in the support of $G_j = G_{\varepsilon(j)}$ where $\varepsilon(j) \rightarrow 0$, with limit (t, x, y). In the elliptic case there is always a ray for A_ε linking x_j to y_j and with length $|t_j|$. From the convergence result on rays, Lemma 2.8, there is a ray of A linking x to y and of length $|t|$, so $\delta_A(x, y) \leq |t|$, proving the Proposition.

§ 4. Exponential vanishing of the heat kernel

If A is any non-negative self-adjoint operator on a Hilbert space the heat semigroup $\exp(-tA)$, $t \geq 0$ is well-defined as a strongly continuous semigroup of contractions. Trivializing density bundles for simplicity of notation, $\exp(-tA)$ has Schwartz kernel

$$H(t; x, y) \in C^{-\infty}(\boldsymbol{R}^+ \times X \times X).$$

The integral formula:

$$\exp(-t\lambda^2) = \int \kappa(t, s) \cos(s\lambda) ds, \qquad t > 0,$$

where

(4.1) $$\kappa(t, s) = (\pi/t)^{1/2} \exp(-s^2/4t)$$

shows that the heat kernel can be recovered from the wave kernel by integration over time lines:

(4.2) $$H(t, x, y)=\int \kappa(t, s)G'(s, x, y)ds.$$

(4.3) **Proposition.** *The heat kernel of a self-adjoint non-negative subelliptic differential operator acting on sections of an Hermitian vector bundle over a compact manifold satisfies, for some constants C and k, an estimate:*

$$H(t, x, y)\leq Ct^{-k} \exp(-\delta_A(x, y)^2/4t),$$

where $\delta_A(x, y)$ *is the 'ray distance' between x and y.*

Proof. According to Proposition 3.4 the wave kernel, $G(t, x, y)$ has its support in the region $\delta_A(x, y)\leq|t|$. The unitarity of $U(t)$ and the hypoellipticity of A show G to be a C^∞ function of x and y, jointly, with values in the tempered distributions in $(0, \infty)$, supported in $t\geq\delta_A(x, y)$. Thus the representation formula (4.2) gives a bound for $H(t, x, y)$ in terms of

$$\sup\{(s+s^{-1})^k|\partial_s^p\kappa(t, s)|;\ s\geq\delta_A(x, y),\ k, p\leq m\}$$

for some m. This gives an inequality as claimed.

§ 5. Microlocal time functions

Recall the notation of § 2. If $\bar{m}\in\Sigma$ is a characteristic point for the Hamiltonian p, defined from a positive function a, then we wish to consider the notion of a microlocal time function at $\bar{m}$. In § 2, only forward-pointing rays were analysed; for backward-pointing rays observe that:

$$M_-=\{m\in M;\ p(m)\geq 0,\ \tau\leq 0\}$$

is just the image of M_+ under the involution sending τ to $-\tau$. We shall define the microlocal (forward) propagation cones in M_- using this involution:

(5.1) $$\Gamma_m=\Gamma_{m'},\quad m=(t, \tau, \alpha)\in M_-,\quad m'=(t, -\tau, \alpha)\in M_+$$

(really the new cones are the images of the old under the involution). This normalization has the advantage of consistency at points to which both definitions apply, i.e. in

$$M_+\cap M_-=\Sigma_{(2)}.$$

The backwards propagation cones at $M_\pm$ are then by definition the negatives of these cones, the $-\Gamma_m$.

The definition (5.1) does however hide a sign change, namely for the regular part,

$$m \in [\Sigma_0 \cup \Sigma_1] \cap M_- \Rightarrow \mathscr{H}_m = -\Gamma_m.$$

With this in mind, a C^∞ function ϕ defined near $\bar{m}$ will be called a time function at $\bar{m}$ if in some neighborhood N of $\bar{m}$,

(5.2) ϕ is non-increasing along Γ_m, $m \in N \cap (M_+ \cup M_-)$.

To reiterate, this means that ϕ is weakly decreasing along the Hamilton vector field in M_+ but weakly increasing along the Hamilton foliation in M_-.

Before proceeding to discuss the existence of microlocal time functions with useful additional porperties we first note why it is enough to have conditions on the sign of the derivative in the bicharacteristic directions in M_+ and M_-.

(5.3) **Lemma.** *Let ϕ be a time function near $\bar{m} \in \Sigma_{(2)}$ which is independent of the variable τ. Then there exists a C^∞ function ψ near m such that*

(5.4) $$\tau\{p, \phi\} - \psi p \geq 0$$

in a neighborhood of $\bar{m}$.

Proof. Since $\pm\tau \geq 0$ on $M_\pm$ it follows directly from the definition that:

(5.5) $$q = \tau\{p, \phi\} \geq 0 \qquad \text{on } M_+ \cup M_-.$$

Now q is a quadratic polynomial in τ, vanishing at $\tau = 0$:

$$q = a\tau^2 - b\tau, \quad p = \tau^2 - c, \quad c \geq 0.$$

From (5.5), $a \geq 0$ and, if $a \neq 0$ then the other zero of q, $\tau = b/a$ must lie in the interval $[-c^{1/2}, c^{1/2}]$. Thus,

$$b^2 \leq a^2 c.$$

Then, $\psi = \frac{1}{2}a$ satisfies the requirements, since:

$$q - \tfrac{1}{2}ap = \tfrac{1}{2}a(\tau - b/a)^2 + (a^2c - b^2)/2a.$$

Since Γ_m is always a proper convex cone there is no difficulty in finding microlocal time functions near any given point. If ϕ is C^∞ near m and has differential in the dual cone, in T_m^*M, to Γ_m then ϕ is obviously a time function near m, because of the inner continuity of the propagation cones. We need more particular microlocal time functions.

(5.6) **Lemma.** *Suppose that $\mathscr{K}_0$, $\mathscr{K}_1$, $\mathscr{K}_2$ are three closed proper convex cones with C^∞ bases and contained in $\{x \in \boldsymbol{R}^k; x_1 > 0\} \cup \{0\}$, with $(1, 0, \cdots, 0) \in \mathscr{K}_i$, $i=0, 1, 2$ and*

$$\mathscr{K}_i \setminus 0 \subset \operatorname{int}(\mathscr{K}_{i+1}) \qquad i=0, 1.$$

Then there exists $\varepsilon > 0$ and a C^∞ function, ϕ, defined in $x_1 \geq -\varepsilon$ which is decreasing in the directions of $\mathscr{K}_0$, has $\partial\phi/\partial x_1(0) = -1$ and

$$\operatorname{supp}(\partial\phi/\partial x_1) = \operatorname{supp}(\phi) \subset \mathscr{F}$$

where

(5.7) $$\mathscr{F} = \{x \in \boldsymbol{R}^k; (x - \mathscr{K}_1) \cap \{x_1 = -\varepsilon\} \subset (-\mathscr{K}_1) \cap \{x_i = -\varepsilon\}\}.$$

The direct construction of such a function ϕ is straightforward. Using it one can construct more sophisticated functions of the same type. For example, by integrating the square of such a function with respect to t it can be arranged that:

$$\phi_t' = -(\psi)^2.$$

In the circumstances where such functions are required below there is an overall $\boldsymbol{R}^+$-action, a homogeneity. Clearly ϕ can be taken homogeneous of any degree and still satisfy the support and positivity conditions in the base.

§ 6. Outline or proof of Theorem 1.8

Suppose that $\Phi(t, x, D_x)$ is a compactly supported self-adjoint pseudodifferential operator, with respect to the measure ν. The symbol of Φ will be a homogeneous time functions as described following Lemma 5.6, cut off near $t = -\varepsilon$. Then if $u \in C^\infty(\boldsymbol{R} \times X)$ and P is given by (1.1) with A self-adjoint with respect to the same measure:

$$\operatorname{Im}\langle \Phi D_t u, Pu \rangle = \langle \Phi C u, u \rangle$$

where,

(6.1) $$C = D_t \Phi_t' D_t - \tfrac{1}{2} i([A, \Phi] D_t + D_t [A, \Phi]) + \tfrac{1}{2}(\Phi_t' A + A \Phi_t').$$

The principal symbol of the self-adjoint operator C is clearly:

(6.2) $$\sigma_2(C)=\tau^2\Phi'_t-\tau H_a\Phi+\Phi'_t a.$$

Suppose that Φ is constructed as discussed following Lemma 5.6, so is a time function except in some closed conic region Ω', of which Ω, invariant under τ translation, is a conic neighborhood. In Lemma 5.6 we take $\mathscr{K}_0$ to be a cone containing Γ_m for all m in some small region. Using a microlocal cutoff in x depending on t as a parameter, equal to the identity on Ω' and essentially supported in Ω, it follows that:

(6.3) $$C=C_1+R,$$

where R has essential support in Ω and C_1 is still a second order differential operator in t now with:

$$\sigma_2(C_1)\leq 0 \qquad \text{in } M_+\cup M_-,$$

by the definition of a time function. Using Lemma 5.3 or rather its homogeneous analogue one can write:

$$C_1=C_2+R'P+PR',$$

where R' is a compactly supported self-adjoint pseudodifferential operator and C_2 has non-positive principal symbol. In fact by direct computation of the symbols of order two and one C_2 can be written as a sum:

(6.4) $$C_2=C_3+R'PAPR'+C'_3-\delta(\Psi A\Psi+D_t\Psi^2D_t),$$

where C_3 has non-positive full symbol and C'_3, of order one, can be controlled in terms of the L^2 norm and the last terms in (6.4). Here Ψ is essentially the square root of $-\Phi'_t$, and $\delta>0$ depends on the cones $\mathscr{K}_i$ is Lemma 5.6. It is at this point that the alternative additional assumptions that A be real or an Hermitian form or hypoelliptic with loss of one derivative are used.

Applying the Fefferman-Phong inequality ([1]) shows that C_3 has an upper bound in L^2. Thus, the commutator identity shows that the L^2 norm of $D_t\Psi u$, and the A-norm of Ψu can be controlled in terms of the L^2 norm of u in any microlocal neighborhood of the essential support of Ψ, some norm on u and regularity of u near $t=-\varepsilon$ in $\mathscr{K}_1$.

To prove Theorem 1.8 from these estimates it is only necessary to iterate. Any degree of regularity can be propagated locally by estimating higher, or lower, D_t derivatives of the solution u which can be regularized by t-convolution. The subellipticity of A implies that this does indeed give complete regularity. By a covering argument with the microlocal propagation cones, using continuity on $\Sigma_{(2)}$ and inner continuity, these microlocal regularity results can be pieced together to show that singularities only propagate along the rays as defined in § 2.

References

[1] C. Fefferman and D. Phong, On positivity of pseudo-differential operators. Proc. Nat. Acad. Sci. USA **75** (1978), 4673–4674.

[2] L. Hörmander, The spectral function of an elliptic operator. Acta Math. **121** (1968), 193–218.

[3] B. Lascar, Propagation des singularités pour des équations hyperboliques à caracteristiques de multiplicité au plus double et singularités Masloviennes. Amer. J. Math. **104** (1982), 227–285.

[4] B. and R. Lascar, do. II, J. Analyse Math. **41** (1982), 1–38.

[5] R. B. Melrose, The wave equation for a hypoelliptic operator with symplectic characteristics of codimension two. J. Analyse Math. **44** (1984/5), 134–182.

MASSACHUSETTS INSTITUTE OF TECHNOLOGY
CAMBRIDGE, MASS. 02139
USA

Taniguchi Symp. HERT
Katata 1984, pp. 193–233

On the Cauchy Problem for Hyperbolic Equations and Related Problems
—Micro-local Energy Method—

Sigeru MIZOHATA

§ 1. Introduction

In 1961, the author showed a theorem, which is now called Lax-Mizohata theorem [11]. This method could be named micro-local energy method. We explain briefly it and show that several basic facts can be explained from this view-point. Our method uses asymptotic expressions of commutators of micro-localizers and pseudo-differential operators. This argument seems to be complicated, but the idea is simple. We use only fairly elementary energy inequalities, so that our method can be applied to fairly wide range problems. We show that this method can be applied to the treatment of micro-local analysis in Gevrey and analytic classes.

§ 2. Lax-Mizohata theorem and Levi condition

In order to explain the method clearly, we consider the following Cauchy problem

$$\begin{cases} \partial_t u(x, t) - a(x, t; D)u = 0, & t \in [0, T] \\ u|_{t=0} = u_0(x) \in H^\infty \end{cases} \tag{2.1}$$

where $a(x, t; D)$ is a pseudo-differential operator of order m, *m being a positive number* (not necessarily integer).

We assume

$$a(x, t; \xi) \in S^m_{1,0},$$

and t is considered as parameter, and that $t \to a(x, t; \xi) \in S^m_{1,0}$ is smooth in $t \in [0, T]$. In order to discuss the H^∞-wellposedness, it is convenient to assume

$$a = a_m + b \tag{2.2}$$

where $a_m(x, t; \xi)$ is homogeneous of degree m in ξ, and

$$b(x, t; \xi) \in S^{m'}_{1,0} \quad \text{with} \quad m' < m.$$

We show

Theorem 2.1. *In order that* (2.1) *be H^∞-wellposed, the condition*

$$\operatorname{Re} a_m(x, 0; \xi) \leqq 0, \qquad \text{for all } (x, \xi) \in \boldsymbol{R}^l \times \boldsymbol{R}^l \diagdown 0$$

is necessary.

For the proof, we use as usual

Proposition 2.2. *If* (2.1) *is H^∞-wellposed for $t \in [0, T]$, the linear mapping $u_0(x) \in H^\infty \to u(\cdot, t) \in C^0_t([0, T]; H^\infty)$ is continuous. In particular, there exist a constant $C(T)$ and a non-negative integer k such that*

$$\sup_{t \in [0,T]} \|u(\cdot, t)\|_{L^2} \leqq C(T) \|u(\cdot, 0)\|_{H^k} \tag{2.3}$$

Micro-localizers $\alpha_n(D)$, $\beta(x)$.

Let (x_0, ξ^0) be a point $(|\xi^0| = 1)$. Given r_0 (we call it *the size of micro-localizer*), we define first $\beta(x) \in C_0^\infty$ $(0 \leqq \beta(x) \leqq 1)$ with support contained in $\{x; |x - x_0| \leqq r_0\}$, and $= 1$ on $\{x; |x - x_0| \leqq r_0/2\}$. Next, we define $\alpha(\xi) \in C_0^\infty$ $(0 \leqq \alpha(\xi) \leqq 1)$ with its support in $\{\xi; |\xi - \xi^0| \leqq r_0\}$ and $= 1$ on $\{\xi; |\xi - \xi^0| \leqq r_0/2\}$. Finally put

$$\alpha_n(\xi) = \alpha(\xi/n) \tag{2.4}$$

where n are positive integers (or more generally positive parameter) tending to ∞. Observe that $\alpha_n(\xi)$ has its support in the r_0-conic neighborhood of ξ^0, and that

$$|\alpha_n^{(\nu)}(\xi)| \leqq c_\nu / n^{|\nu|}. \tag{2.5}$$

We define $\alpha_n(D)v(x)$ by $\widehat{\alpha_n v}(\xi) = \alpha_n(\xi)\hat{v}(\xi)$. We consider $\alpha_n(D)\beta(x)u$, thus operator $\alpha_n(D)\beta(x)$ is called micro-localizer. We consider also $\alpha_n^{(p)}\beta_{(q)}u$ with multi-index p and q, and $\{\alpha_n^{(p)}(D)\beta_{(q)}(x)\}$ are called micro-localizers. Throughout this paper, for $a(x, \xi)$ we denote $a^{(\alpha)}_{(\beta)}(x, \xi) = \partial_\xi^\alpha (i^{-1}\partial_x)^\beta a(x, \xi)$.

Properties of $\alpha_n^{(p)}(D)v$.

(1) Let $a(x, \xi)$ be defined by (2.2) with positive m. Suppose

$$\operatorname{Re} a_m(x, \xi) \geqq \delta |\xi|^m \quad (\exists \delta > 0), \quad \forall (x, \xi) \in \boldsymbol{R}^l \times \boldsymbol{R}^l \diagdown 0. \tag{2.6}$$

Then

$$\text{(2.7)}\qquad \operatorname{Re}(a(x, D)\alpha_n(D)v, \alpha_n(D)v) \geqq \frac{\delta}{2}((1-r_0)n)^m \|\alpha_n v\|^2,$$

if n is large.

(2) Let $a(x, \xi) \in S^m_{1,0}$ (with arbitrary m), then

$$\text{(2.8)}\qquad \begin{aligned} &\|a(x, D)\alpha_n(D)v\| \leqq \text{const. } n^m \|v\| \\ &\|\alpha_n(D)a(x, D)v\| \leqq \text{const. } n^m \|v\| \\ &\|\alpha_n^{(p)}(D)a(x, D)v\| \leqq \text{const. } n^{m-|p|} \|v\|. \end{aligned}$$

More generally

$$\|(\alpha_n^{(p)}(D)a(x, D) - \sum_{|\nu|<N} \nu!^{-1} a_{(\nu)}(x, D)\alpha_n^{(p+\nu)})v\| \leqq \text{const. } n^{m-|p|-N} \|v\|,$$

for n large.

(3) Let

$$\beta(x)a(x, D) = \sum_{|\nu|<N} (-1)^{|\nu|} \nu!^{-1} a^{(\nu)}(x, D)\beta_{(\nu)}(x) + r_N(x, D),$$

then $r_N(x, D) \in S^{m-N}_{1,0}$, hence

$$\text{(2.9)}\qquad \|\alpha_n^{(p)} r_N(x, D)v\| \leqq \text{const. } n^{m-|p|-N} \|v\|.$$

Proof of Theorem 2.1.

We assume that there exists an (x_0, ξ^0) $(|\xi^0|=1)$ such that

$$\text{(2.10)}\qquad \operatorname{Re} a_m(x_0, 0; \xi^0) > 0$$

and that the Cauchy problem for (2.1) is H^∞-wellposed. From these two assumptions, by defining suitably a series of initial data, we deduce two inequalities which are not compatible.

Without loss of generality, we assume $x_0=0$, and by taking r_0 small, we can assume

$$\text{(2.11)}\qquad \operatorname{Re} a_m(x, t; \xi) \geqq \delta|\xi|^m \qquad (\exists\delta>0)$$

for $(x, \xi) \in \operatorname{supp}[\beta] \times \text{conic supp}[\alpha]$, and t small, say $t \in [0, t_0]$.

Operating $\alpha_n(D)\beta(x)$ to (2.1), (we assume $\alpha_n(D)$, $\beta(x)$ are of size r_0),

$$\partial_t(\alpha_n \beta u) - \alpha_n \beta a(x, t; D)u = 0.$$

If necessary, modifying the symbol $a(x, t; \xi)$ outside the support of $\beta(x)$, denoted by $\tilde{a}(x, t; \xi)$, we can assume (2.11) with $\delta/2$ for $\tilde{a}(x, t; \xi)$ for all $(x, t; \xi) \in \boldsymbol{R}^l \times [0, t_0] \times V_{\xi^0}$. Observe $\beta a(x, t; D) = \beta \tilde{a}(x, t; D)$. Hence

$$(2.12) \qquad \partial_t(\alpha_n \beta u) - \tilde{a}(\alpha_n \beta u) = f_n$$

with

$$f_n = [\alpha_n \beta, \tilde{a}(x, t; D)]u.$$

Then, taking into account of (2.7), we get

$$(2.13) \qquad \frac{d}{dt}\|\alpha_n \beta u(\cdot, t)\| \geqq \delta' n^m \|\alpha_n \beta u(\cdot, t)\| - \|f_n(\cdot, t)\|$$

if n is large.

Hereafter, *we denote* $\tilde{a}$ by a. Then

$$(2.14) \qquad f_n = \sum_{1 \leqq |\mu+\nu| \leqq N} \mu!^{-1}(-1)^{|\nu|}\nu!^{-1} a_{(\mu)}^{(\nu)}(\alpha_n^{(\mu)} \beta_{(\nu)} u) + r_N u$$

where

$$\|r_N u\| \leqq \text{const. } n^{m-N-1}\|u\|.$$

We denote

$$(2.15) \qquad \alpha_n^{(p)} \beta_{(q)} u = u_{pq}, \qquad |p+q| \leqq N.$$

By (2.8), (2.11), we obtain

$$(2.16) \qquad \|f_n\| \leqq C(N) n^m \sum_{1 \leqq |\mu+\nu| \leqq N} n^{-|\nu|}\|u_{\mu\nu}\| + \text{const. } n^{m-N-1}\|u\|.$$

In view of (2.13) and (2.16), we need to consider $\alpha_n^{(p)} \beta_{(q)} u = u_{pq}$ for $|p+q| \leqq N$. For this purpose, we operate $\alpha_n^{(p)} \beta_{(q)}(x)$ to (2.1). Then

$$\partial_t u_{pq} - a(u_{pq}) = f_{n,p,q} \quad (= [\alpha_n^{(p)} \beta_{(q)}, a]u).$$

$$f_{n,p,q} = \sum_{1 \leqq |\mu+\nu| \leqq N-|p+q|} \mu!^{-1}(-1)^{|\nu|}\nu!^{-1} a_{(\mu)}^{(\nu)}(u_{p+\mu, q+\nu}) + r_{N,p,q} u$$

with

$$\|r_{N,p,q} u\| \leqq \text{const. } n^{m-N+|q|}\|u\|.$$

Thus we get

$$(2.17) \qquad \begin{aligned} \partial_t \|u_{pq}\| \geqq \delta' n^m \|u_{pq}\| - n^m C(N) \sum_{1 \leqq |\mu+\nu| \leqq N-|p+q|} n^{-|\nu|}\|u_{p+\mu, q+\nu}\| \\ - \text{const. } n^{m-N+|q|}\|u(\cdot, t)\|. \end{aligned}$$

We take

$$N=k+[m] \tag{2.18}$$

where k is a positive integer appearing in (2.3).

Now choosing M (large) in such a way that

$$\sum_{1\leqq|\mu+\nu|\leqq N} M^{-|\mu+\nu|}\leqq\frac{1}{2}\delta'/C(N), \tag{2.19}$$

we consider the following energy form

$$S_n(t;u)=\sum_{|p+q|\leqq N} M^{|p+q|}n^{-|q|}\|u_{pq}\|. \tag{2.20}$$

Then, (2.17), (2.19) give

$$\frac{d}{dt}S_n(t;u)\geqq\frac{1}{2}\delta' n^m S_n(t;u)-\text{const.}\, n^{m-N-1}\|u(\cdot,t)\|. \tag{2.21}$$

Finally, let us define a series of solutions $u_n(\cdot,t)$ of (2.1), which we supposed to exist uniquely, by the following initial data $u_n(\cdot,0)$ ($\in H^\infty$): Let $\psi(\xi)\in C_0^0$, with support in $\{\xi;|\xi|\leqq 1\}$. Suppose $\int|\psi(\xi)|^2d\xi=1$. Put

$$\tilde\psi_n(x)=(2\pi)^{-n}\int e^{ix\xi}\psi(\xi-n\xi_0)d\xi=e^{in\xi_0 x}\tilde\psi_0(x).$$

Then

$$u_n(\cdot,0)=\tilde\psi_n(x)\qquad(=e^{in\xi_0 x}\tilde\psi_0(x)). \tag{2.22}$$

We see that

1) $\|u_n(\cdot,0)\|_{H^k}=O(n^k)$,
2) $\|\alpha_n\beta u_n(\cdot,0)\|\geqq c_0$ (>0), for n large.

2) is seen in the following way.

$$\alpha_n\beta\tilde\psi_n=\beta\tilde\psi_n+\sum_{1\leqq|\nu|<N}\nu!^{-1}\beta_{(\nu)}\alpha_n^{(\nu)}\tilde\psi_n+O(n^{-N}),$$

where the middle term disappears because $\alpha_n^{(\nu)}(\xi)\psi_n(\xi)=0$ for $|\nu|\geqq 1$, and that $\beta\tilde\psi_n=e^{in\xi_0x}\beta(x)\tilde\psi_0(x)$, $\tilde\psi_0(x)$ being analytic.

Observe first

$$S_n(0;u_n)\geqq\|\alpha_n\beta u_n(\cdot,0)\|\geqq c_0.$$

1) implies, in view of (2.3),

$$\|u_n(\cdot, t)\| = O(n^k), \qquad t \in [0, t_0]. \tag{2.23}$$

This implies in particular

$$S_n(t; u_n) = O(n^k). \tag{2.24}$$

On the other hand, (2.21) and (2.23) (recalling (2.18)) imply

$$\frac{d}{dt} S_n(t; u_n) \geqq \frac{1}{2} \delta' n^m S_n(t; u_n) - O(1).$$

Hence

$$S_n(t; u_n) \geqq \exp\left(\frac{1}{2}\delta' n^m t\right)(S_n(0; u_n) - o(1)),$$

$$S_n(t; u_n) \geqq \frac{1}{2} c_0 \exp\left(\frac{1}{2}\delta' n^m t\right), \qquad \text{for } t \in [0, t_0]. \tag{2.25}$$

(2.24) and (2.25) are not compatible for $t > 0$, which completes the proof.

We can apply the same argument to the consideration of Levi condition. Let us consider the forward Cauchy problem to

$$\begin{cases} (\partial_t - i\lambda(x, t; D) - c(x, t; D))u(x, t) = f, \\ u|_{t=t_0} = u_0(x) \in H^\infty, \qquad 0 \leqq t_0 < t, \end{cases} \tag{2.26}$$

where

1) $\lambda(x, t; \xi) \in S^1_{1,0}$, essentially homogeneous of degree 1 in ξ and real-valued,
2) $c(x, t; \xi) \in S^\rho_{1,0}$, with $\rho < 1$, and it can be written in the form

$$c(x, t; \xi) = \mathring{c}(x, t; \xi) + \tilde{c}(x, t; \xi),$$

where $\mathring{c}(x, t; \xi)$ is homogeneous of degree ρ, and order $\tilde{c} \leqq \rho - 1$.

Theorem 2.3 (Levi condition). *In order that the forward Cauchy problem for* (2.26) *be uniformly H^∞-wellposed in $t \in [0, T]$, it is necessary and sufficient that, either $\rho \leqq 0$ or if $\rho > 0$, we have*

$$\forall (x, t; \xi), \qquad \operatorname{Re} \mathring{c}(x, t; \xi) \leqq 0.$$

Proof. We are concerned with the necessity. The proof is carried out along the same line as in Theorem 2.1. We explain only main points. As in Theorem 2.1, we assume

(2.27) $$\exists\xi^0\ (|\xi^0|=1),\quad \text{such that } \operatorname{Re}\mathring{c}(0,0;\xi^0)>0.$$

We assume moreover (2.26) to be H^∞-wellposed in $t\in[0,T]$. From these two assumptions we show a contradiction. First, for micro-localized solution $\alpha_n^{(p)}\beta_{(q)}u$ around $(x,\xi)=(0,\xi^0)$, we obtain

(2.28) $$\frac{d}{dt}\|\alpha_n^{(p)}\beta_{(q)}u(\cdot,t)\|\geqq\delta n^\rho\|\alpha_n^{(p)}\beta_{(q)}u(\cdot,t)\|-\|f_{n,p,q}\|,$$

where $0<\rho<1$, and $\delta>0$, $f_{n,p,q}=[\alpha_n^{(p)}\beta_{(q)},\tilde{a}]u$, where $\tilde{a}$ is defined in the same way as in Theorem 2.1.

(2.29) $$f_{n,p,q}\sim\sum_{1\leqq|\mu+\nu|}(-1)^{|\nu|}\mu!^{-1}\nu!^{-1}\tilde{a}^{(\nu)}_{(\mu)}(u_{p+\mu,q+\nu}).$$

The terms with $|\mu+\nu|=1$ play a decisive role.

$$\|\tilde{a}^{(e_j)}(u_{p,q+e_j})\|\leqq\text{const.}\,\|u_{p,q+e_j}\|,$$
$$\|\tilde{a}_{(e_j)}(u_{p+e_j,q})\|\leqq\text{const.}\,n\|u_{p+e_j,q}\|.$$

In view of (2.28) and (2.29), we define the energy form

(2.30) $$S_n(t;u)=\sum_{|p+q|\leqq N}c^n_{pq}\|u_{pq}(\cdot,t)\|,$$

with

$$c^n_{pq}=M^{|p+q|}n^{(1-\rho)|p|-\rho|q|}.$$

Now, by Proposition 2.2, there exists a k. We choose

(2.31) $$N=[k/\rho]+1.$$

By taking the same sequence of initial data as in Theorem 2.1, we get

$$\frac{d}{dt}S_n(t;u_n)\geqq\frac{\delta}{2}n^\rho S_n(t;u_n)-o(1).$$

Hence

$$S_n(t;u_n)\geqq\frac{1}{2}c_0\exp\left(\frac{\delta}{2}n^\rho t\right).$$

§ 3. Wave front sets

By using the micro-localizers explained in § 2, we can characterize $WF(u)$.

Suppose $(x_0, \xi^0) \notin WF(u)$, or if one prefers, $u \in C^\infty_{(x_0,\xi^0)}$, then if we choose the size r_0 of micro-localizer $(\alpha_n(D), \beta(x))$ around (x_0, ξ^0) sufficiently small, then for any p, q, $\|\alpha_n^{(p)}\beta_{(q)}u\|_{L^2}$ is rapidly decreasing sequence. Conversely, if this occurs for a suitably chosen $(\alpha_n(D), \beta(x))$, then $u \in C^\infty_{(x_0,\xi^0)}$.

Wave front set $WF_s(u)$ $(s>1)$ *(wave front set in the sense of Gevrey class s).*

We say that $(x_0, \xi^0) \in WF_s(u)$, if and only if there exists a cut-off function $\zeta(x) \in \gamma^{(s)}$ with compact support, taking the value 1 in a neighborhood of x_0 such that the estimate of the form

$$|\widehat{\zeta u}(\xi)| \leqq \exp(-\varepsilon_0|\xi|^{1/s}) \qquad (\exists \varepsilon_0>0) \tag{3.1}$$

holds, when ξ tends to ∞, remaining in a suitable conic neighborhood V_{ξ^0} of ξ^0. Here $u \in (\gamma^{(s)})'$ is supposed.

We give a criterion of $WF_s(u)$ by using micro-localizers. Detailed accounts to this subject are given in [12], [16].

Definition of $\alpha_n(\xi)$. Let $\chi_N(\xi)$ ba a C_0^∞-function depending on parameter N, having its support in $|\xi-\xi^0|\leqq r_0$ $(0\leqq \chi_N \leqq 1)$, and satisfying the following conditions:

$$\chi_N(\xi)=1, \qquad \text{for } |\xi-\xi^0|\leqq r_0/2,$$
$$|\chi_N^{(\mu+\nu)}(\xi)| \leqq (Ncr_0^{-1})^{|\mu|}\nu!^{1+\varepsilon'}(cr_0^{-1})^{|\nu|} \qquad \text{for } |\mu|\leqq N,$$
$$(s>1,\ 1+2\varepsilon'<s,\ \varepsilon'>0),$$

where c can be considered as an absolute constant (nemely independent of r_0).

Put

$$\alpha_n(\xi)=\chi_N(\xi/n).$$

Therefore

$$\begin{aligned} &\operatorname{supp}[\alpha_n(\xi)] \subset \{\xi;\ |\xi-n\xi_0|\leqq nr_0\},\\ &\alpha_n(\xi)=1, \qquad \text{for } |\xi-n\xi^0|\leqq nr_0/2,\\ &|\alpha_n^{(\mu+\nu)}(\xi)| \leqq \left(\frac{N}{n}cr_0^{-1}\right)^{|\mu|}\nu!^{1+\varepsilon'}(cr_0^{-1})^{|\nu|}/n^{|\nu|}, \qquad \text{for } |\mu|\leqq N, \end{aligned} \tag{3.2}$$

where n and N are related by

$$N=\theta n^{1/s}, \qquad \theta=(ce)^{-1}r_0. \tag{3.3}$$

Definition of $\beta_n(x)$. $\beta_n(x)$ ($\in C_0^\infty$) ($0 \leqq \beta_n(x) \leqq 1$) has its support in $|x-x_0| \leqq r_0$, and $\beta_n(x)=1$ for $|x-x_0| \leqq r_0/2$. Moreover

$$(3.4) \qquad |\beta_{n(\mu+\nu)}(x)| \leqq (Ncr_0^{-1})^{|\mu|}\nu!^{1+\varepsilon'}(cr_0^{-1})^{|\nu|} \qquad \text{for } |\mu| \leqq N.$$

Let us recall that here also n and N are related by (3.3).

Theorem 3.1. *Suppose $(x_0, \xi^0) \notin WF_s(u)$ (with $|\xi^0|=1$), $u \in (\gamma^{(s)})'$ with compact support. Then, for any micro-localizer $\{\alpha_n(D), \beta_n(x)\}$ satisfying the above conditions, if we choose its size r_0 small, we have the following type estimate:*

$$S_n = \sum_{|p+q| \leqq N} c_{pq}^n \|\alpha_n^{(p)}(D)\beta_{n(q)}(x)u\| \leqq \exp(-\varepsilon n^{1/s}) \qquad (\exists \varepsilon > 0),$$

when n is large, where $c_{pq}^n = n^{(1-1/s)|p|-(1/s)|q|}$. We can say more: when r_0 becomes small, we can choose ε as a constant independent of r_0.

Note. Of course ε depends on each u.

The converse is also true. We can state it in a stronger form.

Theorem 3.2. *Suppose $u \in (\gamma^{(s)})'$ with compact support. If the estimate of the form;*

$$\tilde{S}_n = \sum_{|p| \leqq N} c_{p0}^n \|\alpha_n^{(p)}\beta_n u\| \leqq \exp(-\varepsilon_0 n^{1/s}), \qquad (\exists \varepsilon_0 > 0),$$

holds for n large, for some choice of micro-localizer, then $(x_0, \xi^0) \notin WF_s(u)$.

The following fact, which is called pseudo-local property, is not only one of basic properties of Gevrey symbols, but also its proof illustrates the essential part of our arguments when it is concerned with Gevrey symbols.

Let $a(x, \xi)$ be a Gevrey pseudo-analytic symbol of class s (see [6]). To be precise,

$$|a_{(\beta)}^{(\alpha)}(x, \xi)| \leqq A\alpha!\beta!^s C^{|\alpha+\beta|}|\xi|^{m-|\alpha|} \qquad \text{for } |\alpha| \leqq R^{-1}|\xi|,$$

where R is a suitable large number. Then

Theorem 3.3 (Lemma 3 of [12]). *Let $u \in (\gamma^{(s)})'$ with compact support. If $(x_0, \xi^0) \notin WF_s(u)$, then $(x_0, \xi^0) \notin WF_s(a(x, D)u)$.*

Now we are concrened with the $\gamma^{(s)}$-wellposedness of the Cauchy problem for

$$(3.5) \qquad \begin{cases} \partial_t u(x,t) = (i\lambda(x,t;D) + c(x,t;D))u(x,t), & t \in [0, T] \\ u|_{t=0} = u_0(x), \end{cases}$$

where
i) $\lambda(x, t; \xi)$ is homogeneous of degree 1 in ξ, and real-valued,
ii) $c(x, t; \xi) \in S^{\rho}_{1,0}$, with $\rho < 1$,
and these symbols are of Gevrey class;

$$|\lambda^{(\nu)}_{(\mu)}(x, t; \xi)| \leq A'\mu!^s \nu! C_0^{|\mu+\nu|} |\xi|^{1-|\nu|},$$
$$|c^{(\nu)}_{(\mu)}(x, t; \xi)| \leq \tilde{c}\mu!^s \nu! C_0^{|\mu+\nu|} |\xi|^{\rho-|\nu|},$$
$$\text{for } |\nu| \leq R^{-1}|\xi|^{1/s}, \quad (\exists R, \text{ sufficiently large}).$$

As in Theorem 2.3, we assume that

$$c(x, t; \xi) = \mathring{c}(x, t; \xi) + \tilde{c}(x, t; \xi),$$

where $\mathring{c}(x, t; \xi)$ is homogeneous of degree ρ in ξ and order $\tilde{c} < \rho$. Of course, we assume for $\mathring{c}$ and $\tilde{c}$ the corresponding Gevrey pseudo-analytic symbol estimate.

Now, we are concerned with the $\gamma^{(s)}_{L^2}$-wellposedness, where $f \in \gamma^{(s)}_{L^2}$ means $f \in H^\infty$ and that

$$\|\partial^\alpha f\|_{L^2} \leq A\alpha!^s C_1^{|\alpha|}, \qquad (\exists A, \exists C_1 > 0).$$

Theorem 3.4. *In order that* (3.5) *be $\gamma^{(s)}_{L^2}$-wellposed when $\rho > 1/s$, it is necessary that*

$$\mathrm{Re}\, \mathring{c}(x, 0; \xi) \leq 0, \qquad \forall (x, \xi) \in \boldsymbol{R}^l \times \boldsymbol{R}^l.$$

In other words, if $\rho > 1/s$ and that there exists some (x_0, ξ^0) such that

(3.6) $$\mathrm{Re}\, \mathring{c}(x_0, 0; \xi^0) > 0,$$

then the Cauchy problem for (3.5) *is never $\gamma^{(s)}_{L^2}$-wellposed.*

Proof. We give only very rough sketch of the proof. Detailed proof is found in [15]. We assume (3.6), and moreover we assume (3.5) to be $\gamma^{(s)}_{L^2}$-wellposed. Then, from these two assumptions, we deduce two energy inequalities which are not compatible.

First we use, instead of Proposition 2.2 (case in H^∞ space), the following a-priori estimate.

Let $\gamma^{(s)}_{L^2,A}$ be the space of all functions of $\gamma^{(s)}_{L^2}$ satisfying

$$\|f\|_{s,A} = \sup_{\nu \geq 0} \|\partial^\nu f\| / |\nu|!^s A^{|\nu|} < +\infty.$$

$\gamma^{(s)}_{L^2,A}$ is a Banach space with this norm. Then, the assumption yields the following estimate:

(3.7) $$\sup_{t\in[0,T]} \|u(\cdot, t)\|_{L^2} \leqq C(A)\|u(\cdot, 0)\|_{s,A}.$$

We use the same sequence of initial data (see (2.22)): $u_n(\cdot, 0)=\tilde{\psi}_n(x)$. We see easily that

(3.8) $$\|u_n(\cdot, 0)\|_{s,A} = \|\tilde{\psi}_n(x)\|_{s,A} \leqq \exp\left[c(s)\frac{1}{A^{1/s}} n^{1/s}\right].$$

Next we use the same micro-localizer $(\alpha_n(D), \beta_n(x))$, mentioned in this section, around (x_0, ξ^0). We assume $x_0=0$.

First

$$\partial_t(u_{pq}) - \tilde{a}(x, t; D)(u_{pq}) = f_{n,p,q},$$

$$\frac{d}{dt}\|u_{pq}\| \geqq \delta n^\rho \|u_{pq}\| - \|f_{n,p,q}\|, \qquad (\exists\delta>0).$$

Next, we use the energy form:

$$S_n(t; u) = \sum_{\substack{|p|\leqq N \\ |q|\leqq N}} c^n_{pq} \|u_{pq}(\cdot, t)\|_{L^2},$$

where

$$c^n_{pq} = \begin{cases} n^{(1-\rho)|p|} n^{-\rho|q|} M^{|p+q|}, & \text{for } |p|\leqq N_0, \\ |p|!^{s-1} n^{-\rho|q|} M'^{|p|} M^{|q|}, & \text{for } N_0<|p|\leqq N, \end{cases}$$

where N_0 is determined by $N_0^{s-1}=n^{1-\rho}$, and $M'=2(2e)^{s-1}M$, M being sufficiently large. Finally N is chosen by

$$N=(Me^2lc)^{-1} r_0^{1/s} n^{1/s}.$$

Then, we can show that

(3.9) $$\frac{d}{dt} S_n(t; u) \geqq \frac{1}{2}\delta n^\rho S_n(t; u) - e^{-N} \times (\text{some polynomial in } n) \times \|u(\cdot, t)\|.$$

Now, taking into account of (3.7) and (3.8),

(3.10) $$\|u_n(\cdot, t)\| \leqq C(A)\exp\left[\frac{c(s)}{A^{1/s}} n^{1/s}\right].$$

Then, putting u_n in place of u in (3.9), we see that if A is chosen large, this becomes (observe that $N=\theta n^{1/s}$)

$$\frac{d}{dt}S_n(t; u_n) \geqq \frac{1}{2}\delta n^\rho S_n(t; u_n) - O(1), \qquad \text{if } n \text{ is large.}$$

This implies, observing $S_n(0; u_n) \geqq \|u_{0,0}\| \geqq c_0 > 0$,

$$S_n(t; u_n) \geqq \frac{1}{2} c_0 \exp\left(\frac{1}{2}\delta n^\rho t\right), \tag{3.11}$$

if n is large. On the other hand, we see easily that

$$S_n(t; u_n) \leqq \text{const. } \|u_n(\cdot, t)\|_{L^2},$$

where const. is independent of u and n. This implies, in view of (3.10),

$$S_n(t; u_n) \leqq \text{const. } \exp(\delta' n^{1/s}). \tag{3.12}$$

Inequalities (3.11) and (3.12) are not compatible for $t > 0$, because we supposed $\rho > 1/s$.

§ 4. Analytic wave front sets

The analytic wave front set $WF_A(u)$ is characterized also by using micro-localizers $\{\alpha_n(D), \beta_n(x)\}$ around (x_0, ξ^0). Since fairly detailed accounts are given in [14], we explain these briefly.

Definition of $\alpha_n(\xi)$. We define first $\alpha^N(\xi)$. $\chi_N(\xi)$ is defined as: Let $|\xi^0| = 1$,

$$\chi_N(\xi) = \begin{cases} 1, & \text{for } |\xi - \xi^0| \leqq \left(\frac{1}{2} + \frac{1}{8}\right) r_0, \\ 0, & \text{for } |\xi - \xi^0| \geqq \left(1 - \frac{1}{8}\right) r_0, \end{cases}$$

$$|\chi_N^{(\mu)}(\xi)| \leqq (Ncr_0^{-1})^{|\mu|}, \qquad \text{for } |\mu| \leqq 2N,$$

where c is independent of N and r_0.

Let

$$\begin{cases} \psi(\xi) \in C_0^\infty, \quad 0 \leqq \psi(\xi) \leqq 1, \quad \int \psi(\xi) d\xi = 1; \\ \psi(\xi) \text{ has its support in } |\xi| \leqq r_0/8. \end{cases} \tag{4.1}$$

Denoting

$$\alpha^N(\xi) = \chi_N(\xi) * \psi(\xi),$$

put

$$\alpha_n(\xi)=\alpha^N(\xi/n).$$

Then $\alpha_n(\xi)$ —micro-localizer of size r_0 around ξ^0— has the following properties:

$$\text{(4.2)}\quad \begin{aligned} &\text{i)}\quad \alpha_n(\xi)=1, \qquad \text{for } |\xi-n\xi^0|\leqq nr_0/2,\\ &\quad\ \text{and}\quad =0, \qquad \text{for } |\xi-n\xi^0|\geqq nr_0;\quad 0\leqq\alpha_n(\xi)\leqq 1,\\ &\text{ii)}\quad |\alpha_n^{(\mu+\nu)}(\xi)|\leqq\left(\frac{N}{n}cr_0^{-1}\right)^{|\mu|}\|\psi^{(\nu)}(\xi)\|_{L^1}n^{-|\nu|}, \qquad \text{for } |\mu|\leqq 2N, \end{aligned}$$

N and n are related by

$$\text{(4.3)}\qquad N=\frac{1}{2}l^{-1}(ce)^{-2}r_0^2n=\theta n,$$

observe that θ depends on the size in the manner

$$\theta=c_0r_0^2.$$

Note. We did not specify the form $\psi(\xi)$, except for (4.1). $\psi(\xi)$ was introduced only to obtain Theorem 4.2, in this case we specify the form of $\psi(\xi)$. In other cases we use (4.2) only in the case $\nu=0$.

Definition of $\beta_n(x)$. $\beta_n(x)$ is defined by the following properties:

$$\text{(4.4)}\quad \begin{aligned} &\text{i)}\quad \beta_n(x)=1 \qquad \text{for } |x-x_0|\leqq r_0/2,\\ &\quad\ \text{and}\quad =0 \qquad \text{for } |x-x_0|\geqq r_0;\quad 0\leqq\beta_n(x)\leqq 1,\\ &\text{ii)}\quad |\beta_n^{(\mu+\nu)}(x)|\leqq(Ncr_0^{-1})^{|\mu|}\nu!^s(c'r_0^{-1})^{|\nu|}, \qquad \text{for } |\mu|\leqq 2N, \end{aligned}$$

where $s>1$, and c' is independent of r_0 and ν.

We obtain the following criterion of $WF_A(u)$.

Theorem 4.1. *Let $u\in\mathscr{E}'$ and $(x_0, \xi^0)\notin WF_A(u)$, or one prefers $u\in C^\omega_{(x_0,\xi^0)}$. Then if we choose the size r_0 small, for any micro-localizer $\{\alpha_n(D), \beta_n(x)\}$ around (x_0, ξ^0) satisfying (4.1)–(4.4), putting*

$$\text{(4.5)}\qquad S_n=\sum_{|p+q|\leqq N}c_{pq}^n\|\alpha_n^{(p)}(D)\beta_{n(q)}(x)u\|, \qquad c_{pq}^n=(cr_0^{-1})^{|p+q|}n^{-|q|},$$

we have

$$\text{(4.6)}\qquad S_n\leqq\exp(-\varepsilon n), \quad (\exists\varepsilon>0), \quad \textit{for } n \textit{ large}.$$

We can say more. We can replace the right-hand side by

$$S_n \leqq \exp(-N) \times (\textit{some polynomial in } n), \tag{4.7}$$

if r_0 is sufficiently small, and we can take the same polynomial when u runs through an equi-continuous set in $\mathscr{E}'$.

The converse is also true. We can say it in a stronger form:

Theorem 4.2. *Let*

$$\tilde{S}_n = \sum_{|p| \leqq N} c_{p0}^n \| \alpha_n^{(p)} \beta_n u \|.$$

If $\tilde{S}_n$ satisfies (4.6) *when n is large for some $\{\alpha_n, \beta_n\}$ satisfying* (4.1)–(4.4), *then $(x_0, \xi^0) \notin WF_A(u)$.*

We show fairly direct applications of the above theorems. The following pseudo-local property in analytic sense is known (see Treves [20], Chap. V, Theorem 3.2).

Proposition 4.3. *Let $a(x, D)$ be an analytic pseudo-differential operator, or more generally its symbol $a(x, \xi)$ be pseudo-analytic. Then*

$$(x_0, \xi^0) \notin WF_A(u) \textit{ implies } (x_0, \xi^0) \notin WF_A(a(x, D)u).$$

Next, we consider the local version of the analytic-hypoellipticity (more precisely, analytic hypoellipticity in x) of parabolic operators.

Let $a(x, t; D)$ be a differential operator of order m $(\geqq 1)$, C^∞ in (x, t) and analytic in x. Assume, to its principal symbol: At (x_0, ξ^0), $(|\xi^0| = 1)$,

$$\operatorname{Re} a_m(x_0, t; \xi^0) \leqq -\delta \quad (\exists \delta > 0), \quad \text{for } t \in [0, T].$$

We are concerned with a local solution $u(x, t)$ of

$$\partial_t u = a(x, t; D)u + f, \qquad (x, t) \in \Omega \times [0, T].$$

Proposition 4.4. *Suppose $t \to u(\cdot, t)$ to be C^1 with values in $\mathscr{D}'_x(\Omega)$, and that $\forall t$, $t \in [0, T]$, $(x_0, \xi^0) \notin WF_A(f(\cdot, t))$. Then,*

$$\forall t, \quad 0 < t < T, \quad (x_0, \xi^0) \notin WF_A(u(\cdot, t)).$$

Proof. Since the proof is given in [12], we give here the outline. Let $\{\alpha_n^{(p)} \beta_{n(q)}\}$, $(|p+q| \leqq N)$, be a micro-localizer around (x_0, ξ^0). Then denoting

$$\alpha_n^{(p)}(D) \beta_{n(q)} u(\cdot, t) = u_{pq}(\cdot, t), \qquad \alpha_n^{(p)} \beta_{n(q)} f = f_{pq},$$

we have

$$\partial_t(u_{pq}) = a(u_{pq}) + g_{n,pq} + f_{pq}, \qquad g_{n,pq} = [\alpha_n^{(p)}\beta_{n(q)}, a]u.$$

A little delicate argument shows that, if we choose r_0 small,

$$\sum_{|p+q|\leqq N} c_{pq}^n \|g_{n,pq}\|$$

$$\leqq \frac{1}{4}\delta n^m \sum_{1\leqq|p+q|\leqq N} c_{pq}^n\|u_{pq}\| + e^{-N}\times(\text{some fixed polynomial in } n).$$

$$\frac{d}{dt}\|u_{pq}\| \leqq -\frac{1}{2}\delta n^m\|u_{pq}\| + \|g_{n,pq}\| + \|f_{pq}\|.$$

Denoting $S_n(t; u) = \sum_{|p+q|\leqq N} c_{pq}^n \|u_{pq}(\cdot, t)\|$, we get

$$\frac{d}{dt}S_n(t; u) \leqq -\frac{1}{4}\delta n^m S_n(t; u) + e^{-N}\times(\text{some polynomial}) + S_n(t; f).$$

Taking account of Theorem 4.2, if r_0 is small and n is large, we get

$$S_n(t; u) \leqq \exp\left(-\frac{1}{4}\delta n^m t\right) S_n(0; u) + e^{-N}\times(\text{some polynomial in } n). \tag{4.8}$$

We see easily that $S_n(0; u) \leqq n^L$ $(L \geqq 0)$. Now, since $m \geqq 1$, $S_n(t; u)$ satisfies the condition (4.6), which completes the proof.

Observe that in (4.8), if $S_n(0; u) = 0$, and that when $m = 1$, even when the exponential term is replaced by $\exp(\delta nt)$, $(\delta > 0)$, $S_n(t; u)$ still has micro-local analytic estimate provided that $\delta t < \theta$ (see (4.3)). Hence we are led to an extension of Holmgren's Theorem, which was obtained by Yamanaka-Persson [21], and Baouendi-Goulaouic [1].

We are concerned with

$$\partial_t u = \sum_{j=1}^{l} A_j(x, t)\partial_{x_j} u(x, t) + B(x, t)u,$$

where $A_j(x, t)$ and $B(x, t)$ are $k \times k$ matrices whose entries (coefficients) are continuous functions in t with values in holomorphic functions in an open set $\mathcal{O}_x$ containing the origin.

Proposition 4.5. *Every solution $u(x, t)$, defined in a neighborhood of the origin, continuously differentiable in t with values in $\mathscr{D}'(\mathcal{O}_x)$, and satisfying*

$$u|_{t=0} = 0$$

vanishes in a neighborhood $\mathcal{O}_{x,t}$ of the origin.

Proof. It suffices to prove that $u(x, t)$ is analytic in x for $|t| \leqq t_0$ (small). Denote the above equation by

$$\partial_t u = A(x, t; D)u.$$

Taking $\zeta(x) \in C_0^\infty$ with small support around the origin, and taking the value 1 in a small neighborhood of the origin, we consider ζu instead of u itself. We denote this simply by u. Now ξ^0 ($|\xi^0|=1$) is arbitrarily chosen, we take a micro-localizer $\{\alpha_n(D), \beta_n(x)\}$, of small size r_0, around $(x, \xi)=(0, \xi^0)$. First

$$\partial_t(u_{pq}) = A(x, t; D)(u_{pq}) + f_{n,pq},$$

where

$$\begin{aligned} u_{pq}(\cdot, t) &= \alpha_n^{(p)}\beta_{n(q)}u = {}^t(\alpha_n^{(p)}\beta_{n(q)}u_1, \cdots, \alpha_n^{(p)}\beta_{n(q)}u_k) \\ &= {}^t(u_{1,pq}, \cdots, u_{k,pq}), \quad \text{and} \quad f_{n,pq} = [\alpha_n^{(p)}\beta_{n(q)}, A]u. \end{aligned}$$

Denote,

$$\|u_{pq}\| = \sum_{i=1}^k \|u_{i,pq}(\cdot, t)\|.$$

First,

$$\frac{d}{dt}\|u_{pq}\| \leqq An\|u_{pq}\| + \|f_{n,pq}\|.$$

As in the proof of Proposition 4.4, a little delicate argument shows that,

$$\sum_{|p+q|\leqq N} c_{pq}^n \|f_{n,pq}\| \leqq Kr_0 n S_n(t; u) + e^{-N} \times (\text{some fixed polynomial in } n),$$
$$(K>0, \text{ independent of } r_0 \text{ and } n).$$

Hence

$$\frac{d}{dt} S_n(t; u) \leqq (A+Kr_0)n S_n(t; u) + e^{-N} \times (\text{some polynomial in } n).$$

Hence

$$S_n(t; u) \leqq \exp[(A+Kr_0)nt] \cdot e^{-N} \times (\text{some polynomial in } n).$$

This shows that (supposing $A>0$, and supposing $Kr_0 \leqq A$), when

$$t \leqq \theta/4A,$$

$$S_n(t; u) \leqq e^{-N/2} \times (\text{some polynomial in } n).$$

Hence

$$(0, \xi^0) \notin WF_A(u(\cdot, t)).$$

§ 5. Hyperbolic equations with constant multiplicities

We are concerned with

$$P(u) = \partial_t^m u(x, t) + \sum_{j=1}^{m} a_j(x, t; D)\partial_t^{m-j} u(x, t) = f(x, t), \tag{5.1}$$

where order $a_j \leqq j$ and that the principal symbol is of the form

$$i^m \prod_{j=1}^{k} (\tau - \lambda_j(x, t; \xi))^{m_j}. \tag{5.2}$$

We assume λ_j are real and distinct, and m_j are all constant (constant multiplicity). We assume for some j, $m_j \geqq 2$. We consider (5.1) and (5.2) in $(x, t) \in \boldsymbol{R}^l \times [0, T]$, and assume for simplicity all the coefficients to belong to $\mathscr{B}$.

Our starting point is perfect factorization of the operator P. This simplifies remarkably the treatment of the operators with constant multiplicity. Perfect factorization means the following:

$$P = P_k \circ P_{k-1} \circ \cdots \circ P_1 + R, \tag{5.3}$$

where the notation $\circ$ denotes operator product, and P_j $(1 \leqq j \leqq k)$ is a pseudo-differential operator whose principal symbol is $(i\tau - i\lambda_j)^{m_j}$. To be precise,

$$\begin{aligned} P_j = {} & (\partial_t - i\lambda_j)^{m_j} + a_{1,j}(x, t; D_x)(\partial_t - i\lambda_j)^{m_j - 1} \\ & + \cdots + a_{s,j}(\partial_t - i\lambda_j)^{m_j - s} + \cdots + a_{m_j, j}, \end{aligned} \tag{5.4}$$

with order $a_{s,j}(x, t; \xi) \leqq s-1$. We remark that

$$(\partial_t - i\lambda_j)^{m_j} = \underbrace{(\partial_t - i\lambda_j) \circ (\partial_t - i\lambda_j) \circ \cdots \circ (\partial_t - i\lambda_j)}_{m_j}.$$

The operator R in (5.3) is a regularizing operator, which means the following:

$$R = \sum_{j=1}^{m} r_j(x, t; D_x)\partial_t^{m-j},$$

where $r_j(x, t; \xi)$ $(1 \leqq j \leqq m)$ are null symbols, together with all derivatives $D_t^k r_j$.

Put

$$\rho_j = \max_{1 \leqq i \leqq m_j} \text{order } a_{ij}(x, t; \xi)/i \qquad (<1). \tag{5.5}$$

Now, in C^∞-case, Levi condition is formulated as follows:

Definition 5.1 (Levi condition). We say that P satisfies Levi condition if for all j,

$$\rho_j \leqq 0.$$

We have the following theorem, which is known.

Theorem 5.2. *In order that the forward Cauchy problem*

$$(5.6) \qquad \begin{cases} Pu = f \\ \partial_t^i u|_{t=t_0} = u_i(x) \in H^\infty \quad (0 \leqq i \leqq m-1), \quad (0 \leqq t_0 < T) \end{cases}$$

be uniformly wellposed in H^∞, it is necessary and sufficient that P satisfies Levi condition.

Note. Levi condition is formulated in slightly different ways by Flaschka-Strang [5], Chazarain [3], and De Paris [4]. Observe that when Levi condition is formulated in this way, the sufficiency of the above condition becomes almost evident.

First we show that ρ_j is invariant with respect to the order of perfect factorization.

Invariance of ρ_j.

We can prove this in a standard way. Let us observe that ρ_j is a function of (x, t). So, considering this in an open set V, we define

$$\rho_{j,V} = \max_{(x,t) \in V} \rho_j(x, t).$$

Now, as one sees by the proof which we give here, it suffices to prove the invariance in the following simple case:

$$P = P_2 \circ P_1 + R = \tilde{P}_1 \circ \tilde{P}_2 + \tilde{R}$$

where R and $\tilde{R}$ are both regularizing operator. We denote the number $\rho_{1,V}$ corresponding to P_1 and $\tilde{P}_1$ by r_1 and $\tilde{r}_1$ respectively, and we assume

$$r_1 > 0, \qquad r_1 > \tilde{r}_1,$$

and show that this is impossible.

First let us recall the following well-known fact: Let $\phi(x) \in \mathscr{B}$ with

$$|\phi_x(x)| \geqq \delta \ (>0).$$

Let $p(x, \xi)$ be a formal symbol:

$$p(x, \xi) \sim \sum_{j \geqq 0} p_j(x, \xi),$$

$p_j(x, \xi)$ being of homogeneous of degree $m-j$ in ξ, and bounded with all derivatives in x and ξ for $x \in \boldsymbol{R}^l$ and $|\xi|=1$. Let $\tilde{p}(x, \xi)$ be a true symbol corresponding to $p(x, \xi)$. Let $f(x) \in \mathscr{B}$, then when ρ (positive number) tends to infinity, we obtain

$$e^{-i\phi(x)\rho}\tilde{p}(x, D)(e^{i\phi(x)\rho}f(x)) \sim c_m(x)\rho^m + c_{m-1}(x)\rho^{m-1} + \cdots,$$

where $c_j(x) \in \mathscr{B}$, and in particular

$$c_m(x) = p_m(x, \phi_x(x))f(x).$$

Of course if $r(x, \xi)$ is a null symbol, then

$$e^{-i\phi\rho}r(x, D)(e^{i\phi\rho}f(x))$$

is rapidly decreasing when $\rho \to \infty$.

By definition of r_1, we can assume that there exists an $(x_0, t_0) \in V$ such that $\rho_1(x_0, t_0) = r_1$. Without loss of generality, we can assume $(x_0, t_0) = (0, 0)$. We define $\phi(x, t)$ by

$$\begin{cases} \phi_t(x, t) - \lambda_1(x, t; \phi_x(x, t)) = 0 \\ \phi|_{t=0} = x\xi^0, \end{cases}$$

ξ^0 being defined later. We choose $0<\theta<1$, such that

$$\tilde{r}_1 < \theta < r_1.$$

First observe

$$(D_t - \lambda(x, t; D))(e^{i\phi\rho}f(x)) \sim e^{i\phi\rho}(c_1(x, t)\rho + c_0(x, t) + c_{-1}(x, t)\rho^{-1} + \cdots),$$

where

$$c_1(x, t) = (\phi_t - \lambda(x, t; \phi_x))f(x), \quad \text{and} \quad c_j \in \mathscr{B}.$$

Next,

$$(D_t - \lambda_1(x, t; D))(\exp(i\phi\rho + it\rho^\theta)f(x))$$
$$\sim \exp(i\phi\rho + it\rho^\theta)[f(x)\rho^\theta + \tilde{c}_0(x, t) + \tilde{c}_{-1}(x, t)\rho^{-1} + \cdots].$$

Applying these formulas repeatedly, we see that

$$a_{j,1}(x, t; D)(D_t - \lambda_1(x, t; D))^{m_1 - j}(\exp(i\phi\rho + it\rho^\theta)f(x))$$
$$\sim \exp(i\phi\rho + it\rho^\theta)[\mathring{a}_{j,1}(x, t; \phi_x)f(x)\rho^{l_j} + \cdots],$$

where

$$l_j=(m_1-j)\theta+\text{order } \mathring{a}_{j,1}=m_1\theta+(r_1-\theta)j.$$

This implies

$$(5.7)\qquad \begin{aligned} &P_2 \circ P_1(\exp(i\phi\rho+it\rho^\theta)f(x)) \\ &\quad\sim \exp(i\phi\rho+it\rho^\theta)[(\lambda_1-\lambda_2)^{m_2} \sum_j{}' \mathring{a}_{j1}(x, t: \phi_x)f(x)\rho^{m_2+l_j}+\cdots], \end{aligned}$$

where we assumed P_i is expressed in the form

$$P_i=(D_t-\lambda_i)^{m_i}+\sum_j a_{ji}(D_t-\lambda_i)^{m_i-j},$$

and $\mathring{a}_{j1}(x, t; \xi)$ is the term satisfying order $\mathring{a}_{j,1}=r_1 j$. Now by shifting (x_0, t_0) conveniently small, we can assume the following: Among $\mathring{a}_{j,1}(x, t; \xi)$, with homogeneous of degree $r_1 j$, there exists j_0 such that $\mathring{a}_{j_0,1}(x_0, t_0; \xi)\not\equiv 0$ (as a function of ξ), and for $j>j_0$ $\mathring{a}_{j,1}(x, t; \xi)\equiv 0$, for sufficiently small neighborhood of (x_0, t_0) and for any ξ. Since we supposed $(x_0, t_0)=(0, 0)$, we obtain the following result: The top order term in (5.7) at (0, 0) is, by taking ξ^0 in such a way that $\mathring{a}_{j_0,1}(0, 0; \xi^0)\neq 0$,

$$(\lambda_1-\lambda_2)^{m_2}\mathring{a}_{j_0,1}(0, 0; \xi^0)f(0)\rho^{m_1\theta+(r_1-\theta)j_0+m_2}.$$

On the other hand, in the asymptotic expression of

$$\exp(-i\phi\rho-it\rho^\theta)\tilde{P}_1 \circ \tilde{P}_2(\exp(i\phi\rho+it\rho^\theta)f(x)),$$

the top order in ρ is less than or equal to, when $\tilde{r}_1>0$,

$$(m_1-j)\theta+\tilde{r}_1 j+m_2 \qquad (0\leqq j\leqq m_1).$$

Hence this cannot be greater than $m_1\theta+m_2$. This is the same in the case $\tilde{r}_1\leqq 0$. Since $m_1\theta+(r_1-\theta)j_0+m_2>m_1\theta+m_2$, we arrived at a contradiction.

We explain briefly the outline of the proof of Theorem 5.2. Recall that we proved already Theorem 2.3. We show that, in actual case, the same argument can be applied. First let us consider P_1:

$$P_1=(\partial_t-i\lambda_1)^{m_1}+\sum_{j=1}^{m_1} a_{j1}(x, t; D)(\partial_t-i\lambda_1)^{m_1-j}, \qquad t\in[0, T].$$

We assume $\rho_1>0$. Denoting the principal symbol of $a_{j,1}$ by $\mathring{a}_{j,1}(x, t; \xi)$, homogeneous of degree $\rho_1 j$ in ξ, we consider the sub-characteristic equation

$$(5.8)\qquad \mu^{m_1}+\sum_j \mathring{a}_{j,1}(x, t: \xi)\mu^{m_1-j}=0.$$

For simplicity we *assume* that at $(x, t)=(0, 0)$, there exists at least one root, say $\mu_1(\xi)$, and ξ^0 ($|\xi^0|=1$) such that

$$\text{Re } \mu_1(\xi^0)>0. \tag{5.9}$$

Then we show that

Proposition 5.3 (generalization of Theorem 2.3). *Under the assumption* (5.9), *the Cauchy problem*:

$$\begin{cases} P_1 u=0, & t \in [0, T], \\ \partial_t^j u|_{t=0}=u_j(x) & (\in H^\infty) \end{cases} \tag{5.10}$$

is never H^∞-wellposed.

Note 1. About the condition (5.9) (when $\rho_1>0$), we discuss later. Except for the case $\rho_1=1/2$, this condition is always satisfied, and in the case $\rho_1=1/2$, fairly detailed argument is required.

Proof of Proposition 5.3. Since we are concerned only with P_1, we omit suffix 1. So that, λ_1, a_{j1}, m_1, and ρ_1 are denoted simply by λ, a_j, m, and ρ respectively. Denote

$$(\partial_t - i\lambda)^j u = u_j \qquad (0 \leqq j \leqq m-1). \tag{5.11}$$

Then denoting $U={}^t(u_0, u_1, \cdots, u_{m-1})$, (5.10) is written in the form,

$$\partial_t U = i\lambda(x, t; D)IU + A(x, t; D)U, \tag{5.12}$$

where I is the identity matrix, and

$$A(x, t; D)=\begin{pmatrix} & 1 & & & \\ & & 1 & & \\ & & & \ddots & \\ & & & & 1 \\ -a_m & -a_{m-1} & \cdots & & -a_1 \end{pmatrix}.$$

Next, put

$$\tilde{u}_j = \langle \Lambda \rangle^{(m-1-j)\rho} u_j, \qquad (0 \leqq j \leqq m-1). \tag{5.13}$$

Then, denoting again $\tilde{U}={}^t(\tilde{u}_0, \tilde{u}_1, \cdots, \tilde{u}_{m-1})$, (5.12) becomes

$$\partial_t \tilde{U} = i\lambda(x, t; D)I\tilde{U} + B\langle \Lambda \rangle^\rho \tilde{U}, \tag{5.14}$$

where B is of order 0, and its principal symbol $\mathring{B}(x, t; \xi)$ takes the form:

$$\mathring{B}=\begin{pmatrix} 1 & & & & \\ & 1 & & & \\ & & \ddots & & \\ & & & & 1 \\ b_m & b_{m-1} & \cdots & & b_1 \end{pmatrix}$$

where $b_j(x, t; \xi) = -\mathring{a}_j(x, t; \xi)|\xi|^{-\rho j}$.

Observe that the characteristic equation corresponding to $\mathring{B}$ coincides with the sub-characteristic equation (5.8). Thus, by a well-known lemma, for ε (>0) given small, there exists a *constant matrix* N_0 such that

$$N_0\mathring{B}(0, 0; \xi^0)N_0^{-1} = \begin{pmatrix} \mu_1(\xi^0) & & 0 \\ & \ddots & \\ c_{ij} & & \mu_m(\xi^0) \end{pmatrix}$$

where $|c_{ij}| < \varepsilon$.

Suppose

$$\begin{aligned} &\operatorname{Re}\mu_1(\xi^0), \cdots, \operatorname{Re}\mu_p(\xi^0) > 0, \qquad (p \geqq 1), \\ &\operatorname{Re}\mu_{p+1}(\xi^0), \cdots, \operatorname{Re}\mu_m(\xi^0) \leqq 0. \end{aligned}$$

Denote

$$\delta_0 = \frac{1}{2} \min_{1 \leqq i \leqq p} \operatorname{Re}\mu_i(\xi^0).$$

Hence denoting

$$\mathring{C}(x, t; \xi) = N_0\mathring{B}(x, t; \xi)N_0^{-1} = \begin{pmatrix} \mu_1(x, t; \xi) & & 0 \\ & \ddots & \\ 0 & & \mu_m(x, t; \xi) \end{pmatrix} + \begin{pmatrix} \mu_{ij}(x, t; \xi) \end{pmatrix},$$

we can assume the following properties, by taking r_0 and t_0 sufficiently small: For $|x| \leqq 2r_0$, $|\xi - \xi^0| \leqq r_0$, and $t \in [0, t_0]$,

$$(5.15)\qquad \begin{aligned} &\text{i)} \quad \operatorname{Re}\mu_i(x, t; \xi) \geqq \delta_0, \qquad \text{for } 1 \leqq i \leqq p, \\ &\text{ii)} \quad \operatorname{Re}\mu_i(x, t; \xi) \leqq \varepsilon, \qquad \text{for } p+1 \leqq i \leqq m, \\ &\text{iii)} \quad |\mu_{ij}(x, t; \xi)| \leqq 2\varepsilon. \end{aligned}$$

Let us remark that μ_i and μ_{ij} are supposed to be homogeneous of degree 0. Denoting further

$$(5.16)\qquad N_0\tilde{U} = V \ \ (= {}^t(v_1, v_2, \cdots, v_m)),$$

we get

$$\partial_t V = i\lambda(x, t; D)IV + \mathring{C}(x, t; D)\langle\Lambda\rangle^\rho V + C'(x, t; D)V, \tag{5.17}$$

where $\mathring{C}(x, t; \xi)$ is homogeneous of degree 0 in ξ, and C' is also of order 0. To be concrete,

$$\partial_t v_i = (i\lambda + \mu_i\langle\Lambda\rangle^\rho)v_i + \sum_{j=1}^m \mu_{ij}(x, t; D)\langle\Lambda\rangle^\rho v_j + \sum_{j=1}^m c'_{ij}(x, t; D)v_j, \quad (1 \leqq i \leqq m). \tag{5.18}$$

We apply micro-localizers $\alpha_n^{(p)}(D)\beta_{(q)}(x)$ of size r_0 around $(0, \xi^0)$, which yields, denoting

$$\alpha_n^{(p)}(D)\beta_{(q)}(x)v_i = v_{i,pq}(\cdot, t), \qquad (1 \leqq i \leqq m, |p+q| \leqq N),$$

$$\partial_t(v_{i,pq}) = (i\lambda + \mu_i\langle\Lambda\rangle^\rho)(v_{i,pq}) + \sum_j \mu_{ij}\langle\Lambda\rangle^\rho(v_{i,pq}) + \sum_j c'_{ij}(v_{j,pq}) + f_{i,pq},$$

where $f_{i,pq}$ is the error term arising from the commutation of $\alpha_n^{(p)}\beta_{(q)}$ with the operator. Namely

$$f_{i,pq} = [\alpha_n^{(p)}\beta_{(q)}, i\lambda + \mu_i\langle\Lambda\rangle^\rho]v_i + \sum_j [\alpha_n^{(p)}\beta_{(q)}, \mu_{ij}\langle\Lambda\rangle^\rho + c'_{ij}]v_j. \tag{5.19}$$

From (5.15), we get for $1 \leqq i \leqq p$,

$$\frac{d}{dt}\|v_{i,pq}\| \geqq \frac{1}{2}\delta_0 n^\rho \|v_{i,pq}\| - 3\varepsilon n^\rho \sum_j \|v_{j,pq}\| - \|f_{i,pq}\|,$$

and for $p+1 \leqq i \leqq m$,

$$\frac{d}{dt}\|v_{i,pq}\| \leqq 3\varepsilon n^\rho \|v_{i,pq}\| + 3\varepsilon n^\rho \sum_j \|v_{j,pq}\| + \|f_{i,pq}\|.$$

Denote

$$S_{n,pq}(t; V) = \exp(-\delta' n^\rho t)\left[\sum_{i=1}^p \|v_{i,pq}\| - \sum_{i=p+1}^m \|v_{i,pq}\|\right].$$

We choose

$$\delta' = \frac{1}{4}\delta_0, \qquad \varepsilon = \frac{1}{24(m+1)}\delta_0. \tag{5.20}$$

We get

$$(5.21)\qquad \exp(\delta' n^\rho t)\frac{d}{dt}S_{n,pq}(t;V)\geqq\frac{1}{8}\delta_0 n^\rho\sum_{i=1}^m\|v_{i,pq}\|-\sum_{i=1}^m\|f_{i,pq}\|.$$

Now we look at the form $f_{i,pq}$, given by (5.19). As in Theorem 2.3, the dominant term is $[\alpha_n^{(p)}\beta_{(q)},\lambda](v_i)$. Its L^2-norm is estimated in the form:

$$\text{const.}\sum_{1\leqq|\mu+\nu|\leqq N-|p+q|}n^{1-|\nu|}\|v_{i,p+\mu,q+\nu}\|+\text{const.}\,n^{-N+|q|}\|v_i(\cdot,t)\|.$$

Taking into account of this, we consider the following energy form:

$$(5.22)\qquad S_n(t;V)=\sum_{|p+q|\leqq N}c_{pq}^n S_{n,pq}(t;V),$$

where

$$c_{pq}^n=M^{|p+q|}n^{(1-\rho)|p|-\rho|q|},$$

N will be defined later, and M is taken sufficiently large. From (5.21), we deduce ($^\exists\delta''>0$)

$$\exp(\delta' n^\rho t)\frac{d}{dt}S_n(t;V)$$
$$\geqq\delta'' n^\rho\sum_{|p+q|\leqq N}c_{pq}^n\sum_{i=1}^m\|v_{i,pq}\|-\text{const.}\,n^{-\rho N}\sum_{i=1}^m\|v_i(\cdot,t)\|.$$

Hence, a fortiori,

$$(5.23)\qquad \frac{d}{dt}S_n(t;V)\geqq\delta'' n^\rho S_n(t;V)-\exp(-\delta' n^\rho t)n^{-\rho N}\sum_{i=1}^m\|v_i(\cdot,t)\|.$$

Now we define a series of initial data as follows:

$$V_n(\cdot,0)=\begin{pmatrix}\tilde\psi_n(x)\\0\\\vdots\\0\end{pmatrix},$$

where $\tilde\psi_n$ is defined by (2.12). Observe that this determines uniquely the Cauchy data at $t=0$. In fact, first

$$(5.24)\qquad \tilde U_n(\cdot,0)=N_0^{-1}V_n(\cdot,0)=N_0^{-1}\begin{pmatrix}\tilde\psi_n\\0\\\vdots\\0\end{pmatrix}.$$

Observe further

(5.25) $\partial_t^j u = c_j^j u_0 + c_{j-1}^j u_1 + \cdots + c_1^j u_{j-1} + u_j, \qquad (0 \leqq j \leqq m-1),$

where order $c_k^j(x, t; \xi) \leqq k$,

(5.26) $u_j = d_j^j u + d_{j-1}^j \partial_t u + \cdots + d_1^j \partial_t^{j-1} u + \partial_t^j u, \qquad (0 \leqq j \leqq m-1),$

where order $d_k^j(x, t; \xi) \leqq k$.

In view of (5.13), (5.25) determines the initial data.

On the other hand, if we suppose (5.10) to be H^∞-wellposed, we obtain the following type inequality: $\exists k$ and $C(T)$ such that

(5.27) $$\sup_{t \in [0,T]} \sum_{j=0}^{m-1} \|\partial_t^j u(\cdot, t)\|_{H^{m-1-j}} \leqq C(T) \sum_{j=0}^{m-1} \|\partial_t^j u(\cdot, 0)\|_{H^k}.$$

Now if we put $u = u_n$ in (5.23), and in view of (5.25), (5.26), we see easily that the right hand side is estimated by const. n^{k+m-1}. Hence

$$\sum_{j=0}^{m-1} \|\partial_t^j u_n(\cdot, t)\|_{H^{m-1-j}} \leqq \text{const. } n^{k+m-1}.$$

In view of (5.26), this implies

$$\|\tilde{U}_n(\cdot, t)\| \leqq \text{const. } n^{k+m-1},$$

which implies

(5.28) $$\|V_n(\cdot, t)\| \leqq \text{const. } n^{k+m-1}, \qquad t \in [0, T].$$

In view of (5.23), we fix N by

$$N = [(k+m-1)/\rho] + 1.$$

This implies, since $S_n(0; V_n) \geqq c_0$,

(5.29) $$\begin{aligned} S_n(t; V_n) &\geqq \exp(\delta'' n^\rho t)[S_n(0; V_n) - o(1)] \\ &\geqq \frac{1}{2} c_0 \exp(\delta'' n^\rho t). \end{aligned}$$

On the other hand, (5.28) implies

$$S_n(t; V_n) \leqq \text{const. } n^{k+m-1},$$

that is not compatible with (5.21) for any positive $t \leqq t_0$ when n tends to infinity, which proves Proposition 5.3.

On block diagonalization.

We explain why Proposition 5.3 proves essentially Theorem 5.2. For simplicity, we consider the case $k=2$.

$$P=P_2\circ P_1+R,$$

$$\text{(5.30)}\qquad \begin{cases} Pu=0, \\ \partial_t^j u|_{t=0}=u_j(x), & (0\leqq j\leqq m-1). \end{cases}$$

Put

$$\begin{cases} u_j=(\partial_t-i\lambda_1)^j u, & 0\leqq j\leqq m_1-1, \\ u_{m_1}=P_1 u, \\ u_{m_1+j}=(\partial_t-i\lambda_2)^j u_{m_1+j-1}, & 1\leqq j\leqq m_2-1. \end{cases}$$

Observe

$$P_1 u=(\partial_t-i\lambda_1)u_{m_1-1}+\sum_{j=1}^{m_1} a_{j1}u_{m_1-j},$$

$$P_2\circ P_1 u=(\partial_t-i\lambda_2)u_{m-1}+\sum_{j=1}^{m_2} a_{j2}u_{m-j}.$$

We see that

Lemma 5.4.

$$\partial_t^j u=c_j^j u_0+c_{j-1}^j u_1+\cdots+c_1^j u_{j-1}+u_j, \qquad (0\leqq j\leqq m-1),$$

$$u_j=d_j^j u+d_{j-1}^j\partial_t u+\cdots+d_1^j\partial_t^{j-1}u+\partial_t^j u, \qquad (0\leqq j\leqq m-1),$$

where $c_k^j(x, t; \xi)$, $d_k^j(x, t; \xi)$ are of order $\leqq k$.

Now (5.30) is equivalent, denoting $U={}^t(u_0, u_1, \cdots, u_{m-1})$, to

$$\text{(5.31)}\qquad \partial_t U=\begin{bmatrix} D_1 & 1 \\ 0 & D_2 \end{bmatrix} U+R_1 U,$$

where $R_1(x, t; \xi)$ is a null symbol and

$$D_j=\begin{pmatrix} i\lambda_j & 1 & & & 0 \\ & i\lambda_j & 1 & & \\ & & \ddots & \ddots & \\ & 0 & & \ddots & 1 \\ -a_{m_j,j} & -a_{m_j-1,j} & \cdots & & i\lambda_j-a_{1,j} \end{pmatrix}, \qquad (j=1, 2).$$

We can prove

Lemma 5.5 (block diagonalization). *We can construct a pseudo-differential operator M of order* -1 *such that* (5.31) *is transformed into*

$$\partial_t(I+M)U=\begin{bmatrix} D_1 & \vdots & 0 \\ \cdots & \vdots & \cdots \\ 0 & \vdots & D_2 \end{bmatrix}(I+M)U+R_0U, \tag{5.32}$$

where $R_0(x, t; \xi)$ *is a null symbol.*

Note. The proof is a little delicate. We give it in Appendix.

Recall that there exists a pseudo-differential operator M' of order -1, such that

$$(I+M')(I+M)=I+R', \qquad (I+M)(I+M')=I+R'',$$

where $R'(x, t; \xi)$, $R''(x, t; \xi)$ are null symbols. Put

$$(I+M)U=V. \tag{5.33}$$

Then

$$\partial_t V=\begin{bmatrix} D_1 & \\ & D_2 \end{bmatrix} V+R_0U. \tag{5.34}$$

Now we assign first the initial data for the first m_1-components of V, and assign 0 for the remaining components. It determines the initial data of V. But this does not determine completely that of U. In view of this, we assign first the data of approximation V^0, and define

$$(I+M')V^0=U^0 \ \ (={}^t(u_0^0, u_1^0, \cdots, u_{m-1}^0)).$$

This determines completely the Cauchy data of u in view of Lemma 5.4. Then, the corresponding initial data $\tilde{V}^0=(I+M)U^0$ differs from V^0 by

$$\tilde{V}^0-V^0=((I+M)(I+M')-I)V^0=R''(x, t; D)V^0.$$

Recall again the following property: For any null symbol $r(x, \xi)$, $\|\alpha_n\beta\, r(x, D)v\|$ is rapidly decreasing when n tends to infinity. This shows that Proposition 5.3 proves Theorem 5.2, provided we assume (5.9).

On the condition (5.9).

The consideration of this part has been achieved by the collaboration of Dr. W. Matsumoto. The author wishes to thank him for it.

First let $\rho_1=q/p<1$. We can assume p and q are positive integers relatively prime. *Let us show that when $p\geqq 3$, we can assume* (5.9). Let the sub-characteristic equation be

$$\mu^{m'}+\sum_j \mathring{a}_j(\xi)\mu^{m'-j}=0,$$

where degree $\mathring{a}_j=\rho_1 j$. Since $\rho_1 j$ is integer, j should be multiple of p. Hence the above equation becomes

$$\mu^{m'}+\sum_j \mathring{a}_{pj}(\xi)\mu^{m'-pj}=0. \tag{5.35}$$

This shows that, if μ_0 is a root, then $\mu_0\omega^k$ $(0\leqq k\leqq p-1)$ are also roots, where $\omega=\exp(2\pi i/p)$. This shows that (5.9) can be assumed always.

In the case $p=2$, namely $\rho_1=1/2$, the argument seems to be not simple. The trouble is that all the non-zero roots of (5.35) (with $p=2$) might be pure imaginary. We start from the perfect factorization: Let

$$\lambda_1(x, t; \xi)>\lambda_2(x, t; \xi)>\cdots>\lambda_k(x, t; \xi), \qquad (|\xi|\neq 0).$$

We have

$$\lambda_j(x, t; -\xi)=-\lambda_{k+1-j}(x, t; \xi).$$

More concretely, when $k=2p$ (even), we have

$$\lambda_j(x, t; -\xi)=-\lambda_{2p-j+1}(x, t; \xi), \qquad 1\leqq i\leqq p,$$

and when $k=2p+1$, $(p\geqq 1)$, we have

$$\lambda_j(x, t; -\xi)=-\lambda_{2p+2-j}(x, t; \xi),$$
$$\lambda_{p+1}(x, t; -\xi)=-\lambda_{p+1}(x, t; \xi).$$

In this sense, we say that, in the case $k=2p$, λ_j and λ_{2p-j+1} are coupled, and in the case $k=2p+1$, λ_j and λ_{2p+2-j} are coupled, and λ_{p+1} is coupled with itself. Suppose

$$P=r(x, t;, \xi, \tau)\circ p(x, t; \xi, \tau)+r_1,$$

where the principal symbol of p is $(\tau-\lambda)^{m'}$, and r_1 is null symbol. We assume the index corresponding to p is $1/2$. Let $\tilde{\lambda}$ be the coupled root to λ. Let

$$P=\tilde{r}(x, t; \xi, \tau)\circ q(x, t; \xi, \tau)+r_2,$$

where the principal symbol of q is $(\tau-\tilde{\lambda})^{m'}$, and r_2 is a null symbol. Fairly detailed argument shows that the index to q is also 1/2. Moreover, let

$$\mathring{p}=(\tau-\lambda)^{m'}+\sum_{j\geq 1}\mathring{a}_{2j}\circ(\tau-\lambda)^{m'-2j},$$

$$\mathring{q}=(\tau-\tilde{\lambda})^{m'}+\sum_{j\geq 1}\mathring{b}_{2j}\circ(\tau-\tilde{\lambda})^{m'-2j},$$

where $(\tau-\lambda)^{m'-2j}=(\tau-\lambda)\circ(\tau-\lambda)\circ\cdots\circ(\tau-\lambda)$, and $\mathring{a}_{2j}$, $\mathring{b}_{2j}$ are of homogeneous of degree j. Recall that $\lambda(x,t;-\xi)=-\tilde{\lambda}(x,t;\xi)$. Then, by fairly detailed argument, we can show that

$$\mathring{a}_{2j}(x,t;-\xi)=(-1)^j\mathring{b}_{2j}(x,t;\xi). \tag{5.36}$$

The sub-characteristic equation to p is

$$\mu^{m'}+\sum_{j}(-1)^j\mathring{a}_{2j}(x,t;\xi)\mu^{m'-2j}=0, \tag{5.37}$$

and that one to q is

$$\mu^{m'}+\sum_{j}(-1)^j\mathring{b}_{2j}(x,t;\xi)\mu^{m'-2j}=0.$$

Then, in view of (5.36), when we replace ξ to $-\xi$, the equation becomes

$$\mu'^{m'}+\sum_{j}\mathring{a}_{2j}(x,t;\xi)\mu'^{m'-2j}=0. \tag{5.38}$$

From (5.37), (5.38), we see that if μ_0 is a root of (5.37), then $\pm i\mu_0$ are roots of (5.38).

We reached the following conclusion. If eventually all the roots of the sub-characteristic equation do not satisfy (5.9), then replacing the operator by the coupled one and replacing ξ^0 by $-\xi^0$, we can assume (5.9).

Now we return to the case of Gevrey class. We assume all the coefficients appearing in (5.1) to belong to $\gamma^{(s)}$ $(s>1)$. We generalize Theorem 3.4 in the following way:

Proposition 5.4. *Under the assumption* (5.9), *the problem*

$$\begin{cases} P_1u=0, & t\in[0,T], \\ \partial_t^j u|_{t=0}=u_j(x)\in\gamma_{L^2}^{(s)}, & (0\leq j\leq m_1-1) \end{cases}$$

is never wellposed if $1/s<\rho_1$.

The proof is carried out along the same line as Proposition 5.3, taking account of Theorem 3.4. From this proposition, we obtain

Theorem 5.5. *In order that the Cauchy problem*

$$\begin{cases} Pu=f \in C^0([0, T]; \gamma_{L^2}^{(s)}), \\ \partial_t^j u|_{t=t_0}=u_j(x) \in \gamma_{L^2}^{(s)}, \quad (0\leqq i\leqq m-1), \quad (0\leqq t_0\leqq T), \end{cases}$$

be uniformly well-posed in $\gamma_{L^2}^{(s)}$, *it is necessary and sufficient*

$$\frac{1}{s}>\rho, \qquad \rho=\max \rho_j.$$

In the case $1/s=\rho$, *there exists only a local solution (in* t).

Note. This theorem is not entirely new. The original result of Ohya [17] and recent result of Komatsu [10] give already sufficient part of the theorem, and the necessary part is given by Ivrii [8] under a little different form (but equivalent). The outline of our proof is given in [15].

§ 6. Propagations of singularities

Our starting point of this research was to prove the following fact: Let

$$(6.1) \qquad L(u)=\partial_t u+i\lambda(x, t; D)u+c(x, t; D)u=f, \qquad x \in \boldsymbol{R}^l.$$

We assume

i) $\lambda(x, t; \xi) \in S_{1,0}^1$, real-valued and homogeneous of degree 1 in ξ, and of Gevrey class $\gamma^{(s)}$. More precisely,

$$|\lambda_{(\beta)}^{(\alpha)}(x, t; \xi)|\leqq A\alpha!\beta!^s C_0^{|\alpha+\beta|}|\xi|^{1-|\alpha|}, \qquad (\exists A, C_0>0),$$

ii) $c(x, t; \xi) \in S_{1,0}^{\rho}$, Gevrey pseudo-analytic, namely,

$$|c_{(\beta)}^{(\alpha)}(x, t; \xi)|\leqq A'\alpha!\beta!^s C_0^{|\alpha+\beta|}|\xi|^{\rho-|\alpha|}, \qquad (\rho<1),$$

for $|\alpha|\leqq R^{-1}|\xi|^{1/s}$, ($R$ large).

Let Γ be the bicharacteristic strip of (6.1), issuing from (x_0, ξ^0) at $t=0$. Namely, $\Gamma(x(t), \xi(t))$ is the solution of

$$\dot{x}=\lambda_\xi(x, t; \xi), \quad \dot{\xi}=-\lambda_x(x, t; \xi), \quad (x(0), \xi(0))=(x_0, \xi^0).$$

Suppose $1/s>\rho$.

Theorem 6.1. *Let* $(x(t), \xi(t)) \in WF_s(f(\cdot, t))$. *Then* $(x_0, \xi^0) \notin (\in)$ $WF_s(u(\cdot, 0))$ *implies* $(x(t), \xi(t)) \notin (\in) WF_s(u(\cdot, t))$.

Proof. The proof is given in [12], [16] in the case $s>1$, and in [14] in the case $s=1$. We explain only its idea in the case $s>1$. *We use micro-localizer around Γ.* Let $\alpha_n(\xi, t)$, $\beta_n(x, t)$ be defined by

$$\partial_t\alpha_n(\xi, t)-n\lambda_x(t)\partial_\xi\alpha_n(\xi, t)=0,$$
$$\partial_t\beta_n(x, t)+\lambda_\xi(t)\partial_x\beta_n(x, t)=0,$$

where $\lambda_x(t)=\lambda_x(x(t), t; \xi(t))$, $\lambda_\xi(t)=\lambda_\xi(x(t), t; \xi(t))$; $\alpha_n(\xi, 0)=\alpha_n(\xi)$, $\beta_n(x, 0)=\beta_n(x)$. Here $(\alpha_n(\xi), \beta_n(x))$ is a micro-localizer around (x_0, ξ^0) of small size r_0 (see (3.2)–(3.4)). We see easily that $(\alpha_n(\xi, t), \beta_n(x, t))$ has the same property (3.2)–(3.4). We consider the following local energy form:

$$S_n(t; u(\cdot, t))=\sum_{|p+q|\leqq N} c^n_{pq}\|\alpha_n^{(p)}(D, t)\beta_{n(q)}(x, t)u(\cdot, t)\|,$$

and apply to this Theorems 3.1 and 3.2. Indeed, we can show the following type inequality:

$$\frac{d}{dt}S_n(t; u)\leqq[(2lA)r_0+\text{const.}\, n^{-\delta}]n^{1/s}S_n(t; u)+e^{-N/2}+\exp(-\varepsilon' n^{1/s}),$$

where A is the constant appearing in i), and $\delta>0$.

Remark. Observe that direct extensions of standard arguments in the C^∞-case seem to not work in this case. Indeed, the standard argument shows that the assertion is true in the sense $WF_{2s-1}(u)$, instead of $WF_s(u)$ (see for instance, Kajitani [9]). Taniguchi [19] proved the above theorem independently of ours. He uses refined techniques of Fourier integral operators.

Appendix

§ A.1. In our argument the block diagonalization plays an important role. We explain this briefly. First we show how to construct a pseudo-differential operator $M(x, t; D)$ satisfying (5.32). Here we change slightly the notations.

$$D_1=\begin{pmatrix} \lambda & 1 & & & \\ & \lambda & 1 & & 0 \\ & & \ddots & \ddots & \\ 0 & & & \ddots & \\ & & & \ddots & 1 \\ a_{p-1} & a_{p-2} & \cdots & \cdots & \lambda+a_0 \end{pmatrix}, \qquad D_2=\begin{pmatrix} \mu & 1 & & & \\ & \mu & 1 & & 0 \\ & & \ddots & \ddots & \\ 0 & & & \ddots & \\ & & & \ddots & 1 \\ b_{q-1} & b_{q-2} & \cdots & \cdots & \mu+b_0 \end{pmatrix}.$$

Here we assume

i) $\lambda(x, t; \xi)$, $\mu(x, t; \xi)$ are homogeneous of degree 1 in ξ, and $\lambda \neq \mu$ for $|\xi| \neq 0$,

ii) $a_j(x, t; \xi)$, $b_j(x, t; \xi)$ are *classical formal symbols* of order at most j.

Since $\partial_t((I+M)U) - \partial_t M \cdot U = (I+M) \circ (D+J)U$, the condition on M becomes

$$(I+M) \circ (D+J) = D \circ (I+M) + \partial_t M,$$

namely

$$D \circ M - M \circ D + \partial_t M = (I+M)J. \tag{A.1}$$

We seek M under the form

$$M = \begin{bmatrix} 0 & M_{12} \\ 0 & 0 \end{bmatrix},$$

where M_{12} is a $p \times q$ matrix. Then, since $M \circ J = 0$, (A.1) becomes

$$D_1 \circ M_{12} - M_{12} \circ D_2 + \partial_t M_{12} = J_{12}, \tag{A.2}$$

where

$$J_{12} = \begin{bmatrix} 0 & \\ \vdots & \\ 0 & 0 \\ 1 & \end{bmatrix}.$$

Denoting

$$M_{12} = \begin{bmatrix} m_{11} & m_{12} & \cdots & m_{1q} \\ m_{p1} & m_{p2} & \cdots & m_{pq} \end{bmatrix}, \qquad (m_{ij} = m_{ij}(x, t; \xi)),$$

the above relation becomes:

i) For $1 \leqq i \leqq p-1$,

$$\lambda \circ m_{ij} + m_{i+1,j} - m_{i,j-1} - m_{ij} \circ \mu - m_{iq} \circ b_{q-j} + \partial_t m_{ij} = 0,$$

ii) For $i = p$,

$$a_{p-1} \circ m_{1j} + a_{p-2} \circ m_{2j} + \cdots + a_0 \circ m_{pj} + \lambda \circ m_{pj} - m_{p,j-1} - m_{pj} \circ \mu$$
$$- m_{pq} \circ b_{q-j} + \partial_t m_{pj} = \delta_j^1.$$

Denoting

$$\lambda \circ m - m \circ \mu + \partial_t m = L(m), \tag{A.3}$$

the above relation becomes

$$
\text{(A.4)}\quad
\begin{aligned}
&L(m_{ij})+m_{i+1,j}-m_{i,j-1}=m_{iq}\circ b_{q-j}, \qquad (1\leqq i\leqq p-1)\\
&L(m_{pj})-m_{p,j-1}-\delta_j^1=m_{pq}\circ b_{q-j}\\
&\qquad -(a_{p-1}\circ m_{1j}+a_{p-2}\circ m_{2j}+\cdots+a_0\circ m_{pj}).
\end{aligned}
$$

We solve this equation by successive approximations. For this purpose we prepare a lemma.

Lemma A.1. *Let $a(x, t; \xi)$ and $b(x, t; \xi)$ be homogeneous of degree* 1 *in ξ, and $a\neq b$ for $|\xi|\neq 0$. Let*

$$p(x, \xi)=\sum_{j=m}^{-\infty} p_j(x, t; \xi)$$

be a formal symbol, where p_j is homogeneous of degree j in ξ. Then the equation

$$L(q)=a\circ q-q\circ b+\partial_t q=p$$

has a unique formal solution

$$q(x, t; \xi)=\sum_{j=m-1}^{-\infty} q_j(x, t; \xi).$$

Proof. The above equation is equivalent to

$$(a-b)q_{j-1}+\sum_{\substack{k-|\alpha|=j-1\\ |\alpha|\geqq 1}} \alpha!^{-1}a^{(\alpha)}q_{k(\alpha)}-\sum_{\substack{k-|\alpha|=j-1\\ |\alpha|\geqq 1}} \alpha!^{-1}q_k^{(\alpha)}b_{(\alpha)}+\partial_t q_j=p_j,$$

$$j=m, m-1, m-2, \cdots.$$

Then, $q_{m-1}(x, t; \xi)$, $q_{m-2}(x, t; \xi)$, $\cdots$ are determined successively. In particular,

$$
\begin{aligned}
&q_{m-1}=p_m(a-b)^{-1},\\
&q_{m-2}=\Big(p_{m-1}+\sum_j q_{m-1}^{(ej)}b_{(ej)}-\sum_j a^{(ej)}q_{m-1(ej)}-\partial_t q_{m-1}\Big)(a-b)^{-1}\\
&\cdots\cdots\cdots\cdots\cdots\cdots
\end{aligned}
$$

Using this lemma we obtain

Lemma A.2. *The equation* (A.4) *has a solution.*

Proof.

First step. Let us consider

$$L(m_{pj}^{(0)})-m_{p,j-1}^{(0)}-\delta_j^1=0, \qquad (1\leqq j\leqq m).$$

We solve successively $m^0_{p1}, m^0_{p2}, \cdots, m^0_{pq}$. First, $L(m^{(0)}_{p1})=1$, implies order $m^{(0)}_{p1} \leqq -1$. Hence we get, order $m^{(0)}_{pj} \leqq -j$. Next,

$$L(m^0_{p-1,j})=m^{(0)}_{p-1,j-1}-m^{(0)}_{p,j}, \qquad (m^{(0)}_{p-1,0}=0),$$

can be solved in the order $m^{(0)}_{p-1,1}, m^{(0)}_{p-1,2}, \cdots, m^{(0)}_{p-1,q}$, and

$$\text{order } m^{(0)}_{p-1,j} \leqq -j-1, \qquad (1 \leqq j \leqq q).$$

In general

$$L(m^{(0)}_{ij})=m^{(0)}_{i,j-1}-m^{(0)}_{i+1,j}.$$

We get

$$\text{order } (m^{(0)}_{ij}) \leqq (i-p)-j.$$

Second step.

$$L(m^{(1)}_{pj})-m^{(1)}_{p,j-1}=m^{(0)}_{pq}\circ b_{q-j}-(a_{p-1}\circ m^{(0)}_{1j}+a_{p-2}\circ m^{(0)}_{2j}+\cdots+a_0\circ m^{(0)}_{pj}),$$

we solve these recursively in the order $m^{(1)}_{p1}, m^{(1)}_{p2}, \cdots, m^{(1)}_{pq}$. Now since

$$\begin{aligned}
&\text{order } (a_{p-k}\circ m^{(0)}_{kj}) \leqq (p-k)+(k-p-j)=-j,\\
&\text{order } (m^{(0)}_{pq}\circ b_{q-j}) \leqq -j,\\
&\text{order } (m^{(1)}_{pj}) \leqq -j-1 \qquad (1 \leqq j \leqq q).
\end{aligned}$$

Now we can show that

$$\text{order } (m^{(1)}_{ij}) \leqq (i-p)-j-1, \qquad \text{for } 1 \leqq i \leqq p-1.$$

In fact, let us assume this for $i \geqq i_0$, and $j < j_0$. Then from

$$L(m^{(1)}_{i_0 j_0})=m^{(1)}_{i_0,j_0-1}-m^{(1)}_{i_0+1,j_0}+m_{i_0 q}\circ b_{q-j_0},$$

we see that the order of the second member $\leqq (i_0-p)-j_0$, hence order $(m^{(1)}_{i_0,j_0})$ $\leqq (i_0-p)-j_0-1$.

In this way, for general k, we see that

(A.5) $$\text{order } (m^{(k)}_{ij}) \leqq (i-p)-j-k.$$

Hence we see that

$$m_{ij}=\sum_{k=0}^{\infty} m^{(k)}_{ij}$$

gives the desired solution. In fact, recall that each $m^{(k)}_{ij}$ is a formal symbol, and in this sum of formal symbols, the homogeneous part of degree $-l$ originates from $m_{ij}^{(k)}$ satisfying $(i-p)-j-k\geqq -l$. Namely $k\leqq(i-p)-j+l$.

Thus we proved the existence of a formal symbol M_{12} satisfying (A.2), and by a well-known fact, namely by associating true symbols to formal symbols, this proves (5.32).

§ A.2. We consider the general case. In this case,

$$D=\begin{bmatrix} D_1 & & & 0 \\ & D_2 & & \\ & & \ddots & \\ 0 & & & D_k \end{bmatrix}, \qquad J=\begin{bmatrix} 0 & J_{12} & & & 0 \\ & 0 & J_{23} & & \\ & & 0 & \ddots & \\ 0 & & & \ddots & \\ & & & & 0 \end{bmatrix},$$

where D_i is $m_i \times m_i$ matrix. We seek M under the form

$$M=\begin{pmatrix} 0 & M_{12} & M_{13} & \cdots & M_{1k} \\ & 0 & M_{23} & \cdots & M_{2k} \\ & & 0 & & \\ 0 & & & \ddots & \\ & & & & 0 \end{pmatrix},$$

where M_{ij} $(i<j)$ is $m_i \times m_j$ matrix. Then (A.1) becomes

$$D_i \circ M_{ij} - M_{ij} \circ D_j + \partial_t M_{ij} = M_{i,j-1} J_{j-1,j}, \qquad \begin{pmatrix} 1\leqq i\leqq k-1 \\ i+1\leqq j\leqq k \end{pmatrix}, \tag{A.6}$$

here we write under the convention, $M_{ii}=I_{m_i}$ (identity matrix of order m_i).

To solve these equations of M_{ij}, we start from the case $j=i+1$:

$$D_i \circ M_{i,i+1} - M_{i,i+1} \circ D_{i+1} + \partial_t M_{i,i+1} = J_{i,i+1}. \qquad (1\leqq i\leqq k-1).$$

These $(k-1)$-equations are independent each other, and we have already considered it in § A.1. Thus $M_{i,i+1}$ $(1\leqq i\leqq k-1)$ are determined. Next we pass to the case $j=i+2$.

$$D_i \circ M_{i,i+2} - M_{i,i+2} \circ D_{i+2} + \partial_t M_{i,i+2} = M_{i,i+1} J_{i+1,i+2}, \qquad (1\leqq i\leqq k-2).$$

Here observe that the right-hand side is of the form

$$\begin{bmatrix} c_1 & \\ c_2 & \\ \vdots & 0 \\ \vdots & \\ c_{m_i} & \end{bmatrix},$$

and order $(c_j)\leqq -m_{i+1}-m_i+j$, $(1\leqq j\leqq m_i)$. Since $M_{i,i+1}$ is already determined, we obtain $M_{i,i+2}$ by the same method explained in § A.1. This process can be continued up to M_{1k}. Namely, if we obtain $M_{i,i+p-1}$ $(1\leqq i\leqq k-p+1)$, then by

$$D_i \circ M_{i+p}-M_{i,i+p}\circ D_{i+p}+\partial_t M_{i,i+p}=M_{i,i+p-1}J_{i+p-1,i+p},$$

we obtain $M_{i,i+p}$.

§ A.3. We want to show that if all the coefficients $a_{k,j}(x, t; \xi)$ are Gevrey (including analytic) formal symbols, then $M(x, t; \xi)$ belongs to the same class. This is achieved by using the norm $N(a; T)$ introduced by Boutet de Monvel and Krée in [2]. However we make slight modification in order to get enough information on Gevrey dependency in t of symbols.

We denote $t=x_0$. Let $a(x, x_0; \xi)=\sum_{j\geqq 0} a_{k-j}(x, x_0; \xi)$, a_{k-j} being of homogeneous of degree $k-j$ in ξ.

$$N_k(a; T)=\sum_{\substack{j\geqq 0\\ \alpha_0,\alpha,\beta\geqq 0}} c^{\beta}_{j,\alpha_0,\alpha}|a^{(\beta)}_{k-j(\alpha_0,\alpha)}(x, x_0; \xi)|\, T^{2j+\alpha_0+|\alpha+\beta|},$$

where

$$a^{(\beta)}_{k-j(\alpha_0,\alpha)}=D^{\alpha_0}_{x_0}D^{\alpha}_{x}\partial^{\beta}_{\xi}a_{k-j}(x, x_0; \xi),$$

$$c^{\beta}_{j,\alpha_0,\alpha}=2(2n)^{-j}j!^s(j+\alpha_0)!^{-s}j!(j+|\alpha|)!^{-s}(j+|\beta|)!^{-1},$$

$$N'_k(a; T)=\sum_{j,\alpha_0\geqq 0,|\alpha+\beta|\geqq 1} c^{\beta}_{j,\alpha_0,\alpha}|a^{(\beta)}_{k-j(\alpha_0,\alpha)}|\, T^{2j+\alpha_0+|\alpha+\beta|}.$$

Fundamental properties of the norm $N(a; T)$ showed in [2] are conserved. Hereafter, by abuse of notations, we designate (x, x_0) simply by x. Let $a_1(x, \xi)$ be Gevrey symbol of class s $(s\geqq 1)$, and of homogeneous of degree 1 in ξ, and let

$$b(x, \xi)=\sum_{j\geqq 0} b_{k-j}(x, \xi)$$

be a formal symbol in $\gamma^{(s)}$. Then

Lemma A.3.

$$N_{k+1}(a_1\circ b-a_1 b; T)\ll N'_1(a_1; T)N'_k(b; T),$$

$$N_{k+1}(b\circ a_1-ba_1; T)\ll N'_1(a_1; T)N'_k(b; T).$$

Note. Observe $N'_k(b; T)\ll N_k(b; T)$. The above lemma says the following: order $(a_1\circ b-a_1 b)\leqq k$. Bearing this in mind, when we take $(k+1)$-norm of this symbol, in view of $N'_1(a_1; T)=c_1T+c_2T^2+\cdots=TE(T)$, the second member becomes $TE(T)\times N'_k(b; T)$, which shows that if T is small, then $TE(T)$ becomes small. Since the proof is elementary (but tedious), we omit the proof.

Using this lemma, we show

Lemma A.4. *Denote* $\Psi(T)=N_{-1}((\lambda-\mu)^{-1};T)$. *Then there exists* $T_0>0$ *such that the solution m of*

$$\mathscr{C}(m)=\lambda\circ m-m\circ\mu=a, \qquad \lambda\neq\mu$$

satisfies

$$N_{j-1}(m;T)\ll 2\Psi(T)N_j(a;T) \qquad \textit{for } T\leqq T_0. \tag{A.7}$$

Proof.

$$\mathscr{C}(m)=(\lambda-\mu)m+(\lambda\circ m-\lambda m)-(m\circ\mu-m\mu)=a.$$

Hence the equation is equivalent to

$$m=(\lambda-\mu)^{-1}a-(\lambda-\mu)^{-1}(\lambda\circ m-\lambda m)+(\lambda-\mu)^{-1}(m\circ\mu-m\mu).$$

We solve this equation by successive approximations:

$$\begin{aligned} &m=m^{(0)}+m^{(1)}+\cdots+m^{(k)}+\cdots, \\ &m^{(0)}=(\lambda-\mu)^{-1}a,\ m^{(k)}=-(\lambda-\mu)^{-1}(\lambda\circ m^{(k-1)}-\lambda m^{(k-1)}) \\ &\qquad\qquad +(\lambda-\mu)^{-1}(m^{(k-1)}\circ\mu-m^{(k-1)}\mu), \\ &\qquad\qquad\qquad (k=1,2,3,\cdots). \end{aligned}$$

First

$$N_{j-1}(m^{(0)};T)\ll N_{-1}((\lambda-\mu)^{-1};T)N_j(a;T)=\Psi(T)N_j(a;T).$$

Next, denoting

$$N_1'(\lambda;T)+N_1'(\mu;T)=TE(T), \tag{A.8}$$

and applying Lemma A.3, we obtain

$$\begin{aligned} N_{j-1}(m^{(1)};T)&\ll\Psi(T)N_j(\lambda\circ m^{(0)}-\lambda m^{(0)};T)+\Psi(T)N_j(m^{(0)}\circ\mu-m^{(0)}\mu;T) \\ &\ll\Psi(T)TE(T)N_{j-1}(m^{(0)};T). \end{aligned}$$

In general,

$$N_{j-1}(m^{(k)};T)\ll(\Psi(T)TE(T))^kN_{j-1}(m^{(0)};T).$$

Hence

$$N_{j-1}(m; T) \ll \Psi(T)\Big[1 + \sum_{k=1}^{\infty} (T\Psi(T)E(T))^k\Big] N_j(a; T).$$

Next, let $b(x, \xi) = b_{j-1} + b_{j-2} + \cdots$.

Lemma A.5.

$$N_j(\partial_{x_0} b; T) \ll T N_{j-1}(b; T).$$

This lemma is elementary. So we omit the proof.

Now we consider

$$L(m) = \mathscr{C}(m) + \partial_{x_0} m = a. \tag{A.9}$$

Lemma A.6. *m satisfies, if $T \leqq T_0'$ $(\leqq T_0)$,*

$$N_{j-1}(m; T) \ll \Phi(T) N_j(a; T), \tag{A.10}$$

where

$$\Phi(T) = 4\Psi(T).$$

Proof. We solve (A.9) by

$$m = m^{(0)} + m^{(1)} + \cdots + m^{(k)} + \cdots,$$

where

$$\mathscr{C}(m^{(0)}) = a, \quad \mathscr{C}(m^{(1)}) = -\partial_{x_0} m^{(0)}, \ \cdots, \quad \mathscr{C}(m^{(k)}) = -\partial_{x_0} m^{(k-1)}, \ \cdots.$$

First by (A.7),

$$N_{j-1}(m^{(0)}; T) \ll 2\Psi(T) N_j(a; T).$$

In general, using Lemma A.5,

$$\begin{aligned} N_{j-1}(m^{(k)}; T) &\ll 2\Psi(T) N_j(\partial_{x_0} m^{(k-1)}; T) \ll 2T\Psi(T) N_{j-1}(m^{(k-1)}; T) \\ &\ll (2T\Psi(T))^k N_{j-1}(m^{(0)}; T). \end{aligned}$$

Hence

$$N_{j-1}(m; T) \ll 2\Psi(T)\Big[1 + \sum_{k=1}^{\infty} (2T\Psi(T))^k\Big] N_j(a; T).$$

§ A.4. We estimate the solution of (A.4) in the following way. First let us consider the following system

$$\text{(A.11)}\qquad \begin{aligned} &L(m_{ij})+m_{i+1,j}-m_{i,j-1}=a_{ij}, &&(1\leqq i\leqq p-1),\\ &L(m_{pj})-m_{p,j-1}=a_{pj}, &&(1\leqq j\leqq q).\end{aligned}$$

We repeat the argument of Lemma A.2. Then we get

Lemma A.7. *There exists a constant c_{pq}, depending only on p, q, such that*

$$\text{(A.12)}\qquad \sum_{i,j} N_{i-p-j-1}(m_{ij};T)\ll c_{pq}\left(\sum_{k=0}^{p+q-1}\Phi(T)^k\right)\sum_{i,j} N_{i-p-j}(a_{ij};T),$$

where $\Phi(T)$ is the same one as in (A.10).

This is not difficult, we omit the proof. Now, we consider

$$\begin{aligned} &L(m_{ij}^{(0)})=m_{i,j-1}^{(0)}-m_{i+1,j}^{(0)}, &&(1\leqq i\leqq p-1),\\ &L(m_{pj}^{(0)})=m_{p,j-1}^{(0)}+\delta_j^1 &&(1\leqq j\leqq q).\end{aligned}$$

Denoting $c_{pq}\sum_{k=0}^{p+q-1}\Phi(T)^k=\Phi_{pq}(T)$, we get

$$\text{(A.13)}\qquad \sum_{i,j} N_{i-p-j-1}(m_{ij}^{(0)};T)\ll\Phi_{pq}(T).$$

Next, we set

$$\begin{aligned} &a_{ij}^k=m_{iq}^{(k-1)}\circ b_{q-j}, \qquad (1\leqq i\leqq p-1),\\ &a_{pj}^k=m_{pq}^{(k-1)}\circ b_{q-j}-(a_{p-1}\circ m_{1j}^{(k-1)}+a_{p-2}\circ m_{2j}^{(k-1)}+\cdots+a_0\circ m_{pj}^{(k-1)}),\end{aligned}$$

we get

$$\begin{aligned} N_{i-p-j}(a_{ij}^k;T)&\ll N_{i-p-q-1}(m_{iq}^{(k-1)};T)N_{q-j+1}(b_{q-j};T),\\ N_{-j}(a_{pj}^k;T)&\ll N_{-q-1}(m_{pq}^{(k-1)};T)N_{q-j+1}(b_{q-j};T)\\ &\quad+N_p(a_{p-1};T)N_{-p-j}(m_{1j}^{(k-1)};T)\\ &\quad+\cdots+N_1(a_0;T)N_{-j-1}(m_{pj}^{(k-1)};T).\end{aligned}$$

Observe that there exists an $E_{pq}(T)$ such that

$$\sum_{i,j} N_{i-p-j}(a_{ij}^k;T)\ll T^2E_{pq}(T)\sum_{i,j} N_{i-p-j-1}(m_{ij}^{(k-1)};T).$$

Hence

$$\begin{aligned} \sum_{i,j} N_{i-p-j-1}(m_{ij}^{(k)};T)&\ll\Phi_{pq}(T)\sum_{i,j} N_{i-p-j}(a_{ij}^k;T)\\ &\ll\Phi_{pq}(T)T^2E_{pq}(T)\sum_{i,j} N_{i-p-j-1}(m_{ij}^{(k-1)};T)\\ &\ll\cdots\ll(\Phi_{pq}(T)T^2E(T))^k\sum_{i,j} N_{i-p-j-1}(m_{ij}^{(0)};T).\end{aligned}$$

Hence $m_{ij}=m_{ij}^{(0)}+m_{ij}^{(1)}+\cdots$, satisfies

$$\sum_{i,j} N_{i-p-j-1}(m_{ij};\,T)\ll\Phi_{pq}(T)\left[1+\sum_{k=1}^{\infty}(\Phi_{pq}(T)T^2E_{pq}(T))^k\right].$$

Thus we proved

Theorem A.8. *The solution* $M_{12}=(m_{ij}(x,t;\xi))$ *of* (A.2) *is obtained as formal symbol of class* s ($\geqq 1$), *if we assume all the symbols* $a_i(x,t;\xi)$ *and* $b_j(x,t;\xi)$ *to be formal symbols of class* s.

References

[1] M. S. Baouendi and C. Goulaouic, Cauchy problems with characteristic initial hypersurfaces, Comm. Pure Appl. Math., **26** (1973), 455–475.
[2] M. S. Boutet de Monvel et P. Krée, Pseudo-differential operators and Gevrey classes, Ann. Inst. Fourier, **17** (1967), 295–323.
[3] J. Chazarain, Opérateurs hyperboliques à caractéristiques de multiplicité constante, Ann. Inst. Fourier, **24** (1974), 173–202.
[4] J. De Paris, Problème de Cauchy analytique à données singulières pour un opérateur différentiel bien décomposable, J. Math. Pures Appl., **51** (1972), 465–488.
[5] H. Flaschka and G. Strang, The correctness of the Cauchy problems, Adv. in Math., **6** (1971), 347–379.
[6] S. Hashimoto, T. Matsuzawa et Y. Morimoto, Opérateurs pseudo-différentiels et classes de Gevrey, Comm. Partial Differential Equations, **8** (1983), 1277–1289.
[7] L. Hörmander, Uniqueness theorems and wave front sets for solutions of linear differential equations with analytic coefficients, Comm. Pure Appl. Math., **24** (1971), 671–704.
[8] V. Ya. Ivrii, Conditions for correctness in Gevrey classes of the Cauchy problem for weakly hyperbolic equations, Siberian J. Math., **17** (1976), 422–435 (English translation).
[9] K. Kajitani, Leray-Volevich's system and Gevrey class, J. Math. Kyoto Univ., **21** (1981), 547–574.
[10] H. Komatsu, Linear hyperbolic equations with Gevrey coefficients, J. Math. Pures Appl., **59** (1980), 145–185.
[11] S. Mizohata, Some remarks on the Cauchy problem, J. Math. Kyoto Univ., **1** (1961), 107–127.
[12] ——, Propagation de régularité au sens de Gevrey pour les opérateurs différentiels à multiplicité constante, Séminaire sur les équations aux dérivés partielles hyperboliques et holomorphes de J. Vaillant, Hermann, Paris, 1984, 106–133.
[13] ——, On the meromorphic propagation and the Levi condition, Astérisque Soc. Math. France, **134** (1985), 127–135.
[14] ——, On analytic regularities, Séminaire sur Propagation des singularités et opérateurs différentiels de J. Vaillant, Hermann, Paris, 1985, 82–105.
[15] ——, Sur l'indice de Gevrey, Ibid., 106–120.
[16] ——, On the Cauchy problem (delivered at the Wuhan University), to appear from Science Press, Beijing.
[17] Y. Ohya, Le problème de Cauchy pour les équations hyperboliques à caractéristique multiple, J. Math. Soc. Japan, **16** (1964), 268–286.
[18] T. Nishitani, On the Lax-Mizohata theorem in the domain of analytic functions and Gevrey classes, J. Math. Kyoto Univ., **18** (1979), 501–512.

[19] K. Taniguchi, Fourier integral operators in Gevrey class on $\boldsymbol{R}^n$ and fundamental solution for a hyperbolic operator, Publ. RIMS, Kyoto Univ., **20** (1984), 491–542.
[20] F. Treves, Introduction to Pseudo-differential and Fourier Integral Operators, 1, 2, Plenum Press, New York, 1980.
[21] T. Yamanaka and Persson, On an extension of Holmgren's uniqueness theorem, Comment. Math. Univ. St. Paul, **22** (1973), 19–30.

Department of Mathematics
Kyoto University
Kyoto 606, Japan

Taniguchi Symp. HERT
Katata 1984, pp. 235–255

Microlocal Energy Estimates for Hyperbolic Operators with Double Characteristics

Tatsuo NISHITANI

§ 1. Introduction

In this note, microlocal energy estimates for two different types of operators with double characteristics are studied. One is effectively hyperbolic and the other is non effectively hyperbolic operator whose fundamental matrix has a Jordan block of size four.

Let $P=\xi_0^2-Q(x, \xi')$ be a classical pseudodifferential symbol of degree 2 defined near a point ρ in the cotangent bundle. If we assume that there is a real smooth function $\phi(x, \xi')$ near ρ such that $Q(x, \xi')$ vanishes on the hypersurface $\phi(x, \xi')=0$ at most of order 2 and all bicharacteristics of P having limit point ρ are transversal to this surface, then we can obtain microlocal energy estimates which are stable under perturbations of lower order symbols (in § 2). In fact, in § 3, we shall show that the existence of such function ϕ is equivalent to the (microlocal) effective hyperbolicity of P at ρ.

However, if we start from this analytical formulation of effective hyperbolicity, a proof of deriving microlocal energy estimates for P is much more simplified than that of the previous papers [7], [10]. We sketch this proof in §§ 5 through 7. More precisely, in § 5, introducing a small positive parameter, we define the localizations of symbols and introduce some symbol classes of pseudodifferential operators related to these localized symbols. In § 6, using the calculus of pseudodifferential operators in § 5, we give microlocal energy estimates for localized P. In § 7, we give some estimates of wave front sets in terms of Sobolev norms which also show that the parametrix of the Cauchy problem has finite propagation speed of wave front sets.

In case the fundamental matrix of P has a Jordan block of size four, and hence P is necessarily non effectively hyperbolic, it seems that there are not so many papers concerning with such P. In § 4, we generalize the results in [9] and, under some restrictions, we give a necessary and sufficient condition for that P admits a decomposition which allows microlocal energy estimates to be derived easily. In the most simple case, the decomposability is equivalent to that there is no bicharacteristic issuing from a simple characteristic point which has a limit point in the double characteristic set. If P admits this decomposition, then deriving microlocal energy estimates is routine and we omit it in this note.

§ 2. Study on effectively hyperbolic operators

In this section, we consider the following symbol

$$P(x, \xi)=\xi_0^2-Q(x, \xi')$$

where $x=(x_0, x')$, $\xi=(\xi_0, \xi')$, $x'=(x_1, \cdots, x_d)$, $\xi'=(\xi_1, \cdots, \xi_d)$, $Q(x, \xi')$ is a classical pseudodifferential symbol of degree 2 defined in a conic neighborhood of $\rho=(0, \hat{\xi}') \in \boldsymbol{R}\times T^*\boldsymbol{R}^d$, depending smoothly on x_0. We denote by $P_2(x, \xi)$, $Q_2(x, \xi')$ the principal symbol of $P(x, \xi)$, $Q(x, \xi')$ respectively. We assume that $Q_2(x, \xi')$ is non negative in some conic neighborhood of ρ and ρ is a double characteristic point of P. Then the Taylor expansion of P_2 at ρ begins with a hyperbolic quadratic form P_ρ on $T(T^*\boldsymbol{R}^{d+1})$. We denote by $\Gamma(P, \rho)$ the hyperbolic cone of P_ρ containing $\theta=(0, 1, 0, \cdots, 0)$;

$$\Gamma(P, \rho)=\text{connected component of } \{(x, \xi) \in \boldsymbol{R}^{2d+2};\ P_\rho(x, \xi)\neq 0\}$$
$$\text{containing } \theta.$$

We also denote by H_ϕ, the Hamilton vector field of ϕ.

Remark 2.1. The coefficient matrix of the Hamiltonian system with this quadratic Hamiltonian P_ρ is called the fundamental (or Hamilton) matrix of P. The eigenvalues of the fundamental matrix lie on the imaginary axis with possible exception of one pair λ, $-\lambda$, $\lambda>0$ ([1], [2]).

Theorem 2.1. *Let $\phi(x, \xi')$ be a real smooth function near ρ, positively homogeneous of degree 0 in ξ' such that*

$$-H_\phi(\rho) \in \Gamma(P, \rho),\ Q_2(x, \xi')\geqq c\phi(x, \xi')^2|\xi'|^2, \text{ near } \rho,$$

with some positive constant c. Then for sufficiently small conic neighborhood Γ of ρ in $T^*\boldsymbol{R}^d$, $P(x, D)$ has a parametrix in Γ with finite propagation speed of wave front sets.

Here we give a definition of the parametrix of P in Γ with finite propagation speed of WF ([7]). Let I be an open interval in $\boldsymbol{R}$ containing 0. We denote by $C_+^l(I, H^q)$ the set of all $f \in C^l(I, H^q)$ vanishing in $x_0<0$, where H^q stands for the usual Sobolev space $H^q(\boldsymbol{R}^d)$ with the norm $\|\cdot\|_q$.

We say that an operator G acting from $C^0(I, H^p)$ into $C^0(I, H^q)$ is a plus operator if Gf vanishes in $x_0<0$ for any $f \in C^0_+(I, H^p)$.

Now we assume that $Q(x, \xi')$ is defined in $I\times W$ where W is an open conic neighborhood of ρ in $T^*\boldsymbol{R}^d$. We fix a conic open set $\Gamma \Subset W$ and extend $Q(x, \xi')$ to outside Γ. We shall say that a plus operator G is a parametrix of P in Γ with finite propagation speed of WF if G satisfies the following conditions;

For every open conic sets Γ_0, Γ_1 in $T^*\boldsymbol{R}^d$ with $\Gamma_0 \Subset \Gamma_1 \Subset \Gamma$, there is a positive constant $\delta=\delta(\Gamma_0, \Gamma_1)$ such that for any $g(x', \xi'), h(x', \xi') \in S^0_{1,0}$ with cone supports contained in Γ_0, $T^*\boldsymbol{R}^d \setminus \Gamma^1$ respectively, we have

$$g(PG-Id)=R_1+R_2, \qquad hGg=R_3,$$

where R_1 belongs to $S^{-\infty}$ depending smoothly on $x_0 \in I$ and R_i $(i=1, 2)$ are plus operators satisfying the following estimates,

$$\sum_{j=0}^{1} \| D_0^j R_i f\|_p^2(t) \leqq C_{p,q} \int_0^t \|f\|_q^2(x_0)dx_0,$$

in $0\leqq t\leqq \delta(\Gamma_0, \Gamma_1)$, for any $p, q \in \boldsymbol{R}$, $f \in C^0_+(I, H^q)$.

We notice that the existence of parametrix of P in Γ with finite propagation speed of WF does not depend on extensions of P to outside Γ, and is invariant under elliptic Fourier integral operators on $\boldsymbol{R}^d$.

Remark 2.2. Assume that for each $(0, \xi')$, $|\xi'|=1$, there exists an open conic neighborhood in which P has a parametrix with finite propagation speed of WF. Then the Cauchy problem for $P(x, D)$ is locally solvable in a neighborhood of the origin with data on $x_0=0$.

Theorem 2.2. *Let $\phi(x, \xi')$ be the same as in Theorem* 2.1, *and Ω be a sufficiently small conic neighborhood of ρ in $\boldsymbol{R}\times T^*\boldsymbol{R}^d$. Then it follows from*

$$WF(Pu)\cap \Omega \cap \{\phi(x\ \ \xi')\leqq 0\}\cap \text{Char}\, P=\emptyset$$
$$WF(u)\cap \partial\Omega \cap \{\phi(x, \xi')<0\}\cap \text{Char}\, P=\emptyset,$$

that

$$WF(u)\cap \Omega \cap \{\phi(x, \xi')\leqq 0\}\cap \text{Char}\, P=\emptyset,$$

where $u \in \mathscr{D}'$, Char $P=\{(x, \xi); P_2(x, \xi)=0\}$ *and $\partial\Omega$ denotes the boundary of Ω.*

For more precise estimates of wave front sets, the following result is useful.

Theorem 2.3. *Assume the same condition as in Theorem* 2.1. *Let* $\psi(x, \xi')$ *be a real smooth function near* ρ, *positively homogeneous of degree* 0 *in* ξ' *such that* $\psi(\rho)=0$, $-H_\psi(\rho) \in \Gamma(P, \rho)$ *and* Ω *be a sufficiently small conic neighborhood of* ρ *in* $\boldsymbol{R}\times T^*\boldsymbol{R}^d$. *Then it follows from*

$$WF(Pu)\cap\Omega\cap\{\psi(x, \xi')\leqq 0\}\cap \text{Char } P=\emptyset,$$
$$WF(u)\cap\partial\Omega\cap\{\psi(x, \xi')\leqq 0\}\cap \text{Char } P=\emptyset,$$

that

$$WF(u)\cap\Omega\cap\{\psi(x, \xi')\leqq 0\}\cap \text{Char } P=\emptyset,$$

where $u \in \mathscr{D}'$.

Remark 2.3. Admitting Lemmas 3.1, 3.2 in §3, Theorem 2.3 was obtained in [8] and the same result of another formulation is in [6].

§ 3. An analytical characterization of effective hyperbolicity

Let $P(x, \xi)$ be as in §2. We shall say that P is (microlocally) effectively hyperbolic at $\rho=(0, \hat{\xi}')$ if $P_2(x, \xi)$ satisfies the following conditions,

(3.1) $P_2(x, \xi)$ is hyperbolic with respect to dx_0, that is, $Q_2(x, \xi')$ is non negative near ρ,

(3.2) If $Q_2(\rho)=0$, then the fundamental matrix F_{P_2} of P_2 has non zero real eigenvalues at $(0, 0, \hat{\xi}')$.

In the following, { , } denotes the Poisson bracket.

Lemma 3.1. *Assume that* P_2 *is effectively hyperbolic at* ρ. *Then there are a real smooth function* $\phi(x', \xi')$, *positively homogeneous of degree* 0 *with* $\phi(\rho)=0$ *and positive constants* c, κ, $0<\kappa<1$ *such that near* ρ *we have*

$$Q_2(x, \xi')\geqq c(x_0-\phi(x', \xi'))^2|\xi'|^2, \tag{3.3}$$

$$\{\phi, Q_2\}^2(x, \xi')\leqq 4\kappa Q_2(x, \xi'). \tag{3.4}$$

Conversely, if (3.3), (3.4) *hold with* $\phi(x', \xi')$, c, κ *mentioned above, then* P_2 *is effectively hyperbolic at* ρ.

Remark 3.1. The condition (3.4) is equivalent to the following; all bicharacteristics of P having the limit point ρ are transversal to the hypersurface $x_0-\phi(x', \xi')=0$.

Lemma 3.2. *Assume that* $Q_2(\rho)=dQ_2(\rho)=0$ *and* $Q_2(x, \xi')$ *satisfies*

(3,3), (3.4) *near* ρ. *Then there is a real smooth function* $\psi(x, \xi')$ *defined near* ρ, *positively homogeneous of degree* 0 *such that*

$$-H_\psi(\rho) \in \Gamma(P, \rho),\ Q_2(x, \xi') \geqq c\psi(x, \xi')^2|\xi'|^2, \qquad \text{near } \rho,$$

with some positive c. Conversely, if there exists such $\psi(x, \xi')$, *then* $Q_2(x, \xi')$ *satisfies* (3.3), (3.4) *with some real smooth* $\phi(x', \xi')$.

Using a suitable canonical coordinates on $T^*\boldsymbol{R}^d$, we can rewrite P more convenient form.

Lemma 3.3. *Assume that* $Q_2(x, \xi')$ *satisfies* (3.3), (3.4). *Then in a suitable canonical coordinates on* $T^*\boldsymbol{R}^d$, $Q_2(x, \xi')$ *verifies one of the following conditions,*

$$\text{(3.5)}\qquad \begin{aligned} &Q_2(x, \xi') \geqq c(x_0 - \phi(x', \xi'))^2|\xi'|^2 \quad (\text{near } \rho),\\ &\{\phi, \{\phi, Q_2\}\}(\rho) = 0, \quad \rho = (0, e_1), \end{aligned}$$

$$\text{(3.6)}\qquad \begin{aligned} &Q_2(x, \xi') = L(x, \xi')^2 + \tilde{Q}_2(x, \xi'),\\ &\tilde{Q}_2(x, \xi') \geqq c(x_0 - \phi(x', \xi'))^2|\xi'|^2 \qquad (\text{near } \rho),\\ &\{\phi, \{\phi, \tilde{Q}_2\}\}(\rho) = 0, \quad |\{\phi, L\}(\rho)| < 1, \quad \rho = (0, e_2), \end{aligned}$$

with some real smooth $\phi(x', \xi')$, *where* e_p *denotes the unit vector in* $\boldsymbol{R}^d$ *with p-th component* 1.

Proof. In case $d\phi(\rho) = 0$ in (3.3), it is obvious that $\{\phi, \{\phi, Q_2\}\}(\rho) = 0$. Then we assume that $d\phi(\rho) \neq 0$. If $d\phi$ is proportional to $\sum \xi_i dx_i$ at ρ, then making a linear change of coordinates x', if necessary, we may suppose that $\phi = x_1 + \tilde{\phi}$, $d\tilde{\phi}(\rho) = 0$, $\rho = (0, e_1)$. Hence from Euler's identity it follows that

$$\{\phi, \{\phi, Q_2\}\}(\rho) = 0.$$

Then, to prove this lemma, it suffices to show that, in some canonical coordinates on $T^*\boldsymbol{R}^d$, $Q_2(x, \xi')$ satisfies (3.6) if $d\phi$ and $\sum \xi_i dx_i$ are linearly independent at ρ. First, choosing a suitable canonical coordinates on $T^*\boldsymbol{R}^d$, we may assume that

$$\text{(3.7)}\qquad \begin{aligned} &Q_2(x, \xi') \geqq c(x_0 - x_1)^2|\xi'|^2, \quad \{x_1, Q_2\}^2 \leqq 4\kappa Q_2 \quad (\text{near } \rho),\\ &(\partial^2/\partial\xi_1^2)Q_2(\rho) \neq 0, \quad \rho = (0, e_2) \end{aligned}$$

where $0 < \kappa < 1$. Since $(\partial^2/\partial\xi_1^2)Q_2(\rho) > 0$, Malgrange preparation theorem gives that

$$Q_2(x, \xi') = q(x, \xi')\{(\xi_1 - f(x, \hat{\xi}))^2 + g(x, \hat{\xi})\},$$

where $q(\rho)>0$, $g(x, \tilde{\xi})\geqq 0$, $\tilde{\xi}=(\xi_2, \cdots, \xi_d)$. From (3.7) it follows easily that

$$g(x, \tilde{\xi})\geqq c(x_0-x_1)^2|\xi'|^2, \qquad \text{near } \rho.$$

We take $L(x, \xi')=q(x, \xi')^{1/2}(\xi_1-f(x, \tilde{\xi}))$, $\tilde{Q}_2(x, \xi')=q(x, \xi')g(x, \tilde{\xi})$. Since $f(x, \tilde{\xi})$, $g(x, \tilde{\xi})$ do not depend on ξ_1, (3.6) are easily checked for these L, $\tilde{Q}_2$.

Remark 3.2. It is clear from the proof that in case (3.6) one can take $\phi(x', \xi')=x_1$ and in case (3.5) with $d\phi(\rho)\neq 0$, we may suppose that $\phi=x_1+\tilde{\phi}$, $d\tilde{\phi}(\rho)=0$.

§ 4. Study on non effectively hyperbolic operators

Let $P(x, \xi)=-\xi_0^2+Q(x, \xi')$, where $Q(x, \xi')$ is defined in a conic neighborhood of $\tilde{\rho}=(0, \tilde{\xi}')\in \boldsymbol{R}\times T^*\boldsymbol{R}^d$, non negative and homogeneous of degree 2 in ξ'. We assume that P satisfies

$$\text{(4.1)} \qquad \operatorname{Ker} F_P^2(\rho_0)\cap \operatorname{Ran} F_P^2(\rho_0)\neq\{0\}, \qquad \rho_0=(0, 0, \tilde{\xi}'),$$

where $\operatorname{Ker} F_P^2(\rho)$, $\operatorname{Ran} F_P^2(\rho)$ denotes the kernel and range of $F_P^2(\rho)$ respectively. We start with the following lemma.

Lemma 4.1 ([9]). *Assume that* (4.1) *is satisfied at* ρ_0. *Then in a suitable canonical coordinates on* $T^*\boldsymbol{R}^d$, $Q(x, \xi')$ *takes the following form,*

$$\sum_{i=0}^{p-1} q_i(x, \xi')(x_i-x_{i+1})^2+\sum_{i=1}^{p} r_i(x, \xi')\xi_i^2+g(x, \xi')(x_p, x'', \xi'')r_{p+1}(x, \xi'),$$

with

$$\{\xi_p, \{\xi_p, g\}\}(\hat{\rho})=0, \quad \sum_{i=1}^{p} r_i(\hat{\rho})^{-1}=1, \quad (1\leqq p\leqq d-1),$$

where q_i, r_i *are positive, homogeneous of degree* 2, 0 *respectively,* g *is non negative, vanishing at* $\hat{\rho}$, *homogeneous of degree* 2, *and* $x''=(x_{p+1}, \cdots, x_d)$, $\xi''=(\xi_{p+1}, \cdots, \xi_d)$, $\hat{\rho}=(0, e_d)$.

Taking this lemma in mind, we consider the following symbol,

$$\text{(4.2)} \qquad \begin{aligned} &P=-\xi_0^2+Q(x, \xi'),\\ &Q(x, \xi')=\sum_{i=0}^{p-1} q_i(x, \xi')(x_i-x_{i+1})^2+\sum_{i=1}^{p} r_i(x, \xi')\xi_i^2+g(x'', \xi'')r_{p+1}(x, \xi'). \end{aligned}$$

Now we assume for this P that

(4.3) the double characteristic set $\Sigma=\{(x, \xi); P(x, \xi)=dP(x, \xi)=0\}$ is a manifold and the rank of the Hessian of P is equal to the codimension of Σ at every point of Σ, and the rank of the symplectic form $\sum dx_j \wedge d\xi_j$ restricted to Σ is constant,

(4.4) $$\operatorname{Ker} F_P^2(\rho) \cap \operatorname{Ran} F_P^2(\rho) \neq \{0\} \qquad \text{for every } \rho \in \Sigma,$$

(4.5) $$\{(\partial/\partial x_0)+\cdots+(\partial/\partial x_p)\} r_j(x, \xi')=0 \qquad \text{on } \Sigma.$$

Remark 4.1. Clearly we can write the condition (4.5) invariant way, but this expression is more convenient in the following.

Here we introduce the following condition.

(4.6) $H_\phi^3 P=0$ on Σ near $\hat{\rho}$ for any ϕ satisfying $H_\phi(\rho) \in \operatorname{Ran} F_P^2(\rho) \cap \operatorname{Ker} F_P^2(\rho)$ on Σ near $\hat{\rho}$.

Theorem 4.1. *We assume* (4.3)–(4.5). *If* (4.6) *holds then P admits the following decomposition;*

$$P(x, \xi)=-(\xi_0-\lambda(x, \xi'))(\xi_0-\mu(x, \xi'))+Q(x, \xi'),$$

where $\lambda(x, \xi')$, $\mu(x, \xi')$ are real, homogeneous of degree 1, *$Q(x, \xi')$ is non negative, homogeneous of degree* 2 *and satisfy*

(4.7) $$|\{\xi_0-\lambda(x, \xi'), Q(x, \xi')\}| \leqq C Q(x, \xi')$$

near $\hat{\rho}$ with some constant C. Conversely, if P admits such decomposition, then (4.6) *holds.*

Now we consider more special case. Assume that the double characteristic set Σ is a manifold and (4.4) is satisfied. Then it follows from Lemma 4.1 that the codimension of Σ is greater than 3. If the codimension of Σ is 3 then P takes the following form,

$$P=-\xi_0^2+q_0(x, \xi')(x_0-x_1)^2+r_1(x, \xi')\xi_1^2$$
$$\text{with } r_1(x, \xi')=1 \text{ on } \Sigma'=\{x_0=x_1, \xi_1=0\}.$$

In this case the conditions (4.3), (4.5) are a priori verified and we get

Corollary 4.1 ([9]). *Assume that Σ is a manifold of codimension* 3 *and* (4.4) *is satisfied. Then in order that P admits the decomposition in Theorem* 4.1, *it is necessary and sufficient that* (4.6) *holds.*

We can also express the condition (4.6) in terms of bicharacteristics.

Proposition 4.1 ([9]). *Assume the same condition as in Corollary* 4.1.

Then for that the condition (4.6) *holds, it is necessary and sufficient that there is no bicharacteristic* $\gamma(s)\subset \text{Char } P\setminus\Sigma$ *of* P *which has a limit point in* Σ.

Proof of Theorem 4.1. We shall prove only the necessity of (4.6). First we remark that (4.4) implies that

$$\sum_{i=1}^{p}(r_i(x, \xi'))^{-1}=1 \qquad \text{on } \Sigma. \tag{4.8}$$

Also from the assumption (4.3), we may suppose that $g(x'', \xi'')r_{p+1}(x, \xi')$ is a sum of squares,

$$\sum_{i=1}^{k} b_i(x, \xi')n_i(x'', \xi'')^2,$$

where $dn_i(x'', \xi'')$ are linearly independent near $\hat{\rho}$. Making a linear change of coordinates

$$y_0=x_0, \quad y_j=x_{j-1}-x_j, \quad 1\leqq j\leqq p, \quad y''=x'',$$

and remarking that

$$\sum_{i=1}^{p-1} r_i(x, \xi')(\xi_i-\xi_{i+1})^2+r_p(x, \xi')\xi_p^2-\xi_1^2$$
$$=\sum_{i=2}^{p} a_i(x, \xi')m_i(x, \xi')^2+R(x, \xi')m_1(x, \xi')^2,$$

with $m_i=\xi_i-c_i(x, \xi')\xi_{i-1}$, $2\leqq i\leqq p$, $m_1=\xi_1$, P is reduced to

$$P=-\xi_0^2-2\xi_0\xi_1+\langle U(x, \xi')l, l\rangle+\langle V(x, \xi')m', m'\rangle$$
$$+\langle W(x, \xi')n, n\rangle+R(x, \xi')\xi_1^2,$$

where $U=\text{diag}\,(q_1, \cdots, q_p)$, $V=\text{diag}\,(a_2, \cdots, a_p)$, $W=\text{diag}\,(b_1, \cdots, b_k)$ and $\text{diag}\,(q_1, \cdots, q_p)$ denotes the $p\times p$ diagonal matrix with entries $q_i(x, \xi')$ and $\langle\ ,\ \rangle$ stands for the usual scalar product. Here for the simplicity of the notations, we have set $l=(l_1, \cdots, l_p)$, $l_i=x_i$, $m'=(m_2, \cdots, m_p)$ and $n=(n_1, \cdots, n_k)$. With these notations, the double characteristic set Σ is given by

$$\Sigma=\{(x, \xi);\ l(x, \xi')=0,\ m(x, \xi')=0,\ n(x'', \xi'')=0,\ \xi_0=0\}$$

where $m=(m_1, \cdots, m_p)$.

The condition (4.5) implies that $\partial_0 c_i(x, \xi')$ vanishes on Σ where $\partial_0=(\partial/\partial x_0)$ and in view of (4.8), $R(x, \xi')$ vanishes also on Σ. Then $R(x, \xi')$ is a linear combinations of l, m and n,

$$R(x, \xi')=\langle\alpha(x, \xi'), l\rangle+\langle\beta(x, \xi'), m\rangle+\langle\gamma(x, \xi'), n\rangle,$$

with $\alpha=(\alpha_1, \cdots, \alpha_p)$, $\beta=(\beta_1, \cdots, \beta_p)$, $\gamma=(\gamma_1, \cdots, \gamma_k)$.

Let $\phi(x, \xi)$ be as in (4.6). Then we see easily that

$$H_\phi^3 P = H_\phi^3(R(x, \xi')\xi_1^2) \qquad \text{on } \Sigma,$$

and hence $H_\phi^3 P=0$ on Σ is equivalent to that $\beta_1(x, \xi')$ vanishes on Σ. Now we assume that

$$\beta_1(\bar{\rho})\neq 0 \qquad \text{for some} \quad \bar{\rho}\in\Sigma, \tag{4.9}$$

and we show by a contradiction that P does not admit the decomposition stated in this theorem. Suppose that P admits such decomposition, then it is clear that

$$\lambda(x, \xi')+\mu(x, \xi')=-2\xi_1,\ Q(x, \xi')=T(x, \xi')+\lambda(x, \xi')\mu(x, \xi'), \tag{4.10}$$

where $T=\langle U(x, \xi')l, l\rangle+\langle V(x, \xi')m', m'\rangle+\langle W(x, \xi')n, n\rangle+R(x, \xi')\xi_1^2$.

Since $Q(x, \xi')$ is non negative and $T(x, \xi')$ vanishes on Σ at least of order 2, it follows from (4.10) that $\lambda(x, \xi')$ vanishes on Σ. Then one can write

$$\lambda(x, \xi')=\langle\tilde{A}(x, \xi'), l\rangle+\langle\tilde{B}(x, \xi'), m\rangle+\Psi,\ \Psi=\langle\tilde{C}(x, \xi'), n\rangle.$$

On the other hand, it is easy to see that, up to terms vanishing at least of order 2 on Σ, $\{\xi_0-\lambda(x, \xi'), Q(x, \xi')\}$ is equal to be

$$-\{\lambda(x, \xi'), T_0(x, \xi')\}-2\tilde{A}_1(x, \xi')\lambda(x, \xi'),$$

with $T_0(x, \xi')=T(x, \xi')-R(x, \xi')\xi_1^2$. In virtue of (4.10), $Q(x, \xi')$ vanishes at least of order 2 on Σ, then in order that (4.7) holds, we must have

$$F_{T_0}(\rho)\cdot H_\lambda(\rho)=-2\tilde{A}_1(\rho)H_\lambda(\rho) \qquad \text{for} \quad \rho\in\Sigma.$$

By the non negativity of $T_0(\rho)$, $F_{T_0}(\rho)$ has no non zero real eigenvalue, then $H_\lambda(\rho)\neq 0$ implies that $\tilde{A}_1(\rho)=0$ and $H_\lambda(\rho)\in \operatorname{Ker} F_{T_0}(\rho)$. If we remark that

$$\operatorname{Ker} F_{T_0}(\rho)=\{(X, \Xi);\ dl_i(\rho)(X, \Xi)=0,\ 1\leqq i\leqq p,\ dm_i(\rho)(X, \Xi)=0,\ 2\leqq i\leqq p,$$
$$dn_i(\rho)(X, \Xi)=0,\ 1\leqq i\leqq k\},$$

then it follows from $\tilde{A}_1(\rho)=0$ and $H_\lambda(\rho)\in \operatorname{Ker} F_{T_0}(\rho)$ that

$$\tilde{A}(\rho)=\tilde{B}(\rho)=0 \qquad \text{for} \quad \rho\in\Sigma. \tag{4.11}$$

If $H_\lambda(\rho)=0$ then (4.11) is obvious and in any case, we have (4.11).

Next we show that

$$\{\Psi, n_j\}=0 \qquad \text{on } \Sigma,\ 1\leqq j\leqq k. \tag{4.12}$$

Set $G(x, \xi')=\sum b_j(x, \xi')n_j(x'', \xi'')^2$ and denote by G_ρ the Hessian of G at ρ. In virtue of (4.11), $H_\Psi(\rho)$ is in the kernel of $F_G(\rho)$ and then we have

$$G_\rho(H_\Psi(\rho), H_\Psi(\rho))=\{\Psi, \{\Psi, G\}\}(\rho)=0.$$

This gives (4.12) because $b_j(\rho)>0$. Now (4.11) enables us to write

$$\begin{aligned}\lambda(x, \xi')&=\langle A(x, \xi')m, m\rangle+\langle B(x, \xi')m, l\rangle+\langle C(x, \xi')l, l\rangle\\ &\quad+\langle D(x, \xi'), n\rangle+\Psi(x, \xi'),\\ \lambda_0(x, \xi')&=\langle A(x, \xi')m, m\rangle+\langle B(x, \xi')m, l\rangle+\langle C(x, \xi')l, l\rangle,\end{aligned} \tag{4.13}$$

where $A=(A_{ij}(x, \xi'))$, $C=(C_{ij}(x, \xi'))$ are symmetric matrices and $D=(D_1(x, \xi'), \cdots, D_k(x, \xi'))$ is a vector whose component $D_j(x, \xi')$ are linear functions of $l(x, \xi')$ and $m(x, \xi')$.

Here we define the curve $\Gamma(t)=(x(t), \xi'(t))$ in $\boldsymbol{R}\times T^*\boldsymbol{R}^d$ with $\Gamma(0)=\bar{\rho}=(\bar{x}, \bar{\xi}')$ as follows,

$$\begin{aligned}&\xi_1(t)=\Xi_1 t,\quad \xi_j(t)-c_j(\bar{\rho})\xi_{j-1}(t)=\Xi_j t^2,\quad 2\leqq j\leqq p,\quad \xi_j(t)=\bar{\xi}_j+\Xi_j t^3,\\ &p+1\leqq j\leqq d,\quad x_j(t)=X_j t^{3/2},\quad 0\leqq j\leqq p,\quad x_j(t)=\bar{x}_j+X_j t^3,\quad p+1\leqq j\leqq d,\end{aligned}$$

where (X, Ξ) is a constant vector.

Using the expression (4.13) and the coummutation relation (4.12), it is not difficult to see that

$$\{\xi_0-\lambda, Q\}(\Gamma(t))=-\{\lambda_0, T_0\}(\Gamma(t))+O(t^3).$$

Whereas $\{\lambda_0, T_0\}(\Gamma(t))$ is equal to

$$4\Xi_1\left\{A_{1p}q_pX_p+\sum_{j=1}^{p-1}(A_{1j}-a_{j+1}A_{1j+1})q_jX_j\right\}t^{1+3/2},$$

up to the terms of $O(t^3)$. On the other hand, it is clear that $Q(\Gamma(t))=O(t^3)$ and then to have (4.7), we must have

$$A_{1j}(\bar{\rho})=0 \qquad \text{for } 1\leqq j\leqq p. \tag{4.14}$$

But if $\beta_1(\bar{\rho})\neq 0$, one can not have the inequality

$$Q(x, \xi')=T(x, \xi')+\lambda(x, \xi')\mu(x, \xi')\geqq 0$$

near $\bar{\rho}$ with $\lambda(x, \xi')$ satisfying (4.14). Then the proof is complete.

§ 5. Localization of symbols and related pseudodifferential operators

Let p be an integer $1\leqq p\leqq d-1$. To define the localization of symbols, we introduce $y_j(x, \mu)$, $\eta_j(\xi', \mu)$ as follows,

$$y_0(x, \mu)=\mu x_0, \quad y_j(x, \mu)=\mu\chi_2(x_j), \quad 1\leqq j\leqq p,$$
$$y_j(x, \mu)=\mu^{1/2}\chi_3(\mu^{-1/2}x_j)x_j, \quad p+1\leqq j\leqq d,$$
$$\eta_j(\xi', \mu)=\mu^{-1}\chi_3(\mu^{-1}(\xi_j|\xi'|^{-1}-\delta_{jp}))(\xi_j-\delta_{jp}|\xi'|)+\mu^{-1}\delta_{jp}|\xi'|, \quad 1\leqq j\leqq p,$$
$$\eta_j(\xi', \mu)=\mu^{-1/2}\chi_3(\mu^{-1/2}\xi_j|\xi'|^{-1})\xi_j, \quad p+1\leqq j\leqq d,$$

where δ_{ij} is Kronecker's delta, μ is a small positive parameter and $\chi_2(s)$, $\chi_3(s)$ are smooth functions on $\boldsymbol{R}$ such that

$$|\chi_2(s)|=2 \quad \text{if } |s|\geqq 2, \quad \chi_2(s)=s \quad \text{if } |s|\leqq 1, \quad 0\leqq\chi_2'(s)\leqq 1 \quad \text{for } s\in\boldsymbol{R},$$
$$\chi_3(s)=1 \quad \text{if } |s|\leqq 1, \quad \chi_3(s)=0 \quad \text{if } |s|\geqq 2, \quad 0\leqq\chi_3(s)\leqq 1 \quad \text{for } s\in\boldsymbol{R}.$$

Let $Q(x, \xi')$ be a smooth function defined in a conic neighborhood $I\times W$ of $(0, e_p)$, homogeneous of degree m. We define the localization $Q(x, \xi', \mu)$ of $Q(x, \xi')$ by

$$Q(x, \xi', \mu)=\mu^m Q(y(x, \mu), \eta'(\xi', \mu))=Q(y(x, \mu), \mu\eta'(\xi', \mu)).$$

We note that there is a positive $\bar{\mu}>0$ scuch that

$$(y'(x, \mu), \eta'(\xi', \mu))\in W$$

for any $(x', \xi')\in\boldsymbol{R}^d\times(\boldsymbol{R}^d\setminus 0)$ if $0<\mu\leqq\bar{\mu}$. Hence $Q(x, \xi', \mu)$ is globally defined for small μ. Furthermore, if $0<\mu\leqq\bar{\mu}$, $|x_j|<\mu^{1/2}$, $|\xi_j|\xi'|^{-1}-\delta_{jp}|<\mu$, then we have

$$\begin{aligned}Q(x, \xi', \mu)&=Q(\mu x^{(1)}, \mu^{1/2}x^{(2)}, \xi^{(1)}, \mu^{-1/2}\xi^{(2)})\\&=\mu^m Q(\mu x^{(1)}, \mu^{1/2}x^{(2)}, \mu^{-1}\xi^{(1)}, \mu^{-1/2}\xi^{(2)}),\end{aligned}\tag{5.1}$$

where $x^{(1)}=(x_0, \cdots, x_p)$, $x^{(2)}=(x_{p+1}, \cdots, x_d)$, $\xi^{(1)}=(\xi_1, \cdots, \xi_p)$, $\xi^{(2)}=(\xi_{p+1}, \cdots, \xi_d)$. Remark that the last term in (5.1), up to the factor μ^m, is nothing but the symbol obtained from the original one by a change of scales $y^{(1)}=\mu^{-1}x^{(1)}$, $y^{(2)}=\mu^{-1/2}x^{(2)}$.

Assume that $Q_2(x, \xi')$ satisfies (3.1). In this case, from Remark 3.2, we may suppose that $d\phi(\rho)=0$ or $\phi=x_1+\tilde{\phi}$ with $d\tilde{\phi}(\rho)=0$, $\rho=(0, e_1)$ and we localize these $Q_2(x, \xi')$, $x_0-\phi(x', \xi')$ with $p=1$. In case $Q_2(x, \xi')$ satisfies (3.6) with $\phi=x_1$, $\rho=(0, e_2)$, we localize Q_2, $x_0-\phi$ with $p=2$.

Here to state some properties of these localized symbols, we introduce some classes of pseudodifferential operators with parameter μ. Let $a(x, \xi', \mu)$ be a smooth function on $I\times\boldsymbol{R}^{2d}$ with parameter μ. We shall say that

$a(x, \xi', \mu)$ belongs to $(S)^{m,l}_{\rho,\delta}$ if there is $\mu(a)>0$ such that $a(x, \xi', \mu)$ satisfies the following estimates for any $\alpha \in N^d$, $\beta \in N^{d+1}$, $\mu \in (0, \mu(a)]$ with $C_{\alpha,\beta}$ independent of μ,

$$(5.2) \qquad |a^{(\alpha)}_{(\beta)}(x, \xi', \mu)| \leqq C_{\alpha,\beta}\mu^{l-\rho|\alpha|-\delta|\beta|}\langle\xi'\rangle^{m-|\alpha|}.$$

We also say that $a(x, \xi', \mu)$ belongs to $S^{-\infty}(\mu)$ if for any $l \in N$, there is $\mu(l, a)$ being positive such that the estimates

$$(5.3) \qquad |a^{(\alpha)}_{(\beta)}(x, \xi', \mu)| \leqq C_{\alpha,\beta,l,\mu}\langle\xi'\rangle^{-l+|\beta|/2-|\alpha|/2}$$

hold for any $\alpha \in N^d$, $\beta \in N^{d+1}$, $\mu \in (0, \mu(l, a)]$.

When deriving energy estimates in the following sections, operators belonging to $S^{-\infty}(\mu)$ bring only error terms which will be easily absorbed. Therefore we freely omit the term "modulo $S^{-\infty}(\mu)$" throughout this paper.

Lemma 5.1. *Localizations $Q(x, \xi', \mu)$, $\phi(x', \xi', \mu)$ of $Q_2(x, \xi')$, $\phi(x', \xi')$ satisfy*

$$Q(x, \xi', \mu) \geqq c\mu^2(x_0 - \mu^{-1}\phi(x', \xi', \mu))^2|\xi'|^2,$$
$$\partial_0^j Q(x, \xi', \mu) \in (S)^{2,2}_{1/2,1/2} + (S)^{2,3}_{1,1/2}, \qquad j=0, 1, 2.$$

According to the case $d\phi(\rho)=0$ or $\phi=x_1+\tilde{\phi}$, $d\tilde{\phi}(\rho)=0$ or $\phi=x_1$, we have

$$\mu^{-1}\phi(x', \xi', \mu) \in (S)^{0,1}_{1,1/2}, \qquad \mu^{-1}\phi(x', \xi', \mu) - y_1(x_1) \in (S)^{0,1}_{1,1/2},$$
$$\mu^{-1}\phi(x', \xi', \mu) = y_1(x_1).$$

Next we define the localization $L(x, \xi', \mu)$ of $L(x, \xi')$ in (3.6). Set

$$L_1(x, \xi') = \sum_{|\alpha+\beta|=1} x^\beta(\xi')^\alpha L^{(\alpha)}_{(\beta)}(\rho)|\xi'|^{1-|\alpha|}, \quad L_2(x, \xi') = L(x, \xi') - L_1(x, \xi'),$$

and denote by $L_2(x, \xi', \mu)$ the localization of $L_2(x, \xi')$ defined above with $p=2$. Now we define $L(x, \xi', \mu)$ by

$$L(x, \xi', \mu) = L_1(y(x, \mu), \mu^{-1}\xi_1, \tilde{\eta}(\xi', \mu)) + L_2(x, \xi', \mu)$$
$$= L_1(x, \xi', \mu) + L_2(x, \xi', \mu),$$

where $\tilde{\eta}(\xi', \mu) = (\eta_2(\xi', \mu), \cdots, \eta_d(\xi', \mu))$.

Lemma 5.2. *Let $L(x, \xi', \mu) = L_1(x, \xi', \mu) + L_2(x, \xi', \mu)$ be as above. Then we have*

$$L_1(x, \xi', \mu) - \xi_1 L^{(1)}(\rho) \in (S)^{1,1}_{1/2,1/2}, \qquad L_2(x, \xi', \mu) \in (S)^{1,2}_{1,1/2},$$

where $L^{(1)}(\rho) = (\partial/\partial\xi_1)L(\rho)$.

Finally we introduce two more classes of pseudodifferential operators. As a weight function, we take

$$J(x, \xi', \mu)=\{(x_0-\mu^{-1}\phi(x', \xi', \mu))^2+\langle\mu\xi'\rangle^{-1}\}^{1/2}, \quad \langle\xi'\rangle^2=1+\sum_{j=1}^{d}\xi_j^2.$$

We shall say that $a(x, \xi', \mu) \in C^\infty(I\times \boldsymbol{R}^{2d})$ belongs to $J^r\bar{S}^{m,l}$ if there is $\mu(a)>0$ such that

$$(5.4)\qquad |a_{(\beta)}^{(\alpha)}(x, \xi', \mu)|\leqq C_{\alpha,\beta}\langle\xi'\rangle^{m-|\alpha|/2+|\beta|/2}\langle\mu\xi'\rangle^{l}J(x, \xi', \mu)^r,$$

holds for any $\alpha \in \boldsymbol{N}^d$, $\beta \in \boldsymbol{N}^{d+1}$, $\mu \in (0, \mu(a)]$ with $C_{\alpha,\beta}$ independent of μ. We say that $a(x, \xi', \mu)$ belongs to $J^rS^{m,l}$, if there is $\mu(a)>0$ such that

$$(5.5)\qquad |a_{(\beta)}^{(\alpha)}(x, \xi', \mu)|\leqq C_{\alpha,\beta}\langle\xi'\rangle^{m-|\alpha|}\langle\mu\xi'\rangle^{l+|\alpha|/2+|\beta|/2}J(x, \xi', \mu)^r,$$

is valid for all $\alpha \in \boldsymbol{N}^d$, $\beta \in \boldsymbol{N}^{d+1}$, $\mu \in (0, \mu(a)]$ with $C_{\alpha,\beta}$ independent of μ.

We state some properties of these pseudodifferential operators. For the proofs and for other properties, we refer to [10]. In the following $\|\cdot\|$ and $(\ ,\ )$ denote the norm and scalar product in $L^2(\boldsymbol{R}^d)$.

Lemma 5.3. *Let* $a(x, \xi', \mu) \in J^0S^{m,l}$, *and* $\sup|a(x, \xi', \mu)\langle\mu\xi'\rangle^{-l}\langle\xi'\rangle^{-m}| =c$. *Then for any* $k \in \boldsymbol{N}$, *there is* $C(a, k, \mu)$ *such that*

$$\|au\|^2\leqq(c^2+c(a)\mu)\|\langle D'\rangle^m\langle\mu D'\rangle^l u\|^2+C(a, k, \mu)\|u\|_{-k}^2.$$

Lemma 5.4. *Let* $a(x, \xi', \mu) \in J^0S^{m,l}$, $a(x, \xi', \mu)\geqq c\langle\mu\xi'\rangle^l\langle\xi'\rangle^m$, *with positive* c *independent of* μ. *Then for any* $k \in \boldsymbol{N}$, *there is* $C(a, k, \mu)$ *such that*

$$(au, u)\geqq(c-c(a)\mu)\|\langle\mu D'\rangle^{l/2}\langle D'\rangle^{m/2}u\|^2-C(a, k, \mu)\|u\|_{-k}^2.$$

Lemma 5.5. *Assume that* $a_i(x, \xi', \mu) \in J^{r_i}S^{m_i,l_i}$ $(i=1, 2)$ *are elliptic in the sense*

$$a_i(x, \xi', \mu)\geqq c_i J(x, \xi', \mu)^{r_i}\langle\mu\xi'\rangle^{l_i}\langle\xi'\rangle^{m_i},$$

with positive c_i *independent of* μ. *Let* $b(x, \xi', \mu) \in J^rS^{m,l}$, *then one can write*

$$b=a_1(c+r)a_2,$$

where $r \in J^{r-(r_1+r_2)}\bar{S}^{m-(m_1+m_2),\,l-(l_1+l_2)}$, $c(x, \xi', \mu)=b(x, \xi', \mu)a_1(x, \xi', \mu)^{-1} \times a_2(x, \xi', \mu)^{-1}$.

§ 6. Microlocal energy estimates

In this section, using the calculus of pseudodifferential operators introduced in § 5, we derive microlocal energy estimates for the localized operator

$$P(x, D, \mu)=D_0^2-Q(x, D', \mu).$$

First we introduce the following symbols and operators.

$$J_\pm(x, \xi', \mu)=\pm\{2\chi_0(\pm(x_0-\mu^{-1}\phi(x', \xi', \mu))\langle\mu\xi'\rangle^{1/2})-1\}$$
$$\times(x_0-\mu^{-1}\phi(x', \xi', \mu))+\langle\mu\xi'\rangle^{-1/2},$$
$$\alpha_n^\pm(x, \xi', \mu)=\chi(\pm n^{1/2}(x_0-\mu^{-1}\phi(x', \xi', \mu))\langle\mu\xi'\rangle^{1/2}),$$
$$I_-(n, r)=\mathrm{op}(\langle\mu\xi'\rangle^n J_-(x, \xi', \mu)^{n-r}), \quad I_+(n, r)=\mathrm{op}(J_+(x, \xi', \mu)^{-n-r}),$$

where $\chi_0(s)$, $\chi(s)$ are smooth function on $\boldsymbol{R}$ such that

$$\chi_0(s)=0 \quad \text{for } s\leqq -1/2, \quad \chi_0(s)=1 \quad \text{for } s\geqq -1/4, \quad 0\leqq\chi_0(s)\leqq 1 \quad \text{for } s\in \boldsymbol{R}$$
$$\chi(s)=0 \quad \text{for } s\leqq -1, \quad \chi(s)=1 \quad \text{for } s\geqq 1, \quad \chi(s)+\chi(-s)=1 \quad \text{for } s\in \boldsymbol{R},$$

and $\mathrm{op}(a(x, \xi', \mu))$ denotes the pseudodifferential operator with symbol $a(x, \xi', \mu)$. But sometimes, we do not distinguish operators and their symbols.

Using these symbols we define the following semi-norms.

$$|||u|||_{n,r,k}^2=\|\langle\mu D'\rangle^k I_-(n, r)\alpha_n^- u\|^2+\|\langle\mu D'\rangle^k I_+(n, r)\alpha_n^+ u\|^2,$$
$$[u]_{n,r,k}^2=\|\langle\mu D'\rangle^k I_-(n, r)(D_0-i\theta)\alpha_n^- u\|^2+\|\langle\mu D'\rangle^k I_+(n, r)(D_0-i\theta)\alpha_n^+ u\|^2,$$

where θ is a positive parameter.

Lemma 6.1 (Proposition 4.2 in [10]). *For any $\delta>0$, we have*

$$-2\,\mathrm{Im}\sum\int(I_\pm(n, 1/2)(D_0-i\theta)^2 w^\pm, I_\pm(n, 1/2)(D_0-i\theta)w^\pm)dx_0$$
$$\geqq(2n-1)(1-\delta-c(n)\mu)\int[u]_{n,1,0}^2 dx_0+c\theta\int[u]_{n,1/2,0}^2 dx_0$$
$$+c\delta n^3(1-c(n)\mu)\int|||u|||_{n,2,0}^2 dx_0+c\theta n^2(1-c(n)\mu)\int|||u|||_{n,3/2,0}^2 dx_0$$
$$+c\theta^2 n\int|||u|||_{n,1,0}^2 dx_0-c(n)\mu\int|||(D_0-i\theta)u|||_{n,1,0}^2 dx_0,$$

modulo $c(n, k, \mu)\theta\int|u|_{-k,\theta}^2 dx_0$ *with* $|u|_{-k,\theta}^2=\theta^{-1/2}\|(D_0-i\theta)u\|_{-k}^2+\theta^{1/2}\|u\|_{-k}^2$, *where* $w^\pm=\alpha_n^\pm u$, $0<\mu\leqq\mu_0(n)$.

Lemma 6.2. *Let $L(x, \xi', \mu)$ be as in Lemma 5.2. Then for any $\varepsilon>0$, we have*

$$2\,\mathrm{Im}\int(I_\pm(n,1/2)\,\mathrm{op}(L^2)w^\pm,\ I_\pm(n,1/2)(D_0-i\theta)w^\pm)dx_0$$

$$\geqq(2n\pm1)(1-\bar\lambda-\varepsilon-c(n)\mu^{1/2})\int\|I_\pm(n,1)Lw^\pm\|^2dx_0$$

$$+(2\theta-c)\int\|I_\pm(n,1/2)Lw^\pm\|^2dx_0-c\int\|I_\pm(n,1/2)(D_0-i\theta)w^\pm\|^2dx_0$$

$$-(2n\pm1)(\bar\lambda+\varepsilon+c(n)\mu^{1/2})\int\|I_\pm(n,1)(D_0-i\theta)w^\pm\|^2dx_0$$

$$-\varepsilon^{-1}(c+c(n)\mu)\int|||u|||^2_{n,0,1}dx_0,$$

where $\bar\lambda=|L^{(1)}(\rho)|$.

Proof. We start with the following identity,

$$2\,\mathrm{Im}\int(I_\pm(n,1/2)L^2w^\pm,\ I_\pm(n,1/2)(D_0-i\theta)w^\pm)dx_0$$

$$=2\theta\int\|I_\pm(n,1/2)Lw^\pm\|^2dx_0$$

$$-2\,\mathrm{Re}\int(\partial_0I_\pm(n,1/2)Lw^\pm,\ I_\pm(n,1/2)Lw^\pm)dx_0$$

$$-2\,\mathrm{Re}\int(I_\pm(n,1/2)Lw^\pm,\ I_\pm(n,1/2)(\partial_0L)w^\pm)dx_0 \tag{6.1}$$

$$+2\,\mathrm{Im}\int([I_\pm(n,1/2),L]Lw^\pm,\ I_\pm(n,1/2)(D_0-i\theta)w^\pm)dx_0$$

$$+2\,\mathrm{Im}\int([I_\pm(n,1/2),L](D_0-i\theta)w^\pm,\ I_\pm(n,1/2)Lw^\pm)dx_0$$

$$+2\,\mathrm{Im}\int(I_\pm(n,1/2)Lw^\pm,\ (L^*-L)I_\pm(n,1/2)(D_0-i\theta)w^\pm)dx_0.$$

First we estimate the second term in the right-hand of side (6.1). Since $\partial_0I_\pm(n,1/2)L\alpha_n^\pm+(n\pm1/2)I_\pm(n,3/2)L\alpha_n^\pm\equiv0$ for $n\geqq16$, it follows that

$$\begin{aligned}&-2\mathrm{Re}(\partial_0I_\pm(n,1/2)Lw^\pm,\ I_\pm(n,1/2)Lw^\pm)\\&\equiv(2n\pm1)(I_\pm(n,3/2)Lw^\pm,\ I_\pm(n,1/2)Lw^\pm).\end{aligned} \tag{6.2}$$

On the other hand, using Lemma 5.5, one can write

$$I_\pm(n,1/2)^*I_\pm(n,3/2)\equiv I_\pm(n,1)^*(1+b_\pm)I_\pm(n,1),$$

with $b_\pm\in\mu J^0\bar S^{0,0}$, then the right-hand side of (6.2) is estimated from below by

$$(2n\pm 1-c(n)\mu)\|I_{\pm}(n, 1)Lw^{\pm}\|^2.$$

Next setting $L^{(1)}(\rho)=\lambda$, we shall prove the following estimate,

$$\begin{aligned} &|2\,\mathrm{Im}\,([I_{\pm}(n, 1/2), L](D_0-i\theta)w^{\pm}, I_{\pm}(n, 1/2)Lw^{\pm})| \\ &\quad +|2\,\mathrm{Im}\,([I_{\pm}(n, 1/2), L]Lw^{\pm}, I_{\pm}(n, 1/2)(D_0-i\theta)w^{\pm})| \\ &\leqq 2\bar{\lambda}(2n\pm 1)\|I_{\pm}(n, 1)(D_0-i\theta)w^{\pm}\|\cdot\|I_{\pm}(n, 1)Lw^{\pm}\| \\ &\quad +c(n)\mu^{1/2}\|I_{\pm}(n, 1)(D_0-i\theta)w^{\pm}\|^2+c(n)\mu^{1/2}\|I_{\pm}(n, 1)Lw^{\pm}\|^2. \end{aligned} \tag{6.3}$$

From the properties of the localized symbol $L(x, \xi', \mu)$, we see easily that

$$[I_{\pm}(n, 1/2), L]\equiv -\mathrm{op}\,(\lambda I_{\pm}(n, 1/2)_{(1)})+r_{\pm},$$

where $r_{\pm}\in \mu^{1/2}J^{\mp n-3/2}S^{0,\max(\mp n,0)}$, $I_{\pm}(n, 1/2)_{(1)}=-i(\partial/\partial x_1)I_{\pm}(n, 1/2)$. On the other hand, by the definition of $J_{\pm}(x, \xi', \mu)$, it follows that for $n\geqq 16$,

$$(\partial/\partial x_1)J_{\pm}(x, \xi', \mu)=\mp y_1'(x_1) \qquad \text{on the support of } \alpha_n^{\pm}.$$

Then using this fact and Lemma 5.5, we can express

$$\begin{aligned} &I_{\pm}(n, 1/2)^*\,\mathrm{op}\,(\lambda I_{\pm}(n, 1/2)_{(1)})L\alpha_n^{\pm} \\ &\quad \equiv -i\lambda(n\pm 1/2)I_{\pm}(n, 1)^*(y_1'(x_1)+\hat{b}_{\pm})I_{\pm}(n, 1)L\alpha_n^{\pm}, \\ &I_{\pm}(n, 1/2)^*\,\mathrm{op}\,(\lambda I_{\pm}(n, 1/2)_{(1)})(D_0-i\theta)\alpha_n^{\pm} \\ &\quad \equiv -i\lambda(n\pm 1/2)I_{\pm}(n, 1)^*(y_1'(x_1)+\tilde{b}_{\pm})I_{\pm}(n, 1)(D_0-i\theta)\alpha_n^{\pm}, \end{aligned}$$

with $\hat{b}_{\pm}, \tilde{b}_{\pm}\in \mu J^0\bar{S}^{0,0}$. Therefore taking into account that $0\leqq y_1'(x_1)\leqq 1$, the estimate (6.3) follows from these expressions.

As for the last term in the right-hand side of (6.1), remarking that $L-L^*$ belongs to $(S)^{0,0}_{1,1/2}\subset J^0S^{0,0}$, the integrand of this term is estimated from above by

$$c\|I_{\pm}(n, 1/2)Lw^{\pm}\|^2+c\|I_{\pm}(n, 1/2)(D_0-i\theta)w^{\pm}\|^2.$$

Finally, we prove the following estimates and complete the proof of this lemma; for any $\varepsilon>0$, we have

$$\begin{aligned} &|(I_{\pm}(n, 1/2)Lw^{\pm}, I_{\pm}(n, 1/2)(\partial_0 L)w^{\pm})| \\ &\quad \leqq \varepsilon(c+c(n)\mu)\|I_{\pm}(n, 1)Lw^{\pm}\|^2+\varepsilon^{-1}(c+c(n)\mu)|||u|||^2_{n,0,1}, \end{aligned} \tag{6.4}$$

$$\begin{aligned} &2\,\mathrm{Im}\,(I_{\pm}(n, 1/2)\,\mathrm{op}\,(L^2)w^{\pm}, I_{\pm}(n, 1/2)(D_0-i\theta)w^{\pm}) \\ &\quad \geqq 2\,\mathrm{Im}\,(I_{\pm}(n, 1/2)L^2w^{\pm}, I_{\pm}(n, 1/2)(D_0-i\theta)w^{\pm}) \\ &\qquad -\varepsilon(c+c(n)\mu)\|I_{\pm}(n, 1)(D_0-i\theta)w^{\pm}\|^2 \\ &\qquad -\varepsilon^{-1}(c+c(n)\mu)|||u|||^2_{n,0,1}. \end{aligned} \tag{6.5}$$

Since $\partial_0 L \in (S)^{1,1}_{1,1/2} \subset J^0 S^{0,1}$, in view of Lemma 5.5, one can write

$$I_\pm(n, 1/2)^* I_\pm(n, 1/2)(\partial_0 L) \equiv I_\pm(n, 1)^*(A + a_\pm)\langle \mu D' \rangle I_\pm(n, 0),$$

where $A(x, \xi', \mu) = \langle \mu\xi' \rangle^{-1}(\partial_0 L) \in J^0 S^{0,0}$, $a_\pm \in \mu J^0 \bar{S}^{0,0}$. If taking into account that $A(x, \xi', \mu)$ does not depend on n, Lemma 5.3 shows (6.4). To derive the estimate (6.5), it is enough to remark that $\mathrm{op}(L^2) \equiv L^2 + r$ with $r \in (S)^{1,1}_{1,1/2} \subset J^0 S^{0,1}$.

Lemma 6.3 (cf. Propositions 7.1, 7.2 in [10]). *Let $Q(x, \xi', \mu)$ be as in Lemma* 5.1. *Then we have*

$$\begin{aligned}
&2\,\mathrm{Im} \sum \int (I_\pm(n, 1/2)Qw^\pm, I_\pm(n, 1/2)(D_0 - i\theta)w^\pm)dx_0 \\
&\quad \geqq (2n - 1 - c - c(n)\mu^{1/2})\,\mathrm{Re} \sum \int (Qw_n^\pm, w_n^\pm)dx_0 \\
&\qquad + 2\theta\,\mathrm{Re} \sum \int (Qw_{n-1/2}^\pm, w_{n-1/2}^\pm)dx_0 - (c + c(n)\mu^{1/4}) \int |||u|||^2_{n,0,1} dx_0 \\
&\qquad - (c + c(n)\mu^{1/4}) \int |||u|||^2_{n,2,0} dx_0 - (c + c(n)\mu^{1/4}) \int [u]^2_{n,0,1} dx_0 \\
&\qquad - c(n, \mu) \int |||u|||^2_{n,1,0} dx_0 - \theta n^{-1}(c + c(n)\mu) \int |||u|||^2_{n,-1/2,1} dx_0 \\
&\qquad - n\theta(c + c(n)\mu) \int |||u|||^2_{n,3/2,0} dx_0,
\end{aligned}$$

modulo

$$c(n, k, \mu) \int |u|^2_{-k,\theta} dx_0,$$

where

$w_n^\pm = I_\pm(n, 1/2)J_\pm(1/2)w^\pm$, $w_{n-1/2}^\pm = I_\pm(n, 1/2)w^\pm$, $J_\pm(1/2) = \mathrm{op}(J_\pm(x, \xi', \mu)^{-1/2})$.

Lemma 6.4 (cf. Propositions 7.3, 7.4 in [10]). *For any $k \in \boldsymbol{N}$, we have*

$$\begin{aligned}
\mathrm{Re}(Qw_n^\pm, w_n^\pm) + |||u|||^2_{n,2,0} &\geqq (c - c(n)\mu)|||u|||^2_{n,0,1} - c(n)\mu|||u|||^2_{n,2,0} \\
&\quad - c(n, \mu)|||u|||^2_{n,1,0} - C(n, k, \mu)||u||^2_{-k}, \\
\mathrm{Re}(Qw_{n-1/2}^\pm, w_{n-1/2}^\pm) + |||u|||^2_{n,3/2,0} &\geqq (c - c(n)\mu)|||u|||^2_{n,-1/2,1} - c(n)\mu|||u|||^2_{n,3/2,0} \\
&\quad - c(n, \mu)|||u|||^2_{n,1/2,0} - C(n, k, \mu)||u||^2_{-k}.
\end{aligned}$$

From (3.6), we know that $0 < \bar{\lambda} < 1$ and hence we can take $\varepsilon > 0$ in Lemma 6.2 so that

$$\tilde{\lambda} = \bar{\lambda} + \varepsilon < 1.$$

Now estimating the commutators with $\alpha_n^{\pm}$ and absorbing the modulo terms, we get finally the following estimate from Lemmas 6.1 through 6.4.

Lemma 6.5. *In case* (3.6), *we have for any* $\delta>0$ *that*

$$-2\,\mathrm{Im}\sum\int(I_{\pm}(n,1/2)\alpha_n^{\pm}P_{\theta}u,\ I_{\pm}(n,1/2)(D_0-i\theta)\alpha_n^{\pm}u)dx_0$$

$$\geqq(2n-1)(1-\delta-\tilde{\lambda}-cn^{-1/2}-c(n)\mu^{1/4})\int[u]_{n,1,0}^2dx_0$$

$$+cn^{1/2}(1-c(n)\mu^{1/4})\int|||(D_0-i\theta)u|||_{n,1,0}^2dx_0+(c\theta-c(n))\int[u]_{n,1/2,0}^2dx_0$$

$$+c\theta\int|||(D_0-i\theta)u|||_{n,1/2,0}^2dx_0+cn(\delta-cn^{-1/2}-c(n)\mu^{1/4})\int|||u|||_{n,0,1}^2dx_0$$

$$+c\theta(1-cn^{-1}-c(n)\mu)\int|||u|||_{n,-1/2,1}^2dx_0$$

$$+(2n-1)(1-\tilde{\lambda}-cn^{-1/2}-c(n)\mu^{1/2})\sum\int||I_{\pm}(n,1)Lw^{\pm}||^2dx_0$$

$$+(2\theta-c)\sum\int||I_{\pm}(n,1/2)Lw^{\pm}||^2dx_0$$

$$+c\delta n^3(1-c\delta^{-1}n^{-3/2}-c(n)\mu^{1/4})\int|||u|||_{n,2,0}^2dx_0$$

$$+c\theta n^2(1-cn^{-1}-c(n)\theta^{-1}-c(n)\mu)\int|||u|||_{n,3/2,0}^2dx_0$$

$$+(cn\theta^2-c(n,\mu))\int|||u|||_{n,1,0}^2dx_0,$$

where $P_{\theta}=(D_0-i\theta)^2-Q(x,D',\mu)$. *In case* (3.5), *we have the same estimates with* $\tilde{\lambda}=0$, $L(x,D',\mu)=0$.

Since $\tilde{\lambda}<1$, we can take $\delta>0$ so that $1-\delta-\tilde{\lambda}>0$. Then it follows from Lemma 6.5 that

Theorem 6.1. *For any* $k\in N$, *we have*

$$C(n,k,\mu)\int e^{-2\theta x_0}||Pu||_{-2k}^2dx_0+\int e^{-2\theta x_0}|||Pu|||_{n,0,0}^2dx_0$$

$$\geqq c_1n\int e^{-2\theta x_0}|||D_0u|||_{n,1,0}^2\,dx_0+c_2n\int e^{-2\theta x_0}|||u|||_{n,0,1}^2dx_0$$

$$+c_3\theta\int e^{-2\theta x_0}|||D_0u|||_{n,1/2,0}^2dx_0+c_3\theta\int e^{-2\theta x_0}|||u|||_{n,-1/2,1}^2dx_0$$

$$+c_4\theta^3\int e^{-2\theta x_0}||u||_{-k}^2dx_0+c_4\theta^{3/2}\int e^{-2\theta x_0}||D_0u||_{-k}^2dx_0,$$

for $n \geqq C_0$, $0<\mu\leqq\mu_1(n)$, $\theta\geqq\theta_0(n, k, \mu)$, $u \in C_0^\infty(I\times \boldsymbol{R}^d)$.

From this theorem, the existence of parametrix (in a sufficiently small conic neighborhood of ρ) follows easily. The fact that this parametrix has finite propagation speed of wave front sets will be proved in the next section.

§ 7. Estimates of wave front sets

In this section, we follow the arguments in [7], [8], [4]. To simplify the notation, we set

$$E_{n,s}^2(u; t) = |||u|||_{n,0,s}^2 + |||(D_0 - i\theta)u|||_{n,1,s-1}^2 + \theta|||u|||_{n,-1/2,s}^2 + \theta|||(D_0 - i\theta)u|||_{n,1/2,s-1}^2.$$

Let $Q(x, D', \mu)$ be as in § 6. We fix n, $\mu>0$ so that the estimate in Theorem 6.1 holds. The following lemma is easily derived from Theorem 6.1.

Lemma 7.1. *For any* $k \in \boldsymbol{N}$, $s \in \boldsymbol{R}$ *with* $k\geqq 1-s$, *we have*

$$C(s, k)\int ||P_\theta u||_{-k}^2 dx_0 + c(s)\int |||P_\theta u|||_{n,0,s}^2 dx_0$$
$$\geqq c\int E_{n,s+1}^2(u; x_0)dx_0 + c\int \{\theta^{5/2}||u||_{-k}^2 + \theta||(D_0 - i\theta)u||_{-k}^2\} dx_0,$$

for $\theta\geqq\theta_0(s, k)$, $u \in C_0^\infty(I\times \boldsymbol{R}^d)$.

Let $\phi(x, \xi')$ be a real smooth function positively homogeneous of degree 0 such that $\partial_0\phi = -1$. We define $\Phi(x, \xi')$ by

$$\Phi(x, \xi') = \begin{cases} \exp(-1/\phi(x, \xi')) & \text{if } \phi(x, \xi')>0 \\ 0 & \text{if } \phi(x, \xi')\leqq 0. \end{cases} \tag{7.1}$$

It is clear that $\Phi(x, \xi')$ belongs to $S_{1,0}^0$.

Lemma 7.2 (cf. Lemma 2.1 in [7]). *Assume that* $\phi(x, \xi')$ *satisfies*

$$4(1-\nu)Q(x, \xi', \mu) \geqq \{Q, \phi\}(x, \xi', \mu)^2, \qquad \partial_0\phi = -1, \tag{7.2}$$

with some $\nu>0$. *Then we have for any* $k \in \boldsymbol{N}$, $k\geqq 1-s$,

$$-2\,\mathrm{Im}\,(I_\pm(n, 1/2)\alpha_n^\pm\langle\mu D'\rangle^s[P_\theta, \Phi]u, I_\pm(n, 1/2)(D_0 - i\theta)\alpha_n^\pm\langle\mu D'\rangle^s\Phi u)$$
$$\leqq 2\,\mathrm{Re}\,\partial_0(\langle\mu D'\rangle^s I_\pm(n, 1/2)\alpha_n^\pm(D_0\Phi)u, \langle\mu D'\rangle^s I_\pm(n, 1/2)\alpha_n^\pm\Phi(D_0 - i\theta)u)$$
$$+ c(s, \nu)E_{n,s+3/4}^2(u; x_0) + c(s)|||\Phi P_\theta u|||_{n,0,s}^2 + C(s, k)||\Phi P_\theta u||_{-k}^2$$
$$+ C(s, k)\theta|u|_{-k,\theta}^2,$$

where Φ is defined by (7.1) *with* ϕ.

From Lemmas 7.1 and 7.2, it follows that

Lemma 7.3. *Let $\phi(x, \xi')$ be as in Lemma 7.2. Then for any $k \in \boldsymbol{N}$, $s \in \boldsymbol{R}$, we have*

$$
\begin{aligned}
(7.3)\qquad & C(s, k)\int \|\Phi P_\theta u\|_{-k}^2 dx_0 + c(s)\int |||\Phi P_\theta u|||_{n,0,s}^2 dx_0 + \int E_{n,s+3/4}^2(u; x_0)dx_0 \\
& \geqq \int E_{n,s+1}^2(\Phi u; x_0)dx_0 - \theta\int |u|_{-k}^2 dx_0,
\end{aligned}
$$

where $|u|_{-k}^2 = \theta^2\|u\|_{-k}^2 + \theta\|(D_0 - i\theta)u\|_{-k}^2$.

To prove that the parametrix of P constructed in § 6 has finite propagation speed of WF, it is enough to apply Lemma 7.3 with the special $\phi(x, \xi')$,

$$\phi(x, \xi') = -x_0 + T + \varepsilon(|x' - y'|^2 + |\xi'|\xi'|^{-1} - \eta'|\eta'|^{-1}|^2 + \delta)^{1/2}, \qquad \delta > 0$$

which satisfies (7.2) for sufficiently small $\varepsilon > 0$.

Let $\phi(x, \xi')$ be a real smooth function defined near ρ, positively homogeneous of degree 0 satisfying (3.4) near ρ such that

$$\partial_0\phi(x, \xi') = -1, \qquad \phi(\rho) = 0.$$

Let $\phi(x, \xi', \mu)$ be the localization of $\phi(x, \xi')$ (with $p=1$ or $p=2$ according to the case (3.5) or (3.6)) and set

$$\tilde{\phi}(x, \xi', \mu) = \mu^{-1}\phi(x, \xi', \mu).$$

Then $\tilde{\phi}(x, \xi', \mu)$ is globally defined for small $\mu > 0$ and it is easy to see that $\tilde{\phi}(x, \xi', \mu)$ satisfies (7.2). Without loss of generality, we may assume that the last term in the right-hand side of (7.3) is finite for some $k \in \boldsymbol{N}$. Here applying Lemma 7.3, one can improve the regularity of u with 1/4 with respect to s in $E_{n,s}(u; x_0)$ in the region where $\tilde{\phi}$ is positive. This argument proves Theorems 2.2 and 2.3.

References

[1] L. Hörmander, The Cauchy problem for differential equations with double characteristics, J. Analyse Math. **32** (1977), 118–196.

[2] V. Ja. Ivrii and V. M. Petkov, Necessary conditions for the Cauchy problem for non-strictly hyperbolic equations to be well posed, Russian Math. Surveys, **29** (1974), 1–70.

[3] V. Ja. Ivrii, The well-posedness of the Cauchy problem for non-strictly hyperbolic operators III. The energy integrals, Trans. Moscow Math. Soc.

34 (1978), 149–168.
[4] V. Ja. Ivrii, Wave fronts of solutions of certain pseudodifferential equations, Trans. Moscow Math. Soc. **39** (1981), 49–86.
[5] N. Iwasaki, Cauchy problem for effectively hyperbolic equations (a standard type), Preprint.
[6] R. B. Melrose, The Cauchy problem for effectively hyperbolic operators, Hokkaido Math. J. **12** (1983), 371–391.
[7] T. Nishitani, On the finite propagation speed of wave front sets for effectively hyperbolic operators, Sci. Rep. College Gen. Ed. Osaka Univ. **32**-1 (1983), 1–7.
[8] ——, On wave front sets of solutions for effectively hyperbolic operators, Sci. Rep. College Gen. Ed. Osaka Univ. **32**-2 (1983), 1–7.
[9] ——, Note on some non effectively hyperbolic operators, Sci. Rep. College Gen. Ed. Osaka Univ. **32**-2 (1983), 9–17.
[10] ——, Local energy integrals for effectively hyperbolic operators, I, II, J. Math. Kyoto Univ. **24** (1984), 623–658, 659–666.

Department of Mathematics
College of General Education
Osaka University
Toyonaka, Osaka 560, Japan

Taniguchi Symp. HERT
Kyoto 1984, pp. 257–271

Huygens' Principle for a Wave Equation and the Asymptotic Behavior of Solutions along Geodesics*

Kimimasa NISHIWADA

§ 1. Introduction

We shall define a Lorentzian structure in $\boldsymbol{R}^n$ by the pseudo-Riemannian metric

$$\langle dx, dx\rangle = 2dx_1dx_2 - \sum_{3\leqq \alpha,\beta\leqq n} a_{\alpha\beta}(x_1)dx_\alpha dx_\beta, \tag{1.1}$$

where $(a_{\alpha\beta})_{3\leqq\alpha,\beta\leqq n}$ is a symmetric, positive definite, C^∞ matrix of the first variable x_1 only. The signature of the metric is $(+,-,\cdots,-)$ and therefore d'Alembertian $\square$ of this metric is hyperbolic in $\boldsymbol{R}^n$.

Let P denote the operator

$$P = \square + \sum_{j=1}^{k} 2\lambda_j\langle\nabla \log L_j(x), \nabla\rangle,\ \lambda_j \in \boldsymbol{C},\ j=1,\cdots,k. \tag{1.2}$$

Some results will be presented in this note, concerning Huygens' principle and the behavior along geodesics of the fundamental solution $G(x, y)$ of the operator (1.2). We shall make some assumptions on the coefficients in a domain consisting of the geodesics considered.

The metric (1.1) was introduced in connection with the study of empty space-time gravitational equations (see e.g. Petrow [8]). It was then picked up by Günther [3], who showed that with d'Alembertian of this metric Huygens' principle is indeed valid if $n=4$. Later it was proved by McLenaghan [6] that this was the only non-trivial example of Huygens' principle if $n=4$ and the space-time is empty or the Ricci tensor vanishes.

The fact that Huygens' principle occurs with this metric implies that the Hadamard expansion of the fundamental solution terminates after a finite number of terms. This simplicity indeed facilitates the study of $G(x, y)$, local or global, even when the operator has lower order terms as in our case and therefore the Hadamard expansion does not terminate in general.

---

*) This work was started by the author while being a fellow at the Weizmann Institute of Science, Rehovot, Israel.

If the coefficients $a_{\alpha\beta}$ are constant, then d'Alembertian is equivalent to the ordinary wave equation in the Euclidean space. In this case operators of the type (1.2) were treated by many authors. To quote a few, Delache-Leray [1] studied the operator when $k=1$ and L_1 is linear, calling it an "Euler-Poisson-Darboux operator". Stellmacher [9] considered the case $k=n$ and gave the criterion of Huygens' principle, which lies in the same scope as ours. Hopefully, however, our method may offer a more clarified approach to this problem even in the case of constant coefficient metric.

In the last section we shall give an application of our results to a problem discussed by Littman-Lui [5] and Nishiwada-Tintarev [7].

§ 2. Assumptions and Results

It can be easily proved that $\boldsymbol{R}^n$ is geodesically convex with respect to the metric (1.1). Therefore all the geodesics starting from a point y cover the whole $\boldsymbol{R}^n$. These geodesics are grouped into three types, i.e. time-like, null and space-like geodesics. Further, if a time orientation is fixed arbitrarily, then one can define *the future dependence domain* $D^+(y)$ as the set of all points that can be reached by future-directed time-like geodesics from y.

Let Ω be an open domain in $D^+(y)\setminus y$, which itself consists of some of the future-directed, time-like geodesics from y. Our assumptions will be as follows:

(A–1) The functions $a_{\alpha\beta}(x_1)$ and $L_j(x)$ are real valued, C^∞ in $\bar{\Omega}$ and real analytic in Ω. Moreover, $L_j(x)\neq 0$ in Ω.

(A–2) $\Box L_j=0, j=1, \cdots, k$, and ∇L_j are orthogonal to each other in Ω.

(A–3) $P|\nabla L_j|=|\nabla L_j|P$ as operators.

(A–4) Either $|\nabla L_j|=0$ in Ω or $(|\nabla L_j|^{-1}L_j)(x)|_{x=z(r)}$ is a linear function of r for any geodesic $z(r)$, $r\geqq 0$ starting from y and staying in Ω.

Remark 2.1. $|\nabla L_j|^2=\langle\nabla L_j, \nabla L_j\rangle$, where the inner product is induced from the pseudo-Riemannian metric (1.1). If the latter is the case in (A–4), $\arg|\nabla L_j|$ is fixed in such a way that

$$-i\pi\leqq\arg(|\nabla L_j|^{-2})\leqq i\pi. \tag{2.1}$$

If the number in the parenthesis is negative, then we have two choices of its argument. This will be determined later.

Remark 2.2. The last condition (A–4) makes sense, since the parameter of a geodesic can be changed only up to a linear transformation. Therefore, if $f(z(r))$ is linear in one parameter, then so is it in another.

Example 2.3. Suppose that $k=1$ and that $L(x)=h(x_1)x_n$ with non-vanishing real analytic function h and $n\geqq 3$. Then the conditions (A–1), (A–2) and (A–3) are satisfied, if $x_n\neq 0$ in Ω. As we shall see in (2.3), if $a^{\alpha n}$ are constant for $\alpha=3, \cdots, n$, then (A–4) is also satisfied.

When $|\nabla L_j|\neq 0$ in Ω, we shall write

$$(|\nabla L_j|^{-1}L_j)(z(r))=l_j r+|\nabla L_j|^{-1}L_j(y).$$

Let J denote the set of $j=1, \cdots, k$ such that $|\nabla L_j|^2\neq 0$, $J_-(\subset J)$ such that $|\nabla L_j|^2<0$ and J' such that $l_j\neq 0$.

Huygens' principle in Ω can be characterised completely in terms of the numbers n and λ_j.

Theorem 2.4. *Let $G(x, y)$ be the forward fundamental solution of the operator* (1.2). *Then, $G(x, y)\equiv 0$ in Ω if and only if n is even and λ_j is an integer for any $j\in J$ such that*

$$\sum_{j\in J}\max(\lambda_j, 1-\lambda_j)\leqq\frac{n}{2}-1. \tag{2.2}$$

Remark 2.5. If $n=4$, as the only possible cases satisfying (2.2) we have either $|J|=0$ or $|J|=1$ with $\lambda_j=0$ or 1. These cases are essentially equivalent to $\Box$, as we have

$$L_j^{2\lambda_j-1}P_{(\lambda_1,\cdots,\lambda_j,\cdots,\lambda_n)}L_j^{1-2\lambda_j}=P_{(\lambda_1,\cdots,1-\lambda_j,\cdots,\lambda_n)}$$

and

$$L^{-\sigma}\Box L^{\sigma}=\Box+2\sigma\langle\nabla\log L, \nabla\rangle, \quad \text{if } |\nabla L|=0.$$

We shall now take a geodesic $z(r)$, $r\geqq 0$, from y, which stays in Ω. The equation of a geodesic described in the cotangent space can be written as a Hamilton system;

$$\frac{dz_j}{dr}=\frac{1}{2}H_{\xi_j}(z, \xi), \qquad \frac{d\xi_j}{dr}=-\frac{1}{2}H_{x_j}(z, \xi),$$
$$z(0)=y, \qquad \xi(0)=\theta,$$

where

$$H(x, \xi)=2\xi_1\xi_2-\sum a^{\alpha\beta}(x_1)\xi_\alpha\xi_\beta.$$

The system can be explicitly solved and gives the solution

$$\begin{aligned} x_1&=y_1+\theta_2 r_1,\\ x_2&=y_2+\theta_1 r+\frac{1}{2\theta_2}\sum(A^{\alpha\beta}(x_1, y_1)-a^{\alpha\beta}(y_1))\theta_\alpha\theta_\beta r,\\ x_\alpha&=y_\alpha-\sum A^{\alpha\beta}(x_1, y_1)\theta_\beta r,\end{aligned} \tag{2.3}$$

where

$$A^{\alpha\beta}(x_1, y_1)=\frac{1}{x_1-y_1}\int_{y_1}^{x_1} a^{\alpha\beta}(t)dt.$$

Let $\gamma(x, y)$ denote the square of geodesic distance between the two points x and y. As is well known, $\gamma(x, y)=H(y, r\theta)=\Theta r^2$, where

$$\Theta=2\theta_1\theta_2-\sum a^{\alpha\beta}(y_1)\theta_\alpha\theta_\beta.$$

We put $c_n=\pi^{1-n/2}/2$ and define

$$\kappa(x_1, y_1)=|\det A_{\alpha\beta}(x_1, y_1)|^{1/2}/|\det a_{\alpha\beta}(x_1)\det a_{\alpha\beta}(y_1)|^{1/4}. \tag{2.4}$$

Theorem 2.6. *The function* $(c_n\kappa)^{-1}\prod_1^k (L_j(x)/L_j(y))^{\lambda_j}G(x, y)|_{x=z(r), r>0}$ *is analytically continued over* $\mathbf{C}_r$ *except for some singular points located on the real axis. The singular points on the positive real axis are contained in the interval*

$$\begin{aligned}&[\min_{j\in J_-} 2\Theta^{-1}|\mu_j|(|l_j|+(|l_j|^2+\Theta)^{1/2}),\\ &2\Theta^{-1}(\sum_{j\in J}|\mu_j l_j|+((\sum|\mu_j l_j|)^2+\Theta\sum|\mu_j|^2)^{1/2})],\end{aligned} \tag{2.5}$$

where $\mu_j=|\nabla L_j|^{-1}L_j(y)$. *In particular, if* $J_-=\phi$ *(i.e. no space-like* ∇L_j*), there is no singular points on* $(0, \infty)$.

The following example suggests that the above singularity may be generated by null-geodesics (bicharacteristics) emanating from the intersection of the characteristic conoid $C^+(y)$ and the hypersurface $L_j(x)=0$. This statement can easily be proved when $|J|=1$ and thus the interval (2.5) turns out to be a point.

Example 2.7. Let $n=2$ and $L=x_2-x_1+1$. Then

$$P=2\partial_1\partial_2+\frac{2\lambda}{x_2-x_1+1}(\partial_1-\partial_2).$$

We have $D^+(0)=\{x|x_1>0, x_2>0\}\cup\{0\}$ and, with the geodesics $z(r)=(r, kr)$, $r\geqq 0$, $k\geqq 1$, it follows that $\Theta=2k$, $l=(\sqrt{2}i)^{-1}(k-1)$ and $\mu=(\sqrt{2}i)^{-1}$. Hence the interval consists of the point $r=1$.

Theorem 2.8. *Suppose that* $\lambda_j\neq 1/2$ *for* $j\in J$. *Then we have*

$$(c_n\kappa)^{-1}\prod_1^k (L_j(x)/L_j(y))^{\lambda_j}G(x, y)=r^M(C+o(1)) \tag{2.6}$$

along the geodesic $x=z(r)$ *as* $r\to\infty$, *where*

$$M = 2 - n + \textstyle\sum_{J \cap J'} \max(-\lambda_j, \lambda_j - 1) \qquad (2.7)$$
$$+ \textstyle\sum_{J \backslash J'} 2\max(-\lambda_j, \lambda_j - 1),$$

and

$$C = \prod_J \left(\Lambda_j^{\max(-\lambda_j, \lambda_j - 1)} \frac{\Gamma(\max(2\lambda_j - 1, 1 - 2\lambda_j))}{\Gamma(\max(\lambda_j, 1 - \lambda_j))} \right) \qquad (2.8)$$
$$\times \frac{\Theta^{1-(n/2)+\sum_J \max(-\lambda_j, \lambda_j - 1)}}{\Gamma(1 - (n/2) + \sum_J \max(\lambda_j, 1 - \lambda_j))}.$$

Here, $\Lambda_j^{-1} = 4l_j\mu_j$ *if* $j \in J \cap J'$ *or* $= 4\mu_j^2$ *if* $j \in J \backslash J'$. *Moreover,* $-i\pi \leqq \arg \Lambda_j \leqq i\pi$, *and, when* Λ_j *are negative, we fix* $\arg \Lambda_j = i\pi$ *or* $-i\pi$ *according to whether we bypass the singularities on* $(0, \infty)$ *by going "above" or "below" them in* $\mathbf{C}_r$ *respectively.*

In case $\lambda_j = 1/2$ for some j, we have a similar asymptotic formula, in which $\Gamma(\max(2\lambda_j - 1, 1 - 2\lambda_j))$ in (2.8) is replaced by $(\log r) + c$ with a constant c.

The above theorem, together with this remark, tells when the asymptotic order becomes less than M.

Corollary 2.4. *The left-hand side of* (2.6) *becomes* $o(r^M)$ *as* $r \to \infty$, *if and only if* $1 - (n/2) + \sum_J \max(\lambda_j, 1 - \lambda_j)$ *is a non-positive integer.*

Huygens' principle, as stated in Theorem 2.4, consists of the condition of Corollary 2.4 along with the requirement that λ_j be integers and that n be even. The first one follows from the rest if $|J| = 1$.

Corollary 2.5. *If n is even and* $|J| = 1$, *then*

$$(c_n\kappa)^{-1} \prod_1^k (L_j(x)/L_j(y))^{\lambda_j} G(x, y)|_{x = z(r)} = o(r^M) \quad \text{as } r \to \infty$$

if and only if Huygens' principle is valid in Ω.

§ 3. Construction of $G(x, y)$ and outline of the proof

We shall construct the Hadamard expansion of G, which is quite suitable for our purpose. In so doing we have only to apply the general theory (e.g. developed in Friedlander [2]) to our case.

The Hadamard expansion can be written as

$$G(x, y) = \sum_{\nu=0}^{\infty} U^{(\nu)}(x, y)\psi^+_{\nu+1-n/2}(\gamma(x, y)), \qquad (3.1)$$

where $U^{(\nu)}(x, y) \in C^\infty$ if x is close to y. The functions ψ_q are defined as follows; when $\mathrm{Re}\, q > -1$,

$$\psi_q(t)=\begin{cases}\dfrac{t^q}{\Gamma(q+1)}, & t>0,\\ 0 \quad , & t\leqq 0.\end{cases}$$

It is then easy to see that $\psi_q(t)$ can be analytically continued to a distribution-valued entire function of q. We have, for instance,

$$\psi_{-q}(t)=\delta^{(q-1)}(t), \qquad q=1, 2, \cdots. \tag{3.2}$$

Now put, when Re $q>-1$,

$$\psi_q^+(\gamma(x, y))=\begin{cases}\psi_q(\gamma), & x\in D^+(y)\setminus\{y\}\\ 0 \quad , & \text{otherwise.}\end{cases}$$

This time it can be analytically continued to an analytic function of $q\in C$ with poles at $q=-n/2-m$, $m=0, 1, \cdots$. We refer the precise argument to [1] or [2].

The functions $U^{(\nu)}$ are successively determined by the transport equations;

$$\begin{aligned}&2\langle\nabla\gamma, \nabla U^{(\nu)}\rangle+(\square+\langle a, \nabla\gamma\rangle+4\nu-2n)U^{(\nu)}\\ &\qquad =-PU^{(\nu-1)}, \quad \nu=0, 1, \cdots, \quad (U^{(-1)}=0),\\ &U^{(0)}(y, y)=c_n,\end{aligned} \tag{3.3}$$

where $a=\sum 2\lambda_j\nabla\log L_j$.

(3.3) is a first order equation whose bicharacteristics are the geodesics passing through y. The solutions are therefore given by integrations along these geodesics;

$$U^{(0)}(x, y)=c_n\kappa(x, y)\exp\left(-\frac{1}{2}\int_0^1\left\langle a(z(s)), \frac{dz}{ds}\right\rangle ds\right), \tag{3.4}$$

$$U^{(\nu)}(x, y)=-\frac{1}{4}U^{(0)}(x, y)\int_0^1\frac{PU^{(\nu-1)}}{U^{(0)}}\bigg|_{x=z(s)}s^{\nu-1}ds, \tag{3.5}$$

where $z(s)$, $0\leqq s\leqq 1$, is the geodesic such that $z(0)=y$ and $z(1)=x$. And κ is defined by

$$\kappa(x, y)=\frac{|\det\partial^2\gamma/\partial x_i\partial y_j|^{1/2}}{2^{n/2}|\det g(x)g(y)|^{1/4}},$$

where $g=(g_{ij})$ defines the metric (1.1).

We shall now compute $\kappa(x, y)$ more explicitly. It follows from (2.3) that

$$\gamma(x, y) = (2\theta_1\theta_2 - \sum a^{\alpha\beta}(y_1)\theta_\alpha\theta_\beta)r^2$$
$$= 2(x_1-y_1)(x_2-y_2) - \sum A_{\alpha\beta}(x_1, y_1)(x_\alpha - y_\alpha)(x_\beta - y_\beta),$$

which gives

$$\left(\frac{\partial^2\gamma}{\partial x_i \partial y_j}\right) = \left(\begin{array}{cc|c} 0 & -2 & \\ -2 & 0 & 0 \\ \hline & 0 & 2A_{\alpha\beta} \end{array}\right) + \left[\begin{array}{c|c} * & 0 \; * \; \cdots \\ \hline 0 & \\ * & 0 \\ \vdots & \end{array}\right].$$

Here, the first row vector of the second matrix in the right-hand side is a linear combination of the row vectors from the second to the last of the first matrix. Therefore, $|\det \partial^2\gamma/\partial x_i \partial x_j| = 2^n |\det A_{\alpha\beta}|$, which leads to

$$\kappa(x, y) = \kappa(x_1, y_1) = \frac{|\det A_{\alpha\beta}(x_1, y_1)|^{1/2}}{|\det a_{\alpha\beta}(x_1) \det a_{\alpha\beta}(y_1)|^{1/4}}.$$

This agrees with our previous notation (2.4).

Let us consider the case $\lambda_j = 0, j = 1, \cdots, k$. (3.4) implies $U^{(0)} = c_n\kappa$. Since $PU^{(0)} = 0$, we have $U^{(\nu)} = 0$, $\nu = 1, 2, \cdots$, and thus

(3.6) $$G(x, y) = c_n\kappa(x_1, y_1)\psi^+_{1-n/2}(\gamma(x, y)), \qquad \text{if } \lambda_j = 0.$$

This formula, combined with (3.2), proves Günther's result that Huygens principle is valid for $n = 4, 6, \cdots$.

In the general case the integration in (3.4) is carried out easily, due to the gradient a, and gives

$$U^{(0)} = c_n\kappa \prod_1^k (L_j(x)/L_j(y))^{-\lambda_j}.$$

The computation of $U^{(\nu)}$ can be done by repeating the integrations in (3.5).

Proposition 3.1. *For* $\nu = 0, 1, \cdots,$

(3.7) $$U^{(\nu)}(x, y) = U^{(0)}(x, y) \times \sum_{\nu=\nu_1+\cdots+\nu_k} \prod_j \left\{ \frac{(\lambda_j, \nu_j)(1-\lambda_j, \nu_j)}{\nu_j!} (-4\mu_j(x)\mu_j(y))^{-\nu_j} \right\},$$

where $(a, m) = a(a+1)\cdots(a+m-1)$ *and* $\mu_j(x) = |\nabla L_j|^{-1}L_j(x)$. *In case* $|\nabla L_j| = 0$ *in* Ω, *we put* $\mu_j(x)^{-\nu} = 1$, *if* $\nu = 0$, *and* $= 0$, *if otherwise.*

Proof. Let us first recall a certain general calculation rule; let f and S be any C^2 functions, then

$$(3.8)\qquad \square(Uf(S))=f(S)\square U+(2\langle\nabla S,\nabla U\rangle+U\square S)f'(S)+U\langle\nabla S,\nabla S\rangle f''(S).$$

Applying this equality to the case $f=f_1$, $S=L_1$ and $U=\prod_2^k f_j(L_j)$, and using (A–2), we get, by induction,

$$\square\Big(\prod_1^k f_j(L_j)\Big)=\sum_{j=1}^k |\nabla L_j|^2 f_j''(L_j)\prod_{i\neq j} f_i(L_i)$$

and hence

$$(3.9)\qquad P\Big(\prod_1^k f_j(L_j)\Big)=\sum_{j=1}^k |\nabla L_j|^2(f_j''(L_j)+2\lambda_j L_j^{-1}f_j'(L_j))\prod_{i\neq j} f_i(L_i).$$

Now suppose that $U^{(\nu)}$ has the form (3.7). The assumption (A–3) and (3.9) lead to

$$PU^{(\nu)}=U^{(0)}\sum_{\nu=\nu_1+\cdots+\nu_k}\sum_j\prod_{i\neq j}\frac{(\lambda_i,\nu_i)(1-\lambda_i,\nu_i)}{\nu_i!}(-4\mu_i(x)\mu_i(y))^{-\nu_i}\times\frac{(\lambda_j,\nu_j+1)(1-\lambda_j,\nu_j+1)}{\nu_j!}(-4\mu_j(y))^{-\nu_j}(\mu_j(x))^{-\nu_j-2}.$$

Therefore,

$$\Big(-\frac{1}{4}\Big)\frac{PU^{(\nu)}}{U^{(0)}}=\sum_{\nu+1=\nu_1+\cdots+\nu_n}\Big(\prod_i\frac{(\lambda_i,\nu_i)(1-\lambda_i,\nu_i)}{\nu_i!}(-4\mu_i(y))^{-\nu_i}\Big)\times\sum_j \nu_j\mu_j(y)\mu_j(x)^{-1}\prod_1^k \mu_i(x)^{-\nu_i}.$$

What we have to prove is thus the equality

$$(3.10)\qquad \prod_i \mu_i(x)^{-\nu_i}=\sum_j\int_0^1 \nu_j\mu_j(y)\mu_j(x)^{-1}\prod \mu_i(x)^{-\nu_i}\big|_{x=z(s)}s^{\nu-1}\,ds,$$

where $\nu=\nu_1+\cdots+\nu_k\geqq 1$.

Putting temporally $\mu_i(z(0))=x_i$ and $\mu_i(z(1))=y_i$ we can rewrite (3.10) as in the following lemma in view of the assumptions (A–1) and (A–4).

Lemma 3.2. *Suppose that $x_j\neq 0$, $y_j\neq 0$, $j=1,\cdots,k$. If ν_j are integers such that $\nu=\nu_1+\cdots+\nu_k\geqq 1$, then*

$$(3.11)\qquad \prod_1^k (x_i)^{-\nu_i}=\sum_j\int_0^1\frac{\nu_j y_j s^{\nu-1}ds}{(y_j+s(x_j-y_j))\prod(y_i+s(x_i-y_i))^{\nu_i}}.$$

Proof. If $\nu=\nu_j$, (3.11) says about the formula

$$x^{-\nu}=\int_0^1 \frac{\nu y s^{\nu-1}ds}{(y+s(x-y))^{\nu+1}},$$

which can be easily proved by induction.

We now assume that (3.11) is valid for any $\nu'<\nu$ and also that both $\nu_1, \nu_2>0$. Moreover, suppose that $d=y_1^{-1}x_1-y_2^{-1}x_2\neq 0$. Since

$$s=d^{-1}(y_1^{-1}(y_1+s(x_1-y_1))-y_2^{-1}(y_2+s(x_2-y_2))),$$

the right-hand side of (3.11) turns out to be

$$\sum d^{-1}\Big(y_1^{-1}\int_0^1 \frac{\nu_j y_j s^{\nu-2}ds}{(y_j+s(x_j-y_j))(y_1+s(x_1-y_1))^{\nu_1-1}(y_2+s(x_2-y_2))^{\nu_2}\prod''}$$

$$-y_2^{-1}\int_0^1 \frac{\nu_j y_j s^{\nu-2}ds}{(y_j+s(x_j-y_j))(y_1+s(x_1-y_1))^{\nu_1}(y_2+s(x_2-y_2))^{\nu_2-1}\prod''}\Big)$$

$$=\sum d^{-1}(y_1^{-1}x_1^{-\nu_1+1}x_2^{-\nu_2}-y_2^{-1}x_1^{-\nu_1}x_2^{-\nu_2+1})\prod_{i=3}^{k}x_i^{-\nu_i}$$

$$=\prod_1^k x_i^{-\nu_i}.$$

By analytic continuation (3.11) is valid also when $d=0$. The proof of the lemma and hence that of Prop. 3.1 are completed. Q.E.D.

We have thus obtained a precise expression of the fundamental solution G in a neighborhood of y in $\overline{\Omega}$. In order to know the convergence domain of the Hadamard series and to get a global extension we shall need some of the generalized multi-variable hypergeometric functions defined by Lauricella [4]. Namely, we shall use

$$F_B(a_1,\cdots,a_k,b_1,\cdots,b_k,c,x_1,\cdots,x_k)$$
$$=\sum\frac{(a_1,\nu_1)\cdots(a_k,\nu_k)(b_1,\nu_1)\cdots(b_k,\nu_k)}{(c,\nu_1+\cdots+\nu_k)\nu_1!\cdots\nu_k!}x_1^{\nu_1}\cdots x_k^{\nu_k}$$

and

$$F_A(a,b_1,\cdots,b_k,c_1,\cdots,c_k,x_1,\cdots,x_k)$$
$$=\sum\frac{(a,\nu_1+\cdots+\nu_k)(b_1,\nu_1)\cdots(b_k,\nu_k)}{(c_1,\nu_1)\cdots(c_k,\nu_k)\nu_1!\cdots\nu_k!}x_1^{\nu_1}\cdots x_k^{\nu_k}.$$

Furthermore, one can define, for $m=0, 1, \cdots,$

$$F_B^{(m)}(a_1,\cdots,b_1,\cdots,c,x_1,\cdots)=\left(\frac{d}{dr}\right)^m r^{c-1}F_B(\cdots,c,rx)|_{r=1}.$$

In particular,

$$F_B^{(m)}(a_1, \cdots, b_1, \cdots, 1, x_1, \cdots)$$
$$= \sum_{\nu=\nu_1+\cdots+\nu_k \geqq m} \frac{(a_1, \nu_1)\cdots(b_1, \nu_1)\cdots}{(\nu-m)!\,\nu_1!\cdots} x_1^{\nu_1}\cdots x_k^{\nu_k}.$$

It is known that F_B converges in the domain $|x_1|<1, \cdots, |x_k|<1$, while F_A does in the domain $|x_1|+\cdots+|x_k|<1$.

Note that F_B and F_A are direct extensions of Appell's double variable hypergeometric functions $f_2(a_1, a_2, b_1, b_2, c, x_1, x_2)$ and $f_3(a, b_1, b_2, c_1, c_2, x_1, x_2)$ respectively. These special functions will allow us to rewrite $G(x, y)$ in (3.1).

Proposition 3.3. a) *If n is odd,*

$$G(x, y) = c_n\kappa \prod_1^k (L_j(y)/L_j(x))^{\lambda_j} \psi^+_{1-n/2}(\gamma)$$
$$\times F_B\Big(\lambda_1, \cdots, \lambda_k, 1-\lambda_1, \cdots, 1-\lambda_k, 2-\frac{n}{2},$$
$$\frac{-\gamma}{4\mu_1(y)\mu_1(x)}, \cdots, \frac{-\gamma}{4\mu_k(y)\mu_k(x)}\Big).$$

b) *If n is even,*

$$G(x, y) = c_n\kappa \prod_1^k (L_j(y)/L_j(x))^{\lambda_j} \sum_{\nu=0}^{n/2-2} \psi^+_{1+\nu-n/2}(\gamma)$$
$$\times \sum_{\nu=\nu_1+\cdots+\nu_k} \prod_j \Big(\frac{(\lambda_j, \nu_j)(1-\lambda_j, \nu_j)}{\nu_j!}(-4\mu_j(y)\mu_j(x))^{-\nu_j}\Big)$$
$$+ c_n\kappa \prod_1^k (L_j(y)/L_j(x))^{\lambda_j} \psi_0^+(\gamma)\gamma^{1-n/2}$$
$$\times F_B^{(n/2-1)}\Big(\lambda_1, \cdots, \lambda_k, 1-\lambda_1, \cdots, 1-\lambda_k, 1,$$
$$\frac{-\gamma}{\mu_1(y)\mu_1(x)}, \cdots, \frac{-\gamma}{\mu_k(y)\mu_k(x)}\Big).$$

Proof. The proposition follows easily from (3.1), (3.7) and the obvious identity

$$\psi^+_{q+\nu}(\gamma) = \psi^+_q(\gamma)\frac{\gamma^\nu}{(q+1, \nu)},$$

which holds when q is not a negative integer. Q.E.D.

Proof of Theorem 2.4. Since

$$\psi_q^+(\gamma) = \frac{\gamma^q}{\Gamma(q+1)} \qquad \text{in } D^+(y)\backslash y,$$

$\psi_{1+\nu-n/2}(\gamma)$ vanishes in $D^+(y)\backslash y$ if and only if n is even and $\nu=0, \cdots, (n/2)-2$. Hence, Huygens' principle is valid if and only if n is even and

$$F^{(n/2-1)}\left(\lambda_1, \cdots, 1-\lambda_1, \cdots, 1, -\frac{\gamma^2}{4\mu_1(y)\mu_1(x)}, \right)=0.$$

The latter is clearly equivalent to

$$\prod_{j\in J}(\lambda_j, \nu_j)(1-\lambda_j, \nu_j)=0 \qquad \text{for any}$$

$$\nu_1, \cdots, \nu_k \quad \text{such that} \sum_{j\in J}\nu_j \geqq \frac{n}{2}-1.$$

The statement of the theorem follows immediately from this observation. Q.E.D.

The proofs of Theorem 2.6 and Theorem 2.8 are based on the following connection formula relating F_B near $x=0$ to F_A near $x=\infty$ (see Lauricella [4]).

$$\begin{aligned} &F_B(a_1, \cdots, a_k, b_1, \cdots, b_k, c, x_1, \cdots, x_k) \\ &\quad = \frac{\Gamma(c)\Gamma(b_1-a_1)\cdots\Gamma(b_k-c_k)}{\Gamma(c-a_1-\cdots-a_k)\Gamma(b_1)\cdots\Gamma(b_k)}(-x_1)^{-a_1}\cdots(-x_k)^{-a_k} \\ &\quad \times F_A(a_1+\cdots+a_k+1-c, a_1, \cdots, a_k, a_1+1-b_1, \cdots, \\ &\qquad\qquad a_k+1-b_k, x_1^{-1}, \cdots, x_k^{-1}) \\ &\quad +(2^k-1) \text{ other terms.} \end{aligned} \tag{3.12}$$

Here, the other terms are obtained by substituting the pairs (b_i, a_i), $i\in S$, for (a_i, b_i), $i\in S$, in the first term, where S runs over the non-empty subsets of $\{1, \cdots, k\}$.

The above formula is an extension of the formula of Gauss for the hypergeometric function of one variable;

$$\begin{aligned} F(a, b, c, x) &= \frac{\Gamma(c)\Gamma(b-a)}{\Gamma(b)\Gamma(c-a)}(-x)^{-a}F(a, 1+a-c, 1+a-b, x^{-1}) \\ &+ \frac{\Gamma(c)\Gamma(a-b)}{\Gamma(a)\Gamma(c-b)}(-x)^{-b}F(b, 1+b-c, 1+b-a, x^{-1}). \end{aligned}$$

The formula of Lauricella needs a small modification, as does the Gauss formula, when some of a_i-b_i, $i=1, \cdots, k$, are integers.

The details of how the above formula is applied to the proof of our theorems are routine and omitted in this note.

§ 4. Asymptotic behavior of solutions with compact support Cauchy data

In this section we shall give an application of our previous results to the following problem. Let S be a smooth, space-like hypersurface in $\boldsymbol{R}^n$ and let $u \in C^\infty(\boldsymbol{R}^n)$ be a solution of $Pu=0$ such that the Cauchy data of u in S are of compact support. Take a geodesic $z(r)$ in $D^+(x_0)$ starting from a point $x_0 \in S$.

The statement that,

$$\text{(4.1)} \qquad \begin{aligned} &\text{if} \quad u(z(r))=o(r^{-\infty}), \quad r \to \infty, \\ &\text{then} \quad u(z(r)) \equiv 0, \quad r \geqq 0, \end{aligned}$$

was proved in some cases. Littman-Lui [5] proved (4.1) for some of the homogeneous, constant coefficient, strictly hyperbolic operators when n is odd. The same problem was treated by Nishiwada-Tintarev [7] for operators of the type (1.2) with $a_{\alpha\beta}$ constant, $k=1$ and λ a certain integer.

Our results thus far obtained will enable us to prove (4.1) for a wider class of operators.

In the following theorem we consider the operator P of the type (1.2), where $a_{\alpha\beta}$ are assumed to be constant and L_j to be real valued linear functions in $\boldsymbol{R}^n$ such that ∇L_j is time-like or null for $j=1, \cdots, k$. It is also supposed that $D^+(y)$ lies in the future side S^+ of S for every $y \in S^+$.

Theorem 4.1. *Suppose that n is odd $\geqq 3$ and that $\lambda_j \neq 1/2$ for $j \in J$. Let $u \in C^\infty(\boldsymbol{R}^n)$ be a solution of $Pu=0$ in $\boldsymbol{R}^n$ such that the Cauchy data of u in S have the compact support K and that the set $\bigcup_{y \in K} D^+(y)$ has no common point with the hyperplanes $L_j(x)=0, j=1, \cdots, k$.*

a) *Then $u(z(r))$ is asymptotically*

$$u(z(r))=r^M(C+o(1)), \tag{4.2}$$

where M and C are constants.

b) *Furthermore, if $u(z(r))=o(r^{-\infty})$ in (4.2), then it follows that $u(z(r)) \equiv 0$, $r \geqq 0$.*

Remark 4.2. (4.2) is valid even when n is even, but the second statement no longer holds in this case, as is well illustrated by Huygens' principle. In this case supp u propagates along the bicharacteristics and therefore $u(z(r))$ vanishes for sufficiently large r but does not vanish in general up to $r=0$.

Proof. Take a smooth function φ such that $\varphi > 0$ in the future side of S and $\varphi < 0$ in the past side of S. With the Heaviside function H we have

$$P(H(\varphi)u) = u_0\delta(\varphi) + u_1\delta'(\varphi) = f, \quad \operatorname{supp} f \subset S,$$
$$u = \langle G(x, y), f(y) \rangle \qquad \text{for } \varphi(x) > 0.$$

The expression of G given in Prop. 3.3 and Lauricella's formula (3.12) show that $G(z(r), y)$, $y \in \operatorname{supp} f$, is analytic in a fixed region $R < |r| < \infty$, with at most a pole or branch point at $r = \infty$. (4.2) is therefore just the puiseaux expansion of $u(z(r)) = \langle G(x, y), f(y) \rangle|_{x=z(r)}$ at $r = \infty$.

For the same reason, if $u(z(r)) = o(r^{-\infty})$, $u(z(r))$ vanishes identically in a neighborhood of ∞, namely

$$u(z(r)) = 0, \qquad r > r_0 \tag{4.3}$$

for a certain r_0. Indeed, $u(r)$ vanishes until r, moving from ∞, hits the singular support of the distributions $G(x, y)$, $y \in \operatorname{supp} f$.

Note that the singular points supplied by F_B are all located in $(-\infty, 0)$ due to the assumption $|\nabla L_j|^2 \geqq 0$, $j = 1, \cdots, k$. Therefore in view of Prop. 3.3 the singular support of G is described by $c'\psi^+_{1-n/2}(\gamma)$ with $c' = c_n\kappa$, which, by (3.6), is the fundamental solution of $\square$.

We shall now change the coordinate system to the effect that

$$\square = \partial_1^2 - \sum_2^n \partial_j^2, \qquad \gamma(x, y) = (x_1 - y_1)^2 - \sum_2^n (x_j - y_j)^2$$

and that x_0 becomes the origin of $\boldsymbol{R}^n$.

Let $E(x, y)$ be the solution of the Cauchy problem

$$\square E(x, y) = 0,$$
$$E|_{x_1 = y_1} = 0, \qquad \partial_1 E|_{x_1 = y_1} = \delta(x' - y').$$

$E(x, y)$ is then related to $G(x, y)$ by the equality

$$H(x_1 - y_1)E(x, y) = c'\psi^+_{1-n/2}(\gamma),$$

so that

$$E(x, y) = c'\psi^+_{1-n/2}(\gamma) \qquad \text{if } y \notin \overline{D^+(x)}. \tag{4.4}$$

The expression of E in terms of Fourier integral operators is well known and, making a radial integration, we get

$$E(x, y)=\frac{1}{2}(2\pi i)^{-n+1}\int_{|\xi|=1} \chi_{2-n}(x_1-y_1+\langle x', \xi'\rangle+i0)$$
$$-\chi_{2-n}(-x_1+y_1+\langle x', \xi'\rangle+i0)\omega(\xi'),$$

where

$$\chi_{-q}(z)=\Gamma(q)e^{i\pi q}z^{-q}, \qquad q=1, 2, \cdots,$$

and

$$\omega(\xi')=\sum_{j=2}^{n} (-1)^j \xi_j d\xi_2\wedge\cdots\wedge d\xi_{j-1}\wedge d\xi_{j+1}\wedge\cdots\wedge d\xi_n.$$

Changing ξ' to $-\xi'$ in the second term of the integrand gives

$$E=\frac{1}{2}(2\pi i)^{-n+1}\int_{|\xi|=1} \chi_{2-n}(x_1-y_1+\langle x'-y', \xi'\rangle+i0) \tag{4.5}$$
$$-(-1)^n\chi_{2-n}(x_1-y_1+\langle x'-y', \xi'\rangle-i0)\omega(\xi').$$

Since $z(r)\in D^+(0)$, one can write $z(r)=r(\theta_1, \theta')$ with $\theta_1>|\theta'|$ and hence

$$x_1-y_1+\langle x'-y', \xi'\rangle=r(\theta_1+\langle\theta', \xi'\rangle)-y_1-\langle y', \xi'\rangle, \tag{4.6}$$

which is positive if r is large enough and $y\in \operatorname{supp} f$.

Since n is odd, along with the formulas (4.5) and (4.4), the expression of $G(x, y)$ given in Prop. 3.3 implies that one can write

$$u(z(r))=\langle G(z(r), y), f(y)\rangle$$
$$=g_1(r+i0)+g_2(r+i0), \qquad r>0,$$

where g_1 (resp. g_2) is holomorphic in the upper (resp. lower) half plane of $\boldsymbol{C}$. Moreover they satisfy

$$g_1(r+i0)=g_2(r-i0) \tag{4.7}$$

for large r so that (4.6) is positive. However, (4.3) and (4.7) combine to mean that both g_1 and g_2 vanish identically. Therefore, $u(z(r))=0$, $r>0$. Since u is C^∞, we have $u(z(r))=0$ also at $r=0$. Q.E.D.

References

[1] S. Delache et J. Leray, Calcul de la solution élementaire de l'opérateur d'Euler-Poisson-Darboux et de l'opérateur de Tricomi-Clairaut, hyperbolique, d'ordre 2. Bull. Soc. Math. France, **99** (1971), 313–336.
[2] F. G. Friedlander, The Wave Equation on a Curved Spacetime, Cambridge University Press, London, 1975.

[3] P. Günther, Ein Beispiel einer nichttrivialen Huygensschen Differentialgleichung mit vier unabhängigen Variablen, Arch. Rational Mech. Anal., **18** (1965), 103–106.
[4] G. Lauricella, Sulle funzioni ipergeometriche a piu variabli, Rend. Circ. Mat. Palermo, **7** (1893), 111–158.
[5] W. Littman and R. Lui, Asymptotic behavior of solutions to the wave equation and other hyperbolic equations in an even number of space dimension, Univ. Minnesota Math. Report, 81–107.
[6] R. G. McLenaghan, An explicit determination of the empty space-times on which the wave equation satisfies Huygens' principle, Proc. Camb. Phil. Soc., **65** (1969), 139–155.
[7] K. Nishiwada and K. Tintarev. Lacunas and saturation properties for hyperbolic differential operators with analytic coefficients, Israel J. Math., **49** (1984), 281–306.
[8] A. S. Petrow, Einstein-Räume. Akad. Verlag, Berlin, 1964 (Übersetzung aus dem Russischen).
[9] K. L. Stellmacher, Eine Klasse Huyghensscher Differentialgleichungen und ihre Integration, Math. Ann., **130** (1955), 219–233.

INSTITUTE OF MATHEMATICS
YOSHIDA COLLEGE
KYOTO UNIVERSITY
KYOTO 606, JAPAN

Taniguchi Symp. HERT
Katata 1984, pp. 273–306

Le Problème de Cauchy à Caractéristiques Multiples dans la Classe de Gevrey
—coefficients hölderiens en t—

Yujiro OHYA et Shigeo TARAMA

Dédié au Professeur S. Mizohata pour son soixantième anniversaire

Abstract

Let us consider the Cauchy problem for weakly hyperbolic operator $P(t, x; \partial/\partial t, \partial/\partial x)$ with κ-Hölder continuous coefficients in t and coefficients in $x \in \boldsymbol{R}^l$ of Gevrey class s. We construct a right parametrix for P under the assumption $1 < s < \min(1 + \kappa/r, r/(r-1))$ for $0 < \kappa \leqq 2$, where r is the maximal multiplicity of characteristic roots of P.

Introduction

1. *Historique*

Le problème de Cauchy dans la classe de Gevrey a été étudié par Y. Ohya [12], J. Leray-Y. Ohya $[8]_{1,2}$ sous l'hypothèse que le polynôme caractéristique ait des racines réelles et multiples de multiplicité constante pour les coefficients assez réguliers même par rapport au temps, en employant les inégalites énergétiques. M.D. Bronshtein $[1]_{1,2}$ a étudié la situation plus générale dans le cas de multiplicité variable, en construisant une paramétrice du problème de Cauchy avec des coefficients suffisamment réguliers par rapport à t. Il y a beaucoup de resultats dans le cas intermédiaire. On ne va citer que les articles de V. Ya Ivriĭ [6] et de J.-M. Trépreau [13]. Ils ont traité ce problème sous l'hypothèse que les coefficients de la partie principale soient analytiques en x. Plus récemment, on a des résultats dûs à F. Colombini-E. De Giorgi-S. Spagnolo [3] et à F. Colombini-E. Jannelli-S. Spagnolo [4] concernant des coefficients moins réguliers par rapport à t (voir aussi S. Spagnolo [15]). Le résultat plus proche de nôtre est des articles de T. Nishitani [9] [10].

2. *Notations*

Dans cet article, le problème de Cauchy pour les équations hyperboliques à caractéristiques multiples de multiplicité variable sera résolu dans la classe de fonctions de Gevrey; cette classe sera déterminée selon l'indice de continuité d'Hölder concernant la régularité par rapport au temps de coefficients et la multiplicité la plus haute.

Plus précisément, nous considérons le problème de Cauchy pour

$$\begin{cases} P(t, x; D_t, D_x)u(t, x)=f(t, x) \\ D_t^j u(0, x)=\phi_j(x), \quad j=0, 1, \cdots, m-1, \end{cases} \tag{2.1}$$

dans une bande $\Omega=[0, T]\times \boldsymbol{R}^l$, où $P(t, x; D_t, D_x)$ est un opérateur différentiel du type Kowalewskien;

$$P(t, x; D_t, D_x)=D_t^m+\sum_{\substack{j+|\nu|\leqq m \\ j<m}} a_{j\nu}(t, x)D_t^j D_x^\nu, \tag{2.2}$$

étant $D_t=(1/i)(\partial/\partial t)$ et $D_{x_j}=(1/i)(\partial/\partial x_j)$ $(1\leqq j\leqq l)$.

Avant d'énoncer notre résultat, nous allons donner quelques définitions.

Définition 2.1. On entend que le polynôme caractéristique

$$p_m(t, x; \tau, \xi)=0 \tag{2.3}$$

a des racines réelles par rapport à τ pour tout $(t, x; \xi)\in \Omega\times \boldsymbol{R}_\xi^l$ par l'hyperbolicité.

Définition 2.2. On désigne par $\gamma_{\text{loc}}^s(\boldsymbol{R}^l)$ $(s>1)$ l'ensemble de fonctions $g(x)$ indéfiniment dérivables telles que, pour tout ensemble compact K de $\boldsymbol{R}^l$, il existe deux constantes positives A et C telles que

$$\sup_K |D_x^\alpha g(x)|\leqq AC^{|\alpha|}|\alpha|!^s \tag{2.4}$$

pour n'importe quel multi-indice α.

Définition 2.3. On désigne par $C^\kappa([0, T]; \gamma_{\text{loc}}^s(\boldsymbol{R}^l))$ l'ensemble de fonctions $g(t, x)$ appartenantes à $\gamma_{\text{loc}}^s(\boldsymbol{R}^l)$ pour chaque $t\in[0, T]$ qui satisfont à l'inégalité, K étant un ensemble compact de $\boldsymbol{R}^l$,

$$\sup_{[0,T]\times K} |D_x^\alpha g(t, x)-D_x^\alpha g(s, x)|\leqq A|t-s|^\kappa C^{|\alpha|}|\alpha|!^s \tag{2.5}$$

avec des constantes A et C de (2.4), pour tout $t, s\in[0, T]$ et pour $0<\kappa\leqq 1$.

3. *Enoncé de résultat*

Dans la suite, nous considérons le cas où $0<\kappa\leqq 2$. Pour $1<\kappa\leqq 2$, nous entendons que $g(t, x)$ est une fois continûment dérivable par rapport à t, et que $D_t g(t, x)$ appartient à $C^{\kappa-1}([0, T]; \gamma_{\text{loc}}^s(\boldsymbol{R}^l))$.

Théorème. *Soit r la multiplicité la plus haute de racines caractéristiques. Si l'on suppose que, pour* $j+|\nu|=m$ *(resp.* $j+|\nu|<m$*),* $a_{j\nu}(t, x)$ *appartiennent à* $C^{\kappa}([0, T]; \gamma^s_{loc}(\boldsymbol{R}^l))$ *(resp.* $C^0([0, T]; \gamma^s_{loc}(\boldsymbol{R}^l))$*) bornés dans* Ω*, et que* $f(t, x) \in C^0([0, T]; \gamma^s_{loc}(\boldsymbol{R}^l))$*,* $\phi_j(x) \in \gamma^s_{loc}(\boldsymbol{R}^l)$ *pour* $j=0, 1, \cdots, m-1$*, alors le problème de Cauchy* (2.1) *hyperbolique est bien posé dans* $C^0([0, T]; \gamma^s_{loc}(\boldsymbol{R}^l))$ *pour* $1<s<\min(1+\kappa/r, r/(r-1))$ *où* $0<\kappa\leqq 2$.

§ 1. Construction de parametrice

4. *Propriété de polynômes hyperboliques*

La propriété suivante du ploynôme hyperbolique est bien connue; si l'on suppose que le polynôme caractéristique $p_m(t, x; \tau, \xi)$ d'opérateur différentiel hyperbolique $P(t, x; D_t, D_x)$ ait les racines caractéristiques de multiplicité au plus r, alors les racines de $p(t, x; \tau, \xi)=0$ par rapport à τ se situent en

$$(4.1)\qquad \begin{cases} |\tau|\leqq C(1+|\xi|) \\ |I_m\tau|\leqq C(1+|\xi|)^{(r-1)/r} \end{cases}$$

dans le plan complexe, C étant une constante positive.

En introduisant une notation par

(4.2) $\langle\xi\rangle=(1+|\xi|^2)^{1/2}$, on emploie, C_i $(i=1, 2)$ étant des constantes,

$$(4.1)'\qquad \begin{cases} |\tau|\leqq C_1\langle\xi\rangle \\ |I_m\tau|\leqq C_2\langle\xi\rangle^{(r-1)/r} \end{cases}$$

à la place de (4.1) dans la suite.

Compte tenu de cette propriété, on va employer la transformation de Fourier par

$$(4.3)\qquad \hat{f}(\tau-iH\langle\xi\rangle^\delta, \xi)=(2\pi)^{-(l+1)/2}\int_{-\infty}^{\infty}\int_{\boldsymbol{R}^l} e^{-it(\tau-iH\langle\xi\rangle^\delta)-i\langle x,\xi\rangle}f(t, x)dt\,dx$$

pour $f(t, x) \in C_0^\infty(\boldsymbol{R}^{l+1})$ ou $\mathscr{S}(\boldsymbol{R}^{l+1})$, où H est une constante positive et que δ $(0<\delta<1$ tel que $s\delta=1)$ sera déterminé tout à l'heure. Aussi, la transformation inverse est donnée par

$$(4.4)\qquad f(t, x)=(2\pi)^{-(l+1)/2}\iint e^{it(\tau-iH\langle\xi\rangle^\delta)+i\langle x,\xi\rangle}\hat{f}(\tau-iH\langle\xi\rangle^\delta, \xi)d\tau\, d\xi.$$

5. *Plan de démonstration*

Pour résoudre le problème de Cauchy (2.1), nous nous appuyons essentiellement sur la méthode due à M. D. Bronshtein [1]. Evidemment, il nous faut quelques modifications à faire, puisqu'il nous manque de la régularité par rapport à t.

Note. Pour ne faire la démonstration pas trop compliquée, on suppose que $a_{j\nu}(t, x)\,(j+|\nu|=m)$ soient constants pour $|x|$ assez grand. Voici le déssin de notre démonstration. Tout d'abord, pour construire une paramétrice à droite du problème de Cauchy (2.1), nous allons introduire un opéraetur pseudo-différentiel Q.

Définition 5.1. Le symbole $q(t, x; \tau-iH\langle\xi\rangle^\delta, \xi)$ d'opérateur pseudo-différentiel $Q(t, x; D_t, D_x)$ est défini par

$$q(t, x; \tau-iH\langle\xi\rangle^\delta, \xi)=\int_{-\infty}^{\infty} \frac{\rho\chi(\rho(t-s))}{p_m(s, x; \tau-iH\langle\xi\rangle^\delta, \xi)} ds \tag{5.1}$$

où $\chi(s)$ est une fonction indéfiniment dérivable à support compact tell que

$$\chi(s)=\begin{cases} 1 & \frac{1}{2}\leqq s\leqq 1 \\ 0 & s\leqq\frac{1}{3},\ s\geqq\frac{4}{3}, \end{cases} \tag{5.2}$$

et que $\int_{-\infty}^{\infty}\chi(s)ds=1$, ρ étant choisi ultérieurement ne dépendant que de ξ.

Note. Il est évident que l'on a

$$\int_{-\infty}^{\infty}\rho\chi(\rho s)ds=1. \tag{5.3}$$

En posant $u=Qv$, on met (2.1) en $P[Qv]=f$.

Définition 5.2. On définit un opérateur preudo-différentiel $R(t, x; D_t, D_x)$ par

(5.4) $PQ=I+R$, I étant un opérateur d'identité.

Le calcul élémentaire des opérateurs pseudo-différentiels implique

$$r(t, x; \tau-iH\langle\xi\rangle^\delta, \xi)$$
$$=\sum\frac{1}{\alpha!}\partial_\xi^\alpha p(t, x; \tau-iH\langle\xi\rangle^\delta, \xi)D_x^\alpha q(t, x; \tau-iH\langle\xi\rangle^\delta, \xi)-1;$$

donc R s'écrit par

$$\begin{aligned} &R(t, x; D_t, D_x)v(t, x) \\ &=(2\pi)^{-(l+1)/2}\iint e^{it(\tau-iH\langle\xi\rangle^\delta)+i\langle x,\xi\rangle}r(t, x; \tau-iH\langle\xi\rangle^\delta, \xi) \\ &\quad\times\hat{v}(\tau-iH\langle\xi\rangle^\delta, \xi)d\tau\, d\xi. \end{aligned} \tag{5.5}$$

En notant

(5.6) $$\tilde{v}(t, \xi)=(2\pi)^{-l/2}\int e^{-i\langle x, \xi\rangle}v(t, x)\,dx,$$

on aura

$$\hat{v}(\tau-iH\langle\xi\rangle^{\delta}, \xi)=\frac{1}{\sqrt{2\pi}}\int e^{-it(\tau-iH\langle\xi\rangle^{\delta})}\tilde{v}(t, \xi)\,dt;$$

donc (5.5) s'écrit encore

$$Rv=(2\pi)^{-(l+1)/2}\iint e^{(t-t_1)H\langle\xi\rangle^{\delta}+i\langle x, \xi\rangle}\tilde{v}(t_1, \xi)\frac{1}{\sqrt{2\pi}}\int e^{i(t-t_1)\tau}r\,d\tau\,dt_1 d\xi,$$

par l'emploie du théorème de Fubini.

Si l'on désigne encore

(5.7) $$\tilde{r}(t, x; t-t_1, \xi)=\frac{1}{\sqrt{2\pi}}\int e^{i(t-t_1)\tau}r\,d\tau$$

et

(5.8) $$\tilde{R}(t, x; t-t_1)v(t_1, x)=(2\pi)^{-l/2}\int e^{i\langle x, \xi\rangle}e^{(t-t_1)H\langle\xi\rangle^{\delta}}\tilde{r}\tilde{v}\,d\xi,$$

alors on aura

(5.9) $$Rv=\frac{1}{\sqrt{2\pi}}\int_0^t \tilde{R}(t, x; t-t_1)v(t_1, x)\,dt_1,$$

qui est valable pour $0\leqq t_1\leqq t$.

Note. En remarquant pour R assez grand

$$\int_{-R}^{R} e^{i(t-t_1)\tau}r(t, x; \tau-iH\langle\xi\rangle^{\delta}, \xi)\,d\tau=\int_{-R-i\alpha}^{R-i\alpha} e^{i(t-t_1)\tau}r\,d\tau \quad (\alpha>0)$$

et (19.5), on constate que

(5.7)′ $$\tilde{r}=\frac{1}{\sqrt{2\pi}}\int_0^t e^{i(t-t_1)\tau}r(t, x; \tau-iH\langle\xi\rangle^{\delta}, \xi)\,d\tau.$$

Le raisonnement ci-dessus nous amène (2.1) à

(5.10) $$v+\frac{1}{\sqrt{2\pi}}\int_0^t \tilde{R}(t, x; t-t_1)v(t_1, x)\,dt_1=f.$$

6. *Une propriété de fonctions de la classe de Gevrey*

Supposons que, pour tout $t \in [0, T]$, $v(t, x)$ appartienne à $\gamma^s_{\mathrm{loc}}(\boldsymbol{R}^l)$ dont le support en x est un compact K;

$$\sup_K |D_x^\alpha v(t, x)| \leqq AC^{|\alpha|}|\alpha|!^s \tag{6.1}$$

où A et C sont des constantes positives.

Alors, (6.1) se caractérise de sa transformée de Fourier, B et K_B existant,

$$|\tilde{v}(t, \xi)| \leqq K_B\, e^{-B|\xi|^\delta} \tag{6.2}$$

où l'on a désigné $s = 1/\delta$ (voir Lemme 5.7.2 de L. Hörmander [6]).

Si l'on applique $e^{(h-tH)\langle D\rangle^\delta}$ à $v(t, x)$, où h est une autre constante positive, alors on constate que

$$e^{(h-tH)\langle D\rangle^\delta} v(t, x) = w(t, x) \in L^2(\boldsymbol{R}^l) \text{ pour tout } t \in [0, h/2H) \tag{6.3}$$

entraîne l'appartenance de $v(t, x)$ à la classe de Gevrey.

En effet, en employant le Lemme de Sobolev où $m > l/2$,

$$\begin{aligned} |v(t, x)| &\leqq \text{Const. } \|\langle\xi\rangle^m \tilde{v}(t, \xi)\|_{L^2(\boldsymbol{R}^l_\xi)} \\ &\leqq \text{Const. } \sup_\xi |\langle\xi\rangle^m e^{-(h-tH)\langle\xi\rangle^\delta}| \|\tilde{w}(t, \xi)\|_{L^2(\boldsymbol{R}^l_\xi)} \\ &\leqq \text{Const. } \|w(t, x)\|_{L^2(\boldsymbol{R}^l_x)}, \end{aligned}$$

pour chaque $t \in [0, h/2H)$, vu la formule de Plancherel.

7. *Plan de démonstration$_{bis}$*

En retournant à (5.10), on pose comme (6.3); alors on aura

$$\begin{aligned} &w + e^{(h-tH)\langle D\rangle^\delta} \frac{1}{\sqrt{2\pi}} \int_0^t \tilde{R}(t, x; t-t_1) e^{-(h-t_1H)\langle D\rangle^\delta} w(t_1, x) dt_1 \\ &= e^{(h-tH)\langle D\rangle^\delta} f. \end{aligned} \tag{7.1}$$

Le second terme du premier membre de (7.1) s'interpréte comme il suit; compte tenu de (5.8),

$$\begin{aligned} &\tilde{R}(t, x; t-t_1) e^{-(h-t_1H)\langle D\rangle^\delta} w(t_1, x) \\ &= (2\pi)^{-l/2} \int e^{i\langle x, \xi\rangle} e^{(t-t_1)H\langle\xi\rangle^\delta} \tilde{r}(t, x; t-t_1, \xi) e^{-(h-t_1H)\langle\xi\rangle^\delta} \tilde{w}(t_1, \xi) d\xi \\ &= (2\pi)^{-l/2} \int e^{i\langle x, \xi\rangle} e^{-(h-tH)\langle\xi\rangle^\delta} \tilde{r}(t, x; t-t_1, \xi) \tilde{w}(t_1, \xi) d\xi. \end{aligned} \tag{7.2}$$

Donc, (7.1) s'écrit encore

$$(7.3)\quad \begin{aligned} &w+e^{(h-tH)\langle D\rangle^\delta}\frac{1}{\sqrt{2\pi}}\int_0^t(2\pi)^{-l/2}\int_{R^l}e^{i\langle x,\xi\rangle}e^{-(h-tH)\langle\xi\rangle^\delta}\tilde{r}\tilde{w}\,d\xi\,dt_1\\ &=e^{(h-tH)\langle D\rangle^\delta}f.\end{aligned}$$

En définissant

$$(7.4)\quad \begin{aligned} &\sqrt{2\pi}\,\boldsymbol{R}(t,x;t-t_1)w(t_1,x)\\ &=e^{(h-tH)\langle D\rangle^\delta}(2\pi)^{-l/2}\int_{R^l}e^{i\langle x,\xi\rangle}e^{-(h-tH)\langle\xi\rangle^\delta}\tilde{r}\tilde{w}\,d\xi,\end{aligned}$$

on aura finalement

$$(7.5)\quad w+\int_0^t\boldsymbol{R}(t,x;t-t_1)w(t_1,x)dt_1=e^{(h-tH)\langle D\rangle^\delta}f.$$

Donc, pour montrer l'existence d'une paramétrice à droite du problème de Cauchy (2.1), il nous suffit de prouver la résolubilité de l'équation intégrale de Volterra (7.4) dans l'espace de $C^0([0,T];L^2(\boldsymbol{R}^l))$, compte tenu de l'interprête du n°6.

§ 2. Calcul symbolique pour $0<\kappa\leqq 1$

8. *Calcul précis*

La définition habituelle d'opérateur pseudo-différentiel associé à la Définition 5.1 implique

$$(8.1)\quad \begin{aligned} &Q(t,x;D_t,D_x)f(t,x)\\ &=(2\pi)^{-(l+1)/2}\iint e^{it(\tau-iH\langle\xi\rangle^\delta)}q(t,x;\tau-iH\langle\xi\rangle^\delta,\xi)\hat{f}(\tau-iH\langle\xi\rangle^\delta,\xi)d\tau d\xi\end{aligned}$$

pour $f(t,x)\in C_0^\infty(\boldsymbol{R}^{l+1})$.

Note. Il est évident que $q(t,x;\tau-iH\langle\xi\rangle^\delta,\xi)$ est bien défini; puisque le polynôme caractéristique $p_m(t,x;\tau,\xi)$ n'a que des racines réelles. De plus, nous sommes autorisé de développer $r(t,x;\tau-iH\langle\xi\rangle^\delta,\xi)$ par le second membre dans la Définition 5.2, car q est suffisamment régulier même par rapport à t grâce à (5.1).

Le second membre de r s'écrit encore

$$(8.2)\quad \begin{aligned} &\Big(\sum_{k=0}^m p_k\Big)q+\frac{\partial}{\partial\tau}\Big(\sum_{k=0}^m p_k\Big)D_tq+\sum_{\alpha=1}^l\frac{\partial}{\partial x_\alpha}\Big(\sum_{k=0}^m p_k\Big)D_{x_\alpha}q\\ &+\sum_{|\alpha|\geqq 2}\frac{1}{\alpha!}\partial_\xi^\alpha\Big(\sum_{k=0}^m p_k\Big)D_x^\alpha q-1,\end{aligned}$$

où $p=\sum_{k=0}^{m} p_k(t, x; \tau, \xi)$, $p_k(t, x; \tau, \xi)$ étant homogène de degré k en (τ, ξ).

9. *Allure de* $\left(\sum_{k=0}^{m} p_k\right)q-1$ *par rapport à* $(\rho, \langle\xi\rangle)$

On décompose $\left(\sum_{k=0}^{m} p_k\right)q-1$ à

$$\int_{-\infty}^{\infty} \frac{p_m(t, x; \tau-iH\langle\xi\rangle^\delta, \xi)-p_m(s, x; \tau-iH\langle\xi\rangle^\delta, \xi)}{p_m(s, x; -iH\langle\xi\rangle^\delta, \xi)} \rho\chi(\rho(t-s))ds$$

$$(9.1) \qquad +\int_{-\infty}^{\infty} \frac{p_m(s, x; \tau-iH\langle\xi\rangle^\delta, \xi)}{p_m(s, x; \tau-iH\langle\xi\rangle^\delta, \xi)} \rho\chi(\rho(t-s))ds$$

$$+\sum_{k=0}^{m-1}\int_{-\infty}^{\infty} \frac{p_k(t, x; \tau-iH\langle\xi\rangle^\delta, \xi)}{p_m(s, x; \tau-iH\langle\xi\rangle^\delta, \xi)} \rho\chi(\rho(t-s))ds-1.$$

Pour majorer le premier terme de (9.1), on emploie le

Lemme 9.1. *Il existe une constante positive C telle que*

$$(9.2) \quad |p_m(s, x; \tau-iH\langle\xi\rangle^\delta, \xi)| \geqq C(H\langle\xi\rangle^\delta)^r(|\tau-iH\langle\xi\rangle^\delta|^2+|\xi|^2)^{(m-r)/2}$$

pour tout $(s, x; \xi) \in \Omega\times \boldsymbol{R}^l_\xi$.

Or, le premier terme se majore en module par

$$C\frac{(|\tau-iH\langle\xi\rangle^\delta|^2+|\xi|^2)^{r/2}}{(H\langle\xi\rangle^\delta)^r}\int_{-\infty}^{\infty} |t-s|^\kappa \rho\chi(\rho(t-s))ds,$$

compte tenu de l'inégalité évidente

$$(9.3) \quad \begin{aligned} &|p_m(t, x; \tau-iH\langle\xi\rangle^\delta, \xi)-p_m(s, x; \tau-iH\langle\xi\rangle^\delta, \xi)| \\ &\leqq C|t-s|^\kappa(|\tau-iH\langle\xi\rangle^\delta|^2+|\xi|^2)^{m/2}\end{aligned}$$

qui s'ensuit de l'hypothèse de $a_{j\nu}(t, x)$ $(j+|\nu|=m)$ appartient à C^κ par rapport à t. Aussi, le troisième terme de (9.1) se majore par

$$\sum_{k=0}^{m-1} C\frac{(|\tau-iH\langle\xi\rangle^\delta|^2+|\xi|^2)^{(k-m+r)/2}}{(H\langle\xi\rangle^\delta)^r}\int_{-\infty}^{\infty} \rho\chi(\rho(t-s))ds.$$

En somme, on obtient

$$(9.4) \qquad |(9.1)| \leqq C\rho^{-\kappa}\langle\xi\rangle^{r-r\delta}+C\sum_{k=0}^{m-1}\langle\xi\rangle^{k-m+r-r\delta}.$$

10. *Allure de* $\dfrac{\partial}{\partial\tau}\left(\sum_{k=0}^{m} p_k\right)D_t q$ *par rapport à* $(\rho, \langle\xi\rangle)$

Ils se décomposent à

$$\int_{-\infty}^{\infty} \frac{\frac{\partial p_m}{\partial \tau}(t, x; \tau - iH\langle\xi\rangle^\delta, \xi) - \frac{\partial p_m}{\partial \tau}(s, x; \tau - iH\langle\xi\rangle^\delta, \xi)}{p_m(s, x; \tau - iH\langle\xi\rangle^\delta, \xi)} \rho^2\chi'(\rho(t-s))ds$$

$$(10.1) \qquad + \int_{-\infty}^{\infty} \frac{\frac{\partial p_m}{\partial \tau}(s, x; \tau - iH\langle\xi\rangle^\delta, \xi)}{p_m(s, x; \tau - iH\langle\xi\rangle^\delta, \xi)} \rho^2\chi'(\rho(t-s))ds$$

$$+ \sum_{k=0}^{m-1} \int_{-\infty}^{\infty} \frac{\frac{\partial p_k}{\partial \tau}(t, x; \tau - iH\langle\xi\rangle^\delta, \xi)}{p_m(s, x; \tau - iH\langle\xi\rangle^\delta, \xi)} \rho^2\chi'(\rho(t-s))ds,$$

où $\chi'(s) = \frac{d\chi}{ds}(s)$.

Le premier terme de (10.1) est majoré en module par

$$(10.2) \qquad C\langle\xi\rangle^{m-1-(m-r)}(H\langle\xi\rangle^\delta)^{-r} \left| \int_{-\infty}^{\infty} |t-s|^\kappa \rho^2\chi'(\rho(t-s))ds \right|,$$

puisque l'on a

$$\left| \frac{\partial p_m}{\partial \tau}(t, x; \tau - iH\langle\xi\rangle^\delta, \xi) - \frac{\partial p_m}{\partial \tau}(s, x; \tau - iH\langle\xi\rangle^\delta, \xi) \right|$$
$$\leqq C|t-s|^\kappa(|\tau - iH\langle\xi\rangle^\delta|^2 + |\xi|^2)^{(m-1)/2}$$

avec l'emploie du Lemme 6.1, et que

$$(10.3) \qquad \left| \int_{-\infty}^{\infty} |t-s|^\kappa \rho^2\chi'(\rho(t-s))ds \right| \leqq \text{const. } \rho^{1-\kappa}.$$

Pour estimer le second terme, on emploie l'identité

$$\frac{1}{p_m(s, x; \tau - iH\langle\xi\rangle^\delta, \xi)} \frac{\partial p_m}{\partial \tau}(s, x; \tau - iH\langle\xi\rangle^\delta, \xi)$$
$$= \sum_{j=1}^{m} \frac{1}{\tau - iH\langle\xi\rangle^\delta - \lambda_j(s, x; \xi)},$$

compte tenu de

$$p_m(s, x; \tau - iH\langle\xi\rangle^\delta, \xi) = \prod_{j=1}^{m} (\tau - iH\langle\xi\rangle^\delta - \lambda_j(s, x; \xi));$$

donc, il est plus petit que

$$(10.4) \qquad C\rho\langle\xi\rangle^{-\delta}.$$

En somme, on obtiendra

$$(10.5)\qquad |(10.1)|\leqq C\langle\xi\rangle^{r-1-r\delta}\rho^{1-\kappa}+C\rho\langle\xi\rangle^{-\delta}+C\rho\sum_{k=0}^{m-1}\langle\xi\rangle^{k-m+r-1-r\delta},$$

en employant l'inégalité

$$(10.6)\qquad \left|\frac{\partial p_k}{\partial\tau}(t, x; \tau-iH\langle\xi\rangle^{\delta}, \xi)\right|\leqq C(|\tau-iH\langle\xi\rangle^{\delta}|^2+|\xi|^2)^{(k-1)/2}.$$

11. *Allure de* $\dfrac{\partial}{\partial\xi_\alpha}\left(\sum_{k=0}^{m}p_k\right)D_{x_\alpha}q$ $(1\leqq\alpha\leqq l)$ *par rapport à* $(\rho, \langle\xi\rangle)$

On les décompose à

$$(11.1)\qquad \begin{aligned}&\int_{-\infty}^{\infty}\left\{\frac{\partial p_m}{\partial\xi_\alpha}(t, x; \tau-iH\langle\xi\rangle^{\delta}, \xi)-\frac{\partial p_m}{\partial\xi_\alpha}(s, x; \tau-iH\langle\xi\rangle^{\delta}, \xi)\right\}\\ &\qquad\times D_{x_\alpha}\left(\frac{1}{p_m(s, x; \tau-iH\langle\xi\rangle^{\delta}, \xi)}\right)\rho\chi(\rho(t-s))ds\\ &+\int_{-\infty}^{\infty}\frac{\partial p_m}{\partial\xi_\alpha}(s, x; \tau-iH\langle\xi\rangle^{\delta}, \xi)\\ &\qquad\times D_{x_\alpha}\left(\frac{1}{p_m(s, x; \tau-iH\langle\xi\rangle^{\delta}, \xi)}\right)\rho\chi(\rho(t-s))ds\\ &+\sum_{k=0}^{m-1}\int_{-\infty}^{\infty}\frac{\partial p_k}{\partial\xi_\alpha}(t, x; \tau-iH\langle\xi\rangle^{\delta}, \xi)\\ &\qquad\times D_{x_\alpha}\left(\frac{1}{p_m(s, x; \tau-iH\langle\xi\rangle^{\delta}, \xi)}\right)\rho\chi(\rho(t-s))ds.\end{aligned}$$

Si l'on emploie l'inégalité (voir aussi Lemme 12.1, ci-après)

$$(11.2)\qquad \left|D_{x_\alpha}\left(\frac{1}{p_m(s, x; \tau-iH\langle\xi\rangle^{\delta}, \xi)}\right)\right|\leqq\frac{C}{|p_m(s, x; \tau-iH\langle\xi\rangle^{\delta}, \xi)|}\frac{|\xi|}{H\langle\xi\rangle^{\delta}},$$

compte tenu de l'identité

$$(11.3)\qquad \begin{aligned}&D_{x_\alpha}\left(\frac{1}{p_m(s, x; \tau-iH\langle\xi\rangle^{\delta}, \xi)}\right)\\ &\quad=\frac{1}{p_m(s, x; \tau-iH\langle\xi\rangle^{\delta}, \xi)}\sum_{j=1}^{m}\frac{D_{x_\alpha}\lambda_j(s, x; \xi)}{\tau-iH\langle\xi\rangle^{\delta}-\lambda_j(s, x; \xi)},\end{aligned}$$

alors on aura la majoration pour le premier terme de (11.1)

$$(11.4)\qquad C\int_{-\infty}^{\infty}|t-s|^{\kappa}\frac{\langle\xi\rangle^{r-1}}{(H\langle\xi\rangle^{\delta})^{r}}\langle\xi\rangle^{1-\delta}\rho\chi((t-s))ds\leqq C\rho^{-\kappa}\langle\xi\rangle^{r-r\delta-\delta},$$

vu

$$\begin{aligned}(11.5)\quad &\frac{1}{|p_m(s,x;\tau-iH\langle\xi\rangle^\delta,\xi)|}\left|\frac{\partial p_m}{\partial\xi_\alpha}(t,x;\tau-iH\langle\xi\rangle^\delta,\xi)\right.\\ &\left.-\frac{\partial p_m}{\partial\xi_\alpha}(s,x;\tau-iH\langle\xi\rangle^\delta,\xi)\right|\leqq\frac{C|t-s|^\kappa\langle\xi\rangle^{r-1}}{(H\langle\xi\rangle^\delta)^r}.\end{aligned}$$

Le second terme se majore par

(11.6) $C\langle\xi\rangle^{r-\delta-r\delta}$, compte tenu de

$$\left|\frac{(\partial p_m/\partial\xi_\alpha)(s,x;\tau-iH\langle\xi\rangle^\delta,\xi)}{p_m(s,x;\tau-iH\langle\xi\rangle^\delta,\xi)}\right|\leqq C\frac{\langle\xi\rangle^{r-1}}{(H\langle\xi\rangle^\delta)^r}.$$

En somme, on obtiendra

$$(11.7)\quad |(11.1)|\leqq C\rho^{-\kappa}\langle\xi\rangle^{r-r\delta-\delta}+C\langle\xi\rangle^{r-\delta-r\delta}+C\sum_{k=0}^{m-1}\langle\xi\rangle^{k-m+r-r\delta-\delta}.$$

12. *Allure de* $\sum_{|\alpha|\geqq 2}\frac{1}{\alpha!}\partial_\xi^\alpha\left(\sum_{k=0}^{m}p_k\right)D_x^\alpha q$ *par rapport à* $(\rho,\langle\xi\rangle)$

On ne considère que pour $k=m$, et on entend $\alpha=(\alpha_0,\alpha)$ pour distinguer les dérivées par rapport à (t,τ) et (x,ξ); c'est-à-dire,

$$\begin{aligned}&\sum_{\alpha_0+|\alpha|\geqq 2}\frac{1}{\alpha_0!\alpha!}\partial_\tau^{\alpha_0}\partial_\xi^\alpha p_m D_t^{\alpha_0}D_x^\alpha q\\ &=\sum_{\alpha_0+|\alpha|\geqq 2}\Bigg[\int_{-\infty}^{\infty}\{\partial_\tau^{\alpha_0}\partial_\xi^\alpha(p_m(t,x;\tau-iH\langle\xi\rangle^\delta,\xi)\\ &\qquad -p_m(s,x;\tau-iH\langle\xi\rangle^\delta,\xi)\}\\ (12.1)\quad &\qquad\times\rho^{\alpha_0+1}\chi^{(\alpha_0)}(\rho(t-s))D_x^\alpha\left(\frac{1}{p_m(s,x;\tau-iH\langle\xi\rangle^\delta,\xi)}\right)ds\\ &+\int_{-\infty}^{\infty}\partial_\tau^{\alpha_0}\partial_\xi^\alpha p_m(s,x;\tau-iH\langle\xi\rangle^\delta,\xi)\\ &\qquad\times\rho^{\alpha_0+1}\chi^{(\alpha_0)}(\rho(t-s))D_x^\alpha\left(\frac{1}{p_m(s,x;\tau-iH\langle\xi\rangle^\delta,\xi)}\right)ds\Bigg].\end{aligned}$$

D'abord, on a

$$\begin{aligned}(12.2)\quad &|\partial_\tau^{\alpha_0}\partial_\xi^\alpha(p_m(t,x;\tau-iH\langle\xi\rangle^\delta,\xi)-p_m(s,x;\tau-iH\langle\xi\rangle^\delta,\xi)|\\ &\leqq C|t-s|^\kappa(|\tau-iH\langle\xi\rangle^\delta|^2+|\xi|^2)^{(m-\alpha_0-|\alpha|)/2},\end{aligned}$$

à cause d'homogénéité.

Ensuite, si l'on emploie le

Lemme 12.1 (Prop. 4. de M.D. Bronshtein $[1]_2$). *Il existe des con-*

stantes positives A et C telles que

$$\left|D_x^\alpha\left(\frac{1}{p_m(s,x;\tau-iH\langle\xi\rangle^\delta,\xi)}\right)\right| \leqq AC^{|\alpha|}\frac{|\alpha|!^s}{|p_m(s,x;\tau-iH\langle\xi\rangle^\delta,\xi)|}\sum_{j=0}^{|\alpha|}\frac{(\langle\xi\rangle^{1-\delta}/H)^j}{j!^{s-1}}, \tag{12.3}$$

pour tout $t\in[0,T]$, $x\in K$ *(ensemble compact* $\boldsymbol{R}^l$*) et* $(\tau,\xi)\in\boldsymbol{R}^{l+1}$, *sous l'hypothèse* $a_{j\nu}(t,x)(j+|\nu|=m)\in C^\kappa([0,T];\gamma^s_{\mathrm{loc}}(\boldsymbol{R}^l))$,

alors on aura

$$\begin{aligned}|(12.3)|\leqq AC^{|\alpha|}|\alpha|!^s(H\langle\xi\rangle^\delta)^{-r}\\ \times(|\tau-iH\langle\xi\rangle^\delta|^2+|\xi|^2)^{-(m-r)/2}\sum_{j=0}^{|\alpha|}\frac{(\langle\xi\rangle^{1-\delta}/H)^j}{j!^{s-1}}.\end{aligned} \tag{12.4}$$

Or, le premier terme se majore par

$$\begin{aligned}&C\sum_{\alpha_0+|\alpha|\geqq2}|\alpha|!^s\langle\xi\rangle^{r-\alpha_0-|\alpha|-r\delta}\sum_{j=0}^{|\alpha|}\frac{(\langle\xi\rangle^{1-\delta}/H)^j}{j!^{s-1}}\\ &\qquad\times\int_{-\infty}^{\infty}\rho^{\alpha_0+1}|\chi^{(\alpha_0)}(\rho(t-s))|\,|t-s|^\kappa ds\\ &=C\sum_{\alpha_0+|\alpha|\geqq2}|\alpha|!^s\rho^{\alpha_0-\kappa}\langle\xi\rangle^{r-\alpha_0-|\alpha|-r\delta}\sum_{j=0}^{|\alpha|}\frac{(\langle\xi\rangle^{1-\delta}/H)^j}{j!^{s-1}}.\end{aligned} \tag{12.5}$$

On obtiendra la majoration pour le second terme

$$C\sum_{\alpha_0+|\alpha|\geqq2}|\alpha|!^s\rho^{\alpha_0}\langle\xi\rangle^{r-\alpha_0-|\alpha|-r\delta}\sum_{j=0}^{|\alpha|}\frac{(\langle\xi\rangle^{1-\delta}/H)^j}{j!^{s-1}}, \tag{12.6}$$

en employant l'inégalité évidente

$$\left|\frac{\partial_\tau^{\alpha_0}\partial_\xi^\alpha p_m(s,x;\tau-iH\langle\xi\rangle^\delta,\xi)}{p_m(s,x;\tau-iH\langle\xi\rangle^\delta,\xi)}\right|\leqq C\langle\xi\rangle^{r-\alpha_0-|\alpha|-r\delta}, \tag{12.7}$$

pour tout $(t,x;\xi)\in\Omega\times\boldsymbol{R}^l_\xi$ et tout $\tau\in\boldsymbol{R}$.

13. *Choix de* δ *pour* $0<\kappa\leqq1$

Nous allons choisir ρ comme fonction de $\langle\xi\rangle$;

$$\rho=\langle\xi\rangle^{\delta_1}\quad\text{pour}\quad 0<\delta_1<1. \tag{13.1}$$

Pour définir bien $r(t,x;\tau-iH\langle\xi\rangle^\delta,\xi)$, on va vérifier que la puissance de r par rapport à $\langle\xi\rangle$ est négative.

(9.4) donne des conditions

$$(13.2)\qquad \begin{cases} -\kappa\delta_1+r-r\delta<0, & \text{et} \\ -1+r-r\delta<0; \end{cases}$$

(10.5) donne des conditions

$$(13.3)\qquad \begin{cases} (1-\kappa)\delta_1+r-1-r\delta<0, \\ \delta_1-\delta<0, \text{ et} \\ \delta_1-1+r-1-r\delta<0; \end{cases}$$

(11.7) donne des conditions

$$(13.4)\qquad \begin{cases} -\kappa\delta_1+r-r\delta-\delta<0, \\ r-\delta-r\delta<0, \quad \text{et} \\ -1+r-r\delta-\delta<0; \end{cases}$$

(12.5) et (12.6) donnent

$$(13.5)\qquad \begin{cases} (\alpha_0-\kappa)\delta_1+r-\alpha_0-|\alpha|-r\delta+(1-\delta)|\alpha|<0, \quad \text{et} \\ \alpha_0\delta_1+r-\alpha_0-|\alpha|-r\delta+(1-\delta)|\alpha|<0, \text{ pour } \alpha_0+|\alpha|\geqq 2: \end{cases}$$

Après avoir comparé des inégalités ci-dessus, on extraira les trois inégalités indépendantes,

$$(13.6)\qquad \frac{r-r\delta}{\kappa}<\delta_1$$

$$(13.7)\qquad \delta_1<\delta$$

(13.8) $\quad r-r\delta-\delta<0,\quad$ compte tenu de $0<\delta<1$ et $r\geqq 2$.

Or, on obtiendra

(13.9) $\quad 1<\dfrac{1}{\delta}<1+\dfrac{\kappa}{r}\quad$ de (13.6) et (13.7);

(13.8) donne $1/\delta<1+1/r$ qui est contenu dans (13.9), car $0<\kappa\leqq 1$.

§ 3. Calcul symbolique pour $1<\kappa\leqq 2$

14. *Allure de* $\left(\sum_{k=0}^{m} p_k\right)q-1$ *par rapport à* $(\rho, \langle\xi\rangle)$

Remarquons d'abord que

$$\begin{aligned} p_m(t, x; \tau-iH\langle\xi\rangle^\delta, \xi) &= p_m(t, x; \tau-iH\langle\xi\rangle^\delta, \xi)-p_m(s, x; \tau-iH\langle\xi\rangle^\delta, \xi) \\ &\quad +p_m(s, x; \tau-iH\langle\xi\rangle^\delta, \xi) \\ &= p'_m(s+\theta(t-s), x; \tau-iH\langle\xi\rangle^\delta, \xi)(t-s) \\ &\quad +p_m(s, x; \tau-iH\langle\xi\rangle^\delta, \xi), \end{aligned}$$

en désignant la dérivée de p_m par rapport à t par p'_m, et θ est un nombre tel que $0<\theta<1$. Nous les mettons encore en

$$\begin{aligned}(14.1)\quad &\{p'_m(s+\theta(t-s), x; \tau-iH\langle\xi\rangle^\delta, \xi)-p'_m(s, x; \tau-iH\langle\xi\rangle^\delta, \xi)\}(t-s)\\ &+p'_m(s, x; \tau-iH\langle\xi\rangle^\delta, \xi)(t-s)+p_m(s, x; \tau-iH\langle\xi\rangle^\delta, \xi).\end{aligned}$$

Alors, nous aurons

$$\begin{aligned}(14.2)\quad \left|\left(\sum_{k=0}^{m} p_k\right)q-1\right| &\leqq \int_{-\infty}^{\infty} \frac{|p'_m(s+\theta(t-s))-p'_m(s)||t-s|}{|p_m(s, x; \tau-iH\langle\xi\rangle^\delta, \xi)|}\rho\chi(\rho(t-s))ds\\ &+\int_{-\infty}^{\infty} \frac{|p'_m(s, x; \tau-iH\langle\xi\rangle^\delta, \xi)|}{|p_m(s, x; \tau-iH\langle\xi\rangle^\delta, \xi)|}|t-s|\rho\chi(\rho(t-s))ds\\ &+\sum_{k=0}^{m-1}\int_{-\infty}^{\infty} \frac{|p_k(t, x; \tau-iH\langle\xi\rangle^\delta, \xi)|}{|p_m(s, x; \tau-iH\langle\xi\rangle^\delta, \xi)|}\rho\chi(\rho(t-s))ds.\end{aligned}$$

Si l'on emploie

$$\begin{aligned}(14.3)\quad &|p'_m(s+\theta(t-s))-p'_m(s)|\,|t-s|\\ &\leqq C|\theta(t-s)|^{\kappa-1}|t-s|(|\tau-iH\langle\xi\rangle^\delta|^2+|\xi|^2)^{m/2}\end{aligned}$$

comme le paragraphe précédent, alors on aura la majoration pour le premier terme de (14.2)

$$(14.4)\quad C\int_{-\infty}^{\infty} |t-s|^\kappa\langle\xi\rangle^r(H\langle\xi\rangle^\delta)^{-r}\rho\chi(\rho(t-s))ds.$$

Pour majorer le second terme de (14.2), on s'appuie sur le Lemme suivant.

Lemme 14.1 (S. Tarama [16]). *Il existe une constante positive C telle que*

$$(14.5)\quad \left|\frac{p'_m(s, x; \tau-iH\langle\xi\rangle^\delta, \xi)}{p_m(s, x; \tau-iH\langle\xi\rangle^\delta, \xi)}\right| \leqq C\left(\frac{\langle\xi\rangle^{1-\delta}}{H}\right)^{\max(1, r/\kappa)}$$

pour tout $(s, x; \xi) \in \Omega\times \boldsymbol{R}^l_\xi$ *et* $\tau \in \boldsymbol{R}$.

En somme, on obtiendra

$$\begin{aligned}(14.6)\quad &\left|\left(\sum_{k=0}^{m} p_k\right)q-1\right|\\ &\leqq C\rho^{-\kappa}\langle\xi\rangle^{r-r\delta}+C\rho^{-1}\langle\xi\rangle^{(1-\delta)\max(1, r/\kappa)}+C\sum_{k=0}^{m-1}\langle\xi\rangle^{k-m+r-r\delta}.\end{aligned}$$

15. *Allure de* $(\partial/\partial\tau)\left(\sum_{k=0}^{m} p_k\right)D_t q$ *par rapport à* $(\rho, \langle\xi\rangle)$

D'abord, on constate que

$$\begin{aligned}\frac{\partial p_m}{\partial \tau}(t, x; \tau - iH\langle\xi\rangle^\delta, \xi) &= \frac{\partial p_m}{\partial \tau}(t) - \frac{\partial p_m}{\partial \tau}(s) + \frac{\partial p_m}{\partial \tau}(s) \\ &= \frac{\partial p'_m}{\partial \tau}(s+\theta(t-s))(t-s) + \frac{\partial p_m}{\partial \tau}(s) \\ &= \left(\frac{\partial p'_m}{\partial \tau}(s+\theta(t-s)) - \frac{\partial p'_m}{\partial \tau}(s)\right)(t-s) + \frac{\partial p'_m}{\partial \tau}(s)(t-s) + \frac{\partial p_m}{\partial \tau}(s),\end{aligned} \tag{15.1}$$

où l'on y remarque

$$\begin{aligned}&\left|\left(\frac{\partial p'_m}{\partial \tau}(s+\theta(t-s)) - \frac{\partial p'_m}{\partial \tau}(s)\right)(t-s)\right| \\ &\qquad \leqq C|\theta(t-s)|^{\kappa-1}|t-s|(|\tau - iH\langle\xi\rangle^\delta|^2 + |\xi|^2)^{(m-1)/2}.\end{aligned} \tag{15.2}$$

Donc, on aura

$$\begin{aligned}\left|\frac{\partial p_m}{\partial \tau} D_t q\right| &\leqq C\int_{-\infty}^{\infty} |t-s|^\kappa \langle\xi\rangle^{r-1}(H\langle\xi\rangle^\delta)^{-r}\rho^2|\chi'(\rho(t-s))|\,ds \\ &\quad + C\int_{-\infty}^{\infty} |t-s|\langle\xi\rangle^{-\delta}\left(\frac{\langle\xi\rangle^{1-\delta}}{H}\right)^{\max(1, r/\kappa)} \rho^2|\chi'(\rho(t-s))|\,ds \\ &\quad + \int_{-\infty}^{\infty} \frac{|(\partial p_m/\partial\tau)(s, x; \tau - iH\langle\xi\rangle^\delta, \xi)|}{|p_m(s, x; \tau - iH\langle\xi\rangle^\delta, \xi|}\rho^2|\chi'(\rho(t-s))|\,ds \\ &\leqq C\rho^{-\kappa+1}\langle\xi\rangle^{r-1-r\delta} + C\langle\xi\rangle^{-\delta}\langle\xi\rangle^{(1-\delta)\max(1, r/\kappa)} + C\rho\langle\xi\rangle^{-\delta}.\end{aligned} \tag{15.3}$$

Ensuite, il est évident que l'on a

$$\left|\frac{\partial}{\partial \tau}\left(\sum_{k=0}^{m-1} p_k\right) D_t q\right| \leqq C\rho \sum_{k=0}^{m-1} \langle\xi\rangle^{k-m+r-1-r\delta}. \tag{15.4}$$

16. *Allure de* $(\partial/\partial\xi_\alpha)\left(\sum_{k=0}^{m} p_k\right)D_{x_\alpha} q$ $(1 \leqq \alpha \leqq l)$ *par rapport à* $(\rho, \langle\xi\rangle)$

On répète le même raisonnement

$$\begin{aligned}\frac{\partial p_m}{\partial \xi_\alpha}(t, x; \tau - iH\langle\xi\rangle^\delta, \xi) &= \left(\frac{\partial p'_m}{\partial \xi_\alpha}(s+\theta(t-s)) - \frac{\partial p'_m}{\partial \xi_\alpha}(s)\right)(t-s) \\ &\quad + \frac{\partial p'_m}{\partial \xi_\alpha}(s)(t-s) + \frac{\partial p_m}{\partial \xi_\alpha}(s),\end{aligned} \tag{16.1}$$

d'où l'on tire

$$\begin{aligned}\left|\frac{\partial p_m}{\partial \xi_\alpha}(t)\right| &\leqq C(\theta|t-s|)^{\kappa-1}|t-s|(|\tau - iH\langle\xi\rangle^\delta|^2 + |\xi|^2)^{(m-1)/2} \\ &\quad + \left|\frac{\partial p'_m}{\partial \xi_\alpha}(s)\right||t-s| + \left|\frac{\partial p_m}{\partial \xi_\alpha}(s)\right|.\end{aligned}$$

Donc, on obtiendra

$$(16.2)\quad \begin{aligned} &\left|\frac{\partial p_m}{\partial \xi_\alpha} D_{x_\alpha} q\right| \\ &\leqq C\int_{-\infty}^{\infty} \theta^{\kappa-1}|t-s|^\kappa(|\tau-iH\langle\xi\rangle^\delta|^2+|\xi|^2)^{(m-1)/2}\left|D_{x_\alpha}\frac{1}{p_m(s)}\right|\rho\chi(\rho(t-s))ds \\ &\qquad +\int_{-\infty}^{\infty}|t-s|\left|\frac{\partial p'_m}{\partial \xi_\alpha}(s)\right|\left|D_{x_\alpha}\frac{1}{p_m(s)}\right|\rho\chi(\rho(t-s))ds \\ &\qquad +\int_{-\infty}^{\infty}\left|\frac{\partial p_m}{\partial \xi_\alpha}(s)\right|\left|D_{x_\alpha}\frac{1}{p_m(s)}\right|\rho\chi(\rho(t-s))ds. \end{aligned}$$

Compte tenu du Lemme 12.1, on aura

$$(16.3)\quad \left|D_{x_\alpha}\frac{1}{p_m(s)}\right| \leqq C\frac{1}{|p_m(s)|}\langle\xi\rangle^{1-\delta}.$$

D'abord, on obtient

$$(16.4)\quad \left|\frac{\partial p_m}{\partial \xi_\alpha}D_{x_\alpha}q\right| \leqq C\rho^{-\kappa}\langle\xi\rangle^{r-r\delta-\delta}+C\rho^{-1}\langle\xi\rangle^{1-2\delta+(1-\delta)r/\kappa}+C\langle\xi\rangle^{1-2\delta}.$$

Ensuite

$$(16.5)\quad \left|\frac{\partial}{\partial \xi_\alpha}\left(\sum_{k=0}^{m-1}p_k\right)D_{x_\alpha}q\right| \leqq C\sum_{k=0}^{m-1}\langle\xi\rangle^{k-m+r}\langle\xi\rangle^{-r\delta}\langle\xi\rangle^{1-\delta}.$$

17. *Allure de* $\sum_{|\alpha|\geqq 2}\frac{1}{\alpha!}\partial_\xi^\alpha\left(\sum_{k=0}^{m}p_k\right)D_x^\alpha q$ *par rapport à* $(\rho, \langle\xi\rangle)$

On raisonne comme au n° 10 avec les mêmes notations, en y remarquant

$$(17.1)\quad p_m(t)-p_m(s)=(p'_m(s+\theta(t-s))-p'_m(s))(t-s)+p'_m(s)(t-s);$$

donc on obtient qu'ils sont moins grand que

$$(17.2)\quad \begin{aligned} &\left|\sum_{\alpha_0+|\alpha|\geqq 2}\frac{1}{\alpha_0!\alpha!}\partial_\tau^{\alpha_0}\partial_\xi^\alpha p_m D_t^{\alpha_0}D_x^\alpha q\right| \\ &\leqq \sum_{\alpha_0+|\alpha|\geqq 2}\int_{-\infty}^{\infty}|\partial_\tau^{\alpha_0}\partial_\xi^\alpha(p'_m(s+\theta(t-s))-p'_m(s))(t-s)| \\ &\qquad \times\rho^{\alpha_0+1}|\chi^{(\alpha_0)}(\rho(t-s))|\left|D_x^\alpha\frac{1}{p_m(s)}\right|ds \\ &\qquad +\int_{-\infty}^{\infty}|\partial_\tau^{\alpha_0}\partial_\xi^\alpha p'_m(s)(t-s)|\rho^{\alpha_0+1}|\chi^{(\alpha_0)}(\rho(t-s))|\left|D_x^\alpha\frac{1}{p_m(s)}\right|ds \\ &\qquad +\int_{-\infty}^{\infty}|\partial_\tau^{\alpha_0}\partial_\xi^\alpha p_m(s)|\rho^{\alpha_0+1}|\chi^{(\alpha_0)}(\rho(t-s))|\left|D_x^\alpha\frac{1}{p_m(s)}\right|ds. \end{aligned}$$

Compte tenu du Lemme 12.1, on aura

$$(17.3)\quad \begin{aligned}(17.2) \leqq\, & C\sum_{\alpha_0+|\alpha|\geqq 2}\langle\xi\rangle^{r-\alpha_0-|\alpha|}(H\langle\xi\rangle^{\delta})^{-r}|\alpha|!^s\sum_{j=0}^{|\alpha|}\frac{(\langle\xi\rangle^{1-\delta}/H)^j}{j!^{s-1}}\\ &\times\int_{-\infty}^{\infty}|t-s|^{\kappa}\rho^{\alpha_0+1}|\chi^{(\alpha_0)}(\rho(t-s))|ds\\ &+C\sum_{\alpha_0+|\alpha|\geqq 2}\langle\xi\rangle^{r-\alpha_0-|\alpha|}(H\langle\xi\rangle^{\delta})^{-r}|\alpha|!^t\sum_{j=0}^{|\alpha|}\frac{(\langle\xi\rangle^{1-\delta}/H)^j}{j!^{s-1}}\\ &\times\int_{-\infty}^{\infty}|t-s|\rho^{\alpha_0+1}|\chi^{(\alpha_0)}(\rho(t-s))|ds\\ &+C\sum_{\alpha_0+|\alpha|\geqq 2}\langle\xi\rangle^{r-\alpha_0-|\alpha|}(H\langle\xi\rangle^{\delta})^{-r}|\alpha|!^s\sum_{j=0}^{|\alpha|}\frac{(\langle\xi\rangle^{1-\delta}/H)^j}{j!^{s-1}}\\ &\times\int_{-\infty}^{\infty}\rho^{\alpha_0+1}|\chi^{(\alpha_0)}(\rho(t-s))|ds;\end{aligned}$$

d'où il s'ensuit

$$(17.4)\quad \begin{aligned}\Big|\sum_{|\alpha|\geqq 2}\frac{1}{\alpha!}\partial_\xi^\alpha p_m D_x^\alpha q\Big| \leqq\, & C\sum_{\alpha_0+|\alpha|\geqq 2}\langle\xi\rangle^{r-\alpha_0-|\alpha|}(H\langle\xi\rangle^{\delta})^{-r}|\alpha|!^s\\ &\times\sum_{j=0}^{|\alpha|}\frac{(\langle\xi\rangle^{1-\delta}/H)^j}{j!^{s-1}}(\rho^{-\kappa+\alpha_0}+\rho^{\alpha_0-1}+\rho^{\alpha_0}).\end{aligned}$$

18. *Choix de δ pour $1<\kappa\leqq 2$*

On va raisonner comme au n° 13, en posant $\rho=\langle\xi\rangle^{\delta_1}$ où $0<\delta_1<1$. La demande que la puissance de $\langle\xi\rangle$ soit négative donnera les quatres inégalités suivantes;

(18.1) $-\kappa\delta_1+r-r\delta<0,$

(18.2) $-\delta_1+(1-\delta)\max(1, r/\kappa)<0,$

(18.3) $-1+r+r\delta<0,$

(18.4) $\delta_1-\delta<0,$

compte tenu des résultats de n°s 14, 15, 16 et 17. En se rappellant de $1<\kappa\leqq 2$ et $r\geqq 2$, (18.2) est une inégalité équivalente à (18.1), car $\max(1, r/\kappa)=r/\kappa$. On obtient d'abord

(18.5) $\dfrac{r-r\delta}{\kappa}<\delta_1<\delta,$ vu des inégalités (18.1) et (18.4);

ce qui donne

(18.6) $\dfrac{1}{\delta}<1+\dfrac{\kappa}{r}.$

Ensuite, on met (18.3) en

$$\frac{1}{\delta}<\frac{r}{r-1}; \tag{18.7}$$

donc, en somme, on aura

$$\frac{1}{\delta}<\min\left(1+\frac{\kappa}{r}, \frac{r}{r-1}\right). \tag{18.8}$$

Note. Compte tenu de (13.8) et (18.8), on peut constater qu'il existe $\varepsilon>0$ tel que l'on a

$$|r(t, x; \tau-iH\langle\xi\rangle^{\delta}, \xi)| \leqq \text{Const. } \langle\xi\rangle^{-\varepsilon} \tag{18.9}$$

pour tout $(t, x; \xi) \in \Omega\times \boldsymbol{R}^l_\xi$ et tout $\tau \in \boldsymbol{R}$.

§ 4. Propriété de $R(t, x; D_t, D_x)$

19. *Propriétés de* $r(t, x; \tau-iH\langle\xi\rangle^{\delta}, \xi)$ *et* $\tilde{r}(t, x; t-t_1, \eta)$
D'abord, on se rappelle de la Définition 5.2;

$$r(t, x; \tau-iH\langle\xi\rangle^{\delta}, \xi)=\sum \frac{1}{\alpha!}\partial^\alpha_\xi p D^\alpha_x q-1.$$

On énonce le premier Lemme qui contient une propriété fondamentale (18.9) de $r(t, x; \tau-iH\langle\xi\rangle^{\delta}, \xi)$ comme le cas spécial.

Lemme 19.1. *Sous les mêmes hypothèses que le Théorème, quelque soient* α *et* β, *il existe* C_β, A *et* C *telles que*

$$|D^\alpha_x \partial^\beta_\xi r(t, x; \tau-iH\langle\xi\rangle^{\delta}, \xi)| \leqq C_\beta A C^{|\alpha|}|\alpha|!^s \langle\xi\rangle^{-\varepsilon-\delta|\beta|}\sum_{j=0}^{|\alpha|}\frac{(\langle\xi\rangle^{1-\delta}/H)^j}{j!^{s-1}} \tag{19.1}$$

pour tout $(t, x; \xi) \in [0, T]\times K\times \boldsymbol{R}^l_\xi$, *K étant un ensemble compact de* $\boldsymbol{R}^l$.

La preuve se trouve dans l'appendice à la fin de cet article.
Concernant l'allure de $\tilde{r}(t, x; t-t_1, \eta)$, on peut énoncer un autre

Lemme 19.2. *Sous les mêmes hypothèses que le Théorème, quelque soit* α, *il existe une constante* σ $(0<\sigma<1)$ *telle que*

$$|D^\alpha_x \tilde{r}(t, x; t-t_1, \eta)| \leqq A|t-t_1|^{\sigma-1}C^{|\alpha|}|\alpha|!^s\langle\eta\rangle^{-\varepsilon}\sum_{j=0}^{|\alpha|}\frac{(\langle\eta\rangle^{1-\delta}/H)^j}{j!^{s-1}} \tag{19.2}$$

pour tout $(t, x; \eta) \in [0, T]\times K\times \boldsymbol{R}^l_\eta$, *K étant un ensemble compact de* $\boldsymbol{R}^l$.

Preuve. Tout d'abord, on explique une partie formelle du calcul. Au (5.7), on applique D_x^α à gauche;

$$D_x^\alpha \tilde{r}(t, x; t-t_1, \eta) = \frac{1}{\sqrt{2\pi}} \int_{-\infty}^{\infty} e^{i(t-t_1)\tau} D_x^\alpha r(t, x; \tau - iH\langle\eta\rangle^\delta, \eta) d\tau. \tag{19.3}$$

D'abord, on emploie l'identité

$$(t-t_1)e^{i(t-t_1)\tau} = \frac{1}{i}\partial_\tau e^{i(-t_1)\tau};^{*)}$$

donc on a

$$\begin{aligned}(t-t_1)\tilde{r}(t, x; t-t_1, \eta) &= \frac{1}{\sqrt{2\pi}} \int_{-\infty}^{\infty} \frac{1}{i} \partial_\tau e^{i(t-t_1)\tau} r(t, x; \tau - iH\langle\eta\rangle^\delta, \eta) d\tau \\ &= \frac{1}{\sqrt{2\pi}} \int_{-\infty}^{\infty} e^{i(t-t_1)\tau} \partial_\tau r(t, x; \tau - iH\langle\eta\rangle^\delta, \eta) d\tau\end{aligned} \tag{19.4}$$

par l'intégration par partie, puisque l'on a

$$|r(t, x; \tau - iH\langle\eta\rangle^\delta, \eta)| \leqq \frac{\text{const.}}{|\tau|} \tag{19.5}$$

pour tout $(t, x; \eta) \in \Omega \times \boldsymbol{R}^l$, vu la partie essentielle de la Définition de

$$\begin{aligned}&r(t, x; \tau - iH\langle\eta\rangle^\delta, \eta) \\ &\sim \int_{-\infty}^{\infty} \frac{p_m(t) - p_m(s)}{p_m(s, x; \tau - iH\langle\eta\rangle^\delta, \eta)} \rho\chi(\rho(t-s)) ds.\end{aligned} \tag{19.5$'$}$$

En employant encore

$$\int_{-\infty}^{\infty} \partial_\tau r(t, x; \tau - iH\langle\eta\rangle^\delta, \eta) d\tau = 0, \tag{19.6}$$

on aura

$$(t-t_1) D_x^\alpha \tilde{r}(t, x; t-t_1, \eta) = \frac{i}{\sqrt{2\pi}} \int_{-\infty}^{\infty} [e^{i(t-t_1)\tau} - 1] \partial_\tau D_x^\alpha r d\tau. \tag{19.7}$$

Compte tenu de l'existence σ $(0 < \sigma < 1)$ tel que l'on a

$$|e^{i(t-t_1)\tau} - 1| \leqq \text{const.} |t-t_1|^\sigma |\tau|^\sigma, \tag{19.8}$$

on obtiendra

*) Voir M. Nagase, Comm. P.D.E., **2** (1977), 1045–1061.

$$(19.9)\qquad |D_x^\alpha \tilde{r}(t, x; t-t_1, \eta)| \leqq \text{const.}\, |t-t_1|^{\sigma-1} \int_{-\infty}^{\infty} |\tau|^\sigma |\partial_\tau D_x^\alpha r|\, d\tau.$$

On va vérifier que $\tilde{r}$ est bien défini même pour $\langle\eta\rangle$ assez grand.

20. *Sur* $\int_{-\infty}^{\infty} |\tau|^\sigma |\partial_\tau r|\, d\tau$

Si l'on départ de (19.5)', alors on aura

$$(20.1)\qquad \partial_\tau r \sim \int_{-\infty}^{\infty} \left\{ \frac{\partial_\tau(p_m(t)-p_m(s))}{p_m(s)} - \frac{p_m(t)-p_m(s)}{p_m(s)} \frac{\partial_\tau p_m(s)}{p_m(s)} \right\} \times \rho\chi(\rho(t-s))ds.$$

D'abord, on déssine la partie délicate par le raisonnement grossier. (20.1) donne évidemment

$$(20.2)\qquad |\partial_\tau r| \leqq \begin{cases} \text{const.}\, \dfrac{\langle\xi\rangle^{-\varepsilon}|\xi|}{(|\tau|+|\xi|)^2} & \text{pour } |\tau| \geqq C\langle\xi\rangle \\ \text{const.}\, \langle\xi\rangle^{-\varepsilon-\delta} & \text{pour } |\tau| \leqq C\langle\xi\rangle \end{cases}$$

compte tenu de (18.9)

$$|r(t, x; \tau - iH\langle\xi\rangle^\delta, \xi)| \leqq C\langle\xi\rangle^{-\varepsilon},$$

qui s'ensuit du Lemme 19.1.

Pour $|\tau| \geqq C\langle\xi\rangle$, il n'y a aucune difficulté; en effet,

$$\begin{aligned} |\tilde{r}| &\leqq \text{const.}\, |t-t_1|^{\sigma-1} \langle\xi\rangle^{-\varepsilon+1} \int_{|\tau|\geqq C\langle\xi\rangle} \frac{|\tau|^\sigma}{|\tau|^2} d\tau \\ &= \text{const.}\, |t-t_1|^{\sigma-1} \langle\xi\rangle^{\sigma-\varepsilon}, \end{aligned}$$

qui est bien défini par le choix de σ tel que $0<\sigma<\varepsilon$.

Mais, pour $|\tau| \leqq C\langle\xi\rangle$, ce raisonnement ne marche plus. Par cette motivation, et en supposant de plus que

$$\partial_\tau r = \frac{\langle\xi\rangle^{-\varepsilon}}{\tau - iH\langle\xi\rangle^\delta - \lambda_j(t, x; \xi)}$$

pour n'alourdir pas le raisonnement (sans perdre généralité, compte tenu de (20.1)), on divise l'intégrale de τ en trois partie,

i) $|\tau - \lambda_j(t, x; \xi)| \geqq C\langle\xi\rangle$

ii) $C\langle\xi\rangle \geqq |\tau - \lambda_j(t, x; \xi)| \geqq C\langle\xi\rangle^\delta$

iii) $C\langle\xi\rangle^\delta \geqq |\tau - \lambda_j(t, x; \xi)|$;

c'est-à-dire,

$$|\tilde{r}| \leqq \text{const. } |t-t_1|^{\sigma-1} \int_{-\infty}^{\infty} |\tau|^{\sigma} |\partial_\tau r| d\tau$$

$$= \text{const. } |t-t_1|^{\sigma-1} \Big\{ \int_{|\tau-\lambda_j| \geqq C\langle\xi\rangle} + \int_{C\langle\xi\rangle \geqq |\tau-\lambda_j| \geqq C\langle\xi\rangle^{\delta}} + \int_{|\tau-\lambda_j| \leqq C\langle\xi\rangle^{\delta}} |\tau|^{\sigma} |\partial_\tau r| d\tau \Big\}$$

dont les intégrales on désigne par I_1, I_2 et I_3.

Or, on obtiendra

$$I_2 \leqq \langle\xi\rangle^{\sigma} \langle\xi\rangle^{-\varepsilon} \int_{C\langle\xi\rangle^{\sigma} \leqq |\tau-\lambda_j| \leqq C\langle\xi\rangle} \frac{d\tau}{|\tau-\lambda_j|}$$
$$\leqq \text{const. } \langle\xi\rangle^{\sigma-\varepsilon} \log \langle\xi\rangle,$$

et aussi

$$I_3 \leqq \langle\xi\rangle^{\sigma} \langle\xi\rangle^{-\varepsilon} \int_{|\tau-\lambda_j| \leqq C\langle\xi\rangle^{\delta}} \frac{d\tau}{H\langle\xi\rangle^{\delta}}$$
$$\leqq \text{const. } \langle\xi\rangle^{\sigma-\varepsilon}.$$

Donc, encore par le choix de σ tel que $0 < \sigma < \varepsilon$, l'intégrale $\int_{-\infty}^{\infty} |\tau|^{\sigma} |\partial_\tau r| d\tau$ est bien définie, ce qui nous autorise de définir $\tilde{r}(t, x; t-t_1, \xi)$ comme au n° 5.

21. *Calcul symbolique du produit* $e^{(h-tH)\langle D\rangle^{\delta}}$ *et* $\tilde{R}(t, x; t-t_1)e^{-(h-tH)\langle D\rangle^{\delta}}$

On se rappelle de la définition (7.2) de $\tilde{R}(t, x; t-t_1)$

$$\tilde{R}(t, x; t-t_1)e^{-(h-tH)\langle D\rangle^{\delta}} w(t_1, x)$$
$$= (2\pi)^{-l/2} \int e^{i\langle x, \xi\rangle} e^{-(h-tH)\langle\xi\rangle^{\delta}} \tilde{r}(t, x; t-t_1, \xi) \tilde{w}(t_1, \xi) d\xi. \tag{7.2}$$

Il nous faut calculer le produit de $e^{(h-tH)\langle D\rangle^{\delta}}$ et de $\tilde{R}(t, x; t-t_1) \times e^{-(h-tH)\langle D\rangle^{\delta}}$ dont le symbole est $e^{-(h-tH)\langle\xi\rangle^{\delta}} \tilde{r}(t, x; t-t_1, \xi)$. Pour cela, nous nous appuyons sur la formule du produit de $A(x, D)$ et $B(x, D)$ dont les symboles sont notés par $a(x, \xi)$ et $b(x, \xi)$;

$$\text{le symbole de } AB = (2\pi)^{-l/2} \int e^{i\langle x, \xi-\eta\rangle} a(x, \xi) \hat{b}(\xi-\eta, \eta) d\xi,$$

où

$$\hat{b}(\zeta, \eta) = (2\pi)^{-l/2} \int e^{-i\langle y, \zeta\rangle} b(y, \zeta) dy.$$

En employant cette formule, on obtiendra

$$(21.1)\qquad (2\pi)^{-l/2}\int e^{i\langle x,\xi-\eta\rangle}e^{(h-tH)\langle\xi\rangle^\delta}e^{-(h-tH)\langle\eta\rangle^\delta}\hat{\tilde{r}}(t,\xi-\eta;t-t_1,\eta)d\xi,$$

qui est égale à

$$(21.2)\qquad (2\pi)^{-l}\iint e^{-i\langle z,\zeta\rangle}e^{(h-tH)\{\langle\eta+\zeta\rangle^\delta-\langle\eta\rangle^\delta\}}\tilde{r}(t,x+z;t-t_1,\eta)\,dzd\zeta,$$

où l'on a posé $\xi-\eta=\zeta$ et $y-x=z$ dans (21.1).

Vu $(-D_z)^\alpha e^{-i\langle z,\zeta\rangle}=\zeta^\alpha e^{-i\langle z,\zeta\rangle}$, on aura encore

$$(21.3)\qquad \begin{aligned}&\sqrt{2\pi}\;\boldsymbol{r}(t,x;t-t_1,\eta)\\ &=\text{le symbole de }\sqrt{2\pi}\;\boldsymbol{R}(t,x;t-t_1)\\ &=(2\pi)^{-l}\iint e^{-i\langle z,\zeta\rangle}\,e^{(h-tH)\{\langle\eta+\zeta\rangle^\delta-\langle\eta\rangle^\delta\}}\frac{1}{\zeta^\alpha}D_z^\alpha\tilde{r}(t,x+z;t-t_1,\eta)\,dzd\zeta\end{aligned}$$

par l'intégration par partie, compte tenu de (7.4). Donc, le module du second membre se majore par

$$(21.4)\qquad (2\pi)^{-l}\iint e^{(h-tH)\{\langle\eta+\zeta\rangle^\delta-\langle\eta\rangle^\delta\}}\frac{1}{|\zeta|^{|\alpha|}}|D_z^\alpha\tilde{r}(t,x+z;t-t_1,\eta)|\,dzd\zeta.$$

22. *Majoration de* (21.4)

Proposition 22.1. *Pour tout α, il existe des constantes c_1 et a telles que*

$$(22.1)\qquad \begin{aligned}&\frac{1}{|\zeta|^{|\alpha|}}|D_z^\alpha\tilde{r}(t,x+z;t-t_1,\eta)|\\ &\leqq A|t-t_1|^{\sigma-1}\langle\eta\rangle^{-\varepsilon}\begin{cases}c_1\exp(-sa^\delta H^{-1/(s-1)}|\zeta|^\delta) & \text{pour } \langle\eta\rangle\leqq a|\zeta|\\ c_1\exp\left(-saH^{-1/(s-1)}\dfrac{|\zeta|}{\langle\eta\rangle^{1-\delta}}\right) & \text{pour } \langle\eta\rangle\geqq a|\zeta|\end{cases}\end{aligned}$$

où

$$(22.2)\qquad \begin{aligned}&a=\frac{1}{e^s}\frac{c}{C}H^{s/(s-1)},\qquad c \text{ étant défini par}\\ &c=\frac{e}{[\sqrt{|\alpha|}\,(|\alpha|+1)]^{1/|\alpha|}}.\end{aligned}$$

Note. c est dépendante de $|\alpha|$, donc a l'est aussi. Les deux membres de (22.1) dépendent de $|\alpha|$.

Preuve. Vu le Lemme 19.2, on aura

$$\left|\frac{1}{\zeta^\alpha} D_z^\alpha \tilde{r}(t, x+z; t-t_1, \eta)\right|$$
$$\leqq A|t-t_1|^{\sigma-1}\left(\frac{C}{|\zeta|}\right)^{|\alpha|}|\alpha|!^s\langle\eta\rangle^{-\varepsilon}\sum_{j=0}^{|\alpha|}\frac{(\langle\eta\rangle^{1-\delta}/H)^j}{j!^{s-1}}.$$

Donc, on montre d'abord le Lemme suivant pour prouver cette Proposition, en posant $|\zeta|/C=B$, $|\alpha|=k$ et $\langle\eta\rangle^{1-\delta}/H=A$.

Lemme 22.1.*) *Si l'on définit*

$$S=\frac{k!^s}{B^k}\sum_{j=0}^{k}\frac{A^j}{j!^{s-1}},$$

alors on a

$$(22.4)\qquad S\leqq\begin{cases} c_1\exp\left(-s\dfrac{c^{1/s}}{e}B^{1/s}\right) & pour\ A^{1/(s-1)}\leqq \dfrac{c^{1/s}}{e}B^{1/s} \\ c_1\exp\left(-s\dfrac{c}{e^s}\dfrac{B}{A}\right) & pour\ A^{1/(s-1)}\geqq \dfrac{c}{e^s}\dfrac{B}{A}\end{cases}$$

en choisissant k convenablement.

Preuve. On le montre séparément deux cas.

i) $A^{1/(s-1)}\leqq k$; en générale, il faut tenir compte plus précisément de $k-i-1<A^{1/(s-1)}\leqq k$, mais on suppose $k-1<A^{1/(s-1)}\leqq k$ (c'est-à-dire, $i=0$) sans perdre généralité. En définissant $f(j)=A^j/j!^{s-1}$, on a $f(j)=(A/j^{s-1})f(j-1)$ qui implique

$$(22.5)\qquad f(k-1)\begin{cases}\geqq f(k)\\ >f(k-2)>\cdots>f(1).\end{cases}$$

Donc, on a

$$S\leqq\frac{k!^s}{B^k}(k+1)\frac{A^{k-1}}{(k-1)!^{s-1}}$$
$$=\frac{(k+1)!}{B^k}k^{s-1}A^{k-1}\leqq\frac{(k+1)!}{B^k}k^{k(s-1)}$$
$$=\frac{k^{ks}}{B^k}\frac{c_1k^{k+1/2}e^{-k}}{k^k}(k+1),$$

vu la formule de Stirling $k!=c_1k^{k+1/2}e^{-k}$; c'est-à-dire

$$(22.6)\qquad S\leqq c_1\left\{\frac{k}{c^{1/s}B^{1/s}}\right\}^{ks},$$

*) Voir aussi S. Wakabayashi [14].

où l'on a posé

$$c=\min_k \frac{e}{[\sqrt{k}\,(k+1)]^{1/k}}.$$

ii) $A^{1/(s-1)}>k$; dans ce cas, on met S en

$$S=\frac{k!}{B^k}A^k\sum_{j=0}^{k}\frac{k!^{s-1}}{j!^{s-1}}A^{j-k}.$$

Or,

$$\frac{k!^{s-1}}{j!^{s-1}}A^{j-k}=\prod_{j=0}^{k-i-1}\frac{(k-i)^{s-1}}{A}<1$$

indique

$$S\leqq\frac{A^k}{B^k}c_1k^{k+1/2}e^{-k}(k+1);$$

donc on aura

$$S\leqq c_1\left\{\frac{k}{c(B/A)}\right\}^k. \tag{22.7}$$

Finalement, il suffit de prendre $k=[c^{1/s}B^{1/s}e^{-1}]+1$ pour i), et $k=[cBA^{-1}e^{-s}]$ pour ii), où $[A]$ signifie le plus grand nombre entier $\leqq A$.

Note. $A^{s/(s-1)}\geqq(c/e^s)B$ équivaut à $A^{1/(s-1)}\geqq(c^{1/s}/e)B^{1/s}$.

Pour terminer la preuve de Proposition, $A^{1/(s-1)}\leqq(c^{1/s}/e)B^{1/s}$ est équivalent à $(\langle\eta\rangle^{1-\delta}/H)^{\delta/(1-\delta)}\leqq(c^\delta/e)(|\zeta|/C)^\delta$; donc on aura $\langle\eta\rangle\leqq(1/e^s)(c/C)\times H^{s/(s-1)}|\zeta|$ que l'on écrit, en définissant $a=(1/e^s)(c/C)H^{s/(s-1)}$,

$$\langle\eta\rangle\leqq a|\zeta|. \tag{22.8}$$

Il est évident que l'on a

$$\frac{c^{1/s}}{e}B^{1/s}=\frac{c^{1/s}}{e}\left(\frac{|\zeta|}{C}\right)^{1/s}=\frac{1}{e}\left(\frac{c}{C}\right)^{\delta}|\zeta|^{\delta}=(aH^{-s/(s-1)})^{1/s}|\zeta|^{\delta}$$
$$=a^{\delta}H^{-1/(s-1)}|\zeta|^{\delta},$$

et

$$\frac{c}{e^s}\frac{B}{A}=\frac{c}{e^s}\frac{|\zeta|/C}{\langle\eta\rangle^{1-\delta}/H}=\frac{1}{e^s}\frac{c}{C}H\frac{|\zeta|}{\langle\eta\rangle^{1-\delta}}=aH^{-1/(s-1)}\frac{|\zeta|}{\langle\eta\rangle^{1-\delta}},$$

vu la Définition (22.2) de a; ce qui complète la preuve de la Proposition 22.1. C.Q.F.D.

23. *Majoration de* $(21.4)_{bis}$

Concernant l'évaluation de $\langle\eta+\zeta\rangle^{\delta}-\langle\eta\rangle^{\delta}$, on emploie la formule de moyenne, et on aura

$$\langle\eta+\zeta\rangle^{\delta}-\langle\eta\rangle^{\delta}=\int_0^1 \frac{\delta\sum(\eta_i+k\zeta_i)\zeta_i}{\langle\eta+k\zeta\rangle^{2-\delta}}\,dk; \tag{23.1}$$

donc, à fortiori

$$\langle\eta+\zeta\rangle^{\delta}-\langle\eta\rangle^{\delta}\leqq\int_0^1 \frac{\delta|\zeta|}{\langle\eta+k\zeta\rangle^{1-\delta}}\,dk, \tag{23.2}$$

puisque l'on a $|\sum_{i=1}^{l}(\eta_i+k\zeta_i)\zeta_i|\leqq|\eta+k\zeta||\zeta|$.

On obtient le

Lemme 23.1. *Il existe deux constantes positives* α *et* β *telles que*

$$\langle\eta+\zeta\rangle^{\delta}-\langle\eta\rangle^{\delta}\leqq\begin{cases}\alpha|\zeta|^{\delta} & \textit{pour } \langle\eta\rangle\leqq a|\zeta|\\ \beta\dfrac{|\zeta|}{\langle\eta\rangle^{1-\delta}} & \textit{pour } \langle\eta\rangle\geqq a|\zeta|.\end{cases} \tag{23.3}$$

Preuve. Pour $\langle\eta\rangle\leqq a|\zeta|$, on met (23.2) en

$$\langle\eta+\zeta\rangle^{\delta}-\langle\eta\rangle^{\delta}\leqq\left\{\int_0^1\delta\left(\frac{|\zeta|}{\langle\eta+k\zeta\rangle}\right)^{1-\delta}dk\right\}|\zeta|^{\delta};$$

donc, il suffit de trouver la borne supérieure de

$$\frac{|\zeta|^2}{\langle\eta\rangle^2+2k\sum\eta_i\zeta_i+k^2|\zeta|^2}\qquad\text{pour } 0\leqq k\leqq 1.$$

Il s'écrit encore

$$\frac{1}{(\langle\eta\rangle/|\zeta|)^2+2k\sum(\eta_i/|\zeta|)(\zeta_i/|\zeta|)+k^2}.$$

Or, $\langle\eta\rangle\leqq a|\zeta|$ implique l'existence de α, pusique $|\sum_{i=1}^{l}(\eta_i/|\zeta|)(\zeta_i/|\zeta|)|\leqq|\eta|/|\zeta|$.

Pour $\langle\eta\rangle\geqq a|\zeta|$, on met (23.2) en

$$\langle\eta+\zeta\rangle^{\delta}-\langle\eta\rangle^{\delta}\leqq\left\{\int_0^1\delta\left(\frac{\langle\eta\rangle}{\langle\eta+k\zeta\rangle}\right)^{1-\delta}dk\right\}\frac{|\zeta|}{\langle\eta\rangle^{1-\delta}}.$$

Or, le même raisonnement comme ci-dessus pour

$$\frac{\langle\eta\rangle^2}{\langle\eta\rangle^2+2k\sum\eta_i\zeta_i+k^2|\zeta|^2}$$

montre l'existence de β, car $\langle\eta\rangle \geqq a|\zeta|$.

Pour discuter l'intégrabilité par rapport à ζ de fonction sous le signe d'intégration (21.4), on va prouver qu'elle décroit exponentiellement. Si l'on s'appuie sur la Proposition 21.1 et le Lemme 23.1, alors il suffira de montrer

$$\alpha(h-tH)-sa^{\delta}H^{-1/(s-1)}<0 \qquad \text{pour } \langle\eta\rangle \leqq a|\zeta| \tag{23.4}$$

et

$$\beta(h-tH)-saH^{-1/(s-1)}<0 \qquad \text{pour } \langle\eta\rangle \geqq a|\zeta|, \tag{23.5}$$

en choisissant convenablement h et H pour $0 \leqq t \leqq h/2H$.

Ces inégalités seront remplies en choisissant

$$h<\frac{s}{\alpha}\frac{1}{e}\left(\frac{c}{C}\right)^{1/s} \tag{23.6}$$

et

$$\frac{h}{H}<\frac{s}{\beta}\frac{1}{e^s}\left(\frac{c}{C}\right); \tag{23.7}$$

en effet, $\alpha h<sa^{\delta}H^{-1/(s-1)}$ et $\beta h<saH^{-1/(s-1)}$ entraînent respectivement (23.6) et (23.7) grâce à la définition (22.2) de a.

Note. Si l'on définit $c_0=(s/\beta)(1/e^s)(c/C)$, alors on aura $h/H<c_0$ et $h<(\alpha/s)(\beta c_0/s)^{\delta}$. Ceci permet de répéter nos raisonnements jusq'au $t=T$.

24. *Résolution de l'équation intégrale de Volterra* (7.5)

On va revenir à la résolution de (7.5)

$$w(t, x)+\int_0^t \boldsymbol{R}(t, x; t-t_1)w(t_1, x)dt_1=F(t, x),$$

en posant $e^{(h-tH)\langle D\rangle^{\delta}}f=F$, où

$$\boldsymbol{R}(t, x; t-t_1)w(t_1, x)=(2\pi)^{-l/2}\int e^{i\langle x,\eta\rangle}\boldsymbol{r}(t, x; t-t_1, \eta)\tilde{w}(t_1, \eta)d\eta, \tag{24.1}$$

et

$$\begin{aligned}&\sqrt{2\pi}\,\boldsymbol{r}(t, x; t-t_1, \eta)\\&=(2\pi)^{-l}\iint e^{-i\langle z,\zeta\rangle}e^{(h-tH)\{\langle\eta+\zeta\rangle^{\delta}-\langle\eta\rangle^{\delta}\}}\frac{1}{\zeta^{\alpha}}D_z^{\alpha}\tilde{r}(t, x+z; t-t_1, \eta)dzd\zeta,\end{aligned} \tag{24.2}$$

vu (21.3).

Voici ce que l'on a obtenu jusqu'aux numéros précédents;

$$|\boldsymbol{r}(t, x; t-t_1, \eta)| \leqq \text{const.}\,|t-t_1|^{\sigma-1}\langle\eta\rangle^{-\varepsilon}. \tag{24.3}$$

Donc, on est déjà permi de supposer qu'il existe une constante positive M telle que

$$\|R(t,\ ;t-t_1)w(t_1,\ )\|_{L^2(\boldsymbol{R}^l)} \leqq M|t-t_1|^{\sigma-1}\|w(t_1,\ )\|_{L^2(\boldsymbol{R}^l)} \tag{24.4}$$

pour $0 \leqq t_1 \leqq t$ et $0<\sigma<1$.

Or, il est classique que l'on peut montrer l'existence unique de $w(t, x)$ dans l'espace $C^0([0, T]; L^2(\boldsymbol{R}^l))$.

En effet, on établit le

Lemme 24.1.

$$\|(-\boldsymbol{R})^n F(t_1,\ )\|_{L^2(\boldsymbol{R}^l)} \leqq M^n|t-t_1|^{n\sigma-1}\frac{[\Gamma(\sigma)]^n}{\Gamma(n\sigma)}\|F(t,\ )\|_{L^2(\boldsymbol{R}^l)}, \tag{24.5}$$

on peut constater qu'il existe $w(t, x)$ telle que

$$\sup_{0\leqq t\leqq T}\|w(t,\ )\|_{L^2(\boldsymbol{R}^l)} < +\infty. \tag{24.6}$$

Preuve du Lemme 24.1. Il suffit de remarquer

$$\begin{aligned}
&\int_0^t |t-t_1|^{\sigma-1}\int_0^{t_1}|t_1-t_2|^{\sigma-1}\|F(t_2,\ )\|_{L^2(\boldsymbol{R}^l)}dt_2dt_1\\
&=\int_0^t \|F(t_2,\ )\|_{L^2(\boldsymbol{R}^l)}dt_2\int_{t_2}^t |t-t_1|^{\sigma-1}|t_1-t_2|^{\sigma-1}dt_1\\
&=\int_0^t \|F(t_2,\ )\|_{L^2(\boldsymbol{R}^l)}|t-t_2|^{2\sigma-2}(t-t_2)\,dt_2\int_0^1(1-u)^{\sigma-1}u^{\sigma-1}du,
\end{aligned}$$

selon le changement de coordonnées $t_1=t_2+u(t-t_2)$.

Voici la démonstration complète du Théorème.

§ 5. Unicité et domaine d'influence

25. *Vitesse finie de propagation*

Dans les paragraphes précédents, nous avons obtenu une solution du problème de Cauchy avec le second membre appartenant à l'espace

$$E_{h,H,T,\delta}=\{f(t,x);\ \sup_{0\leqq t\leqq T}\|e^{(h-tH)\langle\xi\rangle^\delta}\hat{f}(t,\xi)\|_{L^2(\boldsymbol{R}^l)}<+\infty\} \tag{25.1}$$

où $T \leqq h/2H$ et $h \leqq c_0 H$ (voir la Note à la fin du n° 23), dans l'espace

$$E^{(m-1)}_{h,H,T,\delta}=\{u(t,x);\ \sup_{\substack{0\leqq t\leqq T\\ j=0,1,\cdots,m-1}}\|\langle\xi\rangle^{\delta(m-j)}e^{(h-tH)\langle\xi\rangle^\delta}D_t^j u(t,\xi)\|_{L^2(\boldsymbol{R}_\xi^l)} <+\infty\}. \tag{25.2}$$

La solution $u(t, x)$ remplit aussi

$$\langle D_x\rangle^{-(1-\delta)m} D_t^m u(t, x) \in E_{h,H,T,\delta}. \tag{25.3}$$

Comme une propriété, remarquons que, pour tout $u \in E_{h,H,T,\delta}$ et tout m,

$$\langle D_x\rangle^m u(t, x) \in E_{\tilde{h},\tilde{H},\tilde{T},\delta} \tag{25.4}$$

où $\tilde{h}<h$ et $H\leqq\tilde{H}$.

Dans ce paragraphe, nous montrons l'unicité de solutions obtenues ci-dessus et l'existence d'une vitesse finie de propagation.

L'idée de preuve est comme ceci; en considérant une suite d'opérateurs strictement hyperboliques $P^{(n)}(t, x; D_t, D_x)$ qui ont les coefficients réguliers même en t, et qui convergent à l'opérateur étudié dans le Théorème, nous construisons les inverses à droite de $P^{(n)}$ comme au n° 5. Puisque ces inverses convergent fortement à l'inverse à droite de $P(t, x; D_t, D_x)$ comme l'application de $E_{h,H,T,\delta}$ dans $E^{(m-1)}_{h,H,T,\delta}$. Or, notre énoncé resulte de l'unicité de solutions et de l'existence de domaine d'influence pour les opérateurs strictement hyperboliques. Le lemme suivant dû à W. Nuij [11] est essentiel.

Lemme 25.1. *Soit*

$$p_m(t, x; \tau, \xi)=\tau^m+\sum_{\substack{j+|\nu|=m\\ j<m}} a_{j\nu}(t, x)\tau^j\xi^\nu$$

la partie principale de $P(t, x; D_t, D_x)$ traité dans le Théorème. Alors, il existe les suites de fonctions à valeurs réelles $a_{j\nu}^{(n)}(t, x)$, $n=1, 2, \cdots$, indéfiniment dérivables en (t, x) telles que les symboles

$$\tau^m+\sum_{\substack{j+|\nu|=m\\ j<m}} a_{j\nu}^{(n)}(t, x)\tau^j\xi^\nu$$

soient strictement hyperboliques par rapport à τ et que les fonctions $a_{j\nu}^{(n)}(t, x)$ soient uniformément bornées dans $C^\kappa([0, T]; \gamma^s_{\text{loc}}(\boldsymbol{R}^l))$, et pour tous j et ν, la suite de fonctions $\{a_{j\nu}^{(n)}(t, x)\}_{n=1,2,\dots}$ converge uniformément à $a_{j\nu}(t, x)$ avec toutes les dérivées jusqu'à l'ordre supérieur en x pour tout ensemble compact de $[0, T]\times\boldsymbol{R}^l$ (pour $1<\kappa\leqq 2$, au premier ordre en t).

26. *Convergence des opérateurs*

En employant la suite des symboles strictement hyperboliques du Lemme 25.1, nous définissons la suite des opérateurs hyperboliques $P^{(n)}(t, x; D_t, D_x)$, $n=1, 2, \cdots$, telle que la partie principale $p_m^{(n)}$ de $P^{(n)}$ soit

$$\tau^m + \sum_{\substack{j+|\nu|=m \\ j<m}} a_{j\nu}^{(n)}(t, x)\tau^j \xi^\nu$$

et que ses termes d'ordre inférieur à m soient identiques à ceux de l'opérateur P.

Les raisonnements des paragraphes précédents montrent que, pour les opérateurs $P^{(n)}$, les constructions des opérateurs

$$\begin{aligned} &Q^{(n)}(t, x; D_t, D_x)v(t, x) \\ (26.1) \qquad &= (2\pi)^{-(l+1)} \int \cdots \int e^{i(t-t_1)(\tau - iH\langle\xi\rangle^\delta) + i\langle x-y, \xi\rangle} \\ &\times \int_{-\infty}^{\infty} \frac{\rho\chi(\rho(t-s))}{p_m^{(n)}(s, x; \tau - iH\langle\xi\rangle^\delta, \xi)} ds\, v(t_1, y)\, dt_1 dy\, d\tau\, d\xi, \end{aligned}$$

et

$$(26.2) \qquad R^{(n)} = P^{(n)} Q^{(n)} - I$$

sont valables, et que leurs symboles ont les majorations indépendantes de n.

Alors, en définissant

$$\begin{aligned} &\tilde{Q}^{(n)}(t, x; t-t_1, \eta) \\ (26.3) \qquad &= (2\pi)^{-l-1/2}\, \mathrm{Os.}\text{-}\int \cdots \int e^{-i\langle z, \zeta\rangle + i(t-t_1)\tau} \\ &\times e^{(h-tH)\{\langle\eta+\zeta\rangle^\delta - \langle\eta\rangle^\delta\}} q^{(n)}(t, x+z; \tau - iH\langle\eta\rangle^\delta, \eta) dz\, d\zeta\, d\tau, \end{aligned}$$

et

$$\begin{aligned} &\tilde{R}^{(n)}(t, x; t-t_1, \eta) \\ (26.4) \qquad &= (2\pi)^{-l-1/2}\, \mathrm{Os.}\text{-}\int \cdots \int e^{-i\langle z, \zeta\rangle + i(t-t_1)\tau} \\ &\times e^{(h-tH)\{\langle\eta+\zeta\rangle^\delta - \langle\eta\rangle^\delta\}} r^{(n)}(t, x+z; \tau - iH\langle\eta\rangle^\delta, \eta)\, dz\, d\zeta\, d\tau, \end{aligned}$$

nous aurons les propriétés suivantes; pour tous $j=0, 1, \cdots, m-1$, α et β multi-indices, en posant, $\varepsilon' = \varepsilon - \sigma$,

$$(26.5) \qquad \sup_{0 \leq t_1 \leq t \leq T} |t-t_1|^{\sigma-1} |D_x^\alpha \partial_\eta^\beta D_t^j \tilde{Q}^{(n)}| \leq C_{j,\alpha,\beta} \langle\eta\rangle^{\delta(m-j) - \delta|\beta| + (1-\delta)|\alpha| - \varepsilon'}$$

et

$$(26.6) \qquad \sup_{0 \leq t_1 \leq t \leq T} |t-t_1|^{\sigma-1} |D_x^\alpha \partial_\eta^\beta \tilde{R}^{(n)}| \leq C_{\alpha,\beta} \langle\eta\rangle^{-\delta|\beta| + (1-\delta)|\alpha| - \varepsilon'}$$

où $C_{j,\alpha,\beta}$ et $C_{\alpha,\beta}$ sont des constantes indépendantes de n; de plus, pour tout $M > 0$,

$$(26.7)\qquad \sup_{\substack{0\leq t_1\leq t\leq T\\ |x|,|\xi|<M}} |t-t_1|^{\sigma-1}|D_t^j D_x^\alpha \partial_\eta^\beta(\tilde{Q}^{(n)}(t,x;t-t_1,\eta)-\tilde{Q}(t,x;t-t_1,\eta))|\to 0$$

et

$$(26.8)\qquad \sup |t-t_1|^{\sigma-1}|D_x^\alpha \partial_\eta^\beta(\tilde{R}^{(n)}(t,x;t-t_1,\eta)-\tilde{R}(t,x;t-t_1,\eta))|\to 0$$

pour $n\to\infty$.

Les majorations (26.5) et (26.6), et les convergences de $\tilde{Q}^{(n)}$ et de $\tilde{R}^{(n)}$ montrent que, pour tout $u\in L^2(\boldsymbol{R}^l)$,

$$(26.9)\qquad \sup_{\substack{0\leq t_1\leq t\leq T\\ j=0,1,\cdots,m-1}} |t-t_1|^{\sigma-1}\|D_t^j[\tilde{Q}^{(n)}(t,x;t-t_1,D_x)u - \tilde{Q}(t,x;t-t_1,D_x)u]\|_{H^{\delta(m-j)}(\boldsymbol{R}^l)}\to 0$$

et

$$(26.10)\qquad \sup |t-t_1|^{\sigma-1}\|\tilde{R}^{(n)}(t,x;t-t_1,D_x)u-\tilde{R}(t,x;t-t_1,D_x)u\|_{L^2(\boldsymbol{R}^l)}\to 0$$

pour $n\to\infty$, où

$$\|u\|_{H^s(\boldsymbol{R}^l)}=\|\langle\xi\rangle^s\hat{u}(\xi)\|_{L^2(\boldsymbol{R}^l_\xi)}.$$

Note. Voir H. Kumano-go [7], Chap. III, § 7.

En somme, les opérateurs $Q^{(n)}(t,x;D_t,D_x)$ sont uniformément bornés comme les applications de $E_{h,H,T,\delta}$ dans $E^{(m-1)}_{h,H,T,\delta}$ et convergent fortement à l'opérateur $Q(t,x;D_t,D_x)$. Il en est de même pour les opérateurs $R^{(n)}(t,x;D_t,D_x)$ comme les applications de $E_{h,H,T,\delta}$ dans lui-même. De plus, ils remplissent la majoration $\|(R^{(n)})^k\|\leq c_k$ où $\sum_{k=1}^\infty c_k<+\infty$.

Or, nous avons le

Lemme 26.1. *Pour tout* $f\in E_{h,H,T,\delta}$

$$(26.11)\qquad \lim_{n\to\infty} Q^{(n)}(I+R^{(n)})^{-1}f=Q(I+R)^{-1}f.$$

Puisque chaque opérateur strictement hyperbolique a la vitesse finie de propagation qui est majorée par le maximum de racines caractéristiques, les Lemmes 25.1 et 26.1 prouvent qu'il y a une constante Λ telle que l'on ait

(26.12) le support de $Q(I+R)^{-1}f$ est contenu dans

$$\bigcup_{(t_0,x_0)\in\mathrm{supp}.f}\{(t,x);\,|x-x_0|\leq\Lambda(t-t_0)\};$$

c'est-à-dire, l'existence de domaine d'influence est établie.

27. *Unicité de solutions*

Pour montrer l'unicité de solutions, soit une fonction $u(t, x) \in E^{(m-1)}_{h,H,T,\delta}$ telle que $\langle D\rangle^{-(1-\delta)m} D_t^m u \in E_{h,H,T,\delta}$ et qu'elle soit une solution du problème de Cauchy

$$(27.1) \qquad \begin{cases} Pu=0 \\ \text{données nulles sur } t=0. \end{cases}$$

De la même façon qu'au n° précédent, si l'on considère une suite $\{P^{(n)}\}$, alors on aura évidemment

$$(27.2) \qquad \begin{cases} P^{(n)}u=P^{(n)}u-Pu \\ \text{données nulles sur } t=0. \end{cases}$$

D'une part, l'unicité est valable pour les opérateurs strictement hyperboliques.

D'autre part, on obtiendra

$$(27.3) \qquad u=Q^{(n)}(I+R^{(n)})^{-1}(P^{(n)}u-Pu), \qquad \text{vu (27.2).}$$

Or, on voit que $u=0$, puisque

$$(27.4) \qquad P^{(n)}u-Pu \longrightarrow 0 \quad \text{dans } E_{\tilde{h},\tilde{H},\tilde{T},\delta} \text{ où } \tilde{h}<h \text{ et } H\leqq\tilde{H};$$

d'où l'on a l'unicité de solutions.

Alors, en employant la partition de l'unité, on aura la

Proposition 27.1. *Pour le problème de Cauchy* (2.1), *on a l'unicité locale de solutions et l'existence de domaine d'influence dans la classe de Gevrey* γ^s_{loc} *où* $1<s<\min(1+\kappa/r, r/(r-1))$.

Appendice

On va donner la preuve du Lemme 19.1.

Compte tenu du n° 5, on a

$$(1) \qquad \begin{aligned} &D_x^\alpha \partial_\xi^\beta r(t, x; \tau-iH\langle\xi\rangle^\delta, \xi) \\ &= \sum_{(\gamma_0,\gamma)}^{\text{fini}} \sum_{a\leqq\alpha} \sum_{b\leqq\beta} \frac{1}{\gamma_0!\gamma!} \binom{\alpha}{a}\binom{\beta}{b} D_x^{\alpha-a}\partial_\tau^{\gamma_0}\partial_\xi^{\gamma+\beta-b} p \partial_\xi^b D_x^{\gamma+a} D_t^{\gamma_0} q, \end{aligned}$$

où l'on ne considère que le cas $|\alpha+\beta|\geqq 1$.

En se rappellant la définition de q, on obtiendra

$$(2)\qquad \partial_\xi^b D_x^{\gamma+a} D_t^{\gamma_0} q = \sum_{c\leqq b} \binom{b}{c} \int_{-\infty}^{\infty} \partial_\xi^{b-c}[\rho^{\gamma_0+1}\chi^{(\gamma_0)}(\rho(t-s))]\partial_\xi^c D_x^{\gamma+a}\left(\frac{1}{p_m(s)}\right)ds.$$

D'abord, on a

$$(3)\qquad \begin{aligned} &|D_x^{\alpha-a}\partial_\tau^{\gamma_0}\partial_\xi^{\gamma+\beta-b} p(t, x; \tau - iH\langle\xi\rangle^\delta, \xi)| \\ &\leqq AC^{|\alpha-a|}|\alpha-a|!^s \langle\xi\rangle^{-\delta(\gamma_0+|\gamma+\beta-b|)} \sum_{j=0}^{|\alpha-a|} \frac{(\langle\xi\rangle^{1-\delta}/H)^j}{j!^{s-1}} |p_m(t)|, \end{aligned}$$

pusique

$$\left|\frac{p_{(\beta)}^{(\alpha)}(t)}{p_m(t)}\right| \leqq AC^{|\beta|}|\beta|!^s \langle\xi\rangle^{-\delta|\alpha|+(1-\delta)|\beta|}$$

pour $|\alpha+\beta|\leqq r$, et

$$\left|\frac{p_{(\beta)}^{(\alpha)}(t)}{p_m(t)}\right| \leqq AC^{|\beta|}|\beta|!^s \langle\xi\rangle^{(1-\delta)r-|\alpha|}$$

pour $|\alpha+\beta|>r$; mais l'on a évidemment $(1-\delta)r-|\alpha|<(1-\delta)(|\alpha+\beta|)-|\alpha| = -\delta|\alpha|+(1-\delta)|\beta|$.

Ensuite, si l'on emploie la formule de dérivée des fonctions composites (voir N. Bourbaki ou $[1]_2$), alors

$$(4)\qquad \begin{aligned} &\partial_\xi^{b-c}[\rho^{\gamma_0+1}\chi^{(\gamma_0)}(\rho(t-s))] = (b-c)! \sum \frac{1}{i_1!\, i_2! \cdots i_k!} \\ &\times \left(\frac{d}{d\rho}\right)^i [\rho^{\gamma_0+1}\chi^{(\gamma_0)}(\rho(t-s))] \prod_{j=1}^{k} \left(\frac{\partial_\xi^{(\gamma^{(j)})}\rho}{\gamma^{(j)}!}\right)^{i_j}, \end{aligned}$$

où

$$\sum_{j=1}^{k} i_j \gamma^{(j)} = b-c, \quad \sum i_j = i \quad \text{et } k \leqq |b-c|.$$

Le calcul direct donne, en posant

$$\Phi(\rho(t-s)) = (\rho|t-s|+1)^i \max_{0\leqq j\leqq i} |\chi^{(\gamma_0+i-j)}(\rho(t-s))|$$

$$(5)\qquad \left|\left(\frac{d}{d\rho}\right)^i [\rho^{\gamma_0+1}\chi^{(\gamma_0)}(\rho(t-s))]\right| \leqq C(\gamma_0+1)^i \rho^{\gamma_0-i}\rho\Phi(\rho(t-s));$$

si l'on emploie la propriété suivante

$$|\partial_\xi^\alpha \rho| \leqq C\rho\langle\xi\rangle^{-|\alpha|}, \qquad \text{vu } \rho = \langle\xi\rangle^{\delta_1},$$

alors on aura

$$(6) \qquad |\prod (\partial_\xi^{(\gamma(j))}\rho)^{i_j}| \leqq C\rho^i \langle\xi\rangle^{-|b-c|}.$$

(5) et (6) donnent

$$(7) \qquad |\partial_\xi^{b-c}[\rho^{\gamma_0+1}\chi^{(\gamma_0)}(\rho(t-s))]| \leqq C(\gamma_0+1)^i \rho^{\gamma_0}\langle\xi\rangle^{-|b-c|}\rho\Phi(\rho(t-s)).$$

Compte tenu du Lemme 12.1, on obtient aussi

$$(8) \qquad \left|\partial_\xi^c D_x^{\gamma+\alpha}\left(\frac{1}{p_m(s)}\right)\right| \leqq AC^{|\gamma+\alpha|}|\gamma+\alpha|!^s \frac{\langle\xi\rangle^{-c}}{|p_m(s)|}\sum_{j=0}^{|\gamma+\alpha|}\frac{(\langle\xi\rangle^{1-\delta}/H)^j}{j!^{s-1}}.$$

En employant (7) et (8), on a l'évaluation

$$(9) \qquad \begin{aligned} &|\partial_\xi^b D_x^{\gamma+\alpha} D_t^{\gamma_0} q| \\ &\leqq AC^{|\gamma+\alpha|}|\gamma+\alpha|!^s \rho^{\gamma_0}\langle\xi\rangle^{-|b|}\sum_{j=0}^{|\gamma+\alpha|}\frac{(\langle\xi\rangle^{1-\delta}/H)^j}{j!^{s-1}}\int_{-\infty}^{\infty}\frac{\rho\Phi(\rho(t-s))}{|p_m(s)|}ds. \end{aligned}$$

En somme, on obtiendra, compte tenu de (3) et (9),

$$(10) \qquad \begin{aligned} &|D_x^\alpha \partial_\xi^\beta r(t,x;\tau-iH\langle\xi\rangle^\delta,\xi)| \leqq A\sum_{(\gamma_0,\gamma)} C^{|\gamma+\alpha|}|\gamma+\alpha|!^s \rho^{\gamma_0}\langle\xi\rangle^{-\delta(\gamma_0+|\gamma|)} \\ &\times\langle\xi\rangle^{-\delta|\beta|}\sum_{j=0}^{|\gamma+\alpha|}\frac{(\langle\xi\rangle^{1-\delta}/H)^j}{j!^{s-1}}|p_m(t)|\int_{-\infty}^{\infty}\frac{\rho\Phi(\rho(t-s))}{|p_m(s)|}ds. \end{aligned}$$

On majore encore

$$|p_m(t)|\int_{-\infty}^{\infty}\frac{\rho\Phi(\rho(t-s))}{|p_m(s)|}ds$$

par

$$\int_{-\infty}^{\infty}\left\{\frac{|p_m(t)-p_m(s)|}{|p_m(s)|}+1\right\}\rho\Phi(\rho(t-s))ds \leqq C(\rho^{-\kappa}\langle\xi\rangle^{r-r\delta}+1),$$

comme au n° 5.

D'où l'on a finalement

$$(11) \qquad |D_x^\alpha\partial_\xi^\beta r(t,x;\tau-iH\langle\xi\rangle^\delta,\xi)| \leqq AC^{|\alpha|}|\alpha|!^s\langle\xi\rangle^{-\delta|\beta|}\sum_{j=0}^{|\alpha|}\frac{(\langle\xi\rangle^{1-\delta}/H)^j}{j!^{s-1}},$$

en employant

$$\begin{aligned} \sum_{j=0}^{|\gamma+\alpha|}\frac{(\langle\xi\rangle^{1-\delta}/H)^j}{j!^{s-1}} &\leqq \sum_{j=0}^{|\alpha|}\frac{(\langle\xi\rangle^{1-\delta}/H)^j}{j!^{s-1}}\sum_{j=0}^{|\gamma|}\frac{(\langle\xi\rangle^{1-\delta}/H)^j}{j!^{s-1}} \\ &\leqq \text{Const.}\,\langle\xi\rangle^{(1-\delta)(\gamma_0+|\gamma|)}\sum_{j=0}^{|\alpha|}\frac{(\langle\xi\rangle^{1-\delta}/H)^j}{j!^{s-1}}, \end{aligned}$$

et

$$\sum_{(\gamma_0,\gamma)}^{\text{fini}} C^{|\gamma|}\frac{|\gamma+\alpha|!^s}{|\alpha|!^s}\langle\xi\rangle^{-\gamma_0(\delta-\delta_1)}\langle\xi\rangle^{(1-2\delta)|\gamma|}\leqq\text{const.},$$

qui s'ensuit de $1>\delta>1/2$ et $\delta_1<\delta$.

Bibliographie

[1]₁ M. D. Bronshtein, The parametrix of the Cauchy problem for hyperbolic operators with characteristics of variable multiplicity, Funkcional. Anal. i Priložen, **10** (1976), 83–84.

[1]₂ M. D. Bronštein, The Cauchy problem for hyperbolic operators with characteristics of variable multiplicity, Trudy Moskov. Mat. Obšč. **41** (1980), 87–103.

[2] ——, Smoothness of roots of polynomials depending on parameters, Sibirsk. Mat. Ž., **20** (1970), 493–501.

[3] F. Colombini, E. De Giorgi et S. Spagnolo, Sur les équations hyperboliques avec des coefficients qui ne dépendent que du temps, Ann. Scuola Norm. Sup. Pisa, **6** (1979), 511–559.

[4] F. Colombini, E. Jannelli and S. Spagnolo, Well-posedness in the Gevrey classes of the Cauchy problem for a nonstrictly hyperbolic equation with coefficients depending on time, Ann. Scuola Norm. Sup. Pisa, **10** (1983), 291–312.

[5] L. Hörmander, Linear Partial Differential Operators, Springer-Verlag, 1963.

[6] V. Ja. Ivriĭ, Correctness of the Cauchy problem in Gevrey classes for non strictly hyperbolic operators, Math. USSR-Sb., **25** (1975), 365–387.

[7] H. Kumano-go, Pseudo-differential Operators, M.I.T. Press, Cambridge, 1981.

[8]₁ J. Leray et Y. Ohya, Systèmes linéaires, hyperboliques non stricts, Colloq. Anal. Fonct., C.B.R.M., 1964, 105–144.

[8]₂ ——, Equations et systèmes non-linéaires, hyperboliques non-stricts, Math. Ann., **170** (1967), 167–205.

[9] T. Nishitani, Sur les équations hyperboliques à coefficients hölderien en *t* et de classe de Gevrey en *x*, Bull. Soc. Math. France, **107** (1983), 113–138.

[10] ——, Energy inequality for hyperbolic operators in the Gevrey class, J. Math. Kyoto Univ., **23** (1983), 739–773.

[11] W. Nuij, A note on hyperbolic polynomials, Math. Scand., **23** (1968), 69–72.

[12] Y. Ohya, Le problème de Cauchy pour les équations hyperboliques à caractéristique multiple, J. Math. Soc. Japan, **16** (1964), 268–286.

[13] J.-M. Trépreau, Le problème de Cauchy hyperbolique dans les classes d'ultra fonctions et d'ultra-distributions, Comm. Partial Differential Equations, **4** (1979), 339–387.

[14] S. Wakabayashi, Singularities of solutions of the Cauchy problem for hyperbolic systems in Gevrey classes, Japan. J. Math., **11** (1985), 157–201.

[15] S. Spagnolo, Analytic and Gevrey well-posedness of the Cauchy problem for second order weakly hyperbolic equations with coefficients irregular in time, Proceedings of this Workshop.

[16] S. Tarama, Une note sur le théorème de Bronshtein concernant le polynôme hyperbolique, à paraître.

DEPARTMENT OF APPLIED MATHEMATICS
AND PHYSICS, FACULTY OF ENGINEERING,
KYOTO UNIVERSITY, KYOTO 606,
JAPAN

Taniguchi Symp. HERT
Kyoto 1984, pp. 307–316

Solutions with Singularities on a Surface of Linear Partial Differential Equations

Sunao ŌUCHI

Let Ω be a domain in C^{n+1}, and K be a connected nonsingular complex hypersurface through the origin. Let $P(z, \partial)$ be a linear partial differential operator in Ω. Its coefficients are holomorphic in Ω. In the following, let us consider an equation

$$P(z, \partial)u(z)=f(z), \tag{0.1}$$

where $u(z)$ and $f(z)$ may have singularities on K, that is, we consider (0.1) in a function space which admits singularities on K.

So first in order to study (0.1), we define some notions which represent the relation between the operator $P(z, \partial)$ and the hypersurface K. Next we give theorems concerning solutions of the equation (0.1).

§ 1. Characteristic indices and localizations for a surface K

Let us give some notations. $z=(z_0, z_1, \cdots, z_n)=(z_0, z')$ is a point in C^{n+1}. For a domain $U\subset C^{n+1}$, $\mathcal{O}(U)$ means the set of all holomorphic functions in U, $\tilde{\mathcal{O}}(U-K)$ is the set of all holomorphic functions on the universal covering space of $U-K$, and $\mathcal{L}(U)$ is the set of all holomorphic vector fields in U, $\mathcal{L}(U)=\{\sum_{i=0}^{n} a_i(z)\partial/\partial z_i;\ a_i(z)\in\mathcal{O}(U)\}$. For a multi-index $\alpha=(\alpha_0, \alpha_1, \cdots, \alpha_n)=(\alpha_0, \alpha')$, $\alpha_i\in Z_+=N\cup\{0\}$, $|\alpha|=\alpha_0+\alpha_1+\cdots+\alpha_n=\alpha_0+|\alpha'|$ and for $X=(X_0, X_1, \cdots, X_n)=(X_0, X')\in\mathcal{L}(U)^{n+1}$, $X^\alpha=(X_0)^{\alpha_0}(X_1)^{\alpha_1}\cdots(X_n)^{\alpha_n}=(X_0)^{\alpha_0}(X')^{\alpha'}$.

Now let p be a point on K. In a small neighbourhood U of p, since K is nonsingular, we can put $K\cap U=\{\varphi(z)=0\}$, $\varphi(z)\in\mathcal{O}(U)$, $d\varphi(z)\neq 0$ on $K\cap U$.

Take vector fields $X=\{X_0, X_1, \cdots, X_n\}\in\mathcal{L}(U)^{n+1}$ as follows:

(1.1) $\{X_i\}$ $(0\leq i\leq n)$ are linearly independent as each point in U.

$$\begin{cases}\langle d\varphi, X_0\rangle\neq 0 & \text{on } K\cap U,\\ \langle d\varphi, X_i\rangle=0 & \text{on } K\cap U \text{ for } 1\leq i\leq n,\end{cases} \tag{1.2}$$

where $\langle\ ,\ \rangle$ means the product of cotangent and tangent vectors. (1.2) means that X_0 is transversal to K and others are tangent to K. Now by using these vector fields $\{X_i\}$ $(0 \leq i \leq n)$, we can represent $P(z, \partial)$ as follows,

$$(1.3)\qquad \begin{cases} P(z, \partial) = \sum_{k=0}^{m} P_k(z, \partial), \\ P_k(z, \partial) = \sum_{|\alpha|=k} A_\alpha(z) X^\alpha = \sum_{|\alpha|=k} a_\alpha(z) \varphi(z)^{j_\alpha} X^\alpha. \end{cases}$$

Here $P_k(z, \partial)$ is the homogeneous part of $P(z, \partial)$ with degree k with respect to X and $A_\alpha(z) = a_\alpha(z)\varphi(z)^{j_\alpha}$, $a_\alpha(z) \not\equiv 0$ on K, that is, $A_\alpha(z)$ vanishes exactly with the order j_α on K.

Put

$$(1.4)\qquad \begin{cases} s_k = \min\{|\alpha'|;\ A_\alpha(z) \not\equiv 0,\ |\alpha| = k\}, \\ d_k = \min\{|\alpha'| + j_\alpha;\ A_\alpha(z) \not\equiv 0,\ |\alpha| = k\}, \\ J_k = \min\{j_\alpha;\ |\alpha'| + j_\alpha = d_k,\ A_\alpha(z) \not\equiv 0,\ |\alpha| = k\}. \end{cases}$$

From (1.4) we have

$$(1.5)\qquad \begin{aligned} P_k(z, \partial) &= \sum_{l=s_k}^{k} \{\sum_{|\alpha'|=l} A_\alpha(z)(X_0)^{k-l}(X')^{\alpha'}\} \\ &= \sum_{d \geq d_k} \{\sum_{j_\alpha + |\alpha'| = d} a_\alpha(z)\varphi(z)^{j_\alpha}(X_0)^{k-|\alpha'|}(X')^{\alpha'}\}. \end{aligned}$$

Put

$$(1.6)\qquad P_{k, d_k}(z, \partial) = \sum_{j_\alpha = J_k, |\alpha'| = d_k - J_k} a_\alpha(z)(X)^{\alpha'}.$$

We note that we can restrict $P_{k, d_k}(z, \partial)$ to the surface K and obtain an operator on K. If $A_\alpha(z) \equiv 0$ for all α with $|\alpha| = k$, $s_k = d_k = J_k = +\infty$.

Now consider the set $A = \{(k, d_k);\ 0 \leq k \leq m\} \subset \boldsymbol{R}^2$ and its convex hull $\hat{A}$. The lower convex parts of the boundary $\partial\hat{A}$ of $\hat{A}$, in general, are segments $\Sigma(i)$ $(1 \leq i \leq l)$ and put $\Sigma = \bigcup_{i=1}^{l} \Sigma(i)$. The set of vertices of Σ consists of $(l+1)$-points $\{(k_i, d_{k_i});\ 0 \leq i \leq l\}$, $k_0 = m > k_1 > k_2 > \cdots > k_l \geq 0$ (see Fig. 1). Put

$$(1.7)\qquad \sigma_i = \max\{(d_{k_{i-1}} - d_{k_i})/(k_{i-1} - k_i),\ 1\}.$$

$(d_{k_{i-1}} - d_{k_i})/(k_{i-1} - k_i)$ is the slope of $\Sigma(i)$. Hence there is a $p \in N$ such that $\sigma_1 > \sigma_2 > \cdots > \sigma_p = 1$.

If $\sigma_1 = 1$, we put

$$(1.8)\qquad \sigma_{1,1} = \max_{\{0 \leq k < m;\ d_m - d_k = m - k\}} \{(J_m - J_k)/(m - k),\ 1\}.$$

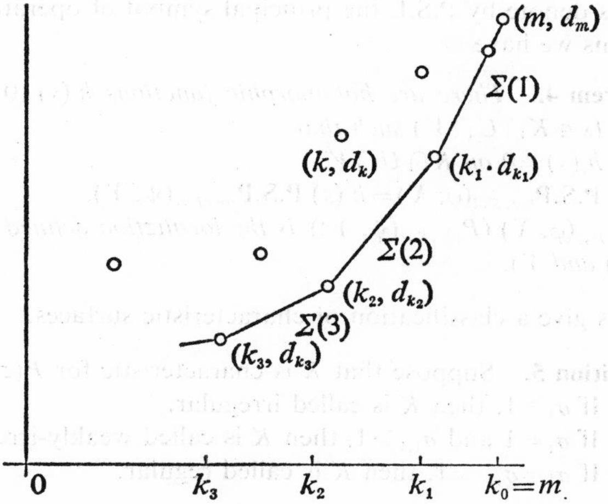

Fig. 1.

As we remarked, we have operators on K, by restricting $P_{k_i, d_{k_i}}(z, \partial)$ $(0 \leq i \leq p-1)$ on K, and put

(1.9) $\qquad P_{\mathrm{loc}, K, i} =$ the restriction of $P_{k_i, d_{k_i}}(z, \partial)$ on K.

Now we give

Definition 1. (1) We call σ_i $(1 \leq i \leq p)$ and $\sigma_{1,1}$ characteristic indices.

(2) We call $P_{\mathrm{loc}, K, i}$ $(0 \leq i \leq p-1)$ the localizations on K. In particular $P_{\mathrm{loc}, K, 0}$ is called the principal lecalization.

Remark 2. For the set $B = \{(k, s_k); 0 \leq k \leq m\}$, by considering convex hull $\hat{B}$ of B, we can define the notions similar to those defined for A. For the later purpose we define $\tilde{\sigma}_1$ and $\tilde{\Sigma}(1)$:

$$\tilde{\sigma}_1 = \max_{0 \leq k < m} \{(s_m - s_k)/(m-k), 1\}, \tag{1.10}$$

and $\tilde{\Sigma}(1)$ is the segment with maximum slope $\tilde{\sigma}_1$ of the lower convex parts of the boundary $\partial\hat{B}$ whose endpoints are (k_0, s_{k_0}) and (k_1, s_{k_1}) $(m = k_0 > k_1)$.

Characteristic indices and localizations are defined by vector fields $\{X_i\}$ $(0 \leq i \leq n)$ and the defining function $\varphi(z)$ of K. We choose another function $\psi(z)$ defining K and other vector fields $Y = \{Y_i\}$ $(0 \leq i \leq n)$ satisfying (1.1)–(1.2) in a neighbourhood V of p. Then we have

Theorem 3. *The characteristic indices σ_i $(0 \leq i \leq p)$ and $\sigma_{1,1}$ do not depend on $\varphi(z)$ defining K and vector fields $X = \{X_i\}$ $(0 \leq i \leq n)$ with* (1.1)–(1.2).

Let us denote by P.S.L the principal symbol of operator L. For the localizations we have

Theorem 4. *There are holomorphic functions* $h_i(s)$ $(0 \leq i \leq p-1)$ *on* $K \cap U \cap V$ $(s \in K \cap U \cap V)$ *such that*

(i) $h_i(s) \neq 0$ *on* $K \cap U \cap V$,

(ii) $\text{P.S.P}_{\text{loc},K,i}(\varphi, X) = h_i(s)\ \text{P.S.P}_{\text{loc},K,i}(\psi, Y)$,

where $P_{\text{loc},K,i}(\varphi, X)$ $(P_{\text{loc},K,i}(\psi, Y))$ *is the localiztion defined by* $\varphi(z)$ *and* X *(resp.* $\psi(z)$ *and* Y).

Let us give a classification of characteristic surfaces.

Definition 5. Suppose that K is characteristic for $P(z, \partial)$.

(i) If $\sigma_1 > 1$, then K is called irregular.

(ii) If $\sigma_1 = 1$ and $\sigma_{1,1} > 1$, then K is called weakly-irregular.

(iii) If $\sigma_1 = \sigma_{1,1} = 1$, then K is called regular.

Remark 6. (i) If K is generically noncharacteristic,

$$\text{P.S.P}\,(z, \xi)|_{z \in K, \xi = d\varphi(z)} \not\equiv 0, \quad \text{then } \sigma_1 = \sigma_{1,1} = 1.$$

If $\sigma_1 > 1$ or if $\sigma_1 = 1$ and $\sigma_{1,1} > 1$, then K is characteristic.

(ii) Let $P(z, \partial)$ be an operator with decomposable principal part and K be its characteristic surface, that is, P.S.P (z, ξ) is decomposed as follows and said to have constant multiple characteristics (see [2], [4]):

$$\text{P.S.P}\,(z, \xi) = g(z, \xi) k(z, \xi)^r \quad (r \in N), \tag{1.11}$$

where $g(z, \xi)$ and $k(z, \xi)$ are homogeneous polynomials of ξ with holomorphic coefficients such that

$$k(z, \xi)|_{z \in K, \xi = d\varphi} = 0, \qquad \text{grad}_\xi k(z, \xi)|_{z \in K, \xi = d\varphi} \not\equiv 0 \tag{1.12}$$

and

$$g(z, \xi)|_{z \in K, \xi = d\varphi} \not\equiv 0. \tag{1.13}$$

Let σ' be the irregularity of characteristic element of $k(z, \xi)$ defined in [4]. We have $\sigma_1 \leq \sigma'$. For $P(z, \partial) = (\partial_1)^3 + z_0(\partial_0)^2$ and $K = \{z_0 = 0\}$, we have $\sigma_1 = 2$, but $\sigma' = 3$. If $P(z, \partial)$ satisfies the Levi's condition, namely $\sigma' = 1$, we have $\sigma_1 = \sigma_{1,1} = 1$.

(iii) If $P(z, \partial)$ is an ordinary differential operator of one variable z_0, the case $\sigma_1 = 1$ and $\sigma_{1,1} > 1$ does not occur.

(iv) If $P(z, \partial)$ is an operator of Fuchsian type for $K = \{z_0 = 0\}$ (see [1]), then $\sigma_1 = \sigma_{1,1} = 1$. Moreover for operators treated in [3] and [11], where characteristic initial value problems or singular Cauchy problems are considered, we have $\sigma_1 = \sigma_{1,1} = 1$ for the initial characteristic surface.

In [1] operators of Fuchsian type are defined by the coordinate chosen so that $K=\{z_0=0\}$. We can give a definition of Fuchsian type by using the notions defined above.

Definition 7. $P(z, \partial)$ is said to be an operator of Fuchsian type with respect to K, if the following conditions hold:

(a) $\sigma_1=\sigma_{1,1}=1$,

(b) $P_{\mathrm{loc},K,0}$ is an operator of order 0, that is, a function.

Under conditions (a) and (b), if we choose the coordinate so that $K=\{z_0=0\}$, $P(z, \partial)$ can be written as follows:

$$P(z, \partial)=(z_0)^{J_m-m}\sum_{i=0}^{m} p_i(z, \partial')(z_0)^{m-i}(\partial_0)^{m-i}, \tag{1.14}$$

where $\operatorname{ord} p_i(z, \partial')\leq i$, $\operatorname{ord} p_i(0, z', \partial')\leq 0$, and $p_0(0, z')$ is the principal localization and $p_0(0, z')\not\equiv 0$.

Finally we give examples. Let $K=\{z_0=0\}$.

$$\begin{cases} P(z, \partial)=z_0(\partial_0)(\partial_1)^3+(z_0)^2(\partial_0)^3+(\partial_1)^3+\partial_0\partial_1, \\ \sigma_1=2, \quad \sigma_2=1, \quad P_{\mathrm{loc},K,0}=(\partial_1)^3, \quad P_{\mathrm{loc},K,1}=I, \end{cases} \tag{1.15}$$

$$\begin{cases} P(z, \partial)=(z_0)^2(\partial_0)^2(\partial_1)^2+(\partial_1)^3+(z_0)(\partial_0)(\partial_1) \\ \qquad\qquad +a(z)(\partial_1)^2+b(z)(\partial_1)+c(z), \\ \sigma_1=1, \quad \sigma_{1,1}=2, \quad P_{\mathrm{loc},K,0}=(\partial_1)^2. \end{cases} \tag{1.16}$$

$K=\{z_0=0\}$ is irregular for (1.15) and weakly irregular for (1.16).

The most part of Section 1 is stated in Ōuchi [9]. In [9] other characteristic indices $\sigma_{p,i}$ and localizations $P_{\mathrm{loc},K,(p,i)}$ are defined.

§ 2. Solutions with singularities on K

In Section 2 let us consider an equation

$$P(z, \partial)u(z)=f(z), \tag{2.1}$$

where $f(z)\in\tilde{\mathcal{O}}(\Omega-K)$, in a neighbourhood of $z=0$. For simplicity we choose the coordinate so that $K=\{z_0=0\}$. Hence $z'=(z_1, z_2, \cdots, z_n)$ is the coordinate in K. For dual variable $\xi=(\xi_0, \xi')$, ξ' is corresponding to z'. By this coordinate $P(z, \partial)$ is expressed as follows:

$$\begin{aligned} P(z, \partial)&=\sum_{k=0}^{m}\Big(\sum_{l=s_k}^{k} A_{k,l}(z, \partial')(\partial_0)^{k-l}\Big) \\ &=\sum_{k=0}^{m}\Big\{\sum_{d\geq d_k}\Big(\sum_{j_{k,l}+l=d}(z_0)^{j_{k,l}}a_{k,l}(z, \partial')(\partial_0)^{k-l}\Big)\Big\}, \end{aligned} \tag{2.2}$$

$$\operatorname{ord} A_{k,l}(z, \partial')\leq l, \quad A_{k,l}(z, \partial')=(z_0)^{j_{k,l}}a_{k,l}(z, \partial').$$

We have

$$P_{\mathrm{loc},K,0}(z', \partial') = a_{m,d_m-J_m}(0, z', \partial'). \tag{2.3}$$

For existence of solutions of (2.1), we have

Theorem 8. *Assume*

$$P_{\mathrm{loc},K,0}(0, \xi') \not\equiv 0. \tag{2.4}$$

Then there is a solution $u(z) \in \tilde{\mathcal{O}}(U-K)$ *of* (2.1), *where the neighbourhood* U *of* $z=0$ *does not depend on* $f(z)$.

Now, in which function space can we find a solution $u(z)$, if $f(z)$ belongs to some function space? To answer this, we give some subclass of $\tilde{\mathcal{O}}(\Omega-K)$:

(2.5) $\tilde{\mathcal{O}}_\gamma(\Omega-K) = \{f(z) \in \tilde{\mathcal{O}}(\Omega-K)$; for any α and β, there are $A_{\alpha,\beta}$ and c such that for $z \in \Omega-K$ with $\alpha < \arg z_0 < \beta$ $|f(z)| \leq A_{\alpha,\beta} \exp(c|z_0|^{-\gamma})\}$,

(2.6) $\tilde{\mathcal{O}}_{(0,\delta)}(\Omega-K) = \{f(z) \in \tilde{\mathcal{O}}(\Omega-K)$; for any α and β, there are $A_{\alpha,\beta}$ and c such that for $z \in \Omega-K$ with $\alpha < \arg z_0 < \beta$ $|f(z)| \leq A_{\alpha,\beta} \exp(c|\log z_0|^{\delta})\}$,

where $\gamma > 0$ and $\delta \geq 1$. If $\delta = 1$, then for $f(z) \in \tilde{\mathcal{O}}_{(0,1)}(\Omega-K)$

$$|f(z)| \leq A'_{\alpha,\beta}|z_0|^{-c} \quad \text{in } z \in \Omega-K \text{ and } \alpha < \arg z_0 < \beta. \tag{2.7}$$

We have

Theorem 9. *Assume* (2.4). (i) *If* $\sigma_1 > 1$ *and* $f(z) \in \tilde{\mathcal{O}}_{(\sigma_1-1)}(\Omega-K)$, *then* $u(z)$ *is found in* $\tilde{\mathcal{O}}_{(\sigma_1-1)}(U-K)$.

(ii) *If* $\sigma_1 = 1$ *and* $f(z) \in \tilde{\mathcal{O}}_{(0,\sigma_1,1)}(\Omega-K)$, *then* $u(z)$ *is found in* $\tilde{\mathcal{O}}_{(0,\sigma_1,1)}(U-K)$.

Theorems 8 and 9 are stated with their application to existence of null solutions in Ōuchi [7]. Their proofs are in Ōuchi [10]. We give an outline. Assume (2.4) is valid for $\xi' = (1, 0, \cdots, 0)$. First we construct a formal solution $v(z, \lambda, \zeta)$ of initial value problem with data given on $z_1 = 0$,

$$\begin{cases} P(z, \partial)v(z, \lambda, \zeta) = (z_0)^{\lambda+d_m-m} f(e^{\zeta}, z'), \\ (\partial_1)^h v(z_0, 0, z'', \lambda, \zeta) = (z_0)^{\lambda} g_h(e^{\zeta}, z'') \quad (0 \leq h \leq L-1), \end{cases} \tag{2.8}$$

where $g_h(z_0, z'') \in \tilde{\mathcal{O}}(K \cap \{z_1=0\})$ and $L = \operatorname{ord} P_{\mathrm{loc},K,0}$. To show Theorems 8 and 9, it is sufficient that we take $g_h(z_0, z'')=0$. A formal solution $v(z, \lambda, \zeta)$ is constructed in the form

$$v(z, \lambda, \zeta)=\sum_{n=0}^{\infty}\left(\sum_{s=-\infty}^{S(n)} v_{n,s}(z', \lambda, \zeta)(z_0)^{\lambda+s}\right). \tag{2.9}$$

Next we integrate $v_{n,s}(z', \lambda, \zeta)$ and get

$$v^*_{n,s}(z, \zeta)=(2\pi i)^{-1}\int_C \exp(-\lambda\zeta)(z_0)^{\lambda+s} v_{n,s}(z', \lambda, \zeta)\log\lambda\, d\lambda. \tag{2.10}$$

It is shown that the infinite sum

$$V(z, \zeta)=\sum_{n=0}^{\infty}\left(\sum_{s=-\infty}^{S(n)} v^*_{n,s}(z, \zeta)\right) \tag{2.11}$$

converges. Then

$$u(z)=(2\pi i)^{-1}\int_\Gamma V(z, \zeta)\, d\zeta. \tag{2.12}$$

C in (2.10) and Γ in (2.12) are some infinite paths.

Next we consider $u(z) \in \tilde{\mathcal{O}}(\Omega - K)$ satisfying

$$P(z, \partial)u(z)=f(z) \in \mathcal{O}(\Omega). \tag{2.13}$$

$u(z)$ is a homogeneous solution modulo holomorphic functions. The purpose is to give an integral representation of $u(z)$. We seek it in the form

$$u(z)=(2\pi i)^{-1}\int_\Gamma \exp(\lambda z_0) W(z, \lambda)\log\lambda\, d\lambda. \tag{2.14}$$

In order to do so we define some constants and operators induced from the given operator $P(z, \partial)$.

Put

$$\alpha=(\tilde{\sigma}_1-1)/\tilde{\sigma}_1. \tag{2.15}$$

From the definition of $\tilde{\sigma}_1$ (see (1.10)), there is a nonnegative rational number β_k such that

$$s_m - s_k + k - m + \beta_k = \alpha(s_m - s_k). \tag{2.16}$$

We define $\tilde{P}(\lambda; z, \partial)$ as follows:

$$P(z, \partial)\{\exp(\lambda z_0) W(z, \lambda)\}=\exp(\lambda z_0)\tilde{P}(\lambda; z, \partial) W(z, \lambda). \tag{2.17}$$

$\tilde{P}(\lambda; z, \partial)$ has a form

$$\tilde{P}(\lambda; z, \partial)=\sum_{k=0}^{m}\left\{\sum_{j=0}^{k-s_k}\lambda^{k-s_k-j}P_{k,s_k+j}(z, \partial)\right\}, \tag{2.18}$$

$$\operatorname{ord} P_{k,s_k+j}(z, \partial)\leq s_k+j.$$

From (2.16), we have

$$\tilde{P}(\lambda; z, \partial)=\lambda^{(1-\alpha)(m-s_m)}\sum_{k=0}^{m}\left\{\sum_{j=0}^{k-s_k}\lambda^{\alpha(m-s_k-j)-(1-\alpha)j-\beta_k}P_{k,s_k+j}(z, \partial)\right\}. \tag{2.19}$$

Thus we define from (2.19)

$$P(\lambda; z, \partial_z, \partial_\zeta)=\sum_{k=0}^{m}\left\{\sum_{j=0}^{k-s_k}\lambda^{-(1-\alpha)j-\beta_k}(\partial_\zeta)^{m-s_k-j}P_{k,s_k+j}(z, \partial)\right\}, \tag{2.20}$$

which is obtained by replacing λ^α by ∂_ζ in (2.19).

We have

Theorem 10. *Assume*

$$\begin{cases} P_{\mathrm{loc},K,0}(0, \xi')|_{\xi'=(1,0,\dots,0)}\neq 0, \\ d_m=s_m. \end{cases} \tag{2.21}$$

Let $|\arg z_0|<\pi/2$. *For* $z\in U-K$ $(U\subset\Omega)$, $u(z)$ *has a representation*

$$u(z)=(2\pi i)^{-1}\int_\Gamma \exp(\lambda z_0)W(z, \lambda)\log\lambda d\lambda+v(z), \tag{2.22}$$

where $v(z)\in\mathcal{O}(U)$ *and*

$$W(z, \lambda)=(2\pi i)^{-1}\int_C \exp(-\lambda^\alpha\zeta)w(z, \zeta, \lambda)d\zeta. \tag{2.23}$$

$w(z, \zeta, \lambda)$ *satisfies*

$$\begin{cases} P(\lambda; z, \partial_z, \partial_\zeta)w(z, \zeta, \lambda)=0, \\ \left(\dfrac{\partial}{\partial z_1}\right)^l w(z_0, 0, z'', \zeta, \lambda)=h_l(\lambda, z'')/\zeta \qquad \text{for } 0\leq l\leq s_m-1, \\ h_l(\lambda, z'')=\displaystyle\int_\gamma \exp(-\lambda t)\left(\frac{\partial}{\partial z_1}\right)^l u(t, 0, z'')dt. \end{cases} \tag{2.24}$$

Γ is an infinite path that starts at $\infty\exp(-\pi i)$ and goes around the origin $\lambda=0$ and ends at $\infty\exp(\pi i)$. C is a circle with radius $d|\lambda|^{1-\alpha}$ that starts at $-d\lambda^{1-\alpha}$ and ends at $-d\lambda^{1-\alpha}\exp(2\pi i)$. γ is a path that starts at $t=b\exp(\pi i)$, goes to $t=\varepsilon\exp(\pi i)$ straightly, goes around once on $|t|=\varepsilon$ and goes from $t=\varepsilon\exp(-\pi i)$ to $t=b\exp(-\pi i)$.

Let us explain Theorem 10. We may assume $\Omega=\{|z|<r\}$. The constant b in γ is chosen so that $0<b<r$. If $u(z)\in\mathcal{O}$, then $h_i(\lambda, z'')\equiv 0$. Let us solve (2.24). We note that the initial values have singularity on $\zeta=0$. We can find $w(z, \zeta, \lambda)$ which is holomorphic in $\{(z, \zeta, \lambda); z\in U, B<|\lambda|<\infty, c<|\zeta|<A|\lambda|^{1-a}\}$. So constant d in path C is chosen $0<d<A$. $W(z, \lambda)$ in (2.23) is holomorphic in $B<|\lambda|<\infty$ and has an estimate, for any $\varepsilon>0$

$$(2.25)\qquad \sup_{z\in U}|W(z, \lambda)|\leq A_\varepsilon \exp(\varepsilon|\lambda|)$$

holds for λ with $|\lambda|\geq B$ and $\operatorname{Re}\lambda\leq 0$. Thus the integral (2.22) converges. It can be holomorphically extensible in a wider domain.

In order to get information of $u(z)$, we have to investigate $w(z, \zeta, \lambda)$. The singularities of $w(z, \zeta, \lambda)$ lie in the disk $\{|\zeta|\leq c\}$. It is expected that they lie on the characteristic surfaces which issue from $\zeta=0$ at $z_1=0$. So the property of $P(\lambda; z, \partial_z, \partial_\zeta)$ is important to analyze the singularities. Hence, adding some conditions, we can investigate characteristic surfaces of $P(\lambda; z, \partial_z, \partial_\zeta)$. They are given by means of sets $A=\{(k, d_k); 0\leq k\leq m\}$ and $B=\{(k, s_k); 0\leq k\leq m\}$. For example, let us assume

$$(2.26)\qquad \Sigma(1)=\tilde{\Sigma}(1)\quad \text{(see Remark 2)}$$

and

$$(2.27)\qquad P_{\text{sub}}(z', \xi')=\sum_{(k,d_k)\in\Sigma(1)} P_{k,d_k}(z', \xi')=0 \quad \text{has } (d_m-d_{k_1})\text{-nonzero}$$

distinct roots at $z'=0$ and $\xi''=0$ with respect to ξ_1.

(2.26) and (2.27) mean that $d_{k_1}=s_{k_1}$ and $P_{\text{sub}}(0, \xi_1, 0)=0$ has $(d_m-d_{k_1})$ $(=s_m-s_{k_1})$-nonzero roots and zero root with multiplicity d_{k_1} $(=s_{k_1})$. Under (2.26) and (2.27), we can study the integral representation in Theorem 10.

Theorem 10 is an improvement of results in [5] and [6], where $P(z, \partial)$ was assumed to be an operator with constant multiple characteristics and the asymptotic behaviour of homogeneous solutions was studied. The proof of Theorem 10 and its applications will be given elsewhere.

Remark 11. In this paper we only use σ_1, $\sigma_{1,1}$, $\Sigma(1)$ and the principal localization. But in Ōuchi [8] other σ_i and localizations are used, where characteristic Cauchy problems are treated and the relation between solutions of formal power series and genuine solutions are investigated.

References

[1] M. S. Baouendi and C. Goulaouic, Cauchy problems with characteristic initial hypersurface, Comm. Pure Appl. Math., **26** (1973), 455–475.

[2] J. C. De Paris, Problème de Cauchy analytique à données singulières pour un opérateur différentiel bien décomposable, J. Math. Pure Appl. **51** (1972), 465–488.

[3] Y. Hasegawa, On the initial value problems with data on a characteristic hypersurface, J. Math. Kyoto Univ., **13** (1973), 579–593.

[4] H. Komatsu, Irregularity of characteristic elements and construction of null solutions, J. Fac. Sci. Univ. Tokyo Sec. IA, Math., **23** (1976), 297–342.

[5] S. Ōuchi, Asymptotic behaviour of singular solutions of linear partial differential equations in the complex domain, J. Fac. Sci. Univ. Tokyo, Sec. IA, Math., **27** (1980), 1–36.

[6] ——, An integral representation of singular solutions of linear partial differential equations in the complex domain, J. Fac. Sci. Univ. Tokyo, Sec. IA, Math., **27** (1980), 37–85.

[7] ——, Characteristic indices and subcharacteristic indices of surfaces for linear partial differential operators, Proc. Japan Acad., **57** (1981), 481–484.

[8] ——, Characteristic Cauchy problems and solutions of formal power series, Ann. Inst. Fourier, **33** (1983), 131–176.

[9] ——, Index, localizations and classification of characteristic surfaces for linear partial differential operators, Proc. Japan Acad., **60** (1984), 189–192.

[10] ——, Existence of singular solutions and null solutions for linear partial differential operators, preprint.

[11] H. Tahara, Singular hyperbolic systems, IV, Remarks on the Cauchy problem for singular hyperbolic partial differential equations, Japan. J. Math., **8** (1982), 297–308.

DEPARTMENT OF MATHEMATICS
FACULTY OF SCIENCE
JŌCHI (SOPHIA) UNIVERSITY
TOKYO 102, JAPAN

Taniguchi Symp. HERT
Katata 1984, pp. 317–327

Poisson Relation for Manifolds with Boundary

Vesselin M. PETKOV

§ 1. Introduction

Let $K \subset \boldsymbol{R}^n$, $n \geq 2$, be a bounded connected domain with C^∞ boundary X. Let Δ_D be a self-adjoint operator in $L^2(K)$, corresponding to the Laplacian with Dirichlet boundary condition on X. Denoting by $\{\lambda_j^2\}_{j=1}^\infty$ the eigenvalues of $-\Delta_D$, set

$$\lambda(t) = \sum_j \cos \lambda_j t \in \mathscr{S}'(\boldsymbol{R}).$$

It is well-known (cf. [1], [14]) that

$$\text{(1)} \qquad \operatorname{singsupp} \lambda(t) \subset -\bigcup_{\gamma \in \mathscr{L}} T_\gamma \cup \{0\} \cup \bigcup_{\gamma \in \mathscr{L}} T_\gamma,$$

where T_γ is the period of a generalized periodic geodesics γ, lying in $\bar{K}$, while $\mathscr{L}$ denotes the union of all such generalized periodic geodesics. A generalized geodesics is a projection on $\bar{K}$ of a generalized periodic bicharacteristics, related to the wave operator $\Box = \partial_t^2 - \Delta$. We refer to [14] for the precise definition of generalized bicharacteristics.

A result, similar to (1), holds if we study the poles μ_j of the scattering matrix $S(\sigma)$, related to the wave operator in the unbounded domain $\boldsymbol{R} \times \Omega$, $\Omega = \boldsymbol{R}^n \backslash K$ with Dirichlet boundary conditions on X. Assuming $n \geq 3$, n odd, consider

$$\mu(t) = \sum_j e^{i\mu_j t} \in \mathscr{D}'(\boldsymbol{R}^+),$$

where the summation is over all poles (see [15]). By using a finite speed of propagation argument, it is easy to deduce from (1) the relation

$$\text{(2)} \qquad \operatorname{singsupp} \mu(t) \subset \bigcup_{\gamma \in \mathscr{L}_e} T_\gamma,$$

where $\mathscr{L}_e$ denotes the union of all generalized geodesics on Ω.

The relations (1), (2) are called Poisson relations. To study some inverse spectral and scattering problems, it is important to examine the following question:

$$\text{(3)} \qquad \text{When } T_\gamma \in \operatorname{singsupp} \lambda(t) \ (\operatorname{singsupp} \mu(t))?$$

The starting point for the analysis of (3) is a suitable trace formula, involving the fundamental solution to the Dirichlet problem for the wave equation. For example, from spectral theory, it follows that

$$\lambda(t) = \operatorname{tr} \cos(\sqrt{-\Delta_D}\, t) = \int_K E(t, x, x)dx. \tag{4}$$

Here $E(t, x, x)$ denotes the solution to the problem

$$\begin{cases} \Box E = 0 \text{ in } \boldsymbol{R} \times K, & E = 0 \text{ in } \boldsymbol{R} \times X, \\ E \restriction_{t=0} = \delta(x-y), & \partial_t E \restriction_{t=0} = 0. \end{cases} \tag{5}$$

A similar trace formula, connecting $\mu(t)$ and the fundamental solution to the exterior Dirichlet problem for the wave equation in Ω, has been established in [15].

To study the singularities of $\lambda(t)$ from (4), some representation of the fundamental solution for t close to T_γ is needed. A global parametrix for $E(t, x, y)$ in a sufficiently small neighborhood of a multiple reflecting ray was constructed in [3], [6]. For this reason we will restrict our attention to multiple reflecting rays called later simply rays. On the other hand, it seems reasonable to conjecture that generically there are no generalized periodic geodesics different from multiple reflecting rays and geodesics lying on the boundary. In this direction a useful geometric information is contained in [11], [20].

Since in general the singularities, related to two rays, could be canceled, two main problems arise:

(R) If generically for any two different rays γ_1, γ_2 with periods T_1, T_2, we have $T_1/T_2 \notin \boldsymbol{Q}$?

(P) If generically the (linear) Poincaré map, related to every ray, has spectrum which does not contain $\sqrt[p]{1}$, $\forall p \in \boldsymbol{N}$? (We refer to [6] for the definition of linear Poincaré map.)

For Riemannian manifolds without boundary for a residual set of Riemannian metrics the answer of these two questions is yes. In our case the situation is more complicated since the metric in $\boldsymbol{R}^n$ is fixed and we could "move" generically only the boundary. The precise definition is given in Section 2. In Section 2 we expose a joint work with L. Stojanov showing that if some "pathological" pairs of rays are excluded, then any two rays, corresponding to a generic situation, have rationally independent periods. In Section 3 we discuss the representation of Poincaré map, given in [18], and we obtain a property of the rays which is more weaker than the condition (P). Finally, in Section 4 we consider some applications concerning spectral invariants for strictly convex planar domains and the distribution of poles of the scattering matrix for trapping obstacles.

§ 2. Rationally independent periods

Let $C^\infty(X, \boldsymbol{R}^n)$ be the space of C^∞ maps with the usual topology. Denote by $C^\infty_{\text{in}}(X, \boldsymbol{R}^n)$ the space of all injective smooth maps. Given $f \in C^\infty_{\text{in}}(X, \boldsymbol{R}^n)$, we may consider the manifold $f(X)$ and the corresponding (multiple reflecting) rays with reflection points on $f(X)$. Consider two such rays γ, δ with different reflection points $\{x_i\}$, $i=1, \cdots, M$, $\{y_i\}$, $i=1, \cdots, N$. Suppose $x_i = y_i$ for some fixed i and consider the unit vectors

$$e_{1,i}, \cdots, e_{k_i,i}, \qquad f_{1,i}, \cdots, f_{l_i,i},$$

corresponding respectively to segments of γ and δ, reflecting at x_i and y_i. To fix an orientation, we make the conditions

$$\langle e_{j,i}, n_{x_i} \rangle > 0, \qquad \langle f_{j,i}, n_{x_i} \rangle > 0,$$

n_{x_i} being the unit normal at x_i pointing into Ω. In general we could have repeating reflection points which mean that $k_i \geq 4$ or $l_i \geq 4$.

Definition 1. A pair (γ, δ) of rays will be called admissible if one of the following conditions holds:

(i) there exists at least one reflection point of $\gamma(\delta)$ which is not reflection point for $\delta(\gamma)$,

(ii) the different reflection points of γ and δ coincide, i.e. $M=N$, $x_i = y_i$ for $i=1, \cdots, M$, but there exists an index i such that for the corresponding unit vectors, introduced above, we have

$$\sum_{m=1}^{k_i} e_{m,i} \neq \sum_{m=1}^{l_i} f_{m,i}.$$

Remark 1. The condition (ii) will be satisfied if there exists at least one reflection point of γ and δ which is not repeating, that is for the corresponding numbers k_i, l_i we have $k_i = l_i = 2$. Therefore, obviously, we get $e_{1,i} + e_{2,i} \neq f_{1,i} + f_{2,i}$.

Remark 2. Let γ be a ray with successive reflection points counting with possible repetition $x_1, \cdots, x_L$. Denoting by M the number of different points x_i, the difference $d = L - M$ is called defect number. Lazutkin ([12], p. 169) proved that for strictly convex planar domains and non-symmetric rays generically we have $d \leq 2$. A simple argument shows that generically for strictly convex planar domains every pair of rays is admissible.

A partial answer to problem (R) is contained in the following

Theorem 1 ([19]). *Let $\mathcal{M}$ be the set of the maps $f \in C^\infty_{\text{in}}(X, \mathbf{R}^n)$ such that every admissible pair of rays, corresponding to the mnaifold $f(X)$ and having reflection points on $f(X)$, is formed by rays with rationally independent (primitive) periods. Then $\mathcal{M}$ contains a countable intersection of open dense sets in $C^\infty_{\text{in}}(X, \mathbf{R}^n)$.*

We will reduce Theorem 1 to the proof of another problem for which a transversality type argument can be adapted. Introduce the set $\Gamma=(k, l, s, \alpha, \beta)$ with $k, l, s \in \mathbf{N}$, $s \leq k$, $s \leq l$, where α, β are maps:

$$\alpha: (1, \cdots, k) \to (1, \cdots, s), \qquad \beta: (1, \cdots, l) \to (1, \cdots, s),$$
$$\alpha(i+1) \neq \alpha(i), \qquad i=1, \cdots, k-1, \qquad \alpha(k) \neq \alpha(1),$$
$$\beta(i+1) \neq \beta(i), \qquad i=1, \cdots, l-1, \qquad \beta(l) \neq \beta(1).$$

Let $X^{(s)}=\{(x_1, \cdots, x_s);\ x_i \in X,\ x_i \neq x_j \text{ for } i \neq j\}$ and let

$$f^{(s)}(x_1, \cdots, x_s)=(f(x_1), \cdots, f(x_s)).$$

Given a pair of rays (γ, δ), we will say that (γ, δ) has type Γ if there exist $(x_1, \cdots, x_s) \in X^{(s)}$ and a map $f \in C^\infty_{\text{in}}(X, \mathbf{R}^n)$ so that, setting $y_i=f(x_i)$, the points

$$y_{\alpha(1)}, \cdots, y_{\alpha(k)}$$

are successive reflection points of γ, while

$$y_{\beta(1)}, \cdots, y_{\beta(l)}$$

are successive reflection points of δ. Recall that in this situation the rays γ, δ have reflection points on the manifold $f(X)$ and the corresponding reflecting rays are related to the map f.

In order to express the lengths of the rays, consider the maps

$$F(y_1, \cdots, y_s)=p(\|y_{\alpha(1)}-y_{\alpha(2)}\|+\cdots+\|y_{\alpha(k)}-y_{\alpha(1)}\|)$$
$$G(y_1, \cdots, y_s)=q(\|y_{\beta(1)}-y_{\beta(2)}\|+\cdots+\|y_{\beta(l)}-y_{\beta(1)}\|)$$

with $p, q \in \mathbf{N}$. It is clear that we can restrict our attention to a fixed configuration Γ, assuming p, q also fixed, which enables one to fix F, G too.

To describe more precisely the geometric situation which occurs, we introduce a set $\mathcal{A}_\Gamma$. First, for $i \in \operatorname{Im} \alpha$ set

$$I_i = \{j \in \operatorname{Im}\alpha;\ \text{there is } t<k, \text{ such that } (i,j) = (\alpha(t), \alpha(t+1)) \text{ or } (i,j)=(\alpha(k), \alpha(1))\}.$$

In the same way for $i \in \operatorname{Im}\beta$ we set

$$J_i = \{j \in \operatorname{Im}\beta;\ \text{there is } t<l, \text{ such that } (i,j) = (\beta(t), \beta(t+1)) \text{ or } (i,j)=(\beta(l), \beta(1))\}.$$

Definition 2. We will say that $\mathscr{A}_\Gamma \subset (R^n)^{(s)}$ is an admissible domain for the pair (γ, δ) if $\mathscr{A}_\Gamma = (y_1, \cdots, y_s)$, where for each $i \in \operatorname{Im}\alpha$, $y_i \notin$ convex hull of the points $\{y_j, y_j \in I_i\}$ and for each $i \in \operatorname{Im}\beta$, $y_i \notin$ convex hull of the points $\{y_j, y_j \in J_i\}$. Moreover, there is $i \leq s$ such that

$$\sum_{k \in I_i} \frac{y_i - y_k}{\|y_i - y_k\|} \neq \sum_{j \in J_i} \frac{y_i - y_j}{\|y_i - y_j\|}.$$

After this preparation work we can define the set of maps

$\mathscr{M}_{\Gamma,p,q} = \{f \in C^\infty_{\text{in}}(X, \boldsymbol{R}^n)$; for every point $(x, \cdots, x_s) \in X^{(s)}$ the following two properties

(a) $(f(x_1), \cdots, f(x_s)) \in \mathscr{A}_\Gamma$,

(b) $(x_1, \cdots, x_s)$ is a critical point for both maps $F \circ f^{(s)}$ and $G \circ f^{(s)}$

imply the property

(c) $F \circ f^{(s)}(x_1, \cdots, x_s) \neq G \circ f^{(s)}(x_1, \cdots, x_s)\}$.

Now, the problem is to show that $\mathscr{M}_{\Gamma,p,q}$ contains a residual set in $C^\infty_{\text{in}}(X, \boldsymbol{R}^n)$.

To prove the above assertion, it is convenient to introduce the manifolds of jets $J^1_s(X, \boldsymbol{R}^n)$.

Consider the standard projections

$$X^s \xleftarrow{\ \alpha^s\ } (J^1_s(X, R^n))^s \xrightarrow{\ \beta^s\ } (R^n)^s.$$

We set $J^1_s(X, R^n) = (\alpha^s)^{-1}(X^{(s)})$, $\mathscr{M}_\Gamma = (\beta^s)^{-1}(\mathscr{A}_\Gamma)$. Next we define a smooth map $H\colon \mathscr{M}_\Gamma \to J^1(X^{(s)}, \boldsymbol{R}^2)$ by

$$H([f_1]_{x_1}, \cdots, [f_s]_{x_s}) = [F \circ (f_1 \times \cdots \times f_s), G \circ (f_1 \times \cdots \times f_s)]_{(x_1, \cdots, x_s)}.$$

Let $\nu\colon J^1(X^{(s)}, \boldsymbol{R}^2) \to \boldsymbol{R}^2$ be given by

$$\nu([F', G']_{(x_1, \cdots, x_s)}) = (F'(x_1, \cdots, x_s), G'(x_1, \cdots, x_s)).$$

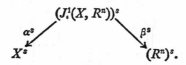

Put

$$S_2=\{\sigma \in J^1(X^{(s)}, \boldsymbol{R}^2);\ \sigma=(F', G')_{(x_1,\cdots,x_s)},$$
$$dF'(x_1, \cdots, x_s)=dG'(x_1, \cdots, x_s)=0\}$$

and set

$$\Sigma=S_2\cap\nu^{-1}(\Delta\boldsymbol{R}^2).$$

It is easy to see that codim $\Sigma=2ns-2s+1$. Our analysis is connected with the set $H^{-1}(\Sigma)$ which in general is not a smooth manifold. The essential point is to construct a finite or countable number of (smooth) manifolds Σ_j with codim $\Sigma_j=(n-1)s+1$ such that

$$H^{-1}(\Sigma)\subset\bigcup_j \Sigma_j.$$

Therefore, applying the multijet transversality theorem in [5], we obtain the assertion concerning $\mathcal{M}_{\Gamma,p,q}$. The details will be given in [19].

§ 3. Poincaré map.

The linear Poincaré map, related to a multiple reflecting ray γ, is a real symplectic map. For the definition of this map the reader should consult for example [6]. It was shown in [18] that in a suitably chosen basis P can be expressed as a product of $(2n-2)\times(2n-2)$ matrices, involving some geometric information. To describe this representation of P, consider a ray γ in Ω with successive reflection points $P_1, \cdots, P_m$, lying on X. Denote by σ_i the matrices corresponding to the symmetry with respect to the tangent planes to K at P_i, and put

$$\lambda_i=\|P_iP_{i+1}\|,\quad i=1, \cdots, m,\quad P_{m+1}=P_1.$$

The form of P, obtained in [18], is given by

$$P=\begin{pmatrix}\prod_{i=1}^m \sigma_i & 0\\ 0 & \prod_{i=1}^m \sigma_i\end{pmatrix}\begin{pmatrix} I & \lambda_m\\ \psi_m & I+\lambda_m\psi_m\end{pmatrix}\cdots\begin{pmatrix} I & \lambda_1\\ \psi_1 & I+\lambda_1\psi_1\end{pmatrix}, \tag{6}$$

where ψ_i are some symmetric matrices related to the differentials of Gauss map at P_i. A more precise information concerning ψ_i is contained in

Proposition 2 ([18]). *The matrix ψ_i is positively definite (semi-definite) if and only if K is strictly convex (convex) at P_i.*

Let Q and R are two arbitrary points with $\|QR\|$ sufficiently small.

Assuming the distance from Q and R to γ sufficiently small, we wish to construct a unique reflecting ray connecting Q and R after just m reflections. This problem is closely related to the map P. In fact, writing

$$P=\begin{pmatrix} A & B \\ C & D \end{pmatrix},$$

it is easy to prove that the existence and uniqueness of a reflecting ray connecting Q and R, will be guaranteed if the matrix B is invertible. (Here we use the basis introduced in [18].) The last condition can be proved if the boundary is convex in all reflection points.

Lemma 3. *Assume* $\psi_i \geq 0$ *for* $i=1, \cdots, m$. *Then* $\det B \neq 0$.

Proof. Setting

$$\begin{pmatrix} A_i & B_i \\ C_i & D_i \end{pmatrix}=\begin{pmatrix} I & \lambda_i \\ \psi_i & I+\lambda_i\psi_i \end{pmatrix}\cdots\begin{pmatrix} I & \lambda_1 \\ \psi_1 & I+\lambda_1\psi_1 \end{pmatrix},$$

we get

$$B_{i+1}=B_i+\lambda_{i+1}D_i, \qquad D_{i+1}=\psi_{i+1}B_{i+1}+D_i.$$

Therefore,

$$\det B_i \neq 0, \quad B_i^* D_i \geq 0 \Rightarrow \det B_{i+1} \neq 0, \quad B_{i+1}^* D_{i+1} \geq 0.$$

Indeed,

$$B_{i+1}^* D_{i+1}=B_{i+1}^*\psi_{i+1}B_{i+1}+B_i^* D_i=\lambda_{i+1}D_i^* D_i \geq 0.$$

On the other hand, if $B_{i+1}u=0$ with $u \neq 0$, we get

$$\|B_i u\|^2+\lambda_{i+1}\langle B_i^* D_i u, u\rangle=0,$$

which yields a contradiction with the assumption $\det B_i \neq 0$.

If a periodic multiple reflecting ray γ starts at $P_1 \in X$ with direction ξ_1 then the billiard map corresponding to m reflections is related to a canonical relation

$$\begin{aligned} C_m=\{&(t, x', \tau, \xi', x'', \xi'') \in T^*(\boldsymbol{R}\times X\times X)\setminus 0; \\ &t=\varphi_m(x', x''),\ \delta_+^m(x'', \xi''/\tau)=(x', \xi'/\tau),\ \tau\neq 0, \\ &\|x'-P_1\|, \|x''-P_1\|<\varepsilon, \|\xi'-\xi_1\|<\varepsilon\tau, \|\xi''-\xi_1\|<\varepsilon\tau\}. \end{aligned}$$

Here $\varphi_m(x', x'')$ is the length of the reflecting ray connecting x' and x'' and making m reflections, while δ_+ denotes the (forward) billiard ball map (see [7], [13]).

Following the arguments in [7], the property of P, discussed above and leading to det $B \neq 0$, enables one to express C_m by a global generating function which is merely

$$\phi_m = (t - \varphi_m(x', x''))\tau.$$

Having this in mind, we can write microlocally the fundamental solution $E(t, x, y)$ as a global Fourier integral operator with a global phase function. Thus, the analysis of the singularities of $\lambda(t)$ is reduced to the investigation of the asymptotics of some oscillatory integral. This approach has been used in [13]. Some other applications will be considered in the next section.

§ 4. Inverse spectral result and distribution of poles of the scattering matrix

Consider a strictly convex domain $K \subset \boldsymbol{R}^2$. According to Theorem 1 and the Remark 2, generically every two multiple reflecting rays have rationally independent periods. On the other hand, Lazutkin proved (Theorem 3, p. 32 in [12]), that generically the spectrum of the Poincaré map of every multiple reflecting ray does not contain $\sqrt[p]{1}$, $\forall p \in \boldsymbol{N}$. Combining these two results with the propagation of singularities of the solution to the problem (5), we conclude that the period of every periodic (multiple reflecting) geodesics is an isolated point in singsupp $\lambda(t)$. To prove that the period of a ray is included in isngsupp $\lambda(t)$, we can apply the (microlocal) construction of a parametrix for the problem (5) near transversal directions, given in [6], and leading to a precise formula for the leading term of the singularity. Here the fact that 1 is not in the spectrum of P is essential for the application of a stationary phase argument. Thus, we are going to the following

Theorem 2. *Let $K \subset \boldsymbol{R}^2$ be a generic strictly convex domain. Then the spectrum $\{\lambda_j\}$ determines the lengths $\{T_\gamma\}$ of all periodic multiple reflecting rays.*

Remark 3. A domain is called generic following the interpretation given in Theorem 1.

Remark 4. The above result has been mentioned in [7], [4], but we have not succeeded to find a complete proof in the literature and it seems that the result announced in [7], [4] could be considered as a conjecture.

Remark 5. Marvizi and Melrose [13] considered striclty convex planar domains without assuming that the Poincaré maps of the rays have suitable spectrum. They proved that some rays have periods included in singsupp $\lambda(t)$. Theorem 2 shows that under a stronger generic assumption we can recover the periods of all geodesics.

Next we will treat the problem of distribution of the poles of scattering matrix for trapping obstacles. As was shown in [16], the existence of a sequence $T_j \nearrow \infty$ of singularities of $\mu(t)$, implies that for every $\varepsilon>0$ there are poles μ_j such that

(7) $$0<\operatorname{Im}\mu_j \leq \varepsilon \log |\mu_j|.$$

To study the singularities of $\mu(t)$, we apply the trace formula

(8) $$\mu(t)=\int_\Omega F(t, x, x)dx,$$

proved in [15]. Here $F(t, x, y)$ is the solution to the problem

(9) $$\begin{cases} \Box F=0 \text{ in } \boldsymbol{R}\times\Omega, & F=0 \text{ in } \boldsymbol{R}\times X, \\ F\restriction_{t=0}=0, & \partial_t F\restriction_{t=0}=\delta(x-y). \end{cases}$$

In [17] we discussed the case when $K=\bigcup_{j=1}^M K_j$, $K_i\cap K_j=\emptyset$ and K_j, $j=1,\cdots,m$ are strictly convex. Set

$$d_{ij}=\operatorname{dist}(K_i, K_j), \quad i\neq j.$$

According to Theorem 1, the (trapping) rays, corresponding to the distance between K_i and K_j, generically have periods rationally independent from the periods of all other rays. On the other hand, the Poincaré map, related to such rays, has a spectrum included in $(0, 1)\cup(1, \infty)$ (see [2], [18]). Thus, the singularities connected with pd_{ij}, $p\in \boldsymbol{N}$, are included in singsupp $\mu(t)$ and we get

Theorem 3. *Let $K=\bigcup_{j=1}^M K_j$ be a generic domain with $K_j\subset\boldsymbol{R}^n$, n odd, $n\geq 3$, $K_i\cap K_j=\emptyset$, K_j strictly convex. Then for every $\varepsilon>0$ there are poles μ_j for which* (7) *holds.*

Remark 6. This result generalizes the situation in [2], [8].

Now consider a multiple reflecting ray γ_0 in $\bar{\Omega}$ with period T_0. Assume T_0 is an isolated point of singsupp $\mu(t)$, that is there exists $\varepsilon_0>0$ such that there are no other periodic generalized geodesics with periods T, $T_0-\varepsilon_0<T<T_0+\varepsilon_0$. Let $P_1,\cdots,P_m$ be reflection points of γ_0 and let K be convex at P_i, $i=1,\cdots,m$. By using the results for propagation of singularities in [14], we can reduce the analysis of the singularity in T_0 to the examination of the microlocal parametrix to the problem (9). Combining Proposition 2 and Lemma 3, we can find a (microlocal) representa-

tion of $F(t, x, y)$ by a Fourier integral operator J with a global phase function $\phi_m(t, x', x'', \tau)$, introduced in Section 3 (see [13] for more details). Therefore, the problem is to study the asymptotics of an oscillatory integral with phase function having isolated minimum. For this purpose we can apply a recent result of Soga [21] concerning the integral

$$I(\tau)=\int_D e^{i\tau\phi(x)}a(x, \tau)dx,$$

with

$$a(x, \tau)\sim\sum_{j=0}^{\infty}(i\tau)^{-j}a_j(x) \quad \text{as } \tau\to\infty,\ a_j(x) \text{ real}.$$

Assume $a_0(x)>0$ on x; $\{x \in D, \phi(x)=\min_{x\in \bar{D}}\phi(x)\}$. Therefore, the result of Soga says that for some $m \in \boldsymbol{R}$ we have $\tau^m I(\tau) \notin L^2(\boldsymbol{R})$. To make use of this result, we must examine the amplitude of the Fourier integral operator J following the global transport equation. This can be done and we conclude that a trapping ray with the properties mentioned above produces a singularity of $\mu(t)$. An application of this argument leads to a generalization of Theorem 3 for the case when K_j are convex. The details will be published elsewhere.

References

[1] K. Andersson and R. Melrose, The propagation of singularities along gliding rays, Invent. Math., **41** (1977), 197–232.

[2] C. Bardos, J. C. Guillot et J. Ralston, La relation de Poisson pour l'équation des ondes dans un ouvert non-borné, Application à la théorie de la diffusion, Comm. Partial Differential Equations, **7** (1982), 905–958.

[3] J. Chazarain, Construction de la paramétrix du problème mixte hyperbolique pour l'équation des ondes, C. R. Acad. Sci. Paris, Ser. A, **276** (1973), 1213–1215.

[4] Y. Colin de Verdiere, Sur les longueurs des trajectoires périodiques d'un billiard, Inst. Fourier Univ. Grenoble I, preprint, October 1983.

[5] M. Golubitsky and V. Guillemin, Stable Mappings and Their Singularities, Springer, 1973.

[6] V. Guillemin and R. Melrose, The Poisson summation formula for manifolds with boundary, Advances in Math., **32** (1979), 204–232.

[7] ——, A Cohomological Invariant of Discrete Dynamical Systems, Christoffel Centennial Volume, Eds. P. L. Butzer and F. Feher, Birkhauser Verlag, Basel, 1981, p. 612–679.

[8] M. Ikawa, On the distribution of the poles of the scattering matrix for two strictly convex obstacles, Hokkaido Math. J., **12** (1983), 343–359.

[9] ——, Trapping obstacles with a sequence of poles of the scattering matrix converging to the real axis, Univ. Osaka, preprint, 1984.

[10] M. Kac, Can one hear the shape of a drum?, Amer. Math. Monthly, **73** (1966), 1–23.

[11] E. E. Landis, Tangential singularities, Functional Anal. Appl., **15**, No. 2, (1981), 36–49 (in Russian).

[12] V. F. Lazutkin, Convex billiard and eigenfunctions of the Laplace operator,

Ed. Lenegrad Univ., 1981, (in Russian).
[13] S. Marvizi and R. Melrose, Spectral invariants of convex planar regions, J. Differential Geom., **17** (1982), 475–502.
[14] R. Melrose and J. Sjöstrand, Singularities in boundary value problems I, II, Comm. Pure Appl. Math., **31** (1978), 593–617 and **35** (1982), 129–168.
[15] R. Melrose, Polynomial bound on the number of scattering poles, J. Funct. Anal., **53** (1983), 287–303.
[16] V. Petkov, Note on the distribution of poles of the scattering matrix, J. Math. Anal. Appl., **101** (1984), 582–587.
[17] ——, La Distribution des Pôles de la Matrice de Diffusion, Séminaire Goulaouic-Meyer-Schwartz 1982–1983, Ecole Polytechnique, Exposé n° VII.
[18] V. Petkov et P. Vogel, La représentation de l'application de Poincaré correspondant aux rayons périodiques reflechissants, C. R. Acad. Sci. Paris, Ser. A, **269** (1983), 633–635.
[19] V. Petkov and L. Stojanov, Periods of multiple reflecting rays and inverse spectral results, preprint.
[20] O. A. Platonova, Singularities of mutual disposition of surface and straight-line, Uspehi Mat. Nauk, **36**, No. 1, (1981), 221–222 (in Russian).
[21] H. Soga, Conditions against rapid decrease of oscillatory integrals and their applications to inverse scattering problems, preprint, 1984.

Institute of Mathematics of
Bulg. Academy of Sciences,
P.O. BOX 373, 1000 Sofia, Bulgaria

Taniguchi Symp. HERT
Kyoto 1984, pp. 329–346

Mixed Problems for Evolution Operators with Dominant Principal Parts in the Volevich-Gindikin Sense

Reiko SAKAMOTO

§ 1. Volevich-Gindikin's results and our problems

Let a polynomial

$$A(\tau, \xi)=A(\tau, \xi_1, \cdots, \xi_n)=\tau^\mu+\sum_{\substack{0\leq j<\mu\\0\leq j+|\nu|\leq m}} a_{j\nu}\tau^j\xi^\nu$$

satisfy

$$\sum_{|\nu|=m} a_{0\nu}\xi^\nu\not\equiv 0,$$

then its *Newton Polygon* N_A is the convex hull of $\Delta_A\cup\{(0, 0)\}$, where

$$\Delta_A=\{(j, |\nu|);\ a_{j,\nu}\neq 0\}.$$

It is figured like the following. We denote $\sigma(N_A)=\{m_1, \cdots, m_l\}$.

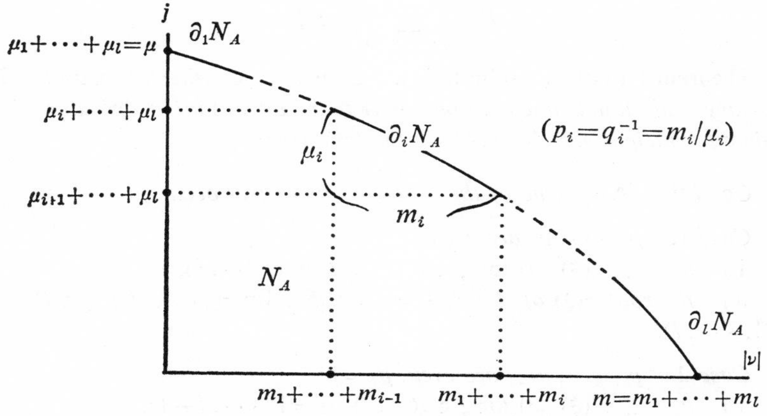

Denoting

$$\partial' N_A=\partial_1 N_A\cup\cdots\cup\partial_l N_A,\quad N_A^1=N_A\cap\{\{N_A-(0, 1)\}\cup\{N_A-(1, 0)\}\},$$

we define the *principal part* of A by

$$A_0 = \sum_{(j,|\nu|)\in\partial' N_A} a_{j\nu}\tau^j\xi^\nu,$$

and define the *lower part* of A by

$$A_1 = \sum_{(j,|\nu|)\in N_A^1} a_{j\nu}\tau^j\xi^\nu.$$

Moreover, we define the *i-th part* of A_0 by

$$A^{(i)} = \sum_{(j,|\nu|)\in\partial_i N_A} a_{j\nu}\tau^j\xi^\nu\tau^{-(\mu_{i+1}+\cdots+\mu_l)}$$
$$= a_{m_1+\cdots+m_{i-1}}(\xi)\tau^{\mu_i}+\cdots+a_{m_1+\cdots+m_i}(\xi).$$

Now, we say that A is an *evolution operator* if the Cauchy problem

$$\text{(C.P.)}\qquad \begin{cases} A(D_t, D_x)u=f & \text{for } 0<t<T,\ x\in R^n,\\ D_t^j u=0\quad (j=0, 1, \cdots, \mu-1) & \text{for } t=0,\ x\in R^n \end{cases}$$

is H^∞-*well posed*, that is, there exists a unique solution $u\in H_0^\infty((0, T)\times R^n)$ for any $f\in H_0^\infty((0, T)\times R^n)$, where $f\in H_0^\infty((0, T)\times R^n)$ means

$$\tilde{f}\in H^\infty((-\infty, T)\times R^n),$$

where $\tilde{f}=f$ for $t>0$ and $\tilde{f}=0$ for $t<0$. We say that A is an *evolution operator with a domainant principal part*, if $A+P$ is an evolution operator for any

$$P = \sum_{(j,|\nu|)\in N_A^1} p_{j\nu}\tau^j\xi^\nu.$$

Theorem I (Volevich-Gindikin). *Constant Coefficient Case. The necessary and sufficient condition for A to be an evolution operator with a dominant principal part is to satisfy the following*

Conditiou (A). *One of the following two cases occurs.*

Case 1. $p_1, \cdots, p_l$ *are even*;

i) $a_{m_1+\cdots+m_i}(\xi)\neq 0$ *for* $\xi\in R^n\setminus\{0\}$ $(i=1, \cdots, l)$,

ii) *the root* $\tau(\xi)$ *of* $A^{(i)}(\tau, \xi)=0$ *satisfies* $\operatorname{Im}\tau(\xi)>0$ *for* $\xi\in R^n\setminus\{0\}$ $(i=1, \cdots, l)$.

Case 2. $p_1, \cdots, p_{l-1}$ *are even*, $p_l=1$;

i) $a_{m_1+\cdots+m_i}(\xi)\neq 0$ *for* $\xi\in R^n\setminus\{0\}$ $(i=1, \cdots, l-1)$,

ii) *the root* $\tau(\xi)$ *of* $A^{(i)}(\tau, \xi)=0$ *satisfies* $\operatorname{Im}\tau(\xi)>0$ *for* $\xi\in R^n\setminus\{0\}$ $(i=1, \cdots, l-1)$,

iii) *the roots* $\tau(\xi)$ *of* $A^{(l)}(\tau, \xi)=0$ *are real and distinct for* $\xi\in R^n\setminus\{0\}$.

When the coefficients of A are dependent of (t, x), Condition (A) is understood as the one with the uniformity with respect to (t, x):

$$a_{m_1+\cdots+m_i}(\xi)\neq 0 \to |a_{m_1+\cdots+m_i}(t, x; \xi)|\geqq c|\xi|^{m_1+\cdots+m_i},$$
$$\operatorname{Im}\tau(\xi)>0 \to \operatorname{Im}\tau(t, x; \xi)\geqq c|\xi|^{p_i},$$
$$\tau_j(\xi)\neq\tau_k(\xi) \to |\tau_j(t, x; \xi)-\tau_k(t, x; \xi)|\geqq c|\xi|.$$

Theorem II (Volevich-Gindikin). *Variable Coefficient Case. Let A satisfy Condition* (A), *then A is an evolution operator with a dominant principal part.*

Now, let us consider the following problem. Let A satisfy Condition (A), where we restrict ourselves to the case 2. Let Ω be R^n_+ or a domain in R^n with a compact and C^∞-boundary, which is non-characteristic to $A(0, \xi)$ (*Condition* (C)). What kind of boundary operators

$$B=\begin{bmatrix} B_1 \\ \vdots \\ B_{m_+} \end{bmatrix}$$

are suitable for the initial-boundary problem (mixed problem)

$$\text{(M.P.)}\qquad \begin{cases} Au=f & \text{in } (0, T)\times\Omega, \\ Bu=g & \text{on } (0, T)\times\partial\Omega, \\ D_t^j u=0 \ (j=0, 1, \cdots, \mu-1) & \text{on } \{t=0\}\times\Omega \end{cases}$$

to be H^∞-well posed? Of course, we say that the mixed problem is H^∞-*well posed* if there exists a unique solution $u\in H_0^\infty((U, T)\times\Omega)$ for any $f\in H_0^\infty((0, T)\times\Omega)$ and $g\in H_0^\infty((0, T)\times\partial\Omega)$. We say that $\{A, B\}$ is an evolution system if (M.P.) is H^∞-well posed, and we say that $\{A, B\}$ is an evolution system with dominant principal parts if $\{A+P, B+Q\}$ is an evolution system for any lower order system $\{P, Q\}$, which will be defined in § 3. Our results are as follows.

Theorem III. *Assume Conditions* (A) *and* (C). *Then Condition* (B), *which will be defined in* § 3, *holds iff there exist positive constants γ_0 and C such that*

$$\gamma^{q_1/2}|||u|||_{N_A^1}+\langle\langle\langle u\rangle\rangle\rangle_{N_A^1}\leqq C\{\gamma^{-q_1/2}\|A_\gamma u\|+\langle B_\gamma^\# u\rangle\}$$

for $\gamma>\gamma_0$, $|||u|||_{N_A}<+\infty$, *where*

$$\|u\|=\|u\|_{L^2(R^1\times\Omega)},\qquad \langle u\rangle=\|u|_{x=0}\|_{L^2(R^1\times\partial\Omega)},$$

$$|||u|||_N = \sum_{(k,|\nu|)\in N} \|(D_t - i\gamma)^k D_x^\nu u\|,$$

$$\langle\langle\langle u\rangle\rangle\rangle_N = \sum_{(k,|\nu|)\in N} \langle(D_t - i\gamma)^k D_x^\nu u\rangle,$$

and

$$A_\gamma = A(D_t - i\gamma, D_x), \qquad B_\gamma^\# = B^\#(D_t - i\gamma, D_x),$$

where $B^\#$ is the standardization of B, which will be defined in § 3.

Remark. Theorem III is valid, when $\{A, B\}$ has variable coefficients, if the Conditions are said in the words of uniformity.

We say that $\{A, B\}$ satisfies Condition (B*) if the adjoint system $\{A^*, B^*\}$ satisfies Condition (B) in the corresponding region, which assures the energy inequality for $\{A^*, B^*\}$. Hence

Theorem IV. *Let $\{A, B\}$ satisfy Conditions* (A), (B), (B*), (C), *then $\{A, B\}$ is an evolution system with dominant principal parts.*

Example 1. Let us consider the solution u satisfying

$$A(D)u = \{D_t - i(D_x^2 + D_y^2)\}\{D_t^2 - (D_x^2 + D_y^2)\}u = f(t, x, y)$$

in $(0, +\infty)\times\Omega = \{t>0\}\times\{x>0, -\infty<y<+\infty\}$, where u satisfiess boundary conditions:

$$u = D_x^2 u = 0 \qquad \text{on } (0, +\infty)\times\partial\Omega,$$

and initial conditions:

$$u = D_t u = D_t^2 u = 0 \qquad \text{on } \{t=0\}\times\Omega.$$

Now, let v, g be the Laplace-Fourier transforms with respect to (t, y), then v, g satisfy

$$\{\tau - i(D_x^2 + \eta^2)\}\{\tau^2 - (D_x^2 + \eta^2)\}v = g \qquad \text{in } \{x>0\},$$

where (τ, η) are parameters in $\Gamma_{\gamma_0} = \{\operatorname{Im}\tau < -\gamma_0, -\infty<\eta<+\infty\}$. Set

$$A_1 = \tau - i(D_x^2 + \eta^2), \qquad A_2 = \tau^2 - (D_x^2 + \eta^2),$$

and

$$A_2 v = v_1, \qquad A_1 v = v_2,$$

then we have

$$\begin{cases} \gamma^{1/2}(\|A_1 v_1\| + \|D_x v_1\|) \leqq C\|g\| & \text{(from the parabolicity of } A_1), \\ \gamma(\|A_2 v_2\| + \|D_x v_2\|) \leqq C\|g\| & \text{(from the hyperbolicity of } A_2), \end{cases}$$

therefore we have

$$\gamma^{1/2}(\|\Lambda_1 v_i\|+\|D_x v_i\|)\leqq C\|g\| \qquad (i=1, 2),$$

where

$$\Lambda_1=|\tau|^{1/2}+|\eta|, \quad \Lambda_2=|\tau|+|\eta|, \quad \|g\|^2=\int_0^\infty |g(x)|^2 dx$$

and $\gamma=-\text{Im}\,\tau>\gamma_0$. Since

$$\begin{cases}\tau^2 v=\tau^2(\tau^2+i\tau)^{-1}(v_1+iv_2),\\(D_x^2+\eta^2)v=2^{-1}\{(\tau^2-i\tau)(\tau^2+i\tau)^{-1}(v_1+iv_2)+(-v_1+iv_2)\},\end{cases}$$

we have

$$\gamma^{1/2}(\|\Lambda_1\Lambda_2^2 v\|+\|\Lambda_2^2 D_x v\|+\|\Lambda_1 D_x^2 v\|+\|D_x^3 v\|)\leqq C\|g\|$$

if $|\tau|>\delta|\eta|$ $(\delta>0)$. On the other hand, since

$$A(\xi)=i(\xi-\xi_1^+)(\xi-\xi_2^+)(\xi-\xi_1^-)(\xi-\xi_2^-)=iA_+(\xi)A_-(\xi),$$
$$|\text{Im}\,\xi_i^\pm|\geqq c|\eta|\geqq c\delta^{-1}|\tau|\geqq c\delta^{-1}\gamma, \qquad |R|\geqq c|\eta|$$

if $|\tau|<\delta|\eta|$, where

$$R=\begin{vmatrix}\dfrac{1}{2\pi i}\oint\dfrac{d\xi}{A_+(\xi)} & \dfrac{1}{2\pi i}\oint\dfrac{\xi d\xi}{A_+(\xi)}\\ \dfrac{1}{2\pi i}\oint\dfrac{\xi^2 d\xi}{A_+(\xi)} & \dfrac{1}{2\pi i}\oint\dfrac{\xi^3 d\xi}{A_+(\xi)}\end{vmatrix}=-(\xi_1^+ +\xi_2^+),$$

we have

$$\gamma(\|\Lambda_2^3 v\|+\|\Lambda_2^2 D_x v\|+\|\Lambda_2 D_x^2 v\|+\|D_x^3 v\|)\leqq C\|g\|$$

if $|\tau|<\delta|\eta|$. Hence we have

$$\gamma^{1/2}(\|\Lambda_1\Lambda_2^2 v\|+\|\Lambda_2^2 D_x v\|+\|\Lambda_2 D_x^2 v\|+\|D_x^3 v\|)\leqq C\|g\|$$

in Γ_{γ_0}.

Example 2. Let

$$A=\tau^3-i(\xi^2+\eta^2)\tau^2+i(\xi^4+\kappa\xi^2\eta^2+\eta^4) \quad (\kappa>1),$$

whose principal part is equal to that in Ex. 1 if $\kappa=2$. Can we get the same type of the energy inequality as in Ex. 1 under the same boundary conditions? The answer is positive. The way of proof is similar to that in Ex. 1. Because, we can divide the parameter space Γ_{γ_0} into some parts, where A can be described as a product of polynomials with respect to ξ, each of which has a good property such as ellipticity, parabolicity, or hyperbolicity. Roughly speaking, A can be considered as follows:

i) $A \sim a^{(1)}a^{(2)}$ in $\Delta^{(0)}=\{|\eta|\leqq\varepsilon|\tau^{1/2}|\}$,

ii) $A \sim A^{(1)}a^{(2)}$ in $\Delta^{(1)}=\{\varepsilon|\tau^{1/2}|\leqq|\eta|\leqq\varepsilon|\tau|\}$,

iii) $A \sim A^{(2)}$ in $\Delta^{(2)}=\{|\eta|\geqq|\tau|\}$,

if ε is small enough and γ_0 is large enough, where

$$A^{(1)}=\tau-i(\xi^2+\eta^2), \quad A^{(2)}=-i(\xi^2+\eta^2)\tau^2+i(\xi^4+\kappa\xi^2\eta^2+\eta^4),$$
$$a^{(1)}=\tau-i\xi^2, \qquad a^{(2)}=\tau^2-\xi^2.$$

§ 2. Properties of the roots of $A_0(\tau, \xi, \eta)=0$ w.r.t. ξ

In the following, let Ω be

$$R^n_+=\{(x, y);\, x>0,\, y=(y_2, \cdots, y_n) \in R^{n-1}\},$$

and change the notations:

$$(\xi_1, \xi_2, \cdots, \xi_n)\to(\xi, \eta_2, \cdots, \eta_n)=(\xi, \eta).$$

Denoting

$$\Gamma_K=\{(\tau, \eta);\, \operatorname{Im}\tau<-K,\, \eta \in R^{n-1}\}$$

and

$$\Delta^{(i)}=\Delta^{(i)}_{K,\varepsilon}=\{(\tau, \eta) \in \Gamma_K;\, \varepsilon|\tau|^{q_i}\leqq|\eta|\leqq\varepsilon|\tau|^{q_{i+1}}\}$$

for $i=0, 1, \cdots, l$, we have $\Gamma_K=\bigcup_{i=0}^{l}\Delta^{(i)}_{K,\varepsilon}$, where

$$\Delta^{(0)}=\{(\tau, \eta);\, \operatorname{Im}\tau<-K,\, \eta \in R^{n-1},\, |\eta|\leqq\varepsilon|\tau|^{q_1}\},$$
$$\Delta^{(l)}=\{(\tau, \eta);\, \operatorname{Im}\tau<-K,\, \eta \in R^{n-1},\, |\eta|\geqq\varepsilon|\tau|^{q_l}\}.$$

Moreover we denote

$$\Lambda_i=|\tau|^{q_i}+|\eta|.$$

We say that a polynomial w.r.t. ξ

$$P=\xi^h+c_1(\tau, \eta)\xi^{h-1}+\cdots+c_h(\tau, \eta), \quad (\tau, \eta) \in \Delta$$

is *p_j-parabolic* in Δ, if

i) $c_k(\tau, \eta) \in S^k_{\Lambda_j}(\Delta)$, i.e.

$$|\partial_\tau^\sigma \partial_\eta^\nu c_k(\tau, \eta)| \leqq C_{\sigma,\nu} \Lambda_j^{k-(\sigma+|\nu|)} \qquad \text{in } \Delta$$

for any σ, ν,

ii) $|P| \geqq c\Lambda_j^h$ for $(\tau, \eta) \in \Delta$, $\xi \in R^1$.

We also say *elliptic* in stead of 1-parabolic. We say that P is *hyperbolic* in Δ, if

i) $c_k(\tau, \eta) \in S_{\Lambda_l}^k(\Delta)$ and

$$|\text{Im}\, c_k(\tau, \eta)| \leqq C\Lambda_l^{k-1} \qquad \text{for } (\tau, \eta) \in R^1 \times R^{n-1},$$

ii) $|\partial_\tau P| \geqq c\Lambda_l^{h-1}$

for $(\tau, \xi, \eta) \in R^1 \times R^1 \times R^{n-1}$ satisfying $P=0$.

Proposition 2.1. *There exist positive constants K_0 and ε_0 such that, for $K > K_0$, $0 < \varepsilon < \varepsilon_0$, we have*

$$A_0(\tau, \xi, \eta) = \mathscr{P}^{(i)}(\tau, \xi, \eta) P^{(i+1)}(\tau, \xi, \eta) \cdots P^{(l-1)}(\tau, \xi, \eta) H(\tau, \xi, \eta)$$

for $(\tau, \eta) \in \Delta_{K,\varepsilon}^{(i)}$ $(i < l)$, where $\mathscr{P}^{(i)}$, $P^{(i+1)}, \cdots, P^{(l-1)}$ are p_i-, $\cdots$, p_l-parabolic and H is hyperbolic.

Proposition 2.2. *Let $(\tau_0, \eta_0) \in \Delta^{(l)}$ and $|\tau_0|^2 + |\eta_0|^2 = 1$, then there exists its neighbourhood $U(\tau_0, \eta_0)$ such that*

$$A_0(\tau, \xi, \eta) = \mathscr{E}(\tau, \xi, \eta) \mathscr{H}(\tau, \xi, \eta)$$

for $(\tau/\Lambda_l, \eta/\Lambda_l) \in U(\tau_0, \eta_0)$, where $\mathscr{E}$ is elliptic and $\mathscr{H}$ is hyperbolic.

Since the proof of Proposition 2.2 is similar to the discussions in the well known hyperbolic mixed problem, we shall only prove Proposition 2.1 hereafter in this section. Let us denote

$$\begin{aligned} A^{(i)}&(\tau, \xi, 0) \\ &= a_{m_1+\cdots+m_{i-1}}(1, 0)\xi^{m_1+\cdots+m_{i-1}}\tau^{\mu_i} + \cdots + a_{m_1+\cdots+m_i}(1, 0)\xi^{m_1+\cdots+m_i} \\ &= a_{m_1+\cdots+m_{i-1}}(1, 0)\xi^{m_1+\cdots+m_{i-1}} a^{(i)}(\tau, \xi), \end{aligned}$$

then we have

Lemma 2.3.

i) *Let*

$$\tilde{A}^{(j)} = A_0 \tau^{-(\mu_{j+1}+\cdots+\mu_l)} - A^{(j)},$$

then we have

$$|\partial_\tau^\sigma \partial_\eta^\nu \tilde{A}^{(j)}| \leqq c_{\sigma\nu}(K, \varepsilon) \Lambda_j^{m_1+\cdots+m_j-(p_j\sigma+|\nu|)}$$

for $(\tau, \eta) \in \Delta_{K,\varepsilon}^{(i)}$ *and* $|\xi| \leqq C\Lambda_j$ $(i \leqq j)$, *where* $c_{\sigma\nu}(K, \varepsilon) \to 0$ *as* $1/K + \varepsilon \to 0$.

ii) *We have*

$$|A^{(j)}(\tau, \xi, \eta) - A^{(j)}(\tau, \xi, 0)| \leqq c(\varepsilon)\Lambda_j^{m_1+\cdots+m_j}$$

for $(\tau, \eta) \in \Delta_{K,\varepsilon}^{(i)}$ $(i < l)$ *and* $|\xi| \leqq C\Lambda_j$ $(i < j)$, *where* $c(\varepsilon) \to 0$ *as* $\varepsilon \to 0$.

Proof of (i). In case of $(\sigma, \nu) = (0, 0)$. We have

$$\begin{aligned} |\tilde{A}^{(j)}| &\leqq |\tau^{\mu_1+\cdots+\mu_j} + \cdots + a_{m_1+\cdots+m_{j-1}-p_{j-1}}\tau^{\mu_j+1}| \\ &\quad + |a_{m_1+\cdots+m_j+p_{j+1}}\tau^{-1} + \cdots + a_{m_1+\cdots+m_l}\tau^{-(\mu_{j+1}+\cdots+\mu_l)}| \\ &\leqq C\Lambda_j^{m_1+\cdots+m_j}\{(|\tau|^{q_{j-1}}/\Lambda_j)^{p_{j-1}} + (\Lambda_j/|\tau|^{q_{j+1}})^{p_{j+1}}\}. \end{aligned}$$

Hence we only remark

$$|\tau|^{q_{j-1}}/\Lambda_j \leqq |\tau|^{q_{j-1}}/|\tau|^{q_j} \leqq K^{-\delta} \quad (\delta = \max_j (q_j - q_{j-1})),$$

$$\Lambda_j/|\tau|^{q_{j+1}} = |\tau|^{q_j}/|\tau|^{q_{j+1}} + |\eta|/|\tau|^{q_{j+1}} \leqq K^{-\delta} + \varepsilon.$$

Proof of (ii). We have

$$\begin{aligned} &|A^{(j)}(\tau, \xi, \eta) - A^{(j)}(\tau, \xi, 0)| \\ &= |(a_{m_1+\cdots+m_{j-1}}(\xi, \eta) - a_{m_1+\cdots+m_{j-1}}(\xi, 0))\tau^{\mu_j} \\ &\qquad + \cdots + (a_{m_1+\cdots+m_j}(\xi, \eta) - a_{m_1+\cdots+m_j}(\xi, 0))| \\ &\leqq C(|\eta|/\Lambda_j)\Lambda_j^{m_1+\cdots+m_j}. \end{aligned}$$

Hence we only remark

$$|\eta|/\Lambda_j \leqq \varepsilon|\tau|^{q_{i+1}}/\Lambda_j \leqq \varepsilon|\tau|^{q_j}/\Lambda_j \leqq \varepsilon.$$

Now, let us consider the root of $a^{(j)}(\tau, \xi) = 0$. Owing to the homogeneity of $a^{(j)}(\tau, \xi)$, $a^{(j)}(1, 0) \neq 0$ and $a^{(j)}(0, 1) \neq 0$, the root ξ of $a^{(j)}(\tau, \xi) = 0$ satisfies

$$c_1|\tau|^{q_j} \leqq |\xi| \leqq c_2|\tau|^{q_j}.$$

Moreover, the root ξ of $a^{(j)}(\tau, \xi) = 0$ is non-real for $\operatorname{Im}\tau \leqq 0$, $\tau \neq 0$ $(j = 1, \cdots, l-1)$. Therefore, it satisfies

$$|\operatorname{Im}\xi| \geqq c_1|\tau|^{q_j}.$$

Let us denote

$$G_{j\pm} = \{\xi \in C^1 : c_1'|\tau|^{q_j} \leqq |\xi| \leqq c_2'|\tau|^{q_j}, \operatorname{Im}\xi \gtrless \pm c_1'|\tau|^{q_j}\} \quad (1 \leqq j \leqq l-1),$$

$$G_l = \{\xi \in C^1 ; c_1'|\tau| \leqq |\xi| \leqq c_2'|\tau|\},$$

where $c_1' < c_1 < c_2 < c_2'$. Since

$$|A^{(j)}(\tau, \xi, 0)| = |\xi|^{m_1 + \cdots + m_{j-1}} |a^{(j)}(\tau, \xi)|,$$

we have

$$\begin{cases} \inf_{\xi \in \partial G_{j\pm}} |A^{(j)}(\tau, \xi, 0)| \geqq c|\tau|^{q_j(m_1 + \cdots + m_j)} & (1 \leqq j \leqq l-1), \\ \inf_{\xi \in \partial G_l} |A^{(l)}(\tau, \xi, 0)| \geqq c|\tau|^{m_1 + \cdots + m_l}. \end{cases}$$

Since

$$|\tau|^{q_j} \leqq \Lambda_j \leqq (1+\varepsilon)|\tau|^{q_j}$$

for $(\tau, \xi) \in \Delta^{(i)}$ $(i < j)$, we have

Lemma 2.4. *We have*

$$\inf_{\xi \in \partial G_{j\pm}} |A^{(j)}(\tau, \xi, 0)| \geqq c\Lambda_j^{m_1 + \cdots + m_j} \quad (i < j < l),$$
$$\inf_{\xi \in \partial G_l} |A^{(l)}(\tau, \xi, 0)| \geqq c\Lambda_l^{m_1 + \cdots + m_l},$$

for $(\tau, \eta) \in \Delta_{K,\varepsilon}^{(i)}$ $(i < l)$, *where* c *is independent of* (K, ε) $(0 < \varepsilon < 1)$.

Next, we consider the root ξ of $A^{(i)}(\tau, \xi, \eta) = 0$ for $(\tau, \eta) \in \Delta^{(i)}$ $(i < l)$. From the homogeneity, we have

$$\begin{aligned} \Lambda_i^{-(m_1 + \cdots + m_i)} A^{(i)}(\tau, \xi, \eta) &= A^{(i)}(\tau/\Lambda_i^{p_i}, \xi/\Lambda_i, \eta/\Lambda_i) \\ &= A^{(i)}(\tau', \xi', \eta'), \end{aligned}$$

where the root ξ' of $A^{(i)}(\tau', \xi', \eta') = 0$ satisfies $\{|\xi'| \leqq C\}$. On the other hand, since

$$(\tau, \eta) \in \Delta^{(i)} = \{(\tau, \eta) \in \Gamma_0;\ \varepsilon|\tau|^{q_i} \leqq |\eta| \leqq \varepsilon|\tau|^{q_{i+1}}\},$$

we have

$$\begin{aligned} (\tau', \eta') &\in \{(\tau', \eta') \in \Gamma_0;\ |\tau'|^{q_i} + |\eta'| = 1,\ \varepsilon|\tau'|^{q_i} \leqq |\eta'| \leqq \varepsilon|\tau'|^{q_{i+1}}\} \\ &\subset \{(\tau', \eta') \in \Gamma_0;\ |\eta'| \geqq \varepsilon/(1+\varepsilon)\}. \end{aligned}$$

Therefore, the root ξ' of $A^{(i)}(\tau', \xi', \eta') = 0$ is non-real, that is, it is contained in $\{|\mathrm{Im}\, \xi'| \geqq c_\varepsilon\}$. Let us denote

$$\mathfrak{G}_{i\pm} = \{\xi \in C^1;\ |\xi| \leqq C'\Lambda_i,\ \mathrm{Im}\, \xi \gtreqless \pm c_\varepsilon' \Lambda_i\},$$

where $c_\varepsilon' < c_\varepsilon < C < C'$, then we have

Lemma 2.5. *We have*

$$\inf_{\xi \in \partial \mathfrak{G}_{i\pm}} |A^{(i)}(\tau, \xi, \eta)| \geqq c_\varepsilon \Lambda_i^{m_1+\cdots m_i}$$

for $(\tau, \eta) \in \Delta^{(i)}$ $(i<l)$.

Finally, we remark that the smoothness of the coefficients is proved by the following lemma, which is proved by the differentiation under the integral sign.

Lemma 2.6. *Let* $P(\tau, \xi, \eta)$, $Q(\tau, \xi, \eta)$ *be polynomials with respect to* ξ, *satisfying*

$$c_1\Lambda^\alpha \leqq |P(\tau, \xi, \eta)| \leqq c_2\Lambda^\alpha,$$
$$|\partial_\tau^\sigma \partial_\eta^\nu P(\tau, \xi, \eta)| \leqq C\Lambda^{\alpha-(\sigma+|\nu|)},$$
$$|\partial_\tau^\sigma \partial_\eta^\nu Q(\tau, \xi, \eta)| \leqq C\Lambda^{\beta-(\sigma+|\nu|)}$$

for $(\tau, \eta) \in \Delta$ *and* $\xi \in \partial G$. *Then we have*

$$|\partial_\tau^\sigma \partial_\eta^\nu c(\tau, \eta)| \leqq C_{\sigma\nu}\Lambda^{\beta-\alpha+1-(\sigma+|\nu|)},$$

where

$$c(\tau, \eta) = \frac{1}{2\pi i} \oint_{\partial G} \frac{Q(\tau, \xi, \eta)}{P(\tau, \xi, \eta)} d\xi.$$

Corollary. *Let* Q *be a polynomial, satisfying* $N_Q \subset N_A^1 + (0, k)$. *Let*

$$\zeta_{i\pm}(\tau, \eta) = \frac{1}{2\pi i} \oint_{\partial \mathfrak{G}_{i\pm}} \frac{Q(\tau, \xi, \eta)}{A_0(\tau, \xi, \eta)} d\xi,$$
$$c_{j\pm}(\tau, \eta) = \frac{1}{2\pi i} \oint_{\partial G_{j\pm}} \frac{Q(\tau, \xi, \eta)}{A_0(\tau, \xi, \eta)} d\xi \quad (i<j<l),$$
$$c_l(\tau, \eta) = \frac{1}{2\pi i} \oint_{\partial G_l} \frac{Q(\tau, \xi, \eta)}{A_0(\tau, \xi, \eta)} d\xi$$

for $i<l$, *then we have* $\zeta_{i\pm} \in S_{\Lambda_i}^k(\Delta^{(i)})$, $c_{j\pm} \in S_{\Lambda_j}^k(\Delta^{(i)})$ $(i<j<l)$, $c_l \in S_{\Lambda_l}^k(\Delta^{(i)})$.

§ 3. Condition (B).

Let us consider a polynomial

$$B = \sum_{(k, |\nu|) \in N_A^1} b_{k\nu} \tau^k \xi^{\nu_1} \eta_2^{\nu_2} \cdots \eta_n^{\nu_2}.$$

Then we define the *enclosure* of N_B with respect to N_A^1 by

$$E_A(N_B)=\bigcap_{N_B\subset N_A^1-(\alpha,\beta)}\{N_A^1-(\alpha,\beta)\}.$$

Now we can give a more constructive definition of $E_A(N_B)$. First, let us denote

$$\partial' N_A^1=\bigcup_{i=1}^{l}\partial_i N_A^1=\{(f(x),x);\ 0\leqq x\leqq m'\},$$

then we have

$$\partial'\{N_A^1-(\alpha,\beta)\}=\{(f(x+\beta)-\alpha,x);\ -\beta\leqq x\leqq m'-\beta\}.$$

Moreover, we have

Lemma 3.1. *Let $\beta<\beta_0$ and $f(x_0+\beta)-\alpha=f(x_0+\beta_0)-\alpha_0$, then we have*

$$f(x+\beta)-\alpha\leqq f(x+\beta_0)-\alpha_0 \qquad \textit{if } x<x_0,$$
$$f(x+\beta)-\alpha\geqq f(x+\beta_0)-\alpha_0 \qquad \textit{if } x>x_0.$$

Proof. $f(x)$ is differentiable without finite points, $f'(x)$ is non-increasing, and

$$(f(x+\beta)-\alpha)-(f(x+\beta_0)-\alpha_0)=\int_\beta^{\beta_0}\{f'(x_0+s)-f'(x+s)\}ds.$$

Now let us define

$$\alpha_0=0, \qquad \beta_0=\sup\{\beta;\ N_B\subset N_A^1-(0,\beta)\},$$
$$x_0=\inf\{x;\ (f(x+\beta_0)-\alpha_0,x)\in N_B\}.$$

Let α_{i-1}, β_{i-1}, x_{i-1} satisfy

$$N_B\subset N_A^1-(\alpha_{i-1},\beta_{i-1}), \qquad x_{i-1}=\inf\{x;\ (f(x+\beta_{i-1})-\alpha_{i-1},x)\in N_B\},$$

then we define

$$\beta_i=\inf\{\beta;\ \beta<\beta_{i-1},\ N_B\subset N_A^1-(f(x_{i-1}+\beta)-f(x_{i-1}+\beta_{i-1})+\alpha_{i-1},\beta)\}$$
$$\alpha_i=f(x_{i-1}+\beta_i)-f(x_{i-1}+\beta_{i-1})+\alpha_{i-1},$$

and

$$x_i=\inf\{x;\ (f(x+\beta_i)-\alpha_i,x)\in N_B\},$$

whenever

$$\{x;\ (f(x+\beta_i)-\alpha_i,x)\in N_B\}\neq\phi.$$

Otherwise, we define $h=i$, $x_h=0$. Finally, we define

$$g(x)=f(x+\beta_i)-\alpha_i \quad \text{for } x_i\leqq x\leqq x_{i-1}\ (i=0,1,\cdots,h) \quad (x_{-1}=m'-\beta_0),$$

then we have

$$\partial' E_A(N_A)=\{(g(x),x);\ 0\leqq x\leqq x_{-1}\}.$$

Here we have

Lemma 3.2. *$E_A(N_B)$ is a polygon with sides parallel to those of N_A^1, where the length of a side of $E_A(N_B)$ is not greater than the corresponding one of N_A^1.*

Let us define the *principal part* of B by

$$B_0=\sum_{(k,|\nu|)\in\partial' E_A(N_B)} b_{k\nu}\tau^k\xi^{\nu_1}\eta_2^{\nu_2}\cdots\eta_n^{\nu_n},$$

and define the *i-th part* of B_0 by

$$B^{(i)}=\sum_{(k,|\nu|)\in\partial_i\{E_A(N_B)\}} b_{k\nu}\tau^k\xi^{\nu_1}\eta_2^{\nu_2}\cdots\eta_n^{\nu_n}\tau^{-h_i},$$

where $h_i=q_{i+1}\sigma_{i+1}+\cdots+q_l\sigma_l$, where $\sigma(E_A(N_B))=\{\sigma_1,\cdots,\sigma_l\}$. Now we define

$$\mathscr{L}_j=\tau^{q_j}-i(|\eta|^2+1)^{1/2}, \qquad \mathscr{L}_j^0=\tau^{q_j}-i|\eta|,$$

and

$$\mathscr{L}_B=\prod_{j=1}^{l}\mathscr{L}_j^{\delta_j} \quad (\delta_j=m'_j-\sigma_j),$$

where τ^{q_j} is a branch with values in the lower half complex plane if τ is there, and $\sigma(N_A^1)=\{m'_1,\cdots,m'_l\}$. Then we define the *standardization* of B by

$$B^{\#}=\mathscr{L}_B B,$$

and the i-th part of $B^{\#}$ by

$$B^{\#(i)}=(-i)^{q_1\delta_1+\cdots+q_{i-1}\delta_{i-1}}|\eta|^{\delta_1+\cdots+\delta_{i-1}}\mathscr{L}_i^{0\delta_i}B^{(i)}.$$

Now, when B is a vector, we define $B^{\#}$ and $B^{\#(i)}$ componentwise:

$$B=\begin{bmatrix}B_1\\ \vdots\\ B_{m+}\end{bmatrix}\longrightarrow B^{\#}=\begin{bmatrix}B_1^{\#}\\ \vdots\\ B_{m+}^{\#}\end{bmatrix}, \qquad B^{\#(i)}=\begin{bmatrix}B_1^{\#(i)}\\ \vdots\\ B_{m+}^{\#(i)}\end{bmatrix},$$

and we define the *i-th Lopatinski determinant* by

$$R^{(i)}(\tau, \eta)=\det[\mathscr{B}^{(i)}(\tau, \eta)b^{(i+1)}(\tau, 0)\cdots b^{(l)}(\tau, 0)] \quad (i=1, \cdots, l-1),$$

where

$$\mathscr{B}^{(i)}(\tau, \eta)$$
$$=\left[\frac{1}{2\pi i}\oint_{\partial\mathfrak{G}_{i+}}\frac{B^{\#(i)}}{A^{(i)}}d\xi\cdots\frac{1}{2\pi i}\oint_{\partial\mathfrak{G}_{i+}}\frac{B^{\#(i)}}{A^{(i)}}\left(\frac{\xi}{\mathscr{L}_i^0}\right)^{m_1^+ +\cdots+m_i^+ -1}d\xi\right],$$

and

$$b^{(i)}(\tau, \eta)=\left[\frac{1}{2\pi i}\oint_{\partial G_{i+}}\frac{B^{\#(i)}}{A^{(i)}}d\xi\cdots\frac{1}{2\pi i}\oint_{\partial G_{i+}}\frac{B^{\#(i)}}{A^{(i)}}\left(\frac{\xi}{\mathscr{L}_i^0}\right)^{m_i^+ -1}d\xi\right].$$

Moreover, the *l-th Lopatinski determinant* (Lopatinski determinant in usual sense) by

$$R^{(l)}(\tau, \eta)=\det\left[\frac{1}{2\pi i}\oint\frac{B^{(l)}}{A_+^{(l)}}d\xi\cdots\frac{1}{2\pi i}\oint\frac{B^{(l)}}{A_+^{(l)}}\xi^{m_+ -1}d\xi\right],$$

then $R^{(1)}$ is holomorphic in $\{\mathrm{Im}\,\tau\leqq 0, \eta\in R^{n-1}, (\tau, \eta)\neq(0, 0)\}$ and $R^{(j)}$ $(j=2, \cdots, l)$ is holomorphic in $\{\mathrm{Im}\,\tau<0, \eta\in R^{n-1}\setminus\{0\}\}$. Here we define

Condition (B) (*Uniform Lopatinski Condition*). Lopatinski determinants satisfy

$$R^{(1)}(\tau, \eta)\neq 0$$

for $\mathrm{Im}\,\tau\leqq 0$, $\eta\in R^{n-1}$, $(\tau, \eta)\neq 0$, and

$$R^{(i)}(\tau, \eta)\neq 0$$

for $\mathrm{Im}\,\tau\leqq 0$, $\eta\in S^{n-2}$ $(i=2, \cdots, l)$.

§ 4. Consideration of Examlpe 2

4.1. Characteristic roots of A w.r.t. ξ. Let us consider the dependence of the roots of

$$\begin{aligned}A&=\tau^3-i(\xi^2+\eta^2)\tau^2+i(\xi^4+\kappa\xi^2\eta^2+\eta^4) \quad (\kappa>1)\\&=\tau^3+A^{(2)}\end{aligned}$$

on the parameters (τ, η) in Γ_{γ_0}. Since A is a polynomial of degree 4 with respect to (τ, ξ, η), $A^{(2)}$ is the principal part of A in usual sense. First, we consider the root of $A^{(2)}=0$:

$$(\xi^2+\eta^2)\tau^2-(\xi^4+\kappa\xi^2\eta^2+\eta^4)=0,$$
$$\text{i.e.}\quad \xi^4+(\kappa\eta^2-\tau^2)\xi^2+\eta^2(\eta^2-\tau^2)=0.$$

Let $(\tau_0, \eta_0) \in S^1$ and $\eta_0 \neq 0$, then all the roots of $A^{(2)}=0$ are non-real if $\tau_0^2<\eta_0^2$, and just two roots of $A^{(2)}=0$ are non-real if $\tau_0^2 \geqq \eta_0^2$. Moreover, we have $\partial_\tau A^{(2)} \neq 0$ at $\xi=\xi_0$, where ξ_0 is a real root of $A=0$. Hence, A is elliptic in $U(\tau_0, \eta_0)$, a conic neighbourhood of (τ_0, η_0) in Γ_{γ_0}, if $\tau_0^2<\eta_0^2$, and

$$A=iE(\tau, \eta; \xi)H(\tau, \eta; \xi),$$

where E is elliptic and H is hyperbolic in $U(\tau_0, \eta_0)$ if $\tau_0^2 \geqq \eta_0^2$.

Let $(\tau_0, \eta_0) \in S^1$ and $\eta_0=0$, then the roots of $A^{(2)}=0$ are ± 1 and 0. Though $\partial_\tau A^{(2)} \neq 0$ at $\xi=\pm 1$, $\partial_\tau A^{(2)}=0$ at $\xi=0$. So, in this case, we need the aid of $A^{(1)}$.

The roots ξ of

$$A^{(1)}=\tau-i(\xi^2+\eta^2)=0, \quad \text{i.e.}\quad \xi^2=-i\tau-\eta^2$$

are in

$$\mathfrak{G}_{1\pm}=\{\xi \in C^1;\ |\xi| \leqq C\Lambda_1,\ \operatorname{Im}\xi \gtrless \pm c\Lambda_1)$$

for $\operatorname{Im}\tau \leqq 0$, $\eta \in R^1$. Since

$$|A\tau^{-2}-A^{(1)}|=|(\xi^4+\kappa\xi^2\eta^2+\eta^4)\tau^{-2}| \leqq C(|\xi|+|\eta|)^2\left(\frac{|\xi|+|\eta|}{|\tau|}\right)^2,$$

we have

$$|A^{(1)}| \geqq c\Lambda_1^2, \qquad |A\tau^{-2}-A^{(1)}| \leqq C'\Lambda_1^2\left(\frac{\Lambda_1}{|\tau|}\right)^2$$

for $\xi \in \partial\mathfrak{G}_{1\pm}$.

Hence the roots ξ of $A=0$ for $(\tau, \eta) \in \Delta^{(0)} \cup \Delta^{(1)}=\{|\eta| \leqq \varepsilon|\tau|\}$ are as figured, if ε and γ_0^{-1} are small enough.

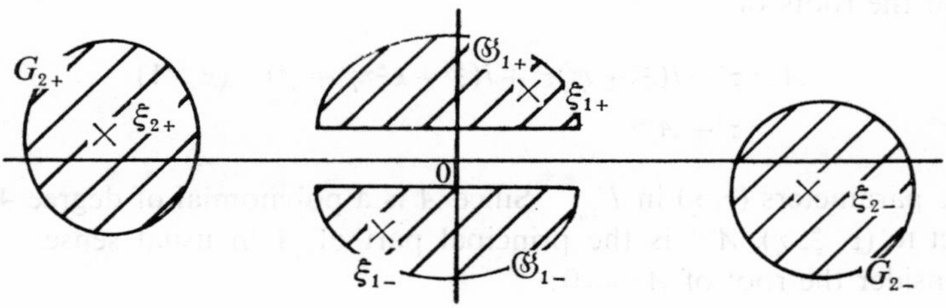

Therefore we have $A=iPH$, where P is parabolic and H is hyperbolic in $\Delta^{(0)}\cup\Delta^{(1)}$. More precisely, we have

$$H=(\xi-\xi_{2+})(\xi-\xi_{2-})=\xi^2+h_1\xi+h_2,$$
$$P=(\xi-\xi_{1+})(\xi-\xi_{1-})=\xi^2+e_1\xi+e_2,$$

where

$$h_1=-c_1,\quad h_2=c_1^2-c_2,\quad e_1=-\varepsilon_1,\quad e_2=\varepsilon_1^2-\varepsilon_2,$$

where

$$c_j=\frac{1}{2\pi i}\oint_{G_2}\frac{\partial_\xi A}{A}\xi^j d\xi,\qquad \varepsilon_j=\frac{1}{2\pi i}\oint_{\mathfrak{G}_1}\frac{\partial_\xi A}{A}\xi^j d\xi.$$

4.2. Uniform Lopatinski condition. Let

$$A=\tau^3-i(\xi^2+\eta^2)\tau^2+i(\xi^4+\kappa\xi^2\eta^2+\eta^4),\quad B_1=1,$$
$$B_2=(a_1\xi+a_2\eta+a)\tau+(\xi^2+b_{12}\xi\eta+b_{22}\eta^2+b_1\xi+b_2\eta+b),$$

we have

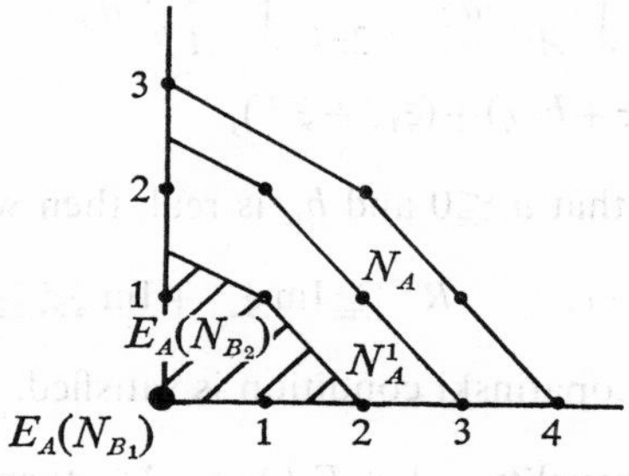

$$A^{(1)}=\tau-i(\xi^2+\eta^2),\qquad A^{(2)}=-i(\xi^2+\eta^2)\tau^2+i(\xi^4+\kappa\xi^2\eta^2+\eta^4),$$
$$B_1^{(1)}=B_1^{(2)}=1,\quad B_2^{(1)}=a_1\xi+a_2\eta,\quad B_2^{(2)}=(a_1\xi+a_2\eta)\tau+(\xi^2+b_{12}\xi\eta+b_{22}\eta^2).$$

Moreover, since

$$B_1^{\#}=\mathscr{L}_1\mathscr{L}_2^2,\qquad B_2^{\#}=\mathscr{L}_2B_2,$$

we have

$$B_1^{\#(1)}=\mathscr{L}_1^0,\qquad B_1^{\#(2)}=-i|\eta|\mathscr{L}_2^{02},$$
$$B_2^{\#(1)}=a_1\xi+a_2\eta,\qquad B_2^{\#(2)}=\mathscr{L}_2^0\{(a_1\xi+a_2\eta)\tau+(\xi^2+b_{12}\xi\eta+b_{22}\eta^2)\}$$

and

$$R^{(1)}=\begin{vmatrix} \frac{1}{2\pi i}\oint_{\partial\mathfrak{G}_{1+}}\frac{B_1^{\#(1)}}{A^{(1)}}d\xi & \frac{1}{2\pi i}\oint_{\partial G_{2+}}\frac{B_1^{\#(2)}}{A^{(2)}}d\xi\Big|_{\eta=0} \\ \frac{1}{2\pi i}\oint_{\partial\mathfrak{G}_{1+}}\frac{B_2^{\#(1)}}{A^{(1)}}d\xi & \frac{1}{2\pi i}\oint_{\partial G_{2+}}\frac{B_2^{\#(2)}}{A^{(2)}}d\xi\Big|_{\eta=0} \end{vmatrix}$$

$$=\begin{vmatrix} \frac{1}{2\pi i}\oint_{\partial\mathfrak{G}_{1+}}\frac{\mathscr{L}_1^0}{\tau-i(\xi^2+\eta^2)}d\xi & 0 \\ \frac{1}{2\pi i}\oint_{\partial\mathfrak{G}_{1+}}\frac{a_1\xi+a_2\eta}{\tau-i(\xi^2+\eta^2)}d\xi & \frac{1}{2\pi i}\oint_{\partial G_{2+}}\frac{\tau\xi(a_1\tau+\xi)}{-i\xi^2(\tau^2-\xi^2)}d\xi \end{vmatrix}$$

$$=\frac{a_1-1}{4(-i)^{1/2}},$$

$$R^{(2)}=\begin{vmatrix} \frac{1}{2\pi i}\oint\frac{B_1^{(2)}}{A_+^{(2)}}d\xi & \frac{1}{2\pi i}\oint\frac{B_1^{(2)}\xi}{A_+^{(2)}}d\xi \\ \frac{1}{2\pi i}\oint\frac{B_2^{(2)}}{A_+^{(2)}}d\xi & \frac{1}{2\pi i}\oint\frac{B_2^{(2)}\xi}{A_+^{(2)}}d\xi \end{vmatrix}$$

$$=\begin{vmatrix} 0 & 1 \\ \frac{1}{2\pi i}\oint\frac{B_2^{(2)}}{A_+^{(2)}}d\xi & \frac{1}{2\pi i}\oint\frac{B_2^{(2)}\xi}{A_+^{(2)}}d\xi \end{vmatrix}$$

$$=-\{(a_1\tau+b_{12}\eta)+(\xi_{1+}^{(2)}+\xi_{2+}^{(2)})\}.$$

Hence, if we assume that $a_1\leqq 0$ and b_{12} is real, then we have

$$|R^{(1)}|\geqq c,\qquad |R^{(2)}|\geqq \operatorname{Im}\xi_{1+}^{(2)}+\operatorname{Im}\xi_{2+}^{(2)}\geqq c|\eta|,$$

that is, the uniform Lopatinski condition is satisfied.

4.3. Energy inequality. Let $E_A(N_Q)=N_A^1$, then we have

$$Q_0=\sum_{(i,j+k)\in\partial_1N\cup\partial_2N} q_{ijk}\tau^i\xi^j\eta^k,$$

$$Q^{(1)}=\sum_{(i,j+k)\in\partial_1N} q_{ijk}\tau^i\xi^j\eta^k\tau^{-2}$$

$$=\sum_{j+k=1} q_{2jk}\xi^j\eta^k,$$

$$Q^{(2)}=\sum_{(i,j+k)\in\partial_2N} q_{ijk}\tau^i\xi^j\eta^k\quad(=Q_0),$$

and

$$|Q\tau^{-2}-Q^{(1)}|\leqq C_0\{|\tau|(|\xi|+|\eta|)^2+(|\xi|+|\eta|)^3\}|\tau|^{-2}$$

$$=C(|\xi|+|\eta|)\left\{\frac{|\xi|+|\eta|}{|\tau|}+\left(\frac{|\xi|+|\eta|}{|\tau|}\right)^2\right\},$$

where we remark

$$\frac{|\xi|+|\eta|}{|\tau|}\leqq C'\frac{\Lambda_1}{|\tau|}=C'\left(|\tau|^{-1/2}+\frac{|\eta|}{|\tau|}\right)\leqq C'(\gamma^{-1/2}+\varepsilon)$$

if $|\xi|\leqq C(|\tau|^{1/2}+|\eta|)$.

In $\Delta^{(2)}=\{|\eta|>\varepsilon|\tau|\}$, we have

$$\begin{aligned}&\langle\Lambda_2^3 v\rangle^2+\langle\Lambda_2 D_x^2 v\rangle^2+\langle\Lambda_2^2 D_x v\rangle^2+\langle D_x^3 v\rangle^2\\&\qquad+\gamma(\|\Lambda_2^3 v\|^2+\|\Lambda_2 D_x^2 v\|^2+\|\Lambda_2^2 D_x v\|^2+\|D_x^3 v\|^2)\\&\quad\leqq C\{\gamma^{-1}\|A(\tau, D_x, \eta)v\|^2+\langle B^{\#}(\tau, D_x, \eta)v\rangle^2\},\end{aligned}$$

owing to the well known result about hyperbolic mixed problems satisfying uniform Lopatinski conditions, because $R^{(2)}$ is the Lopatinski determinant in usual sense.

Now, we consider the energy estimates in $\Delta^{(0)}\cup\Delta^{(1)}=\{|\eta|<\varepsilon|\tau|\}$ and consider the role of $R^{(1)}$. As is seen in 4.1, we have $A=iPH$, where

$$P=(\xi-\xi_{1+})(\xi-\xi_{1-}),\qquad H=(\xi-\xi_{2+})(\xi-\xi_{2-}).$$

Let us denote

$$A_{1\pm}=\frac{A}{\xi-\xi_{1\pm}},\qquad A_{2\pm}=\frac{A}{\xi-\xi_{2\pm}},$$

then we have

$$\begin{bmatrix}B_1^{\#}\\B_2^{\#}\end{bmatrix}=B_+\begin{bmatrix}A_{1+}\\A_{2+}\end{bmatrix}+B_-\begin{bmatrix}A_{1-}\\A_{2-}\end{bmatrix},$$

where

$$B_\pm=\begin{pmatrix}\dfrac{1}{2\pi i}\displaystyle\oint_{\partial\mathfrak{G}_{1\pm}}\frac{B_1^{\#}}{A}d\xi & \dfrac{1}{2\pi i}\displaystyle\oint_{\partial G_{2\pm}}\frac{B_1^{\#}}{A}d\xi\\ \dfrac{1}{2\pi i}\displaystyle\oint_{\partial\mathfrak{G}_{1\pm}}\frac{B_2^{\#}}{A}d\xi & \dfrac{1}{2\pi i}\displaystyle\oint_{\partial G_{2\pm}}\frac{B_2^{\#}}{A}d\xi\end{pmatrix}.$$

Let us denote

$$B_+^{(1)}(\tau,\eta)=\begin{pmatrix}\dfrac{1}{2\pi i}\displaystyle\oint_{\partial\mathfrak{G}_{1+}}\frac{B_1^{\#(1)}(\tau,\xi,\eta)}{A^{(1)}(\tau,\xi,\eta)}d\xi & \dfrac{1}{2\pi i}\displaystyle\oint_{\partial G_{2+}}\frac{B_1^{\#(2)}(\pi,\xi,0)}{A^{(2)}(\tau,\xi,0)}d\xi\\ \dfrac{1}{2\pi i}\displaystyle\oint_{\partial\mathfrak{G}_{1+}}\frac{B_2^{\#(1)}(\tau,\xi,\eta)}{A^{(1)}(\tau,\xi,\eta)}d\xi & \dfrac{1}{2\pi i}\displaystyle\oint_{\partial G_{2+}}\frac{B_2^{\#(2)}(\tau,\xi,0)}{A^{(2)}(\tau,\xi,0)}d\xi\end{pmatrix},$$

then the difference of B_+ and $B_+^{(1)}$ can be small in $\Delta^{(0)}\cup\Delta^{(1)}=\{|\eta|<\varepsilon|\tau|\}$

if ε and γ_0^{-1} is small enough. Since

$$|R^{(1)}(\tau, \eta)| = |\det B_+^{(1)}(\tau, \eta)| > c$$

in $\{\operatorname{Im} \tau \leqq 0, \eta \in R^1, (\tau, \eta) \neq 0\}$, we have

$$|\det B_+| \geqq c'$$

in $\Delta^{(0)} \cup \Delta^{(1)}$ ($\varepsilon, \gamma_0^{-1}$: small enough).

Let Q be a polynomial such that $E_A(N_Q) = N_A^1$, then we have

$$Q = Q_+ \begin{bmatrix} A_{1+} \\ A_{2+} \end{bmatrix} + Q_- \begin{bmatrix} A_{1-} \\ A_{2-} \end{bmatrix},$$

where $Q_\pm = Q_\pm(\tau, \eta)$ is a smooth bounded 1×2-matrix in $\Delta^{(0)} \cup \Delta^{(1)}$, that is,

$$Q = Q_+ B_+^{-1} \begin{bmatrix} B_1^\# \\ B_2^\# \end{bmatrix} + (Q_- - Q_+ B_+^{-1} B_-) \begin{bmatrix} A_{1-} \\ A_{2-} \end{bmatrix}.$$

Since we have the energy estimates for the operators $D_x - \xi_{i\pm}$:

$$\begin{cases} \langle A_{i-} v \rangle^2 + \gamma^{1/2} \|A_{i-} v\|^2 \leqq C \gamma^{-1/2} \|Av\|^2, \\ \gamma^{1/2} \|A_{i+} v\|^2 \leqq C(\gamma^{-1/2} \|Av\|^2 + \langle A_{i+} v \rangle^2), \end{cases}$$

we have

$$\langle Qv \rangle^2 + \gamma^{1/2} \|Qv\|^2 \leqq C(\gamma^{-1/2} \|Av\|^2 + \langle B^\# v \rangle^2)$$

in $\Delta^{(0)} \cup \Delta^{(1)}$. Together with the energy inequality in $\Delta^{(2)}$, we have

$$\langle Qv \rangle^2 + \gamma^{1/2} \|Qv\|^2 \leqq C(\gamma^{-1/2} \|Av\|^2 + \langle B^\# v \rangle^2)$$

in $\{\operatorname{Im} \tau < -\gamma_0, \eta \in R^1\}$.

References

[1] L. R. Volevich and S. G. Gindikin, The problem of Cauchy for differential operators with a dominant principal part, Functional Anal. Appl. **2** (1968), 204–218.

[2] R. Sakamoto, Mixed problems for hyperbolic equations I, II, J. Math. Kyoto Univ. **10** (1970), 349–373, 403–417.

[3] ——, Mixed problems for evolution equations, J. Math. Kyoto Univ. **24** (1984), 473–505.

Department of Mathematics
Nara Women's University
Nara 630, Japan

Taniguchi Symp. HERT
Katata 1984, pp. 347–362

Tunnel Effects for Semiclassical Schrödinger Operators

J. Sjöstrand

§ 0. Introduction

The results of this survey have been obtained jointly with B. Helffer. Naturally, we want to concentrate the exposition on the most recent results, but since our earlier works [4, 5, 6, 7] can not be considered as generally known, we recall in Sections 1 and 2 some of the results of [4, 5]. The results of Section 4 are still very preliminary and that section should be considered as an informal workshop communication.

Let $P=-h^2\Delta+V(x)$ be a semiclassical Schrödinger operator on a compact Riemannian manifold M or on $M=\boldsymbol{R}^n$. Here Δ is the Laplace operator and V is a real-valued C^∞ potential. We are then interested in the eigenvalues of P near some fixed energy $E_0 \in \boldsymbol{R}$, that we can assume to be 0 after replacing V by $V-E_0$. In the case when $M=\boldsymbol{R}^n$ we will assume, except in Section 4, that

$$\lim_{|x|\to\infty} V(x)>0. \tag{0.1}$$

This implies that P has discrete spectrum near 0 (if we denote by P also the Friedrichs extension).

By a potential well, we mean a compact connected component of the set $V(x)\leq 0$. From the Agmon-Simon decay estimates, we know that if $(P-E(h))u=0$, $\|u\|_{L^2}=1$, $E(h)\to 0$ as $h\to 0$, then u is exponentially small outside the set $V(x)\leq 0$. The "tunnel effects" that we are interested in are then the "interactions" between the various wells through the potential barrier, given by $V(x)>0$.

§ 1. General abstract results

A very important role is played by the so called Agmon-metric $\max(V, 0)dx^2$, where dx^2 is the given Riemannian metric on M. One can easily understand the importance of this metric, by considering $e^{-f(x)/h}Pe^{f/h}$ as an h-pseudodifferential operator with principal symbol $\xi^2+(V(x)-(f'(x))^2-2if'(x)\cdot\xi)$ and ask for what real-valued functions f this operator is elliptic. Agmon [1] proved decay results for eigenfunctions of Schrödinger operators at infinity and B. Simon [9] adapted this to the semiclassical situation. We have by integration by parts the following result:

Proposition 1.1. *Let $\Omega \subset M$ be a bounded domain with C^2 boundary and let Φ be a real Lipschitz function on $\bar{\Omega}$. If $V(x)-(\nabla\Phi(x))^2=F_+(x)^2-F_-(x)^2$ a.e. and $F_\pm \geq 0$ are in $L^\infty(\Omega)$, then*

$$(1.1)\qquad \begin{aligned} &h^2\|\nabla(e^{\Phi/h}u)\|^2+(1/2)\|F_+e^{\Phi/h}u\|^2 \\ &\quad \leq \|(F_++F_-)^{-1}e^{\Phi/h}Pu\|^2+(3/2)\|F_-e^{\Phi/h}u\|^2, \end{aligned}$$

for every $u \in C^2(\Omega)$ with $u|_{\partial\Omega}=0$. The norms are the usual ones on $L^2(\Omega)$.

Here is a typical application: Let M be compact for simplicity and let $u \in L^2(M)$ be normalized with $(P-E(h))u=0$, $E(h)\to 0$, $h\to 0$. Let K be the set where $V\leq 0$. Then for every $\varepsilon>0$, we have

$$(1.2)\qquad \|e^{d(x,K)/h}\nabla u\|^2+\|e^{d(x,K)/h}u\|^2\leq C_\varepsilon e^{\varepsilon/h},$$

where d is the Agmon distance. In fact, we just take $\Phi(x)=(1-\delta)d(x, K)$. Then $V-E(h)-(\nabla\Phi)^2\geq 1/C(\delta)$ when $d(x,K)\geq\delta$ and h is small enough and we obtain (1.2) from (1.1), with P replaced by $P-E(h)$, if we make suitable choices of F_+ and F_- and choose $\delta>0$ small enough depending on ε. The estimate (1.2) is essentially due to B. Simon [9]. By making more refined choices of Φ, $F_\pm$, it is often possible to replace the exponential to the right in (1.2) by h^{-N_0} for some large N_0. This improvement is crucial, when we want to compute the asymptotic expanisons of the eigen-functions and of the interaction coefficients in Theorem 1.2 below.

We now assume,

$$(1.3)\qquad \{x \in M;\ V(x)\leq 0\}=U_1\cup\cdots\cup U_N, \text{ where } U_j \text{ are compact disjoint and of diameter 0 for } d.$$

For $\eta>0$ very small, we put $M_j=M\setminus\bigcup_{k\neq j}B(U_k,\eta)$, where $B(U_k,\eta)=\{x \in M;\ d(x, U_k)<\eta\}$. Let P_{M_j} denote the corresponding Dirichlet realization of P on $L^2(M_j)$. Let $J\subset]0, 1]$ have 0 as an accumulation point and let $I(h)\subset \boldsymbol{R}$, $h \in J$ be a compact interval, tending to $\{0\}$ as $h\to 0$. We assume that there is a function $a(h)>0$ with $\log a(h)=o(1/h)$ such that $(I(h)+[-2a, 2a])\setminus I(h)$ has an empty intersection with the spectra of P and P_{M_j}. Let $\mu_{j,k}$ be the eigenvalues of P_{M_j} in $I(h)$ and let $\varphi_{j,k}$, $1\leq k\leq m_j$, be a corresponding orthonormal system in $L^2(M_j)$. (It follows from the analysis below that the $\mu_{j,k}$ change only by $\mathcal{O}_\varepsilon(\exp -2(S_0-\eta-\varepsilon)/h)$, if we further decrease η.) Then by decay estimates

$$(1.4) \qquad \varphi_\alpha(x, h) = \tilde{\mathcal{O}}(\exp - d(x, U_{j(\alpha)})/h),$$

where we write $\alpha = (j, k)$, $j(\alpha) = j$ and $\tilde{\mathcal{O}}$ stands for $\mathcal{O}(\exp(-(d(x, U_{j(\alpha)}) - \varepsilon(\eta))/h))$ uniformly on M_j with $\varepsilon(\eta) \to 0$ as $\eta \to 0$. (Actually, in (1.4) we can take $\varepsilon(\eta)$ arbitrarily small for each η, but the same will not always be true below.)

Let also $\lambda_1, \cdots, \lambda_{\tilde{N}}$ be the eigenvalues of P in $I(h)$ and let $u_1, \cdots, u_{\tilde{N}}$ be a corresponding orthonormal system. We then want to describe $\lambda_1, \cdots, \lambda_{\tilde{N}}$ with the help of φ_α and μ_α. Let $\theta_k \in C_0^\infty(B(U_k, 2\eta))$ be equal to 1 near $\overline{B(U_k, \eta)}$ and put $\chi_j = 1 - \sum_{k \neq j} \theta_k$, $\psi_\alpha = \chi_{j(\alpha)} \varphi_\alpha$. If E and F are the spaces spanned respectively by the ψ_α and the u_k, one shows, combining decay estimates and elementary Hilbert space arguments, that E and F have the same dimension and that the distance between E and F is $\tilde{\mathcal{O}}(\exp(-S_0/h))$. Let $v_\alpha \in F$ be the orthogonal projection of ψ_α. Then the matrix $V = ((v_\alpha \mid v_\beta))$ is the identity plus an exponentially small error, and if we introduce the orthonormal basis $\vec{e} = \vec{v} V^{-1/2}$ in F and the $N \times N$-matrix $\mathcal{D}' = ((1 - \delta_{j,k}) \exp(-d(U_j, U_k)/h))$, where $\delta_{j,k}$ is the Kronecker delta, we have the

Theorem 1.2. *The matrix of* $P|_F$ *with respect to the basis* $\vec{e}$ *is of the form* $\operatorname{diag}(\mu_\alpha) + \hat{W} + R$, *where* $W = ((w_{\alpha,\beta} + w_{\beta,\alpha})/2)$, $w_{\alpha,\beta} = 0$ *if* $j(\alpha) = j(\beta)$ *and otherwise*

$$(1.5) \qquad w_{\alpha,\beta} = h^2 \int \chi_{j(\alpha)} (\varphi_\beta \nabla \varphi_\alpha - \varphi_\alpha \nabla \varphi_\beta) \nabla \chi_{j(\beta)} dx.$$

The remainder $R = (R_{\alpha,\beta})$, *satisfies* $R_{\alpha,\beta} = \tilde{\mathcal{O}}((\mathcal{D}'^2 + \mathcal{D}'^3)_{j(\alpha),j(\beta)})$.

Up to small errors one can also give a surface integral representation of the interaction coefficients $w_{\alpha,\beta}$, which does not depend on the cutoff functions.

§ 2. The case of non-degenerate point wells

In order to obtain more explicit results from Theorem 1.2, we need to look for situations where we can determine the asymptotics of the μ_α and of the φ_α. In the one-dimensional case this is often possible. One case where one gets satisfactory results, also in higher dimensions is the one where all the wells are reduced to points, where V vanishes and has a strict minimum:

$$(2.1) \qquad U_j = \{x_j\}, \quad V''(x_j) > 0, \quad \text{for } j = 1, \cdots, N.$$

We then take $I(h) = [0, \text{const. } h]$ or possibly some subinterval of this interval. In [8] Simon computed rigorously the asymptotics of the μ_α, and in [4] we also obtained the asymptotic behaviour of the φ_α on sufficiently large sets, so that the asymptotics of the interaction coefficients in Theorem 1.2 can be obtained under some very natural extra assumptions.

Let $q(x, \xi) = -p(x, i\xi) = \xi^2 - V(x)$, where $p(x, \xi) = \xi^2 + V(x)$ is the principal symbol of P. We consider first the situation near some fixed well $U_{j_0} = \{x_0\}$. We may choose local coordinates centered at x_0 such that $q = \sum_1^n (\lambda_j/2)(\xi_j^2 - x_j^2) + \mathcal{O}(|(x, \xi)|^3)$. The Hamilton field

$$H_q = \sum_1^n (\partial q/\partial \xi_j)\partial/\partial x_j - (\partial q/\partial x_j)\partial/\partial \xi_j$$

then has a critical point at $(0, 0)$ and its linearisation at that point has the eigenvalues $\lambda_1, \cdots, \lambda_n, -\lambda_1, \cdots, -\lambda_n$. If $e_{\lambda_j}, e_{-\lambda_j}$ are the corresponding eigenvectors, then the stable manifold theorem assures the existence near $(0, 0)$ of two Lagrangian manifolds Λ_+ and Λ_- which are H_q invariant, contain $(0, 0)$ and with tangent space at that point spanned respectively by $e_{\lambda_1}, \cdots, e_{\lambda_n}$ and $e_{-\lambda_1}, \cdots, e_{-\lambda_n}$. It is not hard to see that

$$(2.2) \qquad \Lambda_\pm = \{(x, \pm\varphi'(x));\ x \in \Omega\},$$

where Ω is a suitable neighborhood of 0 and where $\varphi(x) = d(x, 0)$ is a C^∞ function with $\varphi''(0) > 0$.

Associated to U_{j_0} we also have the harmonic oscillator

$$\sum_1^n (\lambda_j/2)(D_{x_j}^2 + x_j^2)$$

whose eigenvalues are of the form $\sum_1^n \lambda_j(\alpha_j + 1/2)$, with $\alpha_j \in N$. Let E_0 be such an eigenvalue and assume for simplicity that E_0 is simple. Then there exist formal symbols $E(h) \sim \sum_0^\infty E_j h^j$ and $a(x, h) \sim h^m \sum_0^\infty a_j(x) h^j$, $x \in \Omega$, where m depends on E_0, such that in the sense of formal asymptotic expansions in h:

$$(2.3) \qquad (P - hE(h))(ae^{-\varphi(x)/h}) \sim 0.$$

Let $E(h)$ also denote an asymptotic sum and let $I(h) = [h(E_0 - \varepsilon_0), h(E_0 + \varepsilon_0)]$. Then if $\varepsilon_0 > 0$ is sufficiently small, one can prove that $Sp(P_{M_{j_0}}) \cap I(h) = \{\mu_{j_0}\}$, where $\mu_{j_0} - hE(h) = \mathcal{O}(h^\infty)$. Moreover, if m is sutiably chosen and $\varphi_{j_0} \in L^2(M_{j_0})$ denotes the corresponding eigenfunction, we have

$$(2.4) \qquad \varphi_{j_0}(x, h) - a(x, h)e^{-\varphi(x)/h} = \mathcal{O}(h^\infty)e^{-\varphi(x)/h}, \quad x \in \Omega,$$

and similarly for all the derivatives. (When V is analytic near 0, all the symbols can be chosen to be analytic symbols and we can replace the error estimates $\mathcal{O}(h^\infty)$ by $\mathcal{O}(e^{-\varepsilon/h})$ for some $\varepsilon>0$.) It turns out that the description (2.4) is valid in quite large sets Ω. If a point $y \in \mathring{M}_{j_0}$ is such that there is a unique minimal geodesic (for the Agmon metric) from 0 to y which in addition is contained in $\mathring{M}_{j_0}$ and non-degenerate in the sense of Riemannian geometry, then we can find an open set $\Omega \subset M_{j_0}$ which contains x_0, y and the geodesic, such that (2.4) is valid. If $z \in M$ is a point which can be linked to x_0 by a minimal geodesic, contained in M_{j_0}, then all the points $y \neq z$ on this geodesic have the properties just quoted.

Now suppose that both U_{j_0} and U_{k_0}, $j_0 \neq k_0$ have E_0 as a simple eigenvalue for their localized harmonic oscillators, and that $d(U_{j_0}, U_{k_0}) < d(U_{j_0}, U_\nu) + d(U_\nu, U_{k_0})$ for all $\nu \notin \{j_0, k_0\}$. Then all the minimal geodesics from U_{j_0} to U_{k_0} avoid the other wells and will therefore be contained in $M_{j_0} \cup M_{k_0}$ if $\eta>0$ is small enough. For the reasons indicated above, we may even find open sets $\Omega_{j_0} \subset \mathring{M}_{j_0}$, $\Omega_{k_0} \subset \mathring{M}_{k_0}$ where a description of the type (2.4) is valid for φ_{j_0} and φ_{k_0} respectively, and such that the intersections between each minimal geodesic from U_{j_0} to U_{k_0} and supp $(\nabla \chi_{j_0})$ and supp $(\nabla \chi_{k_0})$ respectively are contained in $\Omega_{j_0} \cap \Omega_{k_0}$. This means that we know the behaviour of both φ_{j_0} and φ_{k_0} in the regions, which can give contributions to w_{j_0,k_0} of the order of magnitude $\exp(-d(U_{j_0}, U_{k_0})/h - \varepsilon/h)$ or larger, for some $\varepsilon>0$. If in addition the minimal geodesics from U_{j_0} to U_{k_0} are finitely many and non-degenerate then we have a simple asymptotic expansion of w_{j_0,k_0} by the method of stationary phase: $w_{j_0,k_0} \exp(d(U_{j_0}, U_{k_0})/h) \sim h^{m(j_0,k_0)} \sum_0^\infty d_\nu h^\nu$.

Example. Let V be a potential on $\boldsymbol{R}^2$ invariant under the change of sign of any of the coordinates, and with 4 non-degenerate point wells as on the figure:

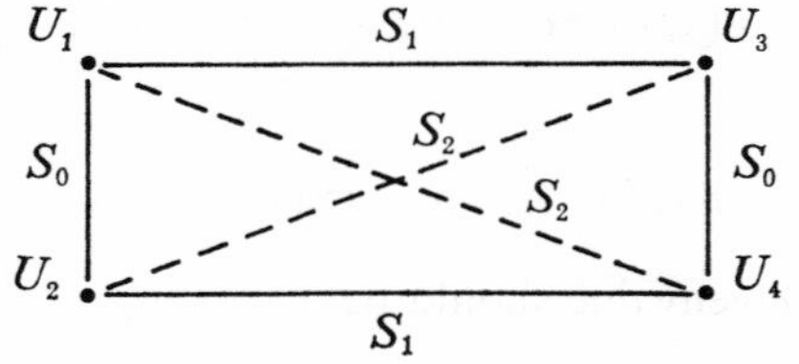

$$d(U_1, U_2) = d(U_3, U_4) = S_0, \qquad d(U_1, U_3) = d(U_2, U_4) = S_1,$$
$$d(U_1, U_4) = d(U_2, U_3) = S_2, \qquad S_0 < S_1 < S_2.$$

We make all the assumptions above concerning the pairs (U_1, U_2), (U_3, U_4), (U_1, U_3), (U_2, U_4). Then with $I(h)$ as above we have $\mu_1 = \mu_2 = \mu_3 = \mu_4$ (if the M_j's are symmetrically chosen). The matrix of $P|_F$ in Theorem 1.2 is then of the form

$$(2.5)\qquad \tilde{\mu}I+\left(\begin{array}{cc|cc}0 & a & b & 0\\ a & 0 & 0 & b\\ \hline b & 0 & 0 & a\\ 0 & b & a & 0\end{array}\right)+\begin{pmatrix}0 & \tilde{\mathcal{O}}(\exp-S_2/h)\\ \tilde{\mathcal{O}}(\exp-S_2/h) & 0\end{pmatrix},$$

where $\tilde{\mu}-\mu=\tilde{\mathcal{O}}(\exp-2S_0/h)$. Here $a=\tilde{\mathcal{O}}(\exp-S_0/h)$ and $b=\tilde{\mathcal{O}}(\exp-S_1/h)$ have asymptotic expansions. We conclude that the eigenvalues of P in $I(h)$ are $\pm a\pm b+\tilde{\mathcal{O}}(\exp-S_2/h)$. The interpretation is here that U_1, U_2 and U_3, U_4 form two molecules and that each molecule has the eigenvalues $\tilde{\mu}+a$, and $\tilde{\mu}-a$, splitted to the distance $2a$ by atomic interaction. The interaction between the molecules gives rise to an additional splitting of $2b$.

§ 3. Wells formed by submanifolds

Results related to the ones of this section have been obtained independently by B. Simon. One motivation for studying this type of wells comes from Witten's paper [11]. (See also [7], where we proved that Witten's ideas can be used to compute the cohomology of a manifold.) We shall restrict ourselves here to the case of one single well, which already presents an interesting interaction phenomenon, namely between certain "sub-wells" of this well.

Let $\Gamma\subset M$ be a closed compact connected submanifold of codimension d. We assume that

(3.1) $V^{-1}(0)=\Gamma$ and V vanishes precisely to the second order on Γ.

We are then looking for asymptotic eigenfunctions of the form

$$(3.2)\qquad u(x,h)=a(x,h)e^{-(\varphi(x)+h^{1/2}\psi(x))/h},$$

where

$$(3.3)\qquad a(x,h)\sim a_0(x)+a_1(x)h^{1/2}+\cdots.$$

The corresponding eigenvalue should be

$$(3.4)\qquad hE(h)\sim h(E_0+E_1h^{1/2}+\cdots).$$

If we put $\Phi=\varphi(x)+h^{1/2}\psi(x)$, then a simple computation gives

$$\begin{aligned}e^{\Phi/h}P(e^{-\Phi/h}a)=&(-(\varphi')^2+V)a-2h^{1/2}(\varphi'\cdot\psi')a\\ &+h(2\nabla\varphi\cdot\nabla+\Delta\varphi-(\psi')^2)a+h^{3/2}(2\nabla\psi\cdot\nabla a+\Delta\psi a)-h^2\Delta a.\end{aligned}$$

Here we work near Γ where $d(x,\Gamma)$ is a smooth function vanishing precisely to the second order on Γ. We take $\varphi = d(x, \Gamma)$, so the first eiconal equation: $-(\varphi')^2 + V = 0$ is satisfied. We shall choose ψ so that

$$\nabla\varphi \cdot \nabla\psi = 0. \tag{3.5}$$

This equation can be solved uniquely, if we prescribe the value of ψ on Γ. It then remains to determine $\psi|_\Gamma$, a and E so that

$$(2\nabla\varphi \cdot \nabla + \Delta\varphi - (\psi')^2)a + h^{1/2}(2\nabla\psi \cdot \nabla a + \Delta\psi a) - h\Delta a - E(h)a = 0. \tag{3.6}$$

Looking at the degrees of homogeneity, we get

$$T_0 a_0 = 0, \tag{3.7}$$

$$2\nabla\psi \cdot \nabla a_0 + \Delta\psi a_0 - E_1 a_0 + T_0 a_1 = 0, \tag{3.8}$$

and further equations$\cdots$. Here $T_0 = 2\nabla\varphi \cdot \nabla + \Delta\varphi - E_0 - (\psi')^2$. We now restrict ourselves to the first eigenvalue, and we therefore expect that a_0 should be non-vanishing. Then in order to have a solution to (3.7), it is necessary that

$$\Delta\varphi - E_0 - (\psi')^2 = 0 \text{ on } \Gamma. \tag{3.9}$$

Now assume that

(3.10) $(\Delta\varphi)|_\Gamma$ has a finite number of minimum points and all these minima are non-degenerate.

We choose E_0 to be the corresponding minimum value. If x_0 is one of the minimum points, and if we restrict the attention to a neighborhood of that point, we see that (3.9) is a new eiconal equation on Γ, associated to the potential $W = (\Delta\varphi)|_\Gamma - E_0$ and as a solution, we let $\psi|_\Gamma(x)$ be the corresponding Agmon distance from x_0 to x. Now (3.7) can be solved with a prescribed value for $a_0|_\Gamma$. To solve (3.8) it is necessary that $2\nabla\psi \cdot \nabla a_0 + \Delta\psi a_0 - E_1 a_0 = 0$ on Γ. This transport equation has a non-vanishing solution a_0 (unique up to a constant factor) if and only if $E_1 = \Delta\psi(x_0)$. Now a_0 is completely determined and (3.8) can be solved for a_1 with a prescribed value on Γ. Naturally, this scheme can be continued and we get an asymptotic solution; $(P - hE(h))u = \mathcal{O}(h^\infty)e^{-(\varphi + h^{1/2}\psi)/h}$. This asymptotic construction works in small neighborhoods in M of open sets in Γ, which have the analogous properties (but now for the new Agmon metric Wdx^2) to the ones, where the *WKB* expansions of Section 2 work.

These formal constructions can be justified, using weighted L^2-estimates, analogous to the ones discussed in Section 1. These estimates are somewhat intricate and can not be described here in detail. Roughly one first proves estimates near Γ, using weight-functions of the type $\exp \tilde{\psi}(x)/h^{1/2}$, and after that, by more simple arguments, we get estimates away from Γ, by using weights $\exp(\tilde{\varphi}(x)/h+\tilde{\psi}(x)/h^{1/2})$.

The final results are completely analogous to the ones of Sections 1 and 2: Let $x_1, \cdots, x_N$ be the minimum points (where $\Delta\varphi=E_0$), let B_k be suitable small neighborhoods of x_k, put $M_j=M\setminus\bigcup_{k\neq j} B_k$. Then the lowest eigenvalue of P_{M_j} is asymptotically equal to $hE^{(j)}(h)$ with $E^{(j)}=E$, obtained as above. This eigenvalue is separated by at least (const.) $h^{3/2}$ from the rest of the spectrum of P_{M_j}. If we let φ_j denote the corresponding eigenfunction, then φ_j has the asymptotic behaviour, that we obtained formally above, in certain "sufficiently large" regions. Moreover, Theorem 1 remains valid with only one modification: $\mathscr{D}'$ is now the $N\times N$ matrix $((1-\delta_{j,k})\exp(-d_W(x_j, x_k)/h^{1/2}))$, where d_W denotes the Agmon distance on Γ, associated to Wdx^2. Again, under mild additional geometric assumptions, we have asymptotic expansions for the interaction coefficients, analogous to the ones in Section 2.

§ 4. Resonances

The results presented in this section are very recent and still in a somewhat preliminary shape, so this should only be considered as an informal workshop communication. We are now working on $\boldsymbol{R}^n$ and we do no more assume that $\underline{\lim}_{|x|\to\infty} V(x)>0$. Then we cannot be sure that P has a self-adjoint realization and even if so, we cannot expect the spectrum of P to be discrete near 0. Instead we look for some other function space where again P should have discrete spectrum near 0. Aguilar and Combes [2] have already done this, and their method consisting in replacing $\boldsymbol{R}^n$ by some other totally real subspace of $\boldsymbol{C}^n$, has been developed by many people, see for instance I. Herbst [3], and is now known as the method of complex scaling.

The method we shall present here is more microlocal, the general machinery is somewhat heavy to build up, but it permits us to treat general classes of potentials, including some, where the method of Aguilar-Combes cannot work. We also think that it gives a little more flexibility in concrete situations.

We start by fixing suitable scales at infinity. Let $r(x)\geq 1$ and $R(x)>0$ be smooth functions on $\boldsymbol{R}^n$ such that $rR\geq 1$, $\partial^\alpha R=\mathcal{O}(R^{1-|\alpha|})$, $\partial^\alpha r=\mathcal{O}(rR^{-|\alpha|})$ for all $\alpha\in\boldsymbol{N}^n$. We put $\tilde{r}=(r^2+\xi^2)^{1/2}$ and we say that a smooth positive function $m(x,\xi)$ on $\boldsymbol{R}^{2n}$ is an order function if

$$\partial_x^\alpha\partial_\xi^\beta m=\mathcal{O}(m\tilde{r}^{-|\beta|}R^{-|\alpha|}),$$

for all α and β in $\boldsymbol{N}^n$. For example $\tilde{r}^j R^k$ is an order function if j and

k are real. If m is an order function we can define a corresponding space of symbols; $S(m)$. These symbols are then defined on $\boldsymbol{R}^{2n}$, some-times they may depend on h and we then require uniformity with respect to h in the defining estimates: $|\partial_x^\alpha \partial_\xi^\beta a| \leq C_{\alpha,\beta} m \tilde{r}^{-|\beta|} R^{-|\alpha|}$. There is also an obvious way to define the symbol spaces $S(mh^k)$, if k is a real number.

Let V be a real-valued potential and assume

(4.1) $\qquad V$ is analytic on $\boldsymbol{R}^n$,

(4.2) $\qquad V$ extends holomorphically to a domain $|\mathrm{Im}\, x| < C^{-1} R(\mathrm{Re}\, x)$ and we have $|V(x)| \leq C(r(\mathrm{Re}\, x))^2$ there.

Then by Cauchy's inequalities, $p = \xi^2 + V$ belongs to $S(\tilde{r}^2)$. In general we shall write $S^{j,k} = S(\tilde{r}^j R^k)$, and we say that $G \in \dot{S}^{1,1}$ if $\partial_{x_j} G \in S^{1,0}$, $\partial_{\xi_j} G \in S^{0,1}$ for all j. If $G \in \dot{S}^{1,1}$, we notice that $H_p(G) \in S^{2,0}$. Here H_p denotes the Hamilton field of p.

Definition 4.1. We say that $G \in \dot{S}^{1,1}$ is an escape function if there is a compact set $K \subset p^{-1}(0)$ and a constant $C > 0$, such that $H_p(G) \geq C^{-1} r^2$ in $p^{-1}(0) \setminus K$.

It is possible to characterize the existence of an escape function in terms of a certain non-trapping condition at infinity. One can say that in the case of Aguilar-Combes the escape function is linear in ξ and often it is just $x \cdot \xi$. For example, suppose that $V(x) = -1 + o(1)$ for $|\mathrm{Im}\, x| \leq C^{-1}(1 + |x|)$. We can then take $R(x) = (1 + x^2)^{1/2}$, $r(x) = 1$, and an escape function is given by $G(x, \xi) = x \cdot \xi$.

If $G \in S^{1,1}$ is sufficiently small, or $t > 0$ is sufficiently small, then the I-lagrangian manifold Λ_t:

(4.4) $$\mathrm{Im}\, \xi = -t \frac{\partial G(\mathrm{Re}\, x, \mathrm{Re}\, \xi)}{\partial\, \mathrm{Re}\, x}, \qquad \mathrm{Im}\, x = t \frac{\partial G}{\partial\, \mathrm{Re}\, \xi},$$

belongs to the domain of p and $p|_{\Lambda_t}$ (considered as a function of $\mathrm{Re}\, x$, $\mathrm{Re}\, \xi$) belongs to the class $S^{2,0}$. If G is an escape function and $t > 0$ is sufficiently small, then outside a compact subset of Λ_t, $p|_{\Lambda_t}$ takes its values in a domain:

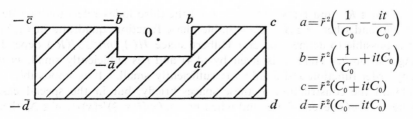

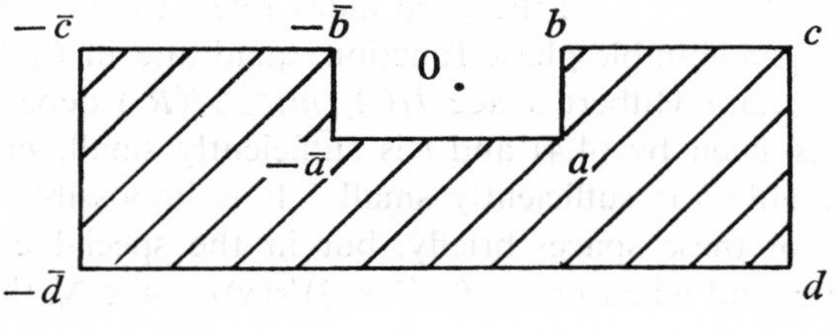

$$a = \tilde{r}^2\left(\frac{1}{C_0} - \frac{it}{C_0}\right)$$

$$b = \tilde{r}^2\left(\frac{1}{C_0} + itC_0\right)$$

$$c = \tilde{r}^2(C_0 + itC_0)$$

$$d = \tilde{r}^2(C_0 - itC_0)$$

This implies that $|p|_{\Lambda_t}|\geq(C_0)^{-1}\tilde{r}(\operatorname{Re}x, \operatorname{Re}\xi)^2$ outside a compact set, so $p|_{\Lambda_t}$ is then an elliptic symbol. We notice that the existence of an escape function implies that

$$\text{(4.5)}\quad \begin{array}{l}\text{There exists } K\Subset \boldsymbol{R}^{2n} \text{ and } C_0>0 \text{ such that}\\ \left|\dfrac{1}{\tilde{r}}\dfrac{\partial p}{\partial\xi}\right|+\left|\dfrac{R}{\tilde{r}^2}\dfrac{\partial p}{\partial x}\right|\geq 1/C_0 \quad \text{on} \quad p^{-1}(0)\backslash K,\end{array}$$

and assuming this necessary condition, we shall now give a necessary and sufficient condition for the existence of an escape function. We notice that a function of class $\dot{S}^{1,1}$ is Lipschitz continuous with respect to the metric $g=(\tilde{r}R)^2 g_0$, where $g_0=(dx/R)^2+(d\xi/\tilde{r})^2$. In view of (4.5), the vector field $\nu=\tilde{r}^{-2}H_p$ has a g-norm of the order of magnitude 1 on $p^{-1}(0)\backslash K$. We have:

There exists an escape function G if and only if there is a compact set $K\subset \boldsymbol{R}^n$ and a constant $C>0$ such that the following condition is satisfied:

(NT) If $[t_1, t_2]\ni t\mapsto\rho(t)\in p^{-1}(0)$ is a piecewise C^1 curve with finitely many jump-discontinuities, and such that $\dot{\rho}(t)=a(t)\nu(\rho(t))+w(t)$, where a and w are bounded and piecewise continuous, $a(t)\geq 0$ with equality when $\rho(t)\in K$, then

$$d_g(\rho(t_1), \rho(t_2))\geq\int_{t_1}^{t_2} a(t)/C-\|w(t)\|_g\, dt.$$

$\Big($Here by convention, we add the sum of the g-lengths of the jumps to $\int_{t_1}^{t_2}\|w(t)\|_g\, dt.\Big)$

If $G\in\dot{S}^{1,1}$, it can always be modified away from $p^{-1}(0)$, so that $G(x, \xi)=g(x)$ is independent of ξ in the region where $|\xi|\geq$ (const.) $r(x)$. We only work with such functions in the following. Using suitable FBI-transforms

$$T_\alpha u(x, h)=\int t(\alpha, h)\, e^{i\tau(x, y, \alpha)/h}u(y)dy,$$

where $\alpha\in\boldsymbol{R}^{2n}$ and $t\in S^{3n/4,\,-3n/4,\,3n/4}$ (and the third index refers to the order in $1/h$) and where $\tau\in S^{1,1}$ is a suitable phase function, quadratic in (x, y), it is possible to define a certain Hilbert space $H(\Lambda_t, m)\subset\mathscr{D}'(\boldsymbol{R}^n)$ depending also on h. Here Λ_t is given by (4.4) and t is sufficiently small, m is any order function and finally h is sufficiently small. It is impossible to explain the construction of these spaces briefly, but in the special case when $G=g(x)$ everywhere, and when $m=m_0(x)(\tilde{r}(x, \xi)/r(x))^N$, $N\in\boldsymbol{N}$, then

$$H(\Lambda_t, m)=\{u \in \mathscr{D}'(\boldsymbol{R}^n);\ \sum_{|\alpha|\le N} \|((h/r(x))D_x)^\alpha m_0(x)e^{-tg(x)/h}u\|_{L^2}^2<\infty\}.$$

If $m\le\tilde{m}$, then $H(\Lambda_t, \tilde{m})\subseteq H(\Lambda_t, m)$ and the inclusion is compact if $m(\alpha)/\tilde{m}(\alpha)\to 0$ when $|\alpha|\to\infty$.

Also using superpositions of Gaussian kernels, if t is sufficiently small and a is a symbol of some class $S(m)$ defined on Λ_t, then for h small enough, we can define a certain pseudodifferential operator $\mathrm{Op}_{\Lambda_t}(a)$ which is uniformly bounded from $H(\Lambda_t, \tilde{m})$ to $H(\Lambda_t, \tilde{m}/m)$, if $\tilde{m}$ is a second order function. We can also compose these pseudodifferential operators with certain differential operators. If $B=\sum_{|\alpha|\le N} b_\alpha(x)(hD_x)^\alpha$ is a differential operator with coefficients holomorphic in a domain $|\mathrm{Im}\, x|$ $<$const. $R(\mathrm{Re}\, x)$, and whose symbol is of class $S(n_0(x)(\tilde{r}/r)^N)$ there, then if $t>0$ is sufficiently small, we have for every order function m and a as above that $\mathrm{Op}_{\Lambda_t}(a)\circ B=\mathrm{Op}_{\Lambda_t}(c)$, $B\circ\mathrm{Op}_{\Lambda_t}(a)=\mathrm{Op}_{\Lambda_t}(\tilde{c})$, where c, $\tilde{c}$ are symbols of class $S(mn_0(\tilde{r}/r)^N)$ and $c-ab$, $\tilde{c}-ab$ are symbols of class $S(mn_0(\tilde{r}/r)^N(h/\tilde{r}R))$.

Let Λ_t be as above with $G\in\dot{S}^{1,1}$ and assume that t is small enough so that the general results above apply.

Definition 4.2. We say that $z\in\boldsymbol{C}$ is Λ_t-resolvent if $p|_{\Lambda_t}-z$ is elliptic. (More precisely $(p|_{\Lambda_t}-z)^{-1}=\mathcal{O}(\tilde{r}^{-2})$ outside a compact set.)

The set of Λ_t-resolvent points is open. Now let G be an escape function. Then we can find a connected open set Ω of Λ_t-resolvent points which contains the closed positive imaginary axis. If $z\in\Omega$ is in some small conic neighborhood of the positive imaginary axis and $\mathrm{Im}\, z$ is large enough, then $p|_{\Lambda_t}-z$ is not only elliptic far away on Λ_t, but $(p|_{\Lambda_t}-z)^{-1}$ is well-defined everywhere on Λ_t and of class $S(\tilde{r}^{-2})$. We then put $Q(\Lambda_t, z)$ $=\mathrm{Op}_{\Lambda_t}((p-z)^{-1})$ and we get

$$(P-z)\circ Q(\Lambda_t, z)=I+R_{-1}(\Lambda_t, z),\ R_{-1}\in\mathrm{Op}_{\Lambda_t}(S^{-1,-1,-1}),$$

so R_{-1} is of norm $\mathcal{O}(h)$ as a bounded operator in $H(\Lambda_t, m)$ for every order function m, when h is small enough. There is a similar result for $Q\circ(P-z)$ and we conclude that for every order function m, $(P-z)$ is bijective: $H(\Lambda_t, m\tilde{r}^2)\to H(\Lambda_t, m)$ and has a uniformly bounded inverse, when h is small enough. For uniformity reasons we now pass to a connected open subset $\Omega'\Subset\Omega$ containing a point $z_0=iy_0$, $y_0>0$, such that the argument above applies to $z=z_0$. If $\chi\in C_0^\infty(\Lambda_t)$ is equal to 1 on a sufficiently large set, we can put

$$\tilde{Q}(z)=\mathrm{Op}_{\Lambda_t}((p-z_0)^{-1}\chi+(p-z)^{-1}(1-\chi))\in\mathrm{Op}_{\Lambda_t}(S^{-2,0}).$$

Then $\tilde{Q}(z_0)=Q(z_0)$ and

$$(4.6) \qquad (P-z)\tilde{Q}(z)=I+K(\Lambda_t, z)+\tilde{R}_{-1}(\Lambda_t, z),$$

where $K=\mathrm{Op}_{\Lambda_t}(k)$, and $k \in S^{0,0,0}$ has compact support, while $\tilde{R}_{-1}$ has the same properties as R_{-1} above. Moreover, K and $\tilde{R}_{-1}$ depend holomorphically on z and $K(\Lambda_t, z_0)=0$.

We have a similar result for $Q \circ (P-z)$ and by applying Fredholm theory, we get:

Theorem 4.3. *Let m be an order function and let Ω, Ω', Λ_t be as above. Then for $h<h_0$, where $h_0>0$ is sufficiently small, there exists a discrete set $\Gamma(h)\subset\Omega'$ such that $(P-z)$: $H(\Lambda_t, m\tilde{r}^2)\to H(\Lambda_t, m)$ is bijective for $z \in \Omega'\backslash\Gamma(h)$, while for $z \in \Gamma(h)$ the same operator is Fredholm of index 0 and splits into a direct sum $F_z\oplus(G_z\cap H(\Lambda_t, m\tilde{r}^2))\to F_z\oplus G_z$, where F_z is finite dimensional, $(P-z)|_{F_z}$ is nilpotent and $(P-z)$: $G_z\cap H(\Lambda_t, m\tilde{r}^2)\to G_z$ is bijective with bounded inverse.*

By a simple deformation argument one can prove that if $\tilde{G}$, t and $\tilde{m}$ is another choice of quantities as above, and if $L\subset\Omega'$ is a subset of the positive imaginary axis, then in some neighborhood of L the corresponding set $\tilde{\Gamma}(h)$ coincides with $\Gamma(h)$, when h is small enough. The elements of $\Gamma(h)$ (with $F_z\neq 0$) will be called resonances. If we restrict the attention to a neighborhood $V(h)$ of 0 which tends to $\{0\}$ as h tends to 0, then the resonances in $V(h)$ are completely well-defined, in the sense that any two choices of (G, t, m) as above, give rise to the same resonances in $V(h)$ when h is sufficiently small.

In describing a first application of all this machinery, we keep the general assumptions above so that the resonances are well-defined near 0. Let $U\subset\ddot{O}\Subset \boldsymbol{R}^n$, where U is compact connected, $\ddot{O}$ is open connected, and assume that $V(x)\leq 0$ on U, $V(x)>0$ on $\ddot{O}\backslash U$, $V(x)=0$ on $\partial\ddot{O}$. We also sharpen the non-trapping condition:

(4.7) There exists $G \in \dot{S}^{1,1}$ such that $H_p(G)\geq(C_0)^{-1}r^2$ on $p^{-1}(0)|_{\complement\ddot{O}}$, $C_0>0$.

Let d be the Agmon distance on $\overline{\ddot{O}}$. For $\eta>0$ small we put $M_0=\{x \in \ddot{O};\ d(x, \complement\ddot{O})>\eta\}$ and we let P_{M_0} be the corresponding Dirichlet realization. Let us consider a simple eigenvalue $\mu=\mu(h)$ of P_{M_0} (for $h \in J$), which tends to 0 as $h\to 0$ and let us assume that this is the only eigenvalue in an interval $[\mu(h)-2a(h), \mu(h)+2a(h)]$ where $a(h)>0$, $\log a(h)=o(1/h)$. Then by combining the techniques outlined in this section with those of Section 1, one can prove:

The closed disc in $\boldsymbol{C}$ with center $\mu(h)$ and radius $a(h)$ contains for h

sufficiently small precisely one resonance $z(h)$. *This resonance is simple in the sense that* $F_{z(h)}$ *is one dimensional. Moreover,*

$$z(h)-\mu(h)=\tilde{\mathcal{O}}(\exp-2S_0/h),$$

where now $S_0=d(U, \complement \ddot{O})$.

We now specialize further, by assuming that $U=\{x^0\}$, $V''(x_0)>0$, and that $\mu(h)\sim h(E_0+E_1h+\cdots)$, where E_0 is a simple eigenvalue of the localized harmonic oscillator at x_0.

We say that a point $x \in \partial\ddot{O}$ is of type 1 if there is a minimal geodesic from x_0 to x in $\ddot{O}\cup\{x\}$ which is of length S_0. Otherwise, we say that x is of type 2. If $x_2 \in \partial\ddot{O}$ is of type 2, then for all $x \in \bar{\ddot{O}}$ in a neighborhood of x_2, we have $d(x_0, x)=S_0+d(x, \partial\ddot{O})$. Let $x_1 \in \partial\ddot{O}$ be of type 1. The (necessarily unique) minimal geodesic γ_0 from x_0 to x_1 is then the projection of the bicharacteristic $\tilde{\gamma}_0$ of $q=\xi^2-V$ from $(x_0, 0)$ to $(x_1, 0)$. In a neighborhood of $\gamma_0\backslash\{x_1\}$ the function $f(x)=d(x_0, x)-S_0$ is analytic and over that neighborhood the Lagrangian manifold Λ_f: $\xi=f'(x)$ coincides with the stable H_q-outgoing Lagrangian manifold through $(x_0, 0)$. Naturally Λ_f can be extended to a smooth H_q-invariant manifold Λ containing $(x_1, 0)$, but the projection π: $\Lambda\rightarrow \boldsymbol{R}^n$ has a singularity at $(x_1, 0)$, since $\pi(\Lambda)\subset\bar{\ddot{O}}$. We assume that

$$\text{(4.8)} \qquad d\pi|_\Lambda(x_1, 0) \text{ is of rank } n-1,$$

(for every $x_1 \in \partial\ddot{O}$ of type 1). Then one can see that π: $\Lambda\rightarrow \boldsymbol{R}^n$ has a fold singularity of the simplest possible type, whose projection is an analytic hypersurface $\mathscr{C}\subset\bar{\ddot{O}}$ which is tangent to $\partial\ddot{O}$ at x_1. f is of class C^1 up to $\mathscr{C}$ and the "boundary value" $\tilde{f}=f|_{\mathscr{C}}$ is analytic and satisfies

$$\text{(4.9)} \qquad \tilde{f}(x)\geq C_0V(x), \qquad C_0>0.$$

If $\mathscr{C}$ is given by $\rho(x)=0$, where ρ is analytic, $d\rho\neq 0$, $\rho|_{\gamma_0}\geq 0$, then

$$\text{(4.10)} \qquad f=F(x, \rho(x)^{1/2}),$$

where F is analytic.

In a neighborhood of $\gamma_0\backslash\{x_1\}$ we can now construct an asymptotic solution of $(P-z(h))v\sim 0$, of the form

$$\text{(4.11)} \qquad v(x, h)\sim a(x, h)\exp-(S_0+f(x))/h,$$

where a is a classical analytic symbol. We can then extend this solution to a neighborhood of x_1 by a siutable "Airy type" oscillatory intergral with a complex contour of integration. On the opposite side of $\mathscr{C}$ from

γ_0 this solution reduces again by stationary phase to the form (4.11), where f is now given by (4.10) and where the choice of branch is given by the choice of contour in the Airy type integral. Using (4.9) and the eiconal equation for f outside $\mathscr{C}$ one shows that $\operatorname{Re} f \geq 0$ in this region. Everything works here with analytic symbols, and if we choose suitable realizations, we get a function $v(x, h)$ defined in a neighborhood of γ_0 such that $v = \mathcal{O}(h^{-N_0} \exp - d(x_0, x)/h)$, $(P - z(h))v = \mathcal{O}(\exp - (\varepsilon_0 + d(x_0, x))/h)$ for $\rho \geq 0$ and $v = \mathcal{O}(h^{-N_0} \exp - S_0/h)$, $(P - z(h))v = \mathcal{O}(\exp - (\varepsilon_0 + S_0)/h)$ for $\rho \leq 0$. One can also make such a choice of contour in the Airy type integral that v is asymptotically outgoing in a suitable sense.

We next compare v with an exact eigen-function u (generator of $F_{z(h)}$). First one shows that with suitable normalizations

$$u - v = \mathcal{O}(\exp - \varepsilon_0/h), \tag{4.12}$$

in a neighborhood of x_0. Let then $\Omega \Subset \boldsymbol{R}^n$ be some large suitable neighborhood of Ö and let $a = a(\eta) \in \boldsymbol{R}$ be the smallest number such that

$$\| u \|_{L^2(\Omega \setminus M_0)} = \hat{\mathcal{O}}(\exp - a/h)$$

(, which here means: $\mathcal{O}(\exp(\varepsilon - a)/h)$ for every $\varepsilon > 0$).

One first shows by general arguments that $a \geq 0$. If $a < S_0 - \eta$, we get from (4.12) and Proposition 1.1, that $u = \hat{\mathcal{O}}(1)$ in $L^2(M_0, e^{2\Phi(x)/h}dx)$, where $\Phi(x) = \min(d(x_0, x), a + d(\complement M_0, x))$. If $\chi \in C_0^\infty(M_0)$ is a suitable cutoff function, equal to 1 near x_0, we get

$$(P - z(h))((1 - \chi)u) = \mathcal{O}(e^{-\alpha/h}), \tag{4.13}$$

for some $\alpha > a$.

Using the invertibility of $(P - z(h))$ away from x_0 in the spaces $H(\Lambda_t, \cdots)$ for arbitrarily small t one then shows that $u = \hat{\mathcal{O}}(e^{-\alpha/h})$ in $L^2(\Omega \setminus M_0)$. This is in contradiction with the assumption about a, so we have proved that $a \geq S_0$, and hence $u = \hat{\mathcal{O}}(\exp - S_0/h)$ in $L^2(\Omega \setminus \text{Ö})$. This result means that u can be analyzed microlocally modulo $\mathcal{O}(\exp - (S_0 + \varepsilon_0)/h)$ by means of FBI-transforms over points of $\Omega \setminus \overline{\text{Ö}}$ and also over points of $\partial\text{Ö}$ which are of type 2. Using the outgoing nature of u one then shows by a deformation argument as in [10], that

$$u = \mathcal{O}(\exp - (S_0 + \varepsilon_0)/h), \qquad \varepsilon_0 > 0, \tag{4.14}$$

near each point of type 2.

Near a point of type 1, the same arguments show that

$$u - v = \mathcal{O}(\exp - (S_0 + \varepsilon_0)/h), \qquad \varepsilon_0 > 0, \tag{4.15}$$

where v is the outgoing asymptotic solution described above.

Let W be a suitable small neighborhood of $\ddot{O}$ with smooth boundary. By Green's formula we have

$$\|u\|^2_{L^2(W)} \operatorname{Im} z(h) = -h^2 \operatorname{Im} \int_{\partial W} (\partial u/\partial n)\bar{u} dS, \tag{4.16}$$

where n is the outward unit normal. We can normalize our choice of v so that $\|u\|_{L^2(W)} = 1 + \mathcal{O}(\exp - \varepsilon_0/h)$. Then from (4.14), (4.15), we know the behaviour of $(\partial u/\partial n)\bar{u}$ on ∂W modulo $\mathcal{O}(\exp - 2(S_0 + \varepsilon_0)/h)$ for some small $\varepsilon_0 > 0$, so up to an error of this size we get $\operatorname{Im} z(h)$ as an oscillatory integral.

When there are only finitely many points of type 1 and $\mathscr{C}$ has only second order contacts with $\partial\ddot{O}$ at these points, the asymptotic expansion of the integral in (4.16) is obtained by simple stationary phase, and we get

$$\operatorname{Im} z(h) = r(h) \exp - 2S_0/h, \tag{4.17}$$

where r is a classical analytic symbol whose order can be determined, but where by accident some or many of the leading terms may vanish when E_0 is not the lowest eigenvalue of the localized harmonic oscillator at x_0. When E_0 is the lowest eigenvalue, then r is a negative elliptic symbol of order 1/2.

When we drop the additional assumption on the points of type 1, we can still estimate $-\operatorname{Im} z(h)$ from above by an expression

$$C_0 h^{C_1} \exp - 2S_0/h, \tag{4.18}$$

where C_0 and C_1 depend on E_0, and when E_0 is the lowest eigenvalue, we also have a lower bound of the form (4.18), with $C_0 > 0$. A recent idea (that we have not yet had time to work out in detail,) indicates that $\operatorname{Im} z(h)$ can be estimated from below by expressions of the form (4.18) with $C_0 > 0$, even when E_0 is not the lowest eigenvalue of the localized harmonic oscillator.

After this conference, we learned about a work of Combes, Duclos and Seiler, who obtain some related results about resonances in the one-dimensional case.

References

[1] S. Agmon, Lectures on exponential decay of solutions of second order elliptic equations, Math. Notes **29**, Princeton University Press, (1982).

[2] J. Aguilar and J. M. Combes, A class of analytic perturbations for one-body Schrödinger Hamiltonians, Comm. Math. Phys. **22** (1971), 269–279.

[3] I. Herbst, Dilation analyticity in a constant electric field, Comm. Math. Phys. **64** (1979), 279–298.
[4] B. Helffer and J. Sjöstrand, Multiple wells in the semiclassical limit I, Comm. Partial Differential Equations **9** (4) (1984), 337–408.
[5] ——, Puits multiples en limite semiclassique II, Ann. Inst. H. Poincaré, à paraître.
[6] ——, Multiple wells . . . III, Math. Nachr., to appear.
[7] ——, Puits multiples . . . IV. Etude du complexe de Witten. Manuscript. (See also the lecture by the second author at the AMS-symposium on pseudo-differential operators, April 2-6, 1984, Notre Dame University, to appear.)
[8] B. Simon, Semiclassical analysis of low lying eigenvalues I, Ann. Inst. H. Poincaré **38** (1983), 295–307.
[9] ——, Semiclassical analysis . . . II. Tunneling, Ann. of Math. **120** (1984), 89–118.
[10] J. Sjöstrand, Singularités analytiques microlocales, Astérisque n° **95** (1982).
[11] E. Witten, Supersymmetry and Morse theory, J. Differential Geom. **17** (1982), 661–692.

DÉPARTEMENT DE MATHÉMATIQUES
UNIVERSITÉ DE PARIS-SUD
91405 ORSAY, FRANCE

Taniguchi Symp. HERT
Katata 1984, pp. 363–380

Analytic and Gevrey Well-Posedness of the Cauchy Problem for Second Order Weakly Hyperbolic Equations with Coefficients Irregular in Time

Sergio SPAGNOLO

Introduction

This paper is divided in three parts.

The first section is a review of the main results of [CDS], $[\mathrm{CS}]_3$ and [CJS] concerning the global solvability in the space of real analytic or Gevrey functions of the Cauchy problem

$$u_{tt}-\sum_{i,j}^{1,n} a_{ij}(t)u_{x_ix_j}=0 \quad (0\leq t\leq T) \tag{1}$$

$$u(x, t)=u_0(x), \qquad u_t(x, t)=u_1(x), \tag{2}$$

where the space-variables $x=(x_1, \cdots, x_n)$ belong to $\boldsymbol{R}^n$ and the coefficients a_{ij} are real integrable functions on the time-interval $[0, T]$ which verify

$$\sum_{i,j}^{1,n} a_{ij}\xi_i\xi_j\geq\nu|\xi|^2 \qquad (\xi\in\boldsymbol{R}^n) \tag{3}$$

for some constant $\nu\geq 0$.

In the second section we report some extensions, due to E. Jannelli, T. Nishitani, Y. Ohya and S. Tarama, of the results of Section 1 to more general hyperbolic equations or system, like

$$u_{tt}-\sum_{i,j}^{1,n}(a_{ij}(x, t)u_{x_i})_{x_j}=0, \qquad \text{with } a_{ij}=a_{ji}, \tag{4}$$

or

$$A_0(x, t)u_t+\sum_{h=1}^{n} A_h(x, t)u_{x_h}+B(x, t)u=0,$$

with coefficients which depend in analytic or Gevrey way on the space-variables.

The third section is devoted to the abstract equation

$$u''+A(t)u=0, \tag{5}$$

where $A(t)$ is a family of self-adjoint operators in Hilbert sapce H, satisfying the weak hyperbolicity condition

$$(A(t)v, v)_H \geq 0.$$

By interplaying between the structure of *Banach scale*, in which it is possible to solve locally the abstract Kovalewskian problems, and the one of *Hilbert triplet*, where the global solvability for the strictly hyperbolic second order equations can be set, we are able to get a result of global well-posedness for Eq. (5) in a certain space of *analytic-like vectors* of H. This result, which is proved in $[AS]_2$, applies in particular to Eq. (4), as well as to some non-Kovalewskian equation such as

$$u_{tt} + c(t)\Delta_x^2 u = 0, \qquad \text{where } c(t) \geq 0.$$

We emphasize that the peculiarity of all these results is in that the coefficients a_{ij}, or the operator $A(t)$, are assumed to have only a *minimum regularity* with respect to the time-variables, namely the Hölder continuity in order to have the Gevrey well-posedness and the mere integrability in order to have the analytic well-posedness (whereas in the classical existence theorems it was assumed at least the Lipschitz continuity).

Finally we remark that the results here exposed, or the techniques employed to prove them, turn out to be useful in other contexts as:

(i) the study of the connections between *weak hyperbolicity* (in the sense that the characteristic roots are real) and *analytic well-posedness*, or between *strict hyperbolicity* (in the sense that the characteristic roots are real and distinct) and C^∞*-well-posedness* (see Theorem 1.2, part (ii), and Theorem 1.4, part (ii) below);

(ii) the study of the *dependence* of the solutions to a hyperbolic equation upon an L^1-perturbation of the coefficients (see $[CS]_1$ and $[CS]_4$);

(iii) the study of the global solvability (in the class of analytic functions) of some non-linear equations of hyperbolic type, like the integro-differential equation

$$u_{tt} - m\left(\int_{\boldsymbol{R}_x^n} |\nabla_x u|^2 dx\right)\Delta_x u = 0,$$

where $m(\rho)$ is any continuous non-negative function (see $[AS]_1$).

Notations

1. $\mathscr{A}(\boldsymbol{R}_x^n)$ is the topological vector space of the *real analytic functions*, i.e. the indefinitely differentiable functions $\phi(x)$ such that, for every compact subset K of $\boldsymbol{R}_x^n$, there exist two positive constants C_K and Λ_K for which

$$|D^\alpha\phi(x)|\le C_K \Lambda_K^{|\alpha|}|\alpha|! \qquad (x\in K,\ \alpha\in N^n).$$

2. $\mathscr{E}^s(\boldsymbol{R}_x^n)$ (s real ≥ 1) is the topological vector space of the *Gevrey functions* with exponent s, i.e. the C^∞ functions $\phi(x)$ such that for every compact $K\subset \boldsymbol{R}_x^n$,

$$|D^\alpha\phi(x)|\le C_K \Lambda_K^{|\alpha|}|\alpha|!^s \qquad (x\in K,\ \alpha\in N^n).$$

For $s=1$ we have $\mathscr{E}^1=\mathscr{A}$.

3. $L^1(0, T; \mathscr{E}^s(\boldsymbol{R}_x^n))$ coincides with the space of the functions $v(x, t)$ on $\boldsymbol{R}_x^n\times[0, T]$, indefinitely differentiable in x for each t and measurable in t for each x, such that for every compact $K\subset\boldsymbol{R}_x^n$ there exist a function $C_K(t)$ in $L^1(0, T)$ and a positive constant Λ_K for which

$$(6)\qquad |D_x^\alpha v(x,t)|\le C_K(t)\Lambda_K^{|\alpha|}|\alpha|!^s \qquad (x\in K,\ \alpha\in N^n).$$

When (6) holds for $C_K(t)$ in $L^\infty(0, T)$, we say that $v(x, t)$ belongs to $L^\infty(0, T; \mathscr{E}^s(\boldsymbol{R}_x^n))$.

4. $C^{k,\alpha}(0, T)$, k integer ≥ 0 and α real number with $0<\alpha\le 1$, denotes the space of the functions of class C^k on $[0, T]$, whose k^{th} derivative is Hölder continuous with exponent α. We also put $C^{k,0}=C^k$.

5. If X is a locally convex topological vector space and $A(t)$ an integrable family of continuous linear operators on X, we say that the Cauchy problem

$$(7)\qquad \begin{cases} u^{(m)}+A(t)u=0 & (0\le t\le T)\\ u^{(j)}(0)=u_j & (j=0, 1, \cdots, m-1)\end{cases}$$

is (*globally*) *well-posed* in X when for every $u_j\in X$ there exists one and only one solution $u(t)$ in $W^{m,1}(0, T; X)$. In the special cases $X=\mathscr{A}(\boldsymbol{R}_x^n)$ or $X=C^\infty(\boldsymbol{R}_x^n)$ we'll briefly say that the Problem is *analytically well-posed* or *C^∞-well-posed.*

When for every u_j in X there exists $\tau=\tau(u_0, u_1, \cdots, u_{m-1})>0$ for which (7) has a unique solution in $W^{m,1}(0, \tau; X)$, we say that the Problem is *locally well-posed* in X (near $t=0$).

§ 1. Coefficients depending only on time

a) *Strictly hyperbolic equations* ($\nu>0$).

Theorem 1.1 ([CDS]). *Let us consider Pb.* {(1), (2)} *where the* $a_{ij}(t)$*'s verify* (3) *for some* $\nu>0$. *Then:*

(i) *If there exist* $C\ge 0$ *and* $0<\alpha<1$ *for which*

(8) $$\int_0^T |a_{ij}(t+\tau)-a_{ij}(t)|dt < C\tau^\alpha \qquad (\tau>0)$$

(where the a_{ij}'s are considered as extended to the whole $[0, +\infty[$ by $a_{ij}(t) \equiv a_{ij}(T)$ for $t>T$), the Problem is well-posed in the Gevrey space $\mathscr{E}^s(\boldsymbol{R}^n_x)$ for

(9) $$1 \leq s < 1+\frac{\alpha}{1-\alpha}.$$

(ii) *If (8) holds for $\alpha=1$, the Problem is well-posed also in $C^\infty(\boldsymbol{R}^n_x)$. More generally there is C^∞-well-posedness as soon as*

$$\int_0^T |a_{ij}(t+\tau)-a_{ij}(t)|\,dt \leq C\cdot\tau(1+|\log\tau|) \qquad (\tau>0).$$

Remark 1.1. The space of the functions which verify (8) is denoted by $BV^\alpha(0, T)$; when $\alpha=1$ this is the space of the functions with *bounded variation* on $[0, T]$. It is immediate that

$$C^{0,\alpha}(0, T) \subseteq BV^\alpha(0, T).$$

Another property of BV^α (see M.H. Taibleson [T]) is the inclusion

$$BV^\alpha(0, T) \subseteq L^s(0, T) \qquad \text{for } 1 \leq s < 1+\frac{\alpha}{1-\alpha}. \quad //$$

Confining ourselves to the simple case

(10) $$\begin{cases} u_{tt}-a(t)u_{xx}=0, \\ u(x, 0)=u_0(x), \qquad u_t(x, 0)=u_1(x), \end{cases}$$

where $x \in \boldsymbol{R}^1$, we can exhibit a class of counter-examples showing the sharpness of Theorem 1.1:

Theorem 1.2 ([CDS]). i) *For every $\alpha \in [0, 1[$ there exists a function $a(t) \geq 1$, belonging to $C^{0,\alpha}(0, T)$, such that Pb. (10) is non well-posed (even locally) in any space $\mathscr{E}^s(\boldsymbol{R}^1_x)$ for $s > 1+\alpha/(1-\alpha)$.*

ii) *There exists a function $a(t) \geq 1$, belonging to $C^{0,\alpha}(0, T)$ for every $\alpha<1$, such that Pb. (10) is non well-posed in $C^\infty(\boldsymbol{R}^1_x)$.*

Remark 1.2. The conclusion (ii) of Theorem 1.2 displays the gap between strict hyperbolicity and C^∞-well-posedness when the coefficients are not sufficiently regular in time. //

b) *Weakly hyperbolic equations* ($\nu=0$).

As in case (a), we have now the analytic well-posedness without any regularity assumption on the a_{ij}'s (but the integrability).

There is also the $\mathscr{E}^s$-well-posedness for $1 \leq s \leq \bar{s}(\alpha)$ when the coefficients belong to $C^{0,\alpha}$, but now $\bar{s}(\alpha)$ is very smaller than in case (a): in particular $\bar{s}(\alpha) \to 2$ for $\alpha \to 1$.

Nevertheless if we confine ourselves to the homogeneous equations (as (1)) we have also the $\mathscr{E}^s$-well-posedness for $s \geq 2$, provided that the coefficients are more than C^1-regular. For instance, when the $a_{ij}(t)$'s are C^∞ functions, Pb. {(1), (2)} turns out to be well-posed in *every* Gevrey space (whereas to get the C^∞-well-posedness we must assume e.g. that the a_{ij}'s are analytic).

Theorem 1.3. *Let the coefficients* $a_{ij}(t)$ *of Pb.* {(1), (2)} *verify* (3) *for* $\nu=0$. *Then*:

(i) ([CDS]) *If the* a_{ij}*'s belong to* $L^1(0, T)$, *the Problem is well-posed in* $\mathscr{A}(\boldsymbol{R}^n_x)$;

(ii) ([CJS]) *If the* a_{ij}*'s belong to* $C^{k,\alpha}(0, T)$ *for some integer* $k \geq 0$ *and* $0 < \alpha \leq 1$, *the Problem is well-posed in* $\mathscr{E}^s(\boldsymbol{R}^n_x)$ *for*

$$1 \leq s < 1 + \frac{k+\alpha}{2}; \tag{11}$$

(iii) *If the* a_{ij}*'s are analytic functions on* $[0, T]$ *the Problem is well-posed in* $C^\infty(\boldsymbol{R}^n_x)$.

The following graph illustrates the different kind of Gevrey well-posedness for the weakly hyperbolic euqations, with respect to the strict hyperbolic ones.

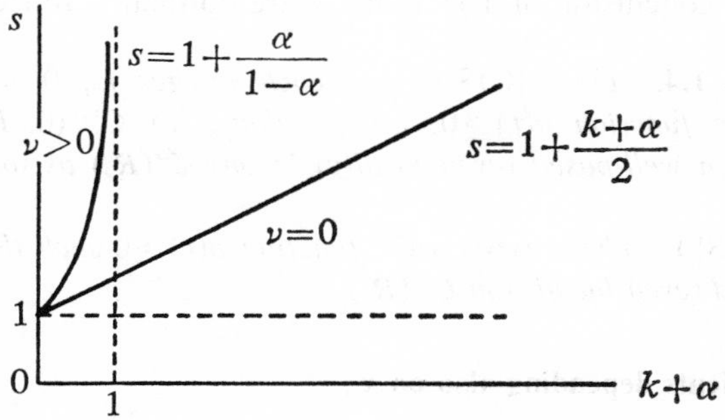

$a_{ij} \in C^{k,\alpha}(0, T) \Longrightarrow$ Pb. {(1), (2)} is $\mathscr{E}^s$-well-posed

Remark 1.3. If we consider, instead of (1), the complete equation

$$u_{tt}-\sum_{i,j}^{1,n} a_{ij}(t)u_{x_ix_j}+\sum_{i=1}^{n} b_i(t)u_{x_i}+c(t)u=0, \tag{12}$$

then the conclusion (i) of Theorem 1.3 holds true, whereas (ii) and (iii) can fail. In fact the Cauchy problem for the equation $u_{tt}-u_x=0$ is well-posed in $\mathscr{E}^s(\boldsymbol{R}_x^n)$ if and only if $1\leq s<2$.

Actually Pb. {(12), (2)} is well-posed in $\mathscr{E}^s(\boldsymbol{R}_x^n)$ for

$$1\leq s<\mathrm{Min}\left\{2,\ 1+\frac{k+\alpha}{2}\right\}.$$

This result can be improved in some particular case; for example, if

$$\left|\sum_{i=1}^{n} b_i(t)\xi_i\right|\leq\lambda(t)\left(\sum_{i,j}^{1,n} a_{ij}(t)\xi_i\xi_j\right)^{\beta}|\xi|^{1-2\beta} \qquad (\xi\in\boldsymbol{R}^n)$$

for some $\lambda(t)$ in $L^1(0, T)$ and some $0\leq\beta\leq 1/2$, then there is the well-posedness in $\mathscr{E}^s$ for

$$1\leq s<\mathrm{Min}\left\{1+\frac{1}{1-2\beta},\ 1+\frac{k+\alpha}{2}\right\}.\quad //$$

Remark 1.4. Conclusion (iii) of Theorem 1.3 is a consequence of a stronger result (easy to prove): a *sufficient* condition for the C^∞-well-posedness of Pb. {(1), (2)} is that there exists some $\delta>0$ in such a way that for every $\xi\in\boldsymbol{R}^n$ the function $\sum a_{ij}(t)\xi_i\xi_j$ makes at most a finite number (independent on ξ) of oscillations with minimum value less than δ.

We do not know if such a condition is also *necessary* for the C^∞-well-posedness (Cf. Th. 1.4 (ii) below). //

Also the conclusion of Theorem 1.3 are optimal. Indeed:

Theorem 1.4. (i) ([CJS]) *For each integer* $k\geq 0$ *and* $0\leq\alpha\leq 1$, *there exists a function* $a(t)\geq 0$, *which belongs to* $C^{k,\alpha}(0, T)$, *such that Pb.* (10) *is non well-posed* (*even locally*) *in any* $\mathscr{E}^s(\boldsymbol{R}_x^1)$ *as soon as* $s>1+(k+\alpha)/2$.

(ii) ([CS]$_3$) *There exists a* C^∞ *function* $a(t)\geq 0$ *such that Pb.* (10) *is non well-posed* (*even locally*) *in* $C^\infty(\boldsymbol{R}_x^1)$.

§ 2. Coefficients depending also on x

We report here some extensions of Theorems 1.1 and 1.3 to the equation

(13) $$u_{tt}-\sum_{i,j}^{1,n} a_{ij}(x, t)u_{x_i x_j}+\sum_{i=1}^{n} b_i(x, t)u_{x_i}+c(x, t)u=0,$$

where the a_{ij}'s are real functions (such that $a_{ij}=a_{ji}$) verifying (3), or to the system

(14) $$A_0(x, t)\underline{u}_t+\sum_{h=1}^{n} A_h(x, t)\underline{u}_{x_h}+B(x, t)\underline{u}=\underline{0}$$

where $A_0, A_1, \cdots, A_n$ and B are $N\times N$ matrices and $\underline{u}(x, t)$ is an N-dimensional vector.

We recall that the system (14) is called *symmetric hyperbolic* if the matrices $A_0, \cdots, A_n$ are hermitian and there exists $\nu>0$ such that

$$(A_0(x, t)\xi, \xi)\geq\nu|\xi|^2 \qquad (\xi \in \boldsymbol{R}_x^n),$$

while it is called *regularly hyperbolic* (following Mizohata) when $A_0(x, t)$ is the identity matrix and, for each $x \in \boldsymbol{R}_x^n$, $t \in [0, T]$, $\xi \in \boldsymbol{R}^n\setminus\{0\}$, the matrix $\sum_{h=1}^n \xi_h\cdot A_h(x, t)$ has N distinct real eigenvalues $\tau_1(x, t; \xi), \cdots, \tau_N(x, t; \xi)$ verifying

$$\operatorname*{Inf}_{\substack{|\xi|=1\\ x,t}} |\tau_i(x, t; \xi)-\tau_j(x, t; \xi)|>0, \qquad \text{if } i\neq j.$$

Theorem 2.1 (*analytic well-posedness*). (i) ([J]$_1$) *Let the coefficientts $a_{ij}(x, t)$, $b_i(x, t)$ and $c(x, t)$ of Eq.* (13) *belong to $L^1(0, T; \mathscr{A}(\boldsymbol{R}_x^n))$ and the a_{ij}'s verify* (3) *for $\nu=0$. Then Pb.* {(13), (2)} *is well-posed in $\mathscr{A}(\boldsymbol{R}_x^n)$.*

(ii) ([J]$_2$) *Let* (14) *be a symmetric hyperbolic system with $A_0(x, t)$ belonging to $L^\infty(0, T; \mathscr{A}(\boldsymbol{R}_x^n))$ and $A_1(x, t), \cdots, A_n(x, t)$, $B(x, t)$ belonging to $L^1(0, T; \mathscr{A}(\boldsymbol{R}_x^n))$. Assume that*

(15) $$\sum_{h=1}^{n} (A_h(x, t)\xi, \xi)\leq\Lambda(t)|\xi|^2 \qquad (\xi \in \boldsymbol{R}^n)$$

for some $\Lambda(t)$ in $L^1(0, T)$.

Then the Cauchy problem for (14) *is well-posed in $\mathscr{A}(\boldsymbol{R}_x^n)$.*

The same conclusion holds if (14) *is a regularly hyperbolic system, provided the matrices $A_h(x, t)$ are continuous on $\boldsymbol{R}_x^n\times[0, T]$ and* (15) *holds for $\Lambda(t)$ constant.*

Remark 2.1. Part (i) of Theorem 2.1 was proved at first in the special case $\nu>0$ in [CS]$_2$. //

Remark 2.2. As far as we know, the analytic well-posedness for the weakly hyperbolic equation of higher order (or for the general weakly hyperbolic systems), with coefficients analytic in x and integrable in t, has not yet been proved. //

As regards the Gevrey well-posedness, there are the following extensions of Theorem 1.1 (i) and Theorem 1.3 (ii) (limitedly to the case $k=0, 1$).

Theorem 2.2 ([N]$_1$, *Gevrey well-posedness*). *Assume that all the coefficients of Pb.* {(13), (2)} *belong to* $\mathscr{E}^s(\boldsymbol{R}^n_x)$ *for each t and are continuous in t uniformly w. r. t. x, and that the leading coefficients* $a_{ij}(x, t)$ *verify* (3) *for some* ν. *Then*:

(i) *if* $\nu>0$ *and the* a_{ij}*'s belong to* $C^{0,\alpha}(0, T)$, *uniformly w. r. t. x, the Problem is well-posed in* $\mathscr{E}^s(\boldsymbol{R}^n_x)$ *for*

$$1<s<1+\frac{\alpha}{1-\alpha},$$

(ii) *if* $\nu=0$ *and the* a_{ij}*'s belong to* $C^{k,\alpha}(0, T)$ *uniformly w. r. t. x, the Problem is well-posed in* $\mathscr{E}^s(\boldsymbol{R}^n_x)$ *for*

$$1<s<1+\frac{k+\alpha}{2}, \qquad k=0, 1.$$

Remark 2.3 (*Equations of higher order*). Recently T. Nishitani (see [N]$_2$) has extended Theorem 2.2 (limitedly to the case $\alpha=1$) to any weakly hyperbolic equation *of order m*. More precisely, assuming that the coefficients of the equation belong to $C^k([0, T]; \mathscr{E}^s(\boldsymbol{R}^n_x))$, he proves the well-posedness in $\mathscr{E}^s$ for

$$1<s<1+\frac{k}{2(r-1)}, \qquad k=1, 2,$$

where r denotes the *maximum multiplicity of the characteristic roots*.

For the same equations, Y. Ohya and S. Tarama (see [O]) assuming that the coefficients belong to $C^{k,\alpha}([0, T]; \mathscr{E}^s(\boldsymbol{R}^n_x))$ have proved the well-posedness in $\mathscr{E}^s$ for

$$1<s<\mathrm{Min}\left\{1+\frac{k+\alpha}{r}, 2\right\}, \qquad k=0, 1, \ 0<\alpha\leq 1. \quad //$$

Concerning the *systems* there are the following result:

Theorem 2.3 ([J]$_3$; *regularly hyperbolic systems*). *Assume that* (14) *is a symmetric or regularly hyperbolic system with coefficients belonging to* $\mathscr{E}^s(\boldsymbol{R}^n_x)$ *for each t and to* $C^{0,\alpha}(0, T)$ *for each x, in a uniform way. Then the Cauchy problem for* (14) *is well-posed in* $\mathscr{E}^s(\boldsymbol{R}^n_x)$ *for*

$$1 < s < 1 + \frac{\alpha}{1-\alpha}.$$

Finally, for equations in *divergence form*, i.e.

$$u_{tt} - \sum_{i,j}^{1,n} (a_{ij}(x, t) u_{x_i})_{x_j} = 0 \qquad (a_{ij} = a_{ji}), \tag{16}$$

there is the following extension of Theorem 1.3 (ii).

Theorem 2.4 ([J]$_4$; *equatious in divergence form*). *Let the coefficients* $a_{ij}(x, t)$ *of Eq.* (16) *verify* (3) *for* $\nu=0$ *and belong to* $L^1(0, T; \mathcal{E}^s(\boldsymbol{R}^n_x))$. *Assume that, for some real* $\sigma > 1$ *and every compact* $K \subset \boldsymbol{R}^n_x$,

(17) $\begin{cases} \xi \longmapsto \left(\sum_{i,j}^{1,n} a_{ij}(x, t) \xi_i \xi_j \right)^{1/\sigma} \\ \text{is a continuous mapping from the unit sphere of } \boldsymbol{R}^n_\xi \text{ into the Banach} \\ \text{space } BV(0, T; L^\infty(K)). \end{cases}$

Then Pb. {(16), (2)} *is well-posed in* $\mathcal{E}^s(\boldsymbol{R}^n_x)$ *for*

$$1 \leq s < 1 + \frac{\sigma}{2}. \tag{18}$$

If we add to Eq. (16) *the lower order terms, we have an analogous conclusion, with* $1 \leq s < \text{Min}\{2, 1+\sigma/2\}$

Remark 2.4. Condition (17) is fulfilled in the following cases:

(i) If

$$a_{ij}(x, t) = c(t) \cdot a^{(0)}_{ij}(x)$$

with $c(t) \geq 0$, $c(t)$ in $C^{k,\alpha}(0, T)$, $a^{(0)}_{ij}(x)$ in $\mathcal{E}^s(\boldsymbol{R}^n_x)$ and

$$0 \leq \sum_{i,j}^{1,n} a^{(0)}_{ij}(x) \xi_i \xi_j \leq \Lambda |\xi|^2 \qquad (\xi \in \boldsymbol{R}^n),$$

then (17) holds for $\sigma = k + \alpha$.

(ii) If the functions $a_{ij}(x, t)$ belong to $\mathcal{E}^s(\boldsymbol{R}^n_x)$ for each t and to $C^{1,\alpha}(0, T)$ uniformly w. r. t. x, then (17) holds for $\sigma = 1 + \alpha$. We observe that such a conclusion is false for $\sigma = k + \alpha$ if $k > 2$. For instance the function $(x-t)^2$ belongs to $C^\infty(0, T)$ uniformly w. r. t. x, but it verifies (17) only for $\sigma < 2$. //

Question. Assume that the $a_{ij}(x, t)$'s are C^∞ in x for each t and analytic in t uniformly w. r. t. x, and that they verify (3) for $\nu = 0$. Is it then possible to conclude that Pb. {(16), (2)} is C^∞-well-posed?

§ 3. Abstract equations

Given a Hilbert space H and a family $\{A(t)\}$ of non-negative self-adjoint operators acting in H, let us consider the Cauchy problem

$$u''+A(t)u=0 \qquad (0<t<T) \tag{19}$$

$$u(0)=u_0, \qquad u'(0)=u_1. \tag{20}$$

Our aim is to solve Pb. {(19), (20)} by taking the initial data u_0 and u_1 is some subspace of H.

a) *Hilbert triplets*

When the non-negative form $(A(t)v, v)_H$ is coercive with respect to some Banach space V, continuously and densely embedded in H, we say that Eq. (19) is *strictly hyperbolic* (w. r. t. V).

The simplest framework for the study of the strict hyperbolic equations is that of the Hilbert triplets. To define the Hilbert triplet $\{V, H, V'\}$ generated by the immersion $V \rightarrow H$, we must only transpose such immersion, by identifying H with its antidual. By assuming that V is *reflexive* and denoting by V' the antidual Banach space of V, we then obtain the chain of continuous and dense immersions

$$V \underset{\rightarrow}{\subsetneq} H \underset{\rightarrow}{\subsetneq} V',$$

with the property that the duality map $\langle \ , \ \rangle$ on $V' \times V$, restricted to $H \times H$, coincides with the scalar product in H.

Let us now assume that the domain of $A(t)$ is a dense subspace of V and that $|(A(t)v, w)_H| \leq C\|v\|_V \cdot \|w\|_V$, in other words let us assume that each $A(t)$ can be extended to a bounded operator form V to V'. More precisely, let us suppose that

$$A(\cdot) \in L^1(0, T; \mathscr{L}(V, V')). \tag{21}$$

The conditions of self-adjointness and coerciveness are then equivalent to

$$\langle A(t)v, w\rangle = \overline{\langle A(t)w, v\rangle} \tag{22}$$

$$\langle A(t)v, v\rangle \geq \nu\|v\|^2 \qquad (\nu>0), \tag{23}$$

for every v and w in V.

Let us finally assume that

$$|\langle A'(t)v, v\rangle| \leq M\|v\|^2 \tag{24}$$

for some positive constant M.

Under all these assumptions it is easy to see that Pb. {(19), (20)} admits a (global) solution u in $L^\infty(0, T; V)$, with $u' \in L^2(0, T; H)$ and $u'' \in L^1(0, T; V')$, for every $u_0 \in V$ and $u_1 \in H$. Indeed, if we consider the *energy function*

$$E(t)=\langle A(t)u, u\rangle+\|u'\|_H^2,$$

by a differentiation in t we get immediately the *a priori estimate*

$$E'(t)\leq \frac{M}{\nu}E(t), \tag{25}$$

for any solution $u(t)$ of Eq. (19).

Unfortunately, (25) becomes meaningless if $M=\infty$ or if $\nu=0$, i.e. when the only regularity of $A(\cdot)$ is the one given by (21), or when $(A(t)v, v)_H$ is merely a non-negative form. Nevertheless one can ask if there is still some global or local existence result for Pb. {(19), (20)}. Now it is easily seen that, in the general case, no existence result can hold, indeed it may occur that the Problem has no local solution (except for initial data both zero). Moreover the concrete case of the weakly hyperbolic differential equations shows that the well-posedness can be achieved only in some complicate topological vector spaces, as $\mathscr{A}(\boldsymbol{R}_x^n)$ or $\mathscr{E}^s(\boldsymbol{R}_x^n)$.

In conclusion, the structure of the Hilbert triplets seems to be inadequate to describe the existence results for weak hyperbolic abstract equations.

b) *Banach scales.*

Neglecting for the moment the hyperbolicity conditions (22) and (23), we can consider Pb. {(19), (20)}, or better any Problem like

$$u^{(m)}+A(t)u=0 \qquad (0\leq t\leq T) \tag{26}$$

$$u^{(j)}(0)=u_j \qquad (j=0, 1, \cdots, m-1), \tag{27}$$

in the framework of the Banach scales. This was firstly made by T. Yamanaka ([Y]) and by L.V. Ovsjannikov ([Ov]) (in the case $m=1$) in order to give a simplified proof of Cauchy-Kovalewsky theorem.

We recall that a Banach scale is a family of Banach spaces $\{X_r, \|\cdot\|_r\}$, where $0<r<\bar{r}$, such that

$$X_r\subseteq X_{r-\delta} \qquad \text{with } \|\cdot\|_{r-\delta}\leq\|\cdot\|_r, \text{ if } 0<\delta<r. \tag{28}$$

Besides the Banach space X_r we can introduce the locally convex *inductive limit*

$$X_{0+} = \bigcup_{r>0} X_r$$

and the Fréchet spaces

$$X_{r-} = \bigcap_{\rho<r} X_\rho \qquad (r>0).$$

Moreover we call *radius of analyticity* of a vector $v \in X_{0+}$ the positive number

$$r(v) = \text{Sup}\,\{\rho>0 : v \in X_\rho\}.$$

Definition 3.1 (*density in itself*). A Banach scale $\{X_r\}$ is said dense in itself when, for every positive r and δ, $X_{r+\delta}$ is dense in the Fréchet space X_{r-}.

Definition 3.2 (*order of an operator*). A linear operator $A: X_{0+} \to X_{0+}$ is said of order $\leq m$ (where m is a real number ≥ 0) in the Banach scale $\{X_r\}$ if, for some positive constant λ,

$$\| Av \|_{r-\delta} \leq \frac{\lambda}{\delta^m} \| v \|_r \qquad (0<\delta<r).$$

When each operator $A(t)$ has order $\leq m$, we say that Eq. (26) is *Kovalewskian* (with respect to the scale $\{X_r\}$).

As a simple extension of Yamanaka and Ovsjannikov theorem, we have:

Theorem 3.1 (*local solvability*). *Let* (26) *be a Kovalewskian equation, more precisely assume that, for each* $v \in X_r$ *and some* $\lambda(t)$ *in* $L^1(0, T)$,

$$A(\cdot)v \qquad \text{is } X_{r-\delta}\text{-measurable} \tag{29}$$

$$\| A(t) \|_{r-\delta} \leq \frac{\lambda(t)}{\delta^m} \| v \|_r, \tag{30}$$

whenever $0<\delta<r$.

Then Pb. {(26), (27)} *is locally well-posed in the space* X_{0+}. *More precisely, for every* $u_j \in X_{r_0}$ *there exists a unique solution* $u(t)$, *defined on the time-interval on which the function*

$$r(t) = r_0 - \left(\frac{e}{(m-1)!}\right)^{1/m} \int_0^t \sqrt[m]{\lambda(s)}\, ds \tag{31}$$

remains positive, the radius of analyticity of $u^{(j)}(t)$, $j=0, 1, \cdots, m-1$, *being at least* $r(t)$.

In view of the hyperbolic equations, it is convenient to particularise the notions of Banach scale and order of an operator, by introducing a *closed operator B* acting in H. Such an operator being fixed, we denote by $D(B^j)$ the domain of the j-th power of B, and we set

$$D(B^\infty)=\bigcap_{j=1}^{\infty} D(B^j).$$

Definition 3.3 (*Banach scale generated by an operator*). The family of Banach spaces

$$X_r(B)=\{v \in D(B^\infty) \colon \|v\|_r<\infty\} \qquad (r>0),$$

where

$$\|v\|_r=\operatorname*{Sup}_{j \in N} \|B^j v\|_H \frac{r^j}{j!}, \tag{32}$$

is called the Banach scale generated by B.

Definition 3.4 (*ω-order of an operator with respect to another operator*). Given a linear operator $A\colon D(B^\infty)\to D(B^\infty)$, we say that A has ω-order (or analyticty order) $\leq m$ with respect to B, where m is an integer ≥ 0, if there exist two positive constants K and Λ, for which

$$\|B^j Av\|_H \leq K(j+m)! \sum_{h=0}^{j+m} \|B^h v\|_H \frac{\Lambda^{j+m-h}}{h!} \tag{33}$$

for every $j \in N$ and $v \in D(B^\infty)$.

It is easy to check that any operator having ω-order $\leq m$ with respect to B has order $\leq m$ w. r. t. the scale $\{X_r(B)\}$ generated by B (see Def. 3.2 and Def. 3.3). More precisely:

If (33) *holds for some constants K and Λ, then* (30) (*with* $\|\cdot\|_r$ *given by* (32)) *holds for* $0<r<1/\Lambda$ *and*

$$\lambda=K\frac{m!}{1-r\Lambda}.$$

On the other hand it is not difficult to verify that condition (29) is a consequence of the following one (easier to satisfy in the concrete cases):

$$B^j A(\cdot)v \quad \text{is } H\text{-measurable} \quad (j \in N,\ v \in D(B^\infty)). \tag{34}$$

Summing up, we have from Theorem 3.1

Theorem 3.2 (*local solvability in Banach scales generated by an operator*). *Let H be a Hilbert space, B a closed operator in H and* $\{X_r(B)\}$ *the Banach scale generated by B (see Def.* 3.3). *Let* $A(t)$ *be a family of linear operators on* $D(B^\infty)$ *verifying* (34) *and such that*

$$\|B^j A(t)v\|_H \leq K(t)(j+m)! \sum_{h=0}^{j+m} \|B^h v\|_H \frac{\Lambda^{j+m-h}}{h!} \tag{35}$$

for every $j \in N$ *and some* $K(t)$ *in* $L^1(0, T)$ *and* $\Lambda > 0$. *Then Pb.* {(26), (27)} *is locally solvable in the space* $X_{0^+}(B)$. *More precisely the same conclusion of Theorem* 3.1 *holds for* $r_0 < 1/\Lambda$ *and*

$$r(t) = r_0 - \left(\frac{me}{1 - r_0\Lambda}\right)^{1/m} \int_0^t \sqrt[m]{K(s)}\, ds.$$

c) *Banach scales coherent with a Hilbert triplet.*

Theorem 3.2 ensures in particular the *global* well-posedness of Pb. {(26), (27)} in the Fréchet space $X_\infty(B) = \bigcap_{r>0} X_r(B)$ of the *B-entire* vectors (provided $A(t)$ satisfies a condition as (35) for *every* $\Lambda > 0$).

In order to have the global well-posedness in the space $X_{0^+}(B)$ of *B-analytic* vectors, some hyperbolicity assumption is indispensable. Hence we go back to the second order Eq. (19), putting ourselves in a composite structure which we describe here below.

Definition 3.5 (*Banach scales coherent with a Hilbert triplet*). A Banach scale $\{X_r\}$ is said coherent with a Hilbert triplet $\{V, H, V'\}$ when it is generated by some operator B (see Def. 3.3) with $D(B) = V$ such that

(36) *the norms* $\|v\|_V$ *and* $\|v\|_H + \|Bv\|_H$ *are equivalent on V.*

We are now in the position to state our main result: the global existence in X_{0^+} of Pb. {(19), (20)} under a supplementary condition on the commutators $[A(t), B^j]$, $j = 1, 2, \cdots$.

Theorem 3.3 ($[AS]_2$; *global solvability*). *Let* $\{V, H, V'\}$ *be a Hilbert triplet and* $A(t)$ *a family of operators, belonging to* $L^1(0, T; L(V, V'))$ *and verifying*

$$\langle A(t)v, w\rangle = \overline{\langle A(t)w, v\rangle}$$
$$\langle A(t)v, v\rangle \geq 0$$

for every v and w in V.

Let moreover $B: V \to H$ *be a linear operator verifying* (36) *and s.t.* $\{X_r(B)\}$, *the Banach scale generated by B, is dense in itself. Assume that*

$A(t)$ has ω-order ≤ 2 w. r. t. B, more precisely that (35) *holds for $m=2$, and that the measurability condition* (34) *is satisfied. Finally, assume that, for some $\alpha(t)$ in $L^1(0, T)$ and every $j \in N$ and $v \in D(B^\infty)$,*

$$\begin{aligned}(37)\quad &\|(B^j A(t) - A(t)B^j)v\|_H \\ &\leq \sqrt{\alpha(t)}(j+2)!\left((A(t)B^j v, B^j v)_H^{1/2} \frac{\Lambda}{(j+1)!} + \sqrt{\alpha(t)} \sum_{h=0}^{j} \|B^h v\|_H \frac{\Lambda^{j+2-h}}{h!}\right).\end{aligned}$$

Then Pb. {(19), (20)} *is globally well-posed in the space $X_{0^+}(B)$. More precisely, for every u_0 and u_1 in $X_{r_0}(B)$ with $r_0 \leq 1/\Lambda$, there exists a unique solution $u(t)$ on* [0, *T*], *and the radii of analyticity of $u(t)$ and $u'(t)$ are $\geq r(t)$, where*

$$(38)\qquad r(t) = r_0 \cdot \exp\left[-\Lambda\left(1 + \frac{2}{\sqrt{1 - r_0 \Lambda}}\right) \cdot \int_0^t \sqrt{\alpha(s)}ds\right].$$

The same conclusion holds true when we add to $A(t)$ any operator $P(t)$ having ω-order ≤ 1 w. r. t. B.

Unlike (31), the equality (38) shows that $r(t)$ is positive for every $t \geq 0$.

Remark 3.1. Taking (21) into account, we see that condition (37) implies that

$$(39)\qquad \|(A(t)B^j - B^j A(t)v\|_H \leq \beta(t)(j+2)! \sum_{h=0}^{j+1} \|B^h v\|_H \frac{\Lambda^{j+2-h}}{h!},$$

for some $\beta(t)$ in $L^1(0, T)$ and every $j \in N$.

We also observe that (39) holds as soon as the commutator $[A(t), B]$ has ω-order ≤ 2 w. r. t. B.

On the other hand, in the special case in which $A(t)$ verifies (23) for some $\nu > 0$, we have immediately that (39) implies (37), so that (39) and (37) are equivalent. //

Remark 3.2. When $A(t)$ commutes with B (in the concrete case this means that $A(t)$ has coefficients depending only on t), then (37) trivially holds for $\alpha(t)$ identically zero. Hence (38) reduces to $r(t) \equiv r_0$, which shows that the radius of analyticity of $(u(t), u'(t))$, where $u(t)$ is a solution of Eq. (19), does not decrease. //

Applications of Theorem 3.2

i) *Initial-periodic problems.*

Given a parallelepiped $P \subset \boldsymbol{R}^n_x$, let us denote by $\mathscr{A}_P(\boldsymbol{R}^n_x)$ the space of the P-periodic functions on $\boldsymbol{R}^n_x$; and let us consider Eq. (13), assuming that every coefficients belong to $L^1(0, T; \mathscr{A}_P(\boldsymbol{R}^n_x))$ while the leading coefficients $a_{ij}(x, t)$ verify (3) for $\nu=0$.

Then Pb. {(13), (2)} is well-posed in $\mathscr{A}_P(\boldsymbol{R}^n_x)$.

An analogous result holds when the coefficients of Eq. (13) are measurable in t, uniformly for $x \in \boldsymbol{R}^n$, and belong to $\mathscr{A}_{L^\infty}(\boldsymbol{R}^n_x)$: in this case Pb. {(13), (2)} is well-posed in $\mathscr{A}_{L^2}(\boldsymbol{R}^n_x)$. We recall that an analytic function $\phi(x)$ is said to belong to $\mathscr{A}_{L^p}(\boldsymbol{R}^n_x)$ when

$$\|D^\alpha\phi\|_{L^p(\boldsymbol{R}^n_x)} \leq C\Lambda^{|\alpha|}|\alpha|! \qquad (\alpha \in \boldsymbol{N}^n).$$

Remark 3.3. The above result follows directly from Theorem 3.2 in the case $n=1$, whereas for $n\geq 2$ it is a consequence of a slight generalization of Theorem 3.2 to the Banach scales generated by an n-tuple of linear operators $B_1, \cdots, B_n$ commuting among them. //

ii) *Initial-boundary problems.*

Let us consider the problem

$$\text{(40)} \qquad \begin{cases} u_{tt}-c(t)\Delta_x u=0 & (x \in \Omega,\ 0\leq t\leq T) \\ u(x, 0)=u_0(x), & u_t(x, 0)=u_1(x) \\ u(x, t)=0 & \text{for } x \in \partial\Omega, \end{cases}$$

where Ω is any open subset of $\boldsymbol{R}^n_x$, and $c(t)$ is a non-negative function belonging to $L^1(0, T)$. Then (40) is well-posed in the space $\mathscr{A}(\Delta_x, \Omega)$ constituted by the functions $\phi(x)$, analytic on Ω, such that

$$\text{(41)} \qquad \Delta^j_x\phi \in W^{1,2}_0(\Omega) \qquad (j \in \boldsymbol{N})$$

and

$$\text{(42)} \qquad \|\Delta^j_x\phi\|_{L^2(\Omega)} \leq C\Lambda^{2j} j!^2 \qquad (j \in \boldsymbol{N})$$

for some C and Λ.

Remark 3.4. Let Ω be bounded. Then condition (42) is fulfilled by any function $\phi(x)$ which is analytic on some neighborhood of Ω. Conversely, any function $\phi(x)$ which satisfies (42) is analytic in a neighborhood of Ω, provided that (41) holds and that the boundary of Ω is sufficiently regular.

iii) *Non-Kovalewskian equations.*

Let us consider the problem

$$\text{(43)} \qquad \begin{cases} u_{tt}+c(t)\Delta^2_x u=0 \\ u(x, 0)=u_0(x), \quad u_t(x, 0)=u_1(x) \end{cases}$$

where $c(t)$ is a non-negative function in $L^1(0, T)$.

Then (43) is well-posed in the space of the analytic functions $\phi(x)$ on $\boldsymbol{R}_x^n$ such that

$$\|D^\alpha\phi\|_{L^2(\boldsymbol{R}_x^n)} \leq C\Lambda^{|\alpha|}(|\alpha|!)^{1/2} \qquad (\alpha \in \boldsymbol{N}^n).$$

Sketch of the proof of Theorem 3.2.

Since the local existence is ensured by Theorem 3.1, we have only to prove that for any local solution $u(t)$ the radius of analyticity of $u(t)$ and $u'(t)$ is greater or equal to $r(t)$ (see (38)). To this end we apply to both terms of Eq. (19) the operator B^j (with $j=1, 2, \cdots$) and we approximate $A(t)$ by a sequence $\{A_j(t)\}$ of coercive and regular operators, to obtain

$$B^j u'' + B^j A_j(t)u = [B^j, A(t)]u + B^j(A_j(t) - A(t))u. \tag{44}$$

We then introduce the *approximate energy of order j* of the solution u, as

$$E_j(t) = (A_j(t)B^{j-1}u, B^{j-1}u) + \|B^{j-1}u'\|_H^2, \tag{45}$$

and the *infinite order energy*

$$\Phi_\rho(t) = \sum_{j=1}^{\infty} \sqrt{E_j(t)}\,\frac{\rho(t)^j}{j!}.$$

If we differentiate (45) and use (44), we can estimate $E_j'(t)$ in terms of $E_1(t), \cdots, E_{j+1}(t)$. Hence, by choosing the approximating sequence $\{A_j(t)\}$ in an appropriate way and taking $\rho(t) = r(t) + \varepsilon$, we obtain an estimate for $\Phi_\rho(t)$ which leads to the wished conclusion. //

References

[AS] A. Arosio and S. Spagnolo

1. "Global solution of the Cauchy problem for a non-linear hyperbolic equation". In "Non Linear Partial Differential Equations and Their Applications" Collège de France Seminars Vol. VI, Eds. H. Brézis and J. L. Lions, Pitman Res. Notes in Math., 1983.
2. "Global existence for abstract evolution equations of weakly hyperbolic type". To appear in J. Math. Pures Appl.

[B] S. Bernstein, "Sur une classe d'équations fonctionnelles aux dérivées partielles". Izv. Akad Nauk SSSR, Ser. Mat. **4** (1940), 17–26.

[CDS] F. Colombini, E. De Giorgi and S. Spagnolo, "Sur les équations hyperboliques avec des coefficients qui ne dépendent que du temps", Ann. Scuola Norm. Sup. Pisa **6** (1979), 511–559.

[CJS] F. Colombini, E. Jannelli and S. Spagnolo, "Well-posednes in the Gevrey class of the Cauchy problem for a non strictly hyperbolic equation with coefficients depending on time". Ann. Scuola Norm. Sup. Pisa, **10** (1983), 291–312.

[CS] F. Colombini and S. Spagnolo
1. "On the convergence of solutions of hyperbolic equations", Comm. Partial Differential Equations **3** (1978), 77–103.
2. "Second order hyperbolic equations with coefficients real analytic in space variables and discontinuous in time", J. Analyse Math. **38** (1980), 1–33.
3. "An example of a weakly hyperbolic Cauchy problem not well posed in C^∞", Acta Math. **148** (1982), 243–253.
4. "Hyperbolic equations with coefficients rapidly oscillating in time: a result of nonstability", J. Differential Equations **52** (1984), 24–38.

[J] E. Jannelli
1. "Weakly hyperbolic equations of second order with coefficients real analytic in space variables", Comm. Partial Differential Equations **7** (1982), 537–558.
2. "Hyperbolic systems with coefficients analytic in space variables", J. Math. Kyoto Univ. **21** (1981), 715–739.
3. "Regularly hyperbolic systems and Gevrey classes" to appear in Ann. Mat. Pura Appl.
4. "Gevrey well-posedness for a class of weakly hyperbolic equations", to appear in J. Math. Kyoto Univ.

[N] T. Nishitani
1. "Sur les équations hyperboliques à coefficients qui sont hölderiens en t et de classe de Gevrey en x", Bull. Sci. Math. **107** (1983), 113–138.
2. "Energy inequality for non strictly hyperbolic operators in the Gevrey class", J. Math. Kyoto Univ. **23** (1983), 739–773.

[O] Y. Ohya and S. Tarama, "Le problème de Cauchy à caractéristiques multiples dans la classe de Gevrey (coefficients hölderiens en t)", Proceedings of this Workshop.

[Ov] L. V. Ovsjannikov, "A singular operator in a scale of Banach spaces", Dokl. Akad. Nauk SSSR **163** (1965), 819–822 (transl. in Soviet Math. Dokl. **6** (1965), 1025–1028).

[T] M. H. Taibleson, "On the theory of Lipschitz spaces of distributions on Euclidean n-space I", J. Math. Mech. **13** (1964), 407–480.

[Y] T. Yamanaka, "Note on Kowalevskaja's system of partial differential equations", Comment. Math. Univ. St. Paul **9** (1960), 7–10.

SCUOLA NORMALE SUPERIORE
7, PIAZZA DEI CAVALIERI
I-56100 PISA, ITALIA

Taniguchi Symp. HERT
Kyoto 1984, pp. 381–394

Fundamental Solution for the Cauchy Problem of Hyperbolic Equation in Gevrey Class and the Propagation of Wave Front Sets

Kazuo TANIGUCHI

Introduction

Consider a hyperbolic operator

$$(1) \qquad L=D_t^m+\sum_{j=0}^{m-1}\sum_{|\alpha|\leq m-j} a_{j,\alpha}(t,x)D_x^\alpha D_t^j \qquad on\ [0,T]\times \boldsymbol{R}_x^n$$

with coefficients $a_{j,\alpha}(t,x)$ in a Gevrey lcass $\gamma^{(\kappa)}([0,T]\times \boldsymbol{R}_x^n)$ of order $\kappa\ (>1)$. Here, we mean that a function $a(t,x)$ belongs to a Gevrey class $\gamma^{(\kappa)}([0,T]\times \boldsymbol{R}_x^n)$ when there exist constants C and M such that

$$|\partial_t^\gamma\partial_x^\alpha a(t,x)|\leq CM^{-(\gamma+|\alpha|)}\gamma!^\kappa\alpha!^\kappa \qquad for\ (t,x)\in[0,T]\times \boldsymbol{R}_x^n.$$

In the recent paper [11] we have constructed the fundamental solution for the Cauchy problem

$$(2) \qquad Lu=0 \quad (0<t\leq T_o), \qquad \partial_t^j u(0)=g_j \quad (j=0,1,\cdots,m-1)$$

when the characteristic roots of L have constant multiplicities and investigated the propagation of wave front sets for the solution $u(t,x)$ of (2). In the present note we summarize the results in [11] and proceed to the case where the characteristic roots of L have variable multiplicities. For the latter case we obtain the same result as the one for the C^∞ case obtained by Kumano-go and the author [5].

In order to investigate these results, we employ the following symbol classes. In what follows we tacitly use the notation in [4].

Definition (cf. [11]). Let κ be a constant satisfying $\kappa>1$.

i) We say that a symbol $p(x,\xi)$ ($\in S^m$) belongs to a class $S^m_{G(\kappa)}$ if

$$(3) \qquad |p^{(\alpha)}_{(\beta)}(x,\xi)|\leq CM^{-(|\alpha|+|\beta|)}\alpha!^\kappa\beta!^\kappa\langle\xi\rangle^{m-|\alpha|}$$

holds with constants C and M independent of α, β.

i)′ We say that a symbol $p(t,x,\xi)$ in $[0,T]\times \boldsymbol{R}^{2n}_{x,\xi}$ belogs to a class $G^{(\kappa)}([0,T];S^m_{G(\kappa)})$ if

$$|\partial_t^\gamma p_{(\beta)}^{(\alpha)}(t, x, \xi)| \leqq CM^{-(\gamma+|\alpha|+|\beta|)}\gamma!^\kappa\alpha!^\kappa\beta!^\kappa\langle\xi\rangle^{m-|\alpha|}$$

holds.

ii) We say that a symbol $p(x, \xi)$ ($\in S^{-\infty}$) bolongs to a class $\mathscr{R}_{G(\kappa)}$ if for any α there exists a constant C_α such that

$$(4) \qquad |p_{(\beta)}^{(\alpha)}(x, \xi)| \leqq C_\alpha M^{-|\beta|}\beta!^\kappa e^{-\varepsilon\langle\xi\rangle^{1/\kappa}}$$

holds for any β with constants M and $\varepsilon>0$ independent of α, β.

ii)′ Let Z be an interval $[0, T]$ or a set $\Delta(T)=\{(t, s); 0\leqq s\leqq t\leqq T\}$. We say that a symbol $p(z, x, \xi)$ in $Z\times\boldsymbol{R}_{x,\xi}^{2n}$ belongs to a class $M(Z; \mathscr{R}_{G(\kappa)})$ if for any $\tilde{\gamma}$ and α there exista a constant $C_{\tilde{\gamma},\alpha}$ such that

$$|\partial_z^{\tilde{\gamma}} p_{(\beta)}^{(\alpha)}(z, x, \xi)| \leqq C_{\tilde{\gamma},\alpha}M^{-|\beta|}\beta!^\kappa e^{-\varepsilon\langle\xi\rangle^{1/\kappa}}$$

holds for any β with constants M and $\varepsilon>0$ independent of $\tilde{\gamma}$, α, β.

For the phase function of Fourier integral operators we employ

Definition. Let $0\leqq\tau<1$. We say that a phase function $\phi(x, \xi)$ belongs to a class $\mathscr{P}_{G(\kappa)}(\tau)$ if $J(x, \xi)\equiv\phi(x, \xi)-x\cdot\xi$ satisfies

$$\sum_{|\alpha+\beta|\leqq 2} |J_{(\beta)}^{(\alpha)}(x, \xi)|/\langle\xi\rangle^{1-|\alpha|} \leqq \tau$$

and

$$|J_{(\beta)}^{(\alpha)}(x, \xi)| \leqq \tau M^{-(|\alpha|+|\beta|)}\alpha!^\kappa\beta!^\kappa\langle\xi\rangle^{1-|\alpha|}$$

for a constant M independent of α, β. We also set $\mathscr{P}_{G(\kappa)}=\bigcup_{0\leqq\tau<1}\mathscr{P}_{G(\kappa)}(\tau)$.

Remark 1. As in [11] we denote $L_{G(\kappa)}^m(\phi)$, for $\phi(x, \xi)\in\mathscr{P}_{G(\kappa)}$, the class of the sum $p_\phi^0(X, D_x)+\tilde{p}_\phi(X, D_x)$ of Fourier integral operators $p_\phi^0(X, D_x)$ and $\tilde{p}_\phi(X, D_x)$ with $p^0(x, \xi)\in S_{G(\kappa)}^m$ and $\tilde{p}(x, \xi)\in\mathscr{R}_{G(\kappa)}$.

Remark 2. For $P_\phi=p_\phi^0(X, D_x)+\tilde{p}_\phi(X, D_x)$ in $L_{G(\kappa)}^m(\phi)$ we call $p^0(x, \xi)$ the main symbol of P_ϕ and denote it by $\sigma_o(P_\phi)$.

Now, consider a hyperbolic operator L of (1). We assume that the characteristic roots of L have constant multiplicities and there exist regularly hyperbolic operators $L_1, L_2, \cdots, L_r$ such that L has a form

$$(5) \qquad L=L_1L_2\cdots L_r+\sum_{j=0}^{m-q}\sum_{|\alpha|\leqq m-q-j} a'_{j,\alpha}(t, x)D_x^\alpha D_t^j$$

with $a'_{j,\alpha}(t, x)$ in $\gamma^{(\kappa)}([0, T]\times\boldsymbol{R}_x^n)$ and $1\leqq q\leqq r$. This condition (5) was first studied by Leray and Ohya [7] and they proved $\gamma^{(\kappa)}$-well-posedness for the Cauchy problem of the operator (5) when $1<\kappa<r/(r-q)$. Let $\lambda_1(t, x, \xi), \cdots, \lambda_h(t, x, \xi)$ be distinct roots of L and we may assume

$\lambda_j(t, x, \xi)$ belong to $G^{(\kappa)}([0, T]; S^1_{G(\kappa)})$. Then, if T is small enough we can solve an eikonal equation

$$\partial_t \phi = \lambda_j(t, x, \nabla_x \phi), \qquad \phi|_{t=s} = x \cdot \xi$$

and get a unique solution $\phi_j(t, s) \equiv \phi_j(t, s; x, \xi)$ belonging to $\mathscr{P}_{G(\kappa)}(c_o|t-s|)$ $(0 \leqq s, t \leqq T)$ for some constant c_o.

Theorem 1. *Assume that the characteristic roots of the operator L in (1) have constant multiplicities and that it has a form (5) for regularly hyperbolic operators $L_1, \cdots, L_r$ with $1 \leqq q \leqq r$. Suppose $1 < \kappa \leqq r/(r-q)$. Then, the fundamental solution $E(t, s) \equiv (E_0(t, s), E_1(t, s), \cdots, E_{m-1}(t, s))$ of L, each component $E_k(t, s)$ of which is an operator satisfying*

$$\begin{cases} LE_k(t, s) = 0 & t > s, \\ D_t^\gamma E_k(s, s) = \delta_{k,\gamma} & (\gamma = 0, 1, \cdots, m-1), \end{cases}$$

is constructed in the form

$$E_k(t, s) = \sum_{j=1}^{h} \sum_{\nu=0}^{\infty} W_{k,j,\nu}(t, s) I_{\phi_j}(t, s) + R(t, s) \quad \text{for } 0 \leqq t, s \leqq T_o \tag{6}$$

for a small T_o. In (6) $\sum_{\nu=0}^{\infty} W_{k,j,\nu}(t, s)$ are series of pseudo-differential operators $W_{k,j,\nu}(t, s)$ with symbols $w_{k,j,\nu}(t, s; x, \xi)$ satisfying for $\gamma = 0, 1, \cdots, m$

$$|\partial_t^\gamma \partial_\xi^\alpha \partial_x^\beta w_{k,j,\nu}(t, s; x, \xi)| \leqq (C_0^\nu |t-s|^\nu \nu!^{-1}) M^{-(|\alpha|+|\beta|)} \alpha!^\kappa \beta!^\kappa \langle \xi \rangle^{m^0 - k + \nu\sigma - |\alpha|}, \tag{7}$$

where $m^0 = m - 1 - \sigma(r-1)$, $\sigma = (r-q)/r$ and constants C_0, M are independent of α, β, ν; $I_{\phi_j}(t, s)$ are Fourier integral operators with the phase functions $\phi_j(t, s)$ $(j = 1, 2, \cdots, h)$; and $R(t, s)$ is a pseudo-differential operator with symbol in $M(\Delta(T_o); \mathscr{R}_{G(\kappa)})$.

From Theorem 1 we obtain

Corollary 1. *In the Cauchy problem (2) we assume*

$$|\partial_x^\alpha g_j(x)| \leqq C M_1^{-|\alpha|} \alpha!^\kappa \qquad (x \in \boldsymbol{R}^n; j = 0, 1, \cdots, m-1).$$

Then, if $1 < \kappa < r/(r-q)$ holds, the solution $u(t, x)$ of (2) exists and satisfies

$$|\partial_t^\gamma \partial_x^\alpha u(t, x)| \leqq C M_2^{-|\alpha|} \alpha!^\kappa \qquad (x \in \boldsymbol{R}^n; \gamma = 0, 1, \cdots, m).$$

For the case $\kappa = r/(r-q)$ we can find a solution $u(t, x)$ of (2) when T_o satisfies $T_o \leqq C_1 M_1$ for a constant C_1 determined only by the constant C_0 in

(7) and independent of M_1.

We note that, when $\kappa=r/(r-q)$, the existence domain $[0, T_o]\times \boldsymbol{R}_x^n$ may depend on the initial data $g_0, \cdots, g_{m-1}$.

Following [3] we define a class $\mathscr{D}_{L^2}^{\{\kappa\}\prime}$ of ultradistributions as

$$\mathscr{D}_{L^2}^{\{\kappa\}\prime}=\operatorname{proj}\lim_{\varepsilon\to 0} \mathscr{D}_{L^2,\varepsilon}^{\{\kappa\}\prime},$$

where $\mathscr{D}_{L^2,\varepsilon}^{\{\kappa\}\prime}=\{u(x);\ e^{-\varepsilon\langle\xi\rangle^{1/\kappa}}\hat{u}(\xi)\in L^2\}$ is the dual space of the Hilbert space $\mathscr{D}_{L^2,\varepsilon}^{\{\kappa\}}=\{u(x)\in L^2;\ e^{\varepsilon\langle\xi\rangle^{1/\kappa}}\hat{u}(\xi)\in L^2\}$ $(\varepsilon>0)$. Then, from Theorem 1 we have

Corollary 2. *Assume* $1<\kappa<r/(r-q)$. *Then, for any* $g_j\in\mathscr{D}_{L^2}^{\{\kappa\}\prime}$ $(j=0, \cdots, m-1)$ *there exists a unique solution* $u(t, x)$ *in* $\mathscr{B}^m([0, T_o];\mathscr{D}_{L^2}^{\{\kappa\}\prime})$ *and satisfies for any* κ_1 *with* $\kappa\leqq\kappa_1<r/(r-q)$

$$\begin{aligned}\sum_{k=0}^{m-1} WF_{G(\kappa_1)}(\partial_t^k u(t))\\ =\bigcup_{j=1}^{h}\Big\{(q_j(t, y, \eta), \rho p_j(t, y, \eta));\ \rho>0, (y, \eta)\in\bigcup_{k=0}^{m-1} WF_{G(\kappa_1)}(g_k)\Big\},\end{aligned}\tag{8}$$

where $WF_{G(\kappa_1)}(u)$ *is a wave front set of* u *in the Gevrey class of order* κ_1 *(see Section* 1) *and* $\{q_j(t, y, \eta), p_j(t, y, \eta)\}$ *is a solution of*

$$\begin{cases}\dfrac{dq}{dt}=-\nabla_\xi\lambda_j(t, q, p), & \dfrac{dp}{dt}=\nabla_x\lambda_j(t, q, p),\\ q|_{t=0}=y, & p|_{t=0}=\eta.\end{cases}$$

The equality (8) is also proved by Mizohata [8], [9], who proves this by a so-called (α, β)-method based on energy inequalities.

Next, we consider a hyperbolic operator (1) with variable multiplicity. Let $\lambda_1(t, x, \xi), \cdots, \lambda_m(t, x, \xi)$ be the characteristic roots of (1) and we may assume $\lambda_j(t, x, \xi)\in G^{(\kappa)}([0, T]; S^1_{G(\kappa)})$. For $\varepsilon>0$, an integer ν and a conic set V in the cotangent bundle $T^*(\boldsymbol{R}_x^n)$ $(\cong\boldsymbol{R}_x^n\times\boldsymbol{R}_\xi^n)$ we define $\Gamma_\varepsilon^\nu(t, V)$ as the set of end points (at t) of all ε-admissible trajectories of, at most, step ν, issuing from the ε-conic neighborhood $\{(x, \xi);\ |x-y|\leqq\varepsilon, |\xi/|\xi|-\eta/|\eta||\leqq\varepsilon, (y, \eta)\in V\}$ of V (concerning the characteristic roots $\lambda_j(t, x, \xi)$, $j=1, \cdots, m$) and define

$$\begin{cases}\Gamma_\varepsilon(t, V)=\textit{the closure of}\ \bigcup_{\nu=0}^{\infty}\Gamma_\varepsilon^\nu(t, V),\\ \Gamma(t, V)=\bigcap_{\varepsilon>0}\Gamma_\varepsilon(t, V).\end{cases}\tag{9}$$

Then, we obtain

Theorem 2. *Let L be a hyperbolic operator* (1) *with variable multiplicity, and assume that L has a form* (5) *with regularly hyperbolic operators*

$L_1, \cdots, L_r$ and that $1<\kappa<r/(r-q)$ $(1 \leqq q \leqq r)$. Then, we have for the solution $u(t)$ of (2)

$$WF_{G(\kappa_1)}(u(t)) \subset \Gamma\left(t, \bigcup_{j=0}^{m-1} WF_{G(\kappa_1)}(g_j)\right), \tag{10}$$

where $\kappa \leqq \kappa_1 < r/(r-q)$.

The theorems 1 and 2 are proved, first, by reducing (2) to the Cauchy problem of a hyperbolic system of first order, and then by constructing the fundamental solution. To construct the fundamental solution, we usually employ a method of solving an eikonal equation and solving transport equations in the case of constant multiplicity. But, in this method the error term in the fundamental solution, $R(t, s)$ in (6), belongs to $\mathscr{R}_{G(2\kappa-1)}$ if the coefficients of L are in the Gevrey class of order κ, and it seems that this is the best result by the method of solving transport equations. In this note we give a sharp estimate of multi-products of Fourier integral operators (Proposition 2.2) and apply it to successive approximation appearing in the fundamental solution of a hyperbolic system. Then, we can construct the fundamental solution with an error term in $\mathscr{R}_{G(\kappa)}$. Here, we emphasize that we do not solve transport equations and instead we use only eikonal equations and a method of successive approximation. This method is found in [5], where we obtained the fundamental solution for a hyperbolic system with variable multiplicity in the C^∞ case and an estimate of wave front sets for its solution.

For the case of the operator (1) with variable multiplicity Bronstein [1] and Wakabayashi [13] constructed a parametrix of (1). They treated only the case $q=1$ but they did not assume the smoothness of the characteristic roots. In [13] Wakabayashi also proved (10) under the above assumption. Restricting ourselves to the case where the characteristic roots are smooth, we apply Wakabayashi's method including the case of q greater than 1. Then, we get (10) when $1<\kappa \leqq \kappa_1 < \min(r/(r-q), 2)$ but we cannot make κ_1 go beyond 2. For example, consider an operator

$$L=(D_t-a_1(x)D_x)(D_t-a_2(x)D_x)(D_t-a_3(x)D_x)+b(x)D_x \quad \text{in } [0, T]\times \boldsymbol{R}_x^1$$

with $a_1(0)=a_2(0)=a_3(0)$ and $b(0)\neq 0$. This is the form (5) with $r=3$ and $q=2$. Then, we obtain (10) under $1<\kappa \leqq \kappa_1 < 3$ by our method, but by Wakabayashi's method we obtain (10) only in the case of $1<\kappa \leqq \kappa_1 < 2$. We also remark that in this note we construct a fundamental solution of L and estimate the wave front sets of the solution of (2) as an equation of evolution. Then, we get the same results of [5] in the case of Gevrey classes.

The plan of this note is the following: In Section 1 we show that pseudo-differential operators and Fourier integral operators with symbols in $S^m_{G(\kappa)}$ act on the class $\mathscr{D}^{\{\kappa\}\prime}_{L^2}$ of ultradistributions and give a formulation of wave front sets $WF_{G(\kappa_1)}(u)$ for $u \in \mathscr{D}^{\{\kappa\}\prime}_{L^2}$ $(\kappa_1 \geqq \kappa)$. In Section 2 we summarize the results obtained in [11] on the fundamental solution of a hyperbolic equation with constant multiplicity and in Section 3 we investigate the propagation of wave front sets for a solution of a hyperbolic equation with variable multiplicity.

§ 1. Pseudo-differential operators on ultradistributions and wave front sets

In the Introduction we introduced the class

$$\mathscr{D}^{\{\kappa\}}_{L^2,\varepsilon}=\{u(x);\ e^{\varepsilon\langle\xi\rangle^{1/\kappa}}\hat{u}(\xi) \in L^2\} \tag{1.1}$$

for $\varepsilon>0$. In the following we define $\mathscr{D}^{\{\kappa\}}_{L^2,\varepsilon}$ for $\varepsilon\leqq 0$ also by (1.1). Then we have $\mathscr{D}^{\{\kappa\}\prime}_{L^2,\varepsilon}=\mathscr{D}^{\{\kappa\}}_{L^2,-\varepsilon}$ for $\kappa\geqq 0$. We denote the norm of $\mathscr{D}^{\{\kappa\}}_{L^2,\varepsilon}$ by

$$\|u\|_\varepsilon=\|e^{\varepsilon\langle\xi\rangle^{1/\kappa}}\hat{u}(\xi)\|_{L^2(R^n_\xi)}.$$

Then, we have

Proposition 1.1. *Let $p(x, \xi)$ be a symbol satisfying* (3) *with $m\geqq 0$, and let ε_0 be a positive number satisfying $0<\varepsilon_0<\kappa M^{1/\kappa}$. Then, for $u \in \mathscr{D}^{\{\kappa\}}_{L^2,\varepsilon}$ with $\langle D_x\rangle^m u \in \mathscr{D}^{\{\kappa\}}_{L^2,\varepsilon}$, where $|\varepsilon|\leqq\varepsilon_0$, we can define an operation $Pu=p(X, D_x)u$ and it satisfies*

$$\|Pu\|_\delta\leqq C\|\langle D_x\rangle^m u\|_\varepsilon$$

for δ satisfying $|\delta|\leqq\varepsilon_0$, $\delta\leqq\varepsilon$. Here, the constant C is independent of ε, δ, m.

Corollary 1. *Let $p(x, \xi)$ and ε_0 be as in the Proposition* 1.1. *Assume now $m>0$. Then, we have for $u \in \mathscr{D}^{\{\kappa\}}_{L^2,\varepsilon}$ and δ with $|\varepsilon|$, $|\delta|\leqq\varepsilon_0$, $\delta<\varepsilon$*

$$\|Pu\|_\delta\leqq C(\varepsilon-\delta)^{-m\kappa}\|u\|_\varepsilon \qquad 0<m<1/\kappa,$$

$$\|Pu\|_\delta\leqq C\Gamma(m\kappa)(\varepsilon-\delta)^{-m\kappa}\|u\|_\delta \qquad m\geqq 1/\kappa.$$

We can prove Proposition 1.1 by using oscillatory integrals.

Let $P=\sum_{\nu=0}^\infty P_\nu$ be a series of pseudo-differential operators P_ν whose symbols $p_\nu(x, \xi)$ satisfy

$$|p_{\nu(\beta)}^{(\alpha)}(x, \xi)|\leqq CA^\nu\nu!^{-1}M^{-(|\alpha|+|\beta|)}\alpha!^\kappa\beta!^\kappa\langle\xi\rangle^{m+\nu\sigma-|\alpha|}$$

with a constant σ satisfying $\sigma\kappa<1$ and constants m, C, A and M independent of α, β, ν. Then, we have from the above Corollary

Corollary 2. *Let $P=\sum_{\nu=0}^{\infty} P_\nu$ be above. Then, the operation*

$$P\colon \mathscr{D}_{L^2}^{\{\kappa\}\prime} \longrightarrow \mathscr{D}_{L^2}^{\{\kappa\}\prime}$$

is well-defined.

The version of Proposition 1.1 for Fourier integral operators is the following

Proposition 1.2. *Let $p(x, \xi)$ be a symbol in $S_{G(\kappa)}^m$ and $\phi(x, \xi)$ be a phase function in $\mathscr{P}_{G(\kappa)}$. Then, the Fourier integral operator P_ϕ acts on $\mathscr{D}_{L^2}^{\{\kappa\}\prime}$ and $P_\phi u$ belongs to $\mathscr{D}_{L^2}^{\{\kappa\}\prime}$ for $u \in \mathscr{D}_{L^2}^{\{\kappa\}\prime}$.*

As in Hörmander [2] we define a wave front set $WF_{G(\kappa_1)}(u)$ for $u \in \mathscr{E}'$ in the Gevrey class of order κ_1 by the following:

We say that (x_o, ξ_o) does not belong to $WF_{G(\kappa_1)}(u)$ when there exist a function $\varphi(x)$ in $\gamma^{(\kappa_1)}(\boldsymbol{R}_x^n)$ with $\varphi(x_o)\neq 0$ and a conic neighborhood Γ of ξ_o such that for positive constants C and ε

$$|\mathscr{F}[\varphi u](\xi)| \leqq Ce^{-\varepsilon\langle\xi\rangle^{1/\kappa_1}} \qquad \textit{for } \xi \in \Gamma,$$

where $\mathscr{F}[u](\xi)$ is the Fourier transform of u.

On the other hand, we have

Proposition 1.3. *Let $p(x, \xi)$ be a symbol in $S_{G(\kappa)}^0$ and there exists a conic set V in $T^*(\boldsymbol{R}_x^n)$ such that*

$$|p(x, \xi)| \geqq C_o > 0 \qquad \textit{for } (x, \xi) \in V.$$

Then, there exists a symbol $q(x, \xi)$ in $S_{G(\kappa)}^0$ such that for any $b(x, \xi) \in S_{G(\kappa)}^0$ with supp $b \subset V_1$

$$B(QP-I) \in \mathscr{R}_{G(\kappa)}, \qquad B(PQ-I) \in \mathscr{R}_{G(\kappa)}, \tag{1.2}$$

where V_1 is a conic set with

$$\{(x, \xi/|\xi|);\, (x, \xi) \in V_1\} \Subset \{(x, \xi/|\xi|);\, (x, \xi) \in V\}. \tag{1.3}$$

Here and in what follows, we denote the class of pseudo-differential operators with symbols in $\mathscr{R}_{G(\kappa)}$ also by $\mathscr{R}_{G(\kappa)}$.

Proof. We fix a conic set V_1 satisfying (1.3) and take a symbol $p_1(x, \xi)$ in $S_{G(\kappa)}^0$ satisfying

$$\begin{cases} p_1(x, \xi) = p(x, \xi) & \textit{for } (x, \xi) \in V_1, \\ |p_1(x, \xi)| \geqq C_o & \textit{for any } (x, \xi) \in \boldsymbol{R}_x^n \times \boldsymbol{R}_\xi^n. \end{cases}$$

Set $p_2(x, \xi) = p_1(x, \xi)^{-1}$ and

$$r(x, \xi)=\sigma_o(p_2(X, D_x)p_1(X, D_x)-I).$$

Then, $r(x, \xi)$ belongs to $S^{-1}_{G(\kappa)}$. Using a function $\chi(\xi)$ in $\gamma^{(\kappa)}(\boldsymbol{R}^n_x)$ with $\chi=1$ for $|\xi|\leqq 1$ and $\chi=0$ for $|\xi|\geqq 2$, we define $r_N(x, \xi)=(1-\chi(\xi/N))r(x, \xi)$. Then, if N is large $r_N(x, \xi)$ satisfies the first inequality of (2.19) in [11] and the inverse operator of $I+r_N(X, D_x)$ exists as the Neumann series $\sum_{\nu=0}^{\infty}(-1)^\nu r_N(X, D_x)^\nu$. Hence, we get a left parametrix Q_1 of $p_1(X, D_x)$ as $Q_1=(I+r_N(X, D_x))^{-1}p_2(X, D_x)$. We note that Q_1 is also a right parametrix of $p_1(X, D_x)$. So, if we define $q(x, \xi)$ as the main symbol of Q_1, we get the desired symbol.

From this proposition the discussions in Section 3 of Chapter 10 in [4] work well and we obtain the following for $u \in \mathscr{E}'$.

(*) $(x_o, \xi_o) \notin WF_{G(\kappa_1)}(u)$ if and only if there exists a symbol $a(x, \xi) \in S^0_{G(\kappa_1)}$ with $a(x_o, \rho\xi_o)\neq 0$ (for $\rho\geqq 1$) such that $a(X, D_x)u \in \gamma^{(\kappa_1)}$.

Since pseudo-differential operators with symbols in $S^m_{G(\kappa)}$ act on $\mathscr{D}^{\{\kappa\}'}_{L^2}$, we may adopt (*) as the definition of the wave front set $WF_{G(\kappa_1)}(u)$ for $u \in \mathscr{D}^{\{\kappa\}'}_{L^2}$ if we modify $a(x, \xi) \in S^0_{G(\kappa_1)}$ as $a(x, \xi) \in S^0_{G(\kappa)}$, that is, we can define $WF_{G(\kappa_1)}(u)$ as

(**) For $u \in \mathscr{D}^{\{\kappa\}'}_{L^2}$, we say that (x_o, ξ_o) does not belong to $WF_{G(\kappa_1)}(u)$, where $\kappa_1\geqq\kappa$, if there exists a symbol $a(x, \xi)$ in $S^0_{G(\kappa)}$ with $a(x_o, \rho\xi_o)\neq 0$ (for $\rho\geqq 1$) such that $a(X, D_x)u$ belongs to $\gamma^{(\kappa_1)}(\boldsymbol{R}^n_x)$.

This makes the discussions of wave front sets in Gevrey classes easy. The detailed description of this section will be given in the forthcoming paper [12].

§ 2. Fundamental solution for a hyperbolic operator with constant multiplicity.

In this section we consider a hyperbolic operator L of (1) in case the characteristic roots of L have constant multiplicities. We assume (5) with regularly hyperbolic operators $L_1, L_2, \cdots, L_r$. Let $\lambda_1(t, x, \xi), \cdots, \lambda_h(t, x, \xi)$ be distinct characteristic roots of L. Then, from Proposition 3.4 of [11] the Cauchy problem (2) is reduced to the Cauchy problem

$$\mathscr{L}_o U(t)=0, \qquad 0<t\leqq T_o, \qquad U(0)=G$$

for the perfectly diagonalized operator

$$\text{(2.1)} \qquad \mathscr{L}_o=D_t-\begin{pmatrix}\lambda_1(t, X, D_x)\mathscr{I}_{l_1} & & 0\\ & \ddots & \\ 0 & & \lambda_h(t, X, D_x)\mathscr{I}_{l_h}\end{pmatrix}+\begin{pmatrix}B_1(t) & & 0\\ & \ddots & \\ 0 & & B_h(t)\end{pmatrix}$$

$$+R_o(t),$$

where $\mathscr{I}_{l_j}$ are $l_j \times l_j$ identity matrices, $B_j(t)$ are $l_j \times l_j$ matrices of pseudo-differential operators with symbols in $G^{(\kappa)}([0, T]; S^{\sigma}_{G(\kappa)})$, $\sigma=(r-q)/r$, and $R_o(t)$ is a matrix of pseudo-differential operators in $\mathscr{R}_{G(\kappa)}$. So, the proof of Theorem 1 is reduced to the investigation of a single operator

$$\mathscr{L}=D_t-\lambda(t, X, D_x)+b(t, X, D_x) \tag{2.2}$$

with $\lambda(t, x, \xi) \in G^{(\kappa)}([0, T]; S^1_{G(\kappa)})$ and $b(t, x, \xi) \in G^{(\kappa)}([0, T]; S^{\sigma}_{G(\kappa)})$ $(0 \leqq \sigma \leqq 1/\kappa)$. For the operator (2.2) we have

Proposition 2.1. *Assume $\lambda(t, x, \xi) \in G^{(\kappa)}([0, T]; S^1_{G(\kappa)})$ is real-valued and $b(t, x, \xi) \in G^{(\kappa)}([0, T]; S^{\sigma}_{G(\kappa)})$ for some $0 \leqq \sigma \leqq 1/\kappa$. Then, the fundamental solution $E(t, s)$ of (2.2) can be written in the form*

$$E(t, s)=\Big\{I+\sum_{\nu=1}^{\infty} W_\nu(t, s)\Big\}I_\phi(t, s)+R(t, s) \qquad \text{for } 0 \leqq t,\ s \leqq T_o \tag{2.3}$$

for a small T_o. In (2.3) $\sum_{\nu=1}^{\infty} W_\nu(t, s)$ is a series of pseudo-differential operators $W_\nu(t, s)$ with symbols $w_\nu(t, s; x, \xi)$ satisfying for $\gamma=0, 1$

$$|\partial_t^\gamma \partial_\xi^\alpha \partial_x^\beta w_\nu(t, s; x, \xi)| \leqq (C_0^\nu |t-s|^\nu \nu!^{-1}) M^{-(|\alpha|+|\beta|)} \alpha!^\kappa \beta!^\kappa \langle \xi \rangle^{\nu\sigma-|\alpha|} \tag{2.4}$$

with constants C_0 and M independent of α, β, ν; $I_\phi(t, s)$ is a Fourier integral operator with phase function $\phi(t, s; x, \xi)$, where $\phi(t, s; x, \xi)$ is a solution of $\partial_t\phi=\lambda(t, x, \nabla_x\phi)$, $\phi|_{t=s}=x\cdot\xi$; and $R(t, s)$ is a pseudo-differential operator in $\mathscr{R}_{G(\kappa)}$.

Remark. From (2.4), Corollary 2 of Proposition 1.1 and Proposition 1.2 the fundamental solution $E(t, s)$ is able to act on $\mathscr{D}_{L^2}^{\{\kappa\}\prime}$.

We give a sketch of the proof. Operate $\mathscr{L}$ in (2.2) to $I_\phi(t, s)$, the Fourier integral operator with symbol 1 and phase function $\phi(t, s; x, \xi)$. Then, we have

$$\mathscr{L}I_\phi(t, s)=P_\phi(t, s)$$

with a Fourier integral operator $P_\phi(t, s)$ in $L^{\sigma}_{G(\kappa)}(\phi(t, s))$ for any t and s. Define $\{\tilde{Q}_\nu(t, s)\}_{\nu=1}^{\infty}$ inductively by

$$\begin{cases} \tilde{Q}_1(t, s)=-iP_\phi(t, s), \\ \tilde{Q}_\nu(t, s)=-i\displaystyle\int_s^t P_\phi(t, \theta)\tilde{Q}_{\nu-1}(\theta, s)d\theta \qquad (\nu \geqq 2). \end{cases} \tag{2.5}$$

Then, we find immediately that

$$E(t, s)=I_\phi(t, s)+\sum_{\nu=1}^{\infty}\int_s^t I_\phi(t, \theta)\tilde{Q}_\nu(\theta, s)d\theta$$

is the fundamental solution of (2.2). From (2.5) we have

$$\begin{aligned}&\int_s^t I_\phi(t,\theta)\tilde{Q}_\nu(\theta,s)d\theta\\ (2.6)\qquad &=(-i)^\nu\int_s^t\int_s^{t_0}\cdots\int_s^{t_{\nu-2}} I_\phi(t,t_0)P_\phi(t_0,t_1)P_\phi(t_1,t_2)\cdots\\ &\qquad\times P_\phi(t_{\nu-1},s)dt_{\nu-1}\cdots dt_0\qquad (t_{-1}=t).\end{aligned}$$

We note that in the integrand of the right hand side of (2.6) a multi-product $I_\phi(t,t_0)P_\phi(t_0,t_1)\cdots P_\phi(t_{\nu-1},s)$ of Fourier integral operators appears. So, if we apply the following proposition, we obtain the desired form (2.3) of the fundamental solution $E(t,s)$.

Proposition 2.2. (Theorem 2.1 in [11]). *Consider multi-products*

$$\tilde{Q}_{\nu+1}=P_{1,\phi_1}P_{2,\phi_2}\cdots P_{\nu+1,\phi_{\nu+1}}$$

of Fourier integral operators P_{j,ϕ_j} in $L^\sigma_{G(\kappa)}(\phi_j)$ $(\sigma\geqq 0)$ for a sequence $\{\phi_j\}$ of phase functions $\phi_j(x,\xi)$ in $\mathscr{P}_{G(\kappa)}(\tau_j)$. We assume

(A.1) *There exists a small constant τ^0 such that $\sum_{j=1}^\infty \tau_j\leqq\tau^0$. If we set $J_j(x,\xi)=\phi_j(x,\xi)-x\cdot\xi$, $J_j(x,\xi)/\tau_j$ satisfies* (3) *with $m=1$ and constants C and M independent of j.*

(A.2) *If we write $P_{j,\phi_j}=p^0_{j,\phi_j}(x,D_x)+\tilde{p}_{j,\phi_j}(X,D_x)$ with $p^0_j(x,\xi)\in S^\sigma_{G(\kappa)}$ and $\tilde{p}_j(x,\xi)\in\mathscr{R}_{G(\kappa)}$, $p^0_j(x,\xi)$ satisfies* (3) *with $m=\sigma$ and constants C and M independent of j, and $\tilde{p}_j(x,\xi)$ satisfies* (4) *with constants C_α, M, ε independent of j.*

Then, the multi-product $\tilde{Q}_{\nu+1}$ is a Fourier integral operator $Q_{\nu+1,\Phi_{\nu+1}}$ in $L^{(\nu+1)\sigma}_{G(\kappa)}(\Phi_{\nu+1})$ with a phase function $\Phi_{\nu+1}(x,\xi)$ in $\mathscr{P}_{G(\kappa)}$ and is represented by the form

$$Q_{\nu+1,\Phi_{\nu+1}}=q^0_{\nu+1}(x,D_x)I_{\Phi_{\nu+1}}+\tilde{q}_{\nu+1}(X,D_x)I_{\Phi_{\nu+1}}$$

for the symbols $q^0_{\nu+1}(x,\xi)$ and $\tilde{q}_{\nu+1}(x,\xi)$ satisfying

$$|q^{0\,(\alpha)}_{\nu+1(\beta)}(x,\xi)|\leqq C_0^\nu M^{-(|\alpha|+|\beta|)}\alpha!^\kappa\beta!^\kappa\langle\xi\rangle^{(\nu+1)\sigma-|\alpha|},$$
$$|\tilde{q}^{(\alpha)}_{\nu+1(\beta)}(x,\xi)|\leqq C_0^\nu C_\alpha M^{-|\beta|}\beta!^\kappa\nu!^{\sigma\kappa}e^{-\varepsilon\langle\xi\rangle^{1/\kappa}}$$

with constants C_0, M, ε independent of ν, α, β and the constant C_α independent of ν, β.

In Proposition 2.2 the phase function $\Phi_{\nu+1}(x,\xi)$ is defined as

$$\Phi_{\nu+1}(x,\xi)=\sum_{j=1}^\infty(\phi_j(X^{j-1}_\nu,\Xi^j_\nu)-X^j_\nu\cdot\Xi^j_\nu)+\phi_{\nu+1}(X^\nu_\nu,\xi)\qquad (X^0_\nu=x),$$

where $\{X^j_\nu,\Xi^j_\nu\}_{j=1}^\nu\equiv\{X^j_\nu,\Xi^j_\nu\}_{j=1}^\nu(x,\xi)$ is a solution of

$$\begin{cases} X_\nu^j = \nabla_\xi \phi_j(X_\nu^{j-1}, \Xi_\nu^j), \\ \Xi_\nu^j = \nabla_x \phi_{j+1}(X_\nu^j, \Xi_\nu^{j+1}), \quad j=1, \cdots, \nu \quad (X_\nu^0 = x, \Xi_\nu^{\nu+1} = \xi). \end{cases}$$

We call $\Phi_{\nu+1}(x, \xi)$ a multi-product of phase functions $\phi_1, \phi_2, \cdots, \phi_{\nu+1}$ and denote it by $\phi_1 \# \phi_2 \# \cdots \# \phi_{\nu+1}$. In the case of (2.6) the corresponding phase function is $\phi(t, t_0) \# \phi(t_0, t_1) \# \cdots \# \phi(t_{\nu-1}, s)$, which is simplified to the phase function $\phi(t, s)$ by 3° of Theorem 2.3 in [6].

For the case of the proof of Theorem 1 we apply the discussions of the proof of Proposition 2.1 to each element $D_t - \lambda_j(t, X, D_x)\mathscr{I}_{l_j} + B_j(t)$ of $\mathscr{L}_o$ in (2.1) and let $E_j(t, s)$ be its fundamental solution. Then,

$$E_o(t, s) = \begin{pmatrix} E_1(t, s) & & 0 \\ & \ddots & \\ 0 & & E_h(t, s) \end{pmatrix}$$

is an approximate fundamental solution of order infinity of $\mathscr{L}_o$. Using $E_0(t, s)$ we follow the same discussions of the proof for Theorem 4.6 in Chapter 10 of [4]. Then, we get the fundamental solution of $\mathscr{L}_o$, from which we immediately derive the fundamental solution $E(t, s) = (E_0(t, s), \cdots, E_{m-1}(t, s))$ of (6) for the operator (1). This proves Theorem 1.

Corollary 1 of Theorem 1 is followed immediately from Theorem 1. Since we give the definition of $WF_{G(\kappa)}(u)$ by (**) in Section 1, for the proof of Corollary 2 of Theorem 1 we have only to prove the following Lemma 2.3. In fact, we obtain (8) by the discussions of the proof of Theorem 3.14 in Chapter 10 of [4] and Lemma 2.3.

Lemma 2.3. i) *Let $\phi(x, \xi) \in \mathscr{P}_{G(\kappa)}$ be homogeneous in ξ for large $|\xi|$ and let $p(x, \xi) \in S^m_{G(\kappa)}$. Let $a(x, \xi)$ and $b(x, \xi)$ be symbols in $S^0_{G(\kappa)}$ and suppose that they satisfy for some $\varepsilon > 0$*

$$|x-y| + |\xi/|\xi| - \nabla_x\phi(y, \eta)/|\nabla_x\phi(y, \eta)|| \geqq \varepsilon$$

if $(x, \xi) \in \operatorname{supp} a$ and $(\nabla_\xi\phi(y, \eta), \eta) \in \operatorname{supp} b$. Then, the operator $AP_\phi B$ belongs to $\mathscr{R}_{G(\kappa)}$.

ii) *Let R be a pseudo-differential operator in $\mathscr{R}_{G(\kappa)}$. Then, for $u \in \mathscr{D}^{\{\kappa\}\prime}_{L^2}$ the function Ru belongs to $\gamma^{(\kappa)}$.*

We can prove this lemma by using the properties of the oscillatory integrals.

§ 3. Propagation of wave front sets for a solution of a hyperbolic equation with variable multiplicity

The end of this section is to study Theorem 2, the propagation of wave front sets, obtained by Y. Morimoto and the author. Let L be a hyperbolic operator (1) and assume that L can be written in the form (5) for regularly hyperbolic operators $L_1, \cdots, L_r$. Let $\lambda_1(t, x, \xi), \cdots, \lambda_m(t, x, \xi)$ be characteristic root of L. We begin with defining an ε-admissible trajectory. Let $\{q_j(t, s; y, \eta), p_j(t, s; y, \eta)\}$ be a solution of

$$\begin{cases} \dfrac{dq}{dt} = -\nabla_\xi \lambda_j(t, q, p), & \dfrac{dp}{dt} = \nabla_x \lambda_j(t, q, p), \\ q|_{t=s} = y, & p|_{t=s} = \eta. \end{cases}$$

Let $\varepsilon \geqq 0$. For $\nu \geqq 1$ and a permutation $J_{\nu+1} = (j_1, j_2, \cdots, j_{\nu+1})$ ($j_k = 1, \cdots, m$, $1 \leqq k \leqq \nu+1$) we define a trajectory $\{q_{J_{\nu+1}}, p_{J_{\nu+1}}\}(\sigma; t, t_1, \cdots, t_\nu; y, \eta)$ for an ordered set $\{t_1, \cdots, t_\nu\}$ $(0 < t_\nu < \cdots < t_1 < t)$ inductively by

$$(3.1) \qquad \{q_{J_{\nu+1}}, p_{J_{\nu+1}}\}(\sigma; t, t_1, \cdots, t_\nu; y, \eta) = \{q_{j_k}, p_{j_k}\}(\sigma, t_k; x^k, \xi^k) \quad \text{for } t_k \leqq \sigma \leqq t_{k-1}, \quad k = 1, \cdots, \nu+1,$$

where $x^{\nu+1} = y$, $\xi^{\nu+1} = \eta$, $t_0 = t$, $t_{\nu+1} = 0$ and

$$(3.2) \qquad (x^k, \xi^k) = \{q_{J_{\nu+1}}, p_{J_{\nu+1}}\}(t_k; t, t_1, \cdots, t_\nu; y, \eta) \qquad (1 \leqq k \leqq \nu).$$

Then, we say that an ordered set $\{t_1, \cdots, t_\nu\}$ is an ε-station chain and that a trajectory $\{q_{J_{\nu+1}}, p_{J_{\nu+1}}\}(\sigma; t, t_1, \cdots, t_\nu; y, \eta)$ is ε-admissible when for (x^k, ξ^k) defined by (3.2) we have

$$|\lambda_{j_k}(t_k, x^k, \xi^k) - \lambda_{j_{k+1}}(t_k, x^k, \xi^k)| \leqq \varepsilon \langle \xi^k \rangle$$

for all k. Let V be a conic set in $T^*(R^n_x)$ and let V_ε be an ε-conic neighborhood $\{(x, \xi); |x-y| \leqq \varepsilon, |\xi/|\xi| - \eta/|\eta|| \leqq \varepsilon, (y, \eta) \in V\}$ of V. Then, we can define $\Gamma^\nu_\varepsilon(t, V)$ explicitly by

$$\begin{aligned} \Gamma^\nu_\varepsilon(t, V) = \{(x, \xi); (x, \xi) = \{q_{J_{\nu'+1}}, p_{J_{\nu'+1}}\}(t; t, t_1, \cdots, t_{\nu'}; y, \eta),\ 1 \leqq \nu' \leqq \nu,\ J_{\nu'+1} \\ = (j_1, \cdots, j_{\nu'+1})\ (j_k = 1, \cdots, m),\ \{t_1, \cdots, t_{\nu'}\} \text{ is an } \varepsilon\text{-station chain} \\ \text{and } (y, \eta) \in V_\varepsilon\} \\ \cup \{(x, \xi); (x, \xi) = \{q_j, p_j\}(t, 0, y, \eta),\ j = 1, \cdots, m,\ (y, \eta) \in V_\varepsilon\}. \end{aligned}$$

We may set

$$\Gamma^0_\varepsilon(t, V) = \{(x, \xi); (x, \xi) = \{q_j, p_j\}(t, 0; y, \eta), j = 1, \cdots, m, (y, \eta) \in V_\varepsilon\}$$

and say that the bicharacteristic curves $\{q_j, p_j\}$ $(t, 0; y, \eta)$ are ε-admissible trajectories for any $\varepsilon \geqq 0$.

Now, we give an outline of the proof of Theorem 2. Using Proposition 3.3 in [11] we can reduce the Cauchy problem (2) to the equivalent Cauchy problem

$$(3.3) \qquad \mathscr{L} U(t) = 0 \qquad 0 \leqq t \leqq T_o, \qquad U(0) = G$$

for a hyperbolic system with an appropriate size l

$$\mathscr{L}=D_t-\begin{pmatrix}\lambda_1'(t,X,D_x) & & 0\\ & \ddots & \\ 0 & & \lambda_l'(t,X,D_x)\end{pmatrix}+(b_{jk}(t,X,D_x)), \tag{3.4}$$

where $\lambda_j'(t,x,\xi)$ $(j=1,\cdots,l)$ is one of $\lambda_k(t,x,\xi)$, $k=1,\cdots,m$, and $b_{jk}(t,x,\xi)$ is a symbol in $G^{(\kappa)}([0,T];S^{\sigma}_{G(\kappa)})$ with $\sigma=(r-q)/r$. In the following we assume $\sigma\kappa<1$. Then, as in the C^∞ case ([5], pp. 185–186) the fundamental solution $E(t,s)$ of $\mathscr{L}$ in (3.4) is constructed in the form

$$\begin{aligned}E(t,s)=\sum_{j=1}^{m}I_{\phi_j}(t,s)+\sum_{\nu=1}^{\infty}\sum_{\substack{j_k=1,\cdots,m\\(k=1,\cdots,\nu+1)}}\int_s^t\int_s^{t_1}\cdots\int_s^{t_{\nu-1}}I_{\phi_{j_1}}(t,t_1)\\ \times W_{j_2,\phi_{j_2}}(t_1,t_2)\cdots W_{j_{\nu+1},\phi_{j_{\nu+1}}}(t_\nu,s)dt_\nu\cdots dt_1 \qquad (t_0=t),\end{aligned} \tag{3.5}$$

where $W_{j,\phi_j}(t,s)$ is a Fourier integral operator with phase function $\phi_j(t,s;x,\xi)$ and with symbol $w_j(t,s;x,\xi)$ in $S^{\sigma}_{G(\kappa)}$ (for any t and s). Here, $\phi_j(t,s;x,\xi)$ is a phase function corresponding to $\lambda_j(t,x,\xi)$ $(j=1,\cdots,m)$. Let V be a closed set in $T^*(\boldsymbol{R}^n_x)$ and let $\Gamma_\varepsilon(t,V)$ be the set defined by (9). Then, we have

Theorem 3.1. *Let $\varepsilon>0$ and let V be a closed conic set in $T^*(\boldsymbol{R}^n_x)$. Let $a(x,\xi)$ and $b(x,\xi)$ be symbols in $S^0_{G(\kappa)}$ satisfying*

$$\begin{cases}\operatorname{supp} b\subset\{(x,\xi);\,|x-y|\leqq\varepsilon/2,\ |\xi/|\xi|-\eta/|\eta||\leqq\varepsilon/2,\ \textit{for } (y,\eta)\in V\},\\ |x-y|\geqq\varepsilon/2\ \textit{or}\ |\xi/|\xi|-\eta/|\eta||\geqq\varepsilon/2\ \textit{if}\ (x,\xi)\in\operatorname{supp} a\\ \qquad\qquad\textit{and}\ (y,\eta)\in\Gamma_\varepsilon(t,V).\end{cases} \tag{3.6}$$

Then, we have for the fundamental solution $E(t,s)$ of (3.5)

$$AE(t,0)B\in\mathscr{R}_{G(\kappa)},$$

where $A=a(X,D_x)$ and $B=b(X,D_x)$.

The proof will be given in the forthcoming joint paper [10] with Y. Morimoto.

Using Theorem 3.1 we can prove Theorem 2 as follows: Let $G={}^t(g_1,\cdots,g_l)$ for $g_j\in\mathscr{D}^{\{\kappa\}\prime}_{L^2}$ and let $V=\bigcup_{j=1}^{l}WF_{G(\kappa_1)}(g_j)$. Let $U(t)={}^t(u_1(t),\cdots,u_l(t))$ be a solution of (3.3). Then, what we have to prove is the following

$$(x_o,\xi_o)\notin WF_{G(\kappa_1)}(U(t)) \qquad \text{if } (x_o,\xi_o)\notin\Gamma(t,V), \tag{3.7}$$

where $WF_{G(\kappa_1)}(U(t))=\bigcup_{j=1}^{l}WF_{G(\kappa_1)}(u_j(t))$. Assume (x_o,ξ_o) does not belong to $\Gamma(t, V)$. Then, there exists an $\varepsilon>0$ such that $(x_o, \xi_o)\neq\Gamma_\varepsilon(t, V)$. Since $\Gamma_\varepsilon(t, V)$ and V are closed conic sets we can find symbols $a(x, \xi)$ and $b(x, \xi)$ in $S^0_{G(\kappa)}$ satisfying (3.6), $b(x, \xi)=1$ in a conic neighborhood of V and $a(x_o, \rho\xi_o)\neq 0$ for $\rho\geqq 1$. Since $V=\bigcup_{j=1}^{l} WF_{G(\kappa_1)}(g_j)$ and $b(x, \xi)\in S^0_{G(\kappa)}$, we have $(I-B)G\in\gamma^{(\kappa_1)}$. On the other hand, from Theorem 3.1 we have $AE(t, 0)BG\in\gamma^{(\kappa)}$. Combining these results we obtain

$$AU(t)=AE(t, 0)G=AE(t, 0)BG+AE(t, 0)(I-B)G\in\gamma^{(\kappa_1)}.$$

This proves (3.7) and (10).

References

[1] M. D. Bronstein, The Cauchy problem for hyperbolic operators with characteristics of variable multiplicity, Trans. Moscow Math. Soc., 1982, 87–103.

[2] L. Hörmander, Uniqueness theorems and wave front sets for solutions of linear differential equations with analytic coefficients, Comm. Pure Appl. Math., **24** (1971), 671–704.

[3] H. Komatsu, Ultradistributions, I, Structure theorems and a characterization, J. Fac. Sci. Univ. Tokyo Sect. IA, **20** (1973), 25–105.

[4] H. Kumano-go, Pseudo-differential Operators, the MIT Press, Cambridge, Massachusetts, and London, England, 1981.

[5] H. Kumano-go and K. Taniguchi, Fourier integral operators of multi-phase and the fundamental solution for a hyperbolic system, Funkcial. Ekvac., **22** (1979), 161–196.

[6] H. Kumano-go, K. Taniguchi and Y. Tozaki, Multi-products of phase functions for Fourier integral operators with an application, Comm. Partial Differential Equations, **3** (1978), 349–380.

[7] J. Leray et Y. Ohya, Systèmes linéaires, hyperbolique non stricts, Colloque de Liège, 1964, C.N.R.B., 105–144.

[8] S. Mizohata, Lecture notes in Chaina, 1983.

[9] ——, Propagation de la régularité au sens de Gevrey pour les opérateures différentiels à multiplicité constante, Séminaire sur les équations aux dérivés partielles hyperboliques et holomorphes à Paris VI, 1984, Hermann, Paris, 1985, pp. 106–133.

[10] Y. Morimoto and K. Taniguchi, Propagation of wave front sets of solutions of the Cauchy problem for hyperbolic equations in Gevrey classes, to appear.

[11] K. Taniguchi, Fourier integral operators in Gevrey class on R^n and the fundamental solution for a hyperbolic operator, Publ. RIMS, Kyoto Univ., **20** (1984), 491–542.

[12] ——, Pseudo-differential operators acting on ultradistribution, Math. Japon., **30** (1985), 719–741.

[13] S. Wakabayashi, Singularities of solutions of the Cauchy problem for hyperbolic systems in Gevrey classes, Japan. J. Math., **11** (1985), 157–201.

DEPARTMENT OF MATHEMATICS
UNIVERSITY OF OSAKA PREFECTURE
SAKAI, OSAKA 591, JAPAN

Taniguchi Symp. HERT
Katata 1984, pp. 395–414

Remification d'intégrales holomorphes

Dédié au Professeur Sigeru Mizohata pour son soixantième anniversaire

Jean VAILLANT

We consider the integral:

$$I(x)=\int_{S_q(x)} \mathcal{U}(x, \tau)d\tau;$$

$\mathcal{U}(x, \tau)$ is a holomorphic germ at the point $(y, 0) \in C^{n+1}\times C^q$; $\mathcal{U}$ is ramified around the hypersurface: $\varphi(x, \tau)=0$, $(x, \tau) \in \Omega_1\times\Omega_2$, Ω_i connected neighbourhood of zero, φ holomorphic, $y \in \Omega_1$ and is arbitrary near zero; $\varphi(0, x', 0)=x^1$, $(x=(x^0, x'))$. $S_q(x)$ is a simplex of dimension q and we integer on a relative cycle defined by this simplex and its faces.

We obtain that $I(x)$ is ramified around the hypersurface: $\Delta(x)=0$, $x\in\mathcal{V}_1$, $\mathcal{V}_1$ connected neighbourhood of zero, $\mathcal{V}_1\subset\Omega_1$; note $V_x=\{\tau;\varphi(x, \tau)=0\}$; if x belongs to $\{x; \Delta(x)=0\}$, either V_x has a singular point, or an edge of the simplex is tangent to V_x.

§ 0. Introduction

φ est un germe de fonction holomorphe à l'origine de $C^{n+1}\times C^q$; plus précisément Ω_1 [resp. Ω_2] est un voisinage connexe de l'origine dans C^{n+1} [resp. C^q] et φ est holomorphe dans $\Omega_1\times\Omega_2$; $x \in \Omega_1$, $\tau \in \Omega_2$,

$$\varphi(0, x', 0)=x^1, \quad (\text{avec } x=(x^0, x')) \text{ et } D_\tau\varphi(0, 0)=0.$$

$\mathcal{U}(x, \tau)$ est un germe de fonction holomorphe au point $(y, 0) \in \Omega_1\times\Omega_2$ $(y^0=0,\ y^1\neq 0)$ ramifié autour de V: $\varphi(x, \tau)=0$; y peut être choisi aussi près que l'on veut de 0.

On considère l'intégrale:

$$\begin{aligned} I(x)&=\int_{S_q(x)} \mathcal{U}(x, \tau)d\tau \\ &=\int_0^{x_0} d\tau_1\cdots\int_0^{x^0-\Sigma_{1\le k\le j-1}\tau^k} d\tau_j\cdots\int_0^{x^0-\Sigma_{1\le k\le q-1}\tau^k} \mathcal{U}(x, \tau)d\tau_q; \end{aligned}$$

elle définit un germe de fonction holomorphe au point y, dont nous nous proposons d'étudier la ramification.

Une telle intégrale apparaît, en particulier dans la représentation des solutions du problème de Cauchy à données singulières pour des opérateurs aux dérivées partielles holomorphes [3], [7], [8]. Dans [7], [8] nous avons étudié le cas de l'intégrale double pour φ polynômiale en τ et holomorphe en x. T. Koyayashi [5] étudie le cas général en employant le théorème d'isotopie de Thom [2] sous une forme qui le conduit à une compactification nécessaire pour rendre la projection propre. Dans [9] nous considérons aussi le cas général sous des hypothèses de "finitude" de Weierstrass qui nous permettent de contrôler les zéros; J. Leray [6] avait traité le cas où φ est polynômiale en (x, τ); nous prolongeons sa méthode au cas où φ est holomorphe en (x, τ). Les propositions essentielles concernent les variétés holomorphes polaires (ou duales); notons $V_x=\{\tau;\ \varphi(x, \tau)=0\}$; si, pour tout x, V_x n'a aucune composante développable, nous exprimons sous forme d'annulation d'un discriminant itéré, (l'appui), le fait qu'un plan soit tangent à V_x ou passe par un point singulier de V_x, (on utilise à ce propos un lemme de Henry, Merle, Sabbah [4] (cf. [1]). Ce résultat "géométrique" ramène, aux récurrences près, la démonstration de notre théorème au cas d'une intégrale simple. On obtient finalement que $I(x)$ se ramifie autour d'une hypersurface $\Delta_{S_q}(x)=0$; dire que x appartient à cette hypersurface implique que V_x a un point singulier ou bien qu'une des arêtes du simplexe $S_q(x)$ sur lequel on intègre est tangente à V_x.

Les détails des démonstrations seront publiés prochainement.

§ 1. Hypersurfaces discriminantes et appuis

1.1. $f(u, \sigma)$ est un germe de fonction holomorphe à l'origine de $\boldsymbol{C}^l \times \boldsymbol{C}$; $f(0, \sigma)\not\equiv 0$, $f(0, 0)=0$; $\pi(u, \sigma)$ est le polynôme de Weierstrass en σ de f: pour que f soit réduit, il faut et il suffit que:

$$\operatorname{discr}_\sigma \pi(u, \hat{\sigma})\not\equiv 0,$$

où $\operatorname{discr}_\sigma$ désigne le discriminant du polynôme en σ.

1.2. $(x, \tau) \in \boldsymbol{C}^{n+1}\times \boldsymbol{C}^q$; l'ensemble des hyperplans affines F de $\boldsymbol{C}^q$ d'équations:

$$\tau_1+a^2\tau_2+\cdots+a^q\tau_q+\sigma=0$$

s'identifie à $\boldsymbol{C}^q$; à F correspond: $(a, \sigma)=(a^2, \cdots, a^q, \sigma) \in \boldsymbol{C}^q$. $f(x, a, \sigma)$ est maintenant un germe de fonction holomorphe à l'origine de $\boldsymbol{C}^{n+1}\times \boldsymbol{C}^q$ et tel que: $f(0, 0, \sigma)\not\equiv 0$, $f(0, 0, 0)=0$; f_r est le germe réduit correspondant, de polynôme de Weierstrass π_r; $\mathrm{discr}_\sigma\, \pi_r(x, a, \hat{\sigma})$ est un germe à l'origine de $\boldsymbol{C}^{n+1}\times\boldsymbol{C}^{q-1}$; il résulte de 1.1 que ce germe n'est pas identiquement nul; il se décompose de façon unique, à un facteur inversible près, en facteurs irréductibles dans l'anneau des germes de fonctions holomorphes à l'origine de $\boldsymbol{C}^{n+1}\times\boldsymbol{C}^{q-1}$; le produit des facteurs irréductibles, distincts, indépendants des a_j, de $\mathrm{discr}_\sigma\, \pi_r(x, a, \hat{\sigma})$ sera noté:

$$\delta f(x, \hat{F});$$

$\delta f(x, \hat{F})=0$ est un germe d'hypersurface analytique réduite à l'origine de $\boldsymbol{C}^{n+1}$ appelé germe d'hypersurface discriminante de f; il résultera des hypothèses que, dans la suite $\delta f(0, \hat{F})=0$, de sorte que cette hypersurface ne sera pas vide.

Dire que: $\delta f(\underset{\sim}{x}, \hat{F})=0$ signifie que $\underset{\sim}{x}$ appartient au plus grand germe d'hypersurface analytique réduit, tel que pour chaque point x de cette hypersurface, le germe $f(x, F)$ réduit génériquement ne soit plus réduit en $\underset{\sim}{x}$. On appliquera cette remarque à une famille holomorphe en x d'ensembles polaires: $f(x, F)=0$.

La définition précédente se généralise facilement en remplaçant $\boldsymbol{C}^{n+1}$ par une variété holomorphe (lisse) de dimension finie et 0 par un point de cette variété.

Il résulte du théorème de Weierstrass qu'il existe des voisinages connexes de 0 dans $\boldsymbol{C}^{n+1}$, $\boldsymbol{C}^{q-1}$, $\boldsymbol{C}$: Ω_1, $\mathscr{A}$, $\mathscr{S}$, tels que: $\forall(x, a) \in \Omega_1\times\mathscr{A}$, le nombre de zéros de f, inclus dans $\mathscr{S}$, comptés avec leur multiplicité, soit constant, égal à l'ordre du zéro nul de $f(0, 0, \sigma)=0$ et que, $\forall(x, a) \in \{(x, a) \in \Omega_1\times\mathscr{A};\ \mathrm{discr}_\sigma\, \pi_r(x, a, \hat{\sigma})\neq 0\}$, le nombre de zéros distincts de f soit constant égal au degré de π_r; de tels voisinages seront appelés voisinages de Weierstrass. Un voisinage Ω_1 tel que δf soit défini sera appelé un voisinage permis.

Enfin on notera: $\tilde{f}(x, \alpha, \sigma)=f(x, \alpha\sigma, \sigma)$ et $\tilde{F}\simeq(\alpha, \sigma) \in \boldsymbol{C}^q$; on a le lemme:

$$\text{si } x \text{ est tel que: } \delta\tilde{f}(x, \hat{\tilde{F}})=0, \text{ alors: } f(x, 0, 0)\cdot\delta f(x, \hat{F})=0.$$

1.3. La grassmannienne des $q-j$-plans affines de $\boldsymbol{C}^q$ est noté Γ^{q-j}, $0\leq j\leq q$. C'est une variété holomorphe de dimension $j(q+1-j)$.

On désigne dans $\boldsymbol{C}^q$, par T_i, $1\leq i\leq q$, l'hyperplan:

$$\tau_i=0$$

et par T_0^0 l'hyperplan:

$$\sum_{k=1}^{k=q}\tau_k=0;$$

pour des: $(i_1, \cdots, i_k, \cdots, i_j)$ tels que:

$$1 \leq i_1 < \cdots i_k \cdots < i_j \leq q,$$

on note:

$$T_{i_1 \cdots i_j} = T_{i_1} \cap T_{i_2} \cap \cdots \cap T_{i_j} \in \Gamma^{q-j},$$
$$T^0_{i_1 \cdots i_j 0} = T_{i_1 \cdots i_j} \cap T^0_0 \in \Gamma^{q-j-1}, \quad j \leq q-1.$$

$f(x, F_{i_1 \cdots i_j})$ est un germe de fonction holomorphe au point:

$(0, T_{i_1 \cdots i_j})$ de $C^{n+1} \times \Gamma^{q-j}$; soit i_l un des indices i_k.

Il existe des hyperplans $F_{i_1}, F_{i_2}, \cdots, F_{i_{l-1}}, F_{i_{l+1}}, \cdots, F_{i_j}, F_{i_l}$, d'équations:

$$F_{i_1}: \tau_{i_1} + \sum_{k \neq i_1} a^k_{i_1} \tau_k + \sigma_{i_1} = 0$$
$$\vdots$$
$$F_{i_{l-1}}: \tau_{i_{l-1}} + \sum_{k \notin \{i_1, \cdots, i_{l-1}\}} a^k_{i_{l-1}} \tau_k + \sigma_{i_{l-1}} = 0$$

$$F_{i_{l+1}}: \tau_{i_{l+1}} + \sum_{k \notin \{i_1, \cdots, i_{l-1}, i_{l+1}\}} a^k_{i_{l+1}} \tau_k + \sigma_{i_{l+1}} = 0$$
$$\vdots$$
$$F_{i_j}: \tau_{i_j} + \sum_{k \notin \{i_1, \cdots, i_{l-1}, i_{l+1}, \cdots, i_j\}} a^k_{i_j} \tau_k + \sigma_{i_j} = 0$$

$$F_{i_l}: \tau_{i_l} + \sum_{k \notin \{i_1, \cdots, i_j\}} a^k_{i_l} \tau_k + \sigma_{i_l} = 0$$

et tels que:

$$F_{i_1 \cdots i_j} = F_{i_1} \cap F_{i_2} \cap \cdots \cap F_{i_{l-1}} \cap F_{i_{l+1}} \cap \cdots \cap F_{i_j} \cap F_{i_l}.$$

De même tout $q-j+1$ plan voisin de $T_{i_1 \cdots i_{l-1} i_{l+1} \cdots i_j}$ peut s'écrire:

$$F_{i_1 \cdots i_{l-1} i_{l+1} \cdots i_j} = F_{i_1} \cap F_{i_2} \cap \cdots \cap F_{i_{l-1}} \cap F_{i_{l+1}} \cap \cdots \cap F_{i_j}.$$

Le germe f induit au point $(0, T_{i_1}, \cdots, T_{i_j})$ un germe $f \circ p$ tel que:

$$f \circ p(x, F_{i_1}, \cdots, F_{i_j}) = f(x, F_{i_1} \cap \cdots \cap F_{i_j});$$

supposons la condition de Weierstrass:

$$f(0, T_{i_1 \cdots i_{l-1} i_{l+1} \cdots i_j} \cap \dot{F}_{i_l}) \not\equiv 0,$$

où $\dot{F}_{i_l}: \tau_{i_l} + \sigma_{i_l} = 0$; le § 2.2 nous permet de définir:

$$\delta(f \circ p)(x, F_{i_1}, \cdots, F_{i_{l-1}}, F_{i_{l+1}}, \cdots, F_{i_j}, \hat{F}_{i_l}).$$

D'autre part le germe f induit au point $(0, T_{i_1, \cdots i_{l-1} i_{l+1} \cdots i_j}, T_{i_l}) \in C^{n+1} \times$

$\Gamma^{q-j+1} \times \boldsymbol{C}^{q-j+1}$ le germe f' tel que:

$$f'(x, F_{i_1 \ldots i_{l-1} i_{l+1} \ldots i_j}, F_{i_l}) = f(x, F_{i_1} \cap \cdots \cap F_{i_j})$$

et:

$$f'(0, T_{i_1 \ldots i_{l-1} i_{l+1} \ldots i_j}, \dot{F}_{i_l}) \not\equiv 0;$$

le § 1.2 nous permet de définir:

$$\delta f'(x, F_{i_1 \ldots i_{l-1} i_{l+1} \ldots i_j}, \hat{F}_{i_l}).$$

On obtient alors l'équivalence:

$$\delta f'(x, F_{i_1 \ldots i_{l-1} i_{l+1} \ldots i_j}, \hat{F}_{i_l}) = 0$$

équivaut à:

$$\delta(f \circ p)(x, F_{i_1}, \cdots, F_{i_{l-1}}, F_{i_{l+1}}, \cdots, F_{i_j}, \hat{F}_{i_l}) = 0,$$

on notera:

$$\delta f(x, F_{i_1 \ldots i_{l-1} i_{l+1} \ldots i_j} \cap \hat{F}_{i_l}) \equiv \delta f'(x, F_{i_1 \ldots i_{l-1} i_{l+1} \ldots i_j}, \hat{F}_{i_l}).$$

On a ainsi fait correspondre au germe $f(x, F_{i_1 \ldots i_j})$ au point $(0, T_{i_1 \ldots i_j}) \in \boldsymbol{C}^{n+1} \times \Gamma^{q-j}$ un germe de fonction holomorphe:

$$\delta f(x, F_{i_1 \ldots i_{l-1} i_{l+1} \ldots i_j} \cap \hat{F}_{i_l})$$

au point:

$$(0, T_{i_1 \ldots i_{l-1} i_{l+1} \ldots i_j}) \in \boldsymbol{C}^{n+1} \times \Gamma^{q-j+1},$$

on dira encore que δf est défini dans un voisinage permis (§ 1.2) et que $(x, F_{i_1 \ldots i_{l-1} i_{l+1} \ldots i_j})$ de ce voisinage est permis.

On fait de même correspondre au germe de fonction holomorphe: $f(x, F_{i_1 \ldots i_j 0})$ au point $(0, T^0_{i_1 \ldots i_j 0}) \in \boldsymbol{C}^{n+1} \times \Gamma^{q-j-1}$, le germe de fonction holomorphe $\delta f(x, F_{i_1 \ldots i_{l-1} i_{l+1} \ldots i_j 0} \cap \hat{F}_{i_l})$ au point:

$$(0, T^0_{i_1 \ldots i_{l-1} i_{l+1} \ldots i_j 0}) \in \boldsymbol{C}^{n+1} \times \Gamma^{q-j},$$

pourvu que soit réalisée la condition de Weierstrass:

$$f(0, T^0_{i_1 \ldots i_{l-1} i_{l+1} \ldots i_j 0} \cap \dot{F}_{i_l}) \not\equiv 0.$$

1.4. $(x, \tau) \in \Omega_1 \times \Omega_2 \mapsto \varphi(x, \tau) \in \boldsymbol{C}$, $(\Omega_1 = \{x \in \boldsymbol{C}^{n+1};\ \sup |x_i| < R_1\}$, $\Omega_2 = \{\tau \in \boldsymbol{C}^q;\ \sup |\tau_j| < R_2\})$ définit un germe de fonction holomorphe à l'origine de $\boldsymbol{C}^{n+1} \times \boldsymbol{C}^q$; on suppose que:

$$\varphi(0, 0) = D_\tau\varphi(0, 0) = 0;$$

on pose: $x = (x^0, x') \in \boldsymbol{C} \times \boldsymbol{C}^n$; on suppose que:

$$\varphi(0, x', 0) = x^1$$

et par suite que:

$$D_{x^1}\varphi(0, 0) = 1 \neq 0.$$

On note:

$$V\text{: le germe d'hypersurface: } \varphi(x, \tau) = 0.$$

Définition. (x, τ) s'appuie sur V si et seulement si

$$\Delta_0(x, \tau) = \varphi(x, \tau) = 0.$$

1ères conditions de Weierstrass

On suppose que: $\forall j,\ \Delta_0(0, T_{1\cdots(j-1)(j+1)\cdots q} \cap \dot{F}_j) \not\equiv 0$

$$\forall j, k \quad j < k,\ \Delta_0(0, T^0_{1\cdots(j-1)(j+1)\cdots(k-1)(k+1)\cdots q0} \cap \dot{F}_j) \not\equiv 0.$$

Définition.

$$(\Delta_1)_{\hat{j}}(x, F_{1\cdots(j-1)(j+1)\cdots q}) = \delta\Delta_0(x, F_{1\cdots(j-1)(j+1)\cdots q} \cap \hat{F}_j)$$

$$(\Delta_1)_{\hat{j}}(x, F_{1\cdots(j-1)(j+1)\cdots(k-1)(k+1)\cdots q0}) = \delta\Delta_0(x, F_{1\cdots(j-1)(j+1)\cdots(k-1)(k+1)\cdots q0} \cap \hat{F}_j).$$

Si la première expression est mulle, on dit que:

$$(x, F_{1\cdots(j-1)(j+1)\cdots q}) \text{ s'appuie sur } V$$

et de même si la deuxième est nulle, on dit que:

$$(x, F_{1\cdots(j-1)(j+1)\cdots(k-1)(k+1)\cdots q0}) \text{ s'appuie sur } V.$$

Le premier germe est un germe au point $(0, T_{1\cdots(j-1)(j+1)\cdots q}) \in \boldsymbol{C}^{n+1} \times \Gamma^1$; le 2ê germe est un germe au point $(0, T_{1\cdots(j-1)(j+1)\cdots(k-1)(k+1)\cdots q0}) \in \boldsymbol{C}^{n+1} \times \Gamma^1$; les voisinages où ils sont définis sont appelés permis, ainsi que les x et les droites appartenant à ces voisinages; les points de V dans le voisinage de Weierstrass correspondant à une droite permise sont dits permis. Les notations ne sont pas ambiguës pourvu que l'on indique les variables dans le Δ_1.

On obtient de même par récurrence des conditions de Weierstrass que l'on suppose satisfaites et on définit les appuis correspondants.

Conditions de Weierstrass et appuis généraux

Pour toutes les suites: $1 \leq l_1 < \cdots < l_k \leq q$, on suppose:

$$(\Delta_{k-1})_{\hat{l}_2 \dots \hat{l}_k}(0, T_{1\dots(l_1-1)(l_1+1)\dots(l_2-1)(l_2+1)\dots(l_k-1)(l_k+1)\dots q} \cap \dot{F}_{l_1}) \not\equiv 0;$$

on définit:

$$(\Delta_k)_{\hat{l}_1 \dots \hat{l}_k}(x, F_{1\dots(l_1-1)(l_1+1)\dots(l_k-1)(l_k+1)\dots q})$$
$$= \delta(\Delta_{k-1})_{\hat{l}_2 \dots \hat{l}_k}(x, F_{1\dots(l_1-1)(l_1+1)\dots(l_k-1)(l_k+1)\dots q} \cap \hat{F}_{l_1});$$

Si cette expression est nulle, (x, le k-plan correspondant) s'appuie sur V; un voisinage de définition de Δ_k est dit permis ainsi que les x et k-plans de ce voisinage.

On remarque que l'on obtient à la fin:

$$(\Delta_q)_{\hat{1}\dots\hat{q}}(x).$$

De même, pour toutes les suites: $1 \leq l_1 < \cdots < l_{k+1} \leq q$, on suppose:

$$(\Delta_{k-1})_{\hat{l}_2 \dots \hat{l}_k}(0, T^0_{1\dots(l_1-1)(l_1+1)\dots(l_{k+1}-1)(l_{k+1}+1)\dots q0} \cap \dot{F}_{l_1}) \not\equiv 0;$$

on définit:

$$(\Delta_k)_{\hat{l}_1 \dots \hat{l}_k}(x, F_{1\dots(l_1-1)(l_1+1)\dots(l_{k+1}-1)(l_{k+1}+1)\dots q0})$$
$$= \delta(\Delta_{k-1})_{\hat{l}_2 \dots \hat{l}_k}(x, F_{1\dots(l_1-1)(l_1+1)\dots(l_{k+1}-1)(l_{k+1}+1)\dots q0} \cap \hat{F}_{l_1}).$$

Si cette expression est nulle, (x, le k-plan correspondant) s'appuie sur V; le voisinage de définition de Δ_k est dit permis ainsi que les x et les k-plans de ce voisinage.

1.5. On note T_0 l'hyperplan de $\boldsymbol{C}^q$:

$$\sum_{k=1}^{k=q} \tau_k - x^0 = 0$$

et on pose, si $1 \leq i_1 < \cdots < i_j \leq q$:

$$T_{i_1 \dots i_j 0} = T_{i_1 \dots i_j} \cap T_0;$$

enfin:

$$F_{[j]} = F_{1\dots j}, \quad \text{où} \quad F_k\colon \tau_k + \sum_{l>k} a_k^l \tau_l + \sigma_k = 0.$$

On définit le simplexe $S_{q-j}(x, F_{[j]})$, $0 \leq j \leq q$, de dimension $q-j$, s'il est non dégénéré, par ses $q-j+1$ sommets:

$$A_j = F_{[j]} \cap T_{j+1\cdots q},$$
$$A_j^l = F_{[j]} \cap T_{(j+1)\cdots(l-1)(l+1)\cdots q0}, \quad j < l \leq q.$$

Une arête de dimension k de S_{q-j} est le k-plan affine défini par $k+1$ sommets.

Si F_j prend la valeur T_j, A_j et A_j^l deviennent A_{j-1} et A_{j-1}^l. Si F_j passe par A_{j-1}^j, $\sigma_j = -\theta_j$, où θ_j est la $j^{\text{ème}}$ coordonnée de A_{j-1}^j:

$$\theta_j = x^0 + \sigma_1 + \cdots + \sigma_{j-1} + h_j(a_k^l, \sigma_k), \quad 1 \leq k \leq j-1,$$

où h_j est une fonction holomorphe des (a, σ) nulle pour $(a) = 0$. Le simplexe $S_q(x)$ ne dépend que de $T_1, \cdots, T_q$, et $T_0(x)$.

Définition. $\Delta_{S_{q-j}}(x, F_{[j]})$ est le germe au point $(0, T_{[j]})$ obtenu en effectuant le produit des germes d'appuis des arêtes distinctes de S_{q-j}. $\Delta_{S_q}(x)$ est un germe à l'origine de $\boldsymbol{C}^{n+1}$.

§ 2. Appuis et contacts

2.1. Pour $x \in \Omega_1$, V_x est l'hypersurface analytique des $\tau \in \Omega_2$ tels que: $\varphi(x, \tau) = 0$.

Supposons dans le 2.1 que $V_{\underline{x}}$ *a un point quadratique* $\underline{\tau}$, *dont l'hyperplan tangent est* $\underline{F}$ *et que* $\underline{F} \cap \underline{G}$ *est un* $q-2$ *plan tangent passant par* $\underline{\tau}$ *et à contact quadratique.*

On suppose aussi que $\underline{F}$ a pour équation:

$$\tau_1 + \langle \underline{a}, \tau' \rangle + \underline{\sigma} = 0, \qquad \tau' = (\tau_2, \cdots, \tau_q),$$

et que $\underline{G}$ a pour équation:

$$\tau_2 + \langle \underline{b}, \tau'' \rangle + \underline{s} = 0, \qquad \tau'' = (\tau_3, \cdots, \tau_q).$$

Il résulte du théorème des fonctions implicites que pour (x, τ) voisin de $(\underline{x}, \underline{\tau})$, V a l'équation:

$$\tau_1 + \psi(x, \tau') = 0,$$

que $\underline{F}$ a les coordonnées:

$$\underline{a} = D_{\tau'}\psi(\underline{x}, \underline{\tau}'), \qquad \underline{\sigma} = \psi(\underline{x}, \underline{\tau}') - \langle \underline{a}', \underline{\tau}' \rangle.$$

Par continuité pour x voisin de $\underline{x}$, V_x a une composante non développable V_x' (c'est-à-dire possédant un point quadratique); si $\tau \in V_x'$ est quadratique, de plan tangent F: $\tau_1 + \langle a, \tau' \rangle + \sigma = 0$, il résulte du théorème d'inversion locale, que l'on a une bijection biholomorphe:

$$(x, \tau') \longmapsto (x, a)$$

(puisque Hess. $_{\tau}\psi(\underline{x}, \underline{\tau}') \neq 0$), d'inverse $(x, T(x, a))$ et que V'_x a localement une hypersurface polaire (duale) $\check{V}'_x$ d'équation

$$P(x, a, \sigma) \equiv \sigma + S(x, a) = 0,$$

où $S(x, a) = -\psi[x, T(x, a)] + \langle a, T(x, a)\rangle$; on a évidemment: $P(\underline{x}, \underline{a}, \underline{\sigma}) = 0$. $F \cap G$ sera un hyperplan de F voisin de $\underline{F} \cap \underline{G}$;

$$G\colon \tau_2 + \langle b, \tau''\rangle + s = 0.$$

On posera: $\xi = (a, \sigma)$, $\eta = (1, b, s)$.

Nous chercherons, *pour F voisin de $\underline{F}$, à quelle condition sur G voisin de $\underline{G}$, $G \cap F$ est-il tangent à V_x.*

Une considération géométrique simple montre que cette condition est la suivante:

(x, ξ, η) est tel qu'il existe $\rho \in \boldsymbol{C}$, vérifiant:

$$P(x, \xi + \rho\eta) = 0, \qquad \frac{d}{d\rho} P(x, \xi + \rho\eta) = 0; \tag{2.1}$$

ce sont les équations du contour apparent de $\check{V}'_x$ vu de ξ. Or P vérifie

$$P(\underline{x}, \underline{\xi}) = 0, \quad \frac{d}{d\rho} P(\underline{x}, \underline{\xi} + \rho\underline{\eta})(0) = 0, \quad \frac{d^2}{d\rho^2} P(\underline{x}, \underline{\xi} + \rho\underline{\eta})(0) \neq 0,$$

la deuxième équation exprimant que $\underline{F} \cap \underline{G}$ contient $\underline{\tau}$ et la dernière que le contact de $\underline{F} \cap \underline{G}$ avec $V_{\underline{x}}$ est quadratique. On peut donc appliquer le théorème de préparation et (avec i inversible):

$$P(x, \xi + \rho\eta) = i[P_0(x, \xi, \eta) + \rho P_1(x, \xi, \eta) + \rho^2];$$

on peut remplacer (2.1) par:

$$P_1^2(x, \xi, \eta) - 4P_0(x, \xi, \eta) = 0; \tag{2.2}$$

si on pose:

$$Q(x, \xi, \eta) = i^2(x, \xi, \eta, 0) P_1(x, \xi, \eta),$$

(2.2) s'écrit encore:

$$Q^2(x, \xi, \eta) - 4i^3(x, \xi, \eta, 0) P(x, \xi) = 0; \tag{2.3}$$

en appliquant encore le théorème de préparation:

$$Q(x, \xi, (1, b, s)) = j[Q_0(x, \xi, b) + s - \underline{s}],$$

avec $Q_0(\underline{x}, \underline{\xi}, \underline{b}) = 0$, on remplace (2.3) par:

$$[Q_0(x, \xi, b) + s - \underline{s}]^2 - i'(x, \xi, \eta)P(x, \xi) = 0, \tag{2.4}$$

i' inversible. On note alors $\gamma P(x, F \cap G)$ *le premier membre de* (2.4).

$$\gamma P(x, F \cap G) = 0,$$

exprime donc que $F \cap G$ *est tangent à* V_x.

On déduit de cette définition le **lemme**.

$$P(x, F) = 0 \ \textit{implique} \ \delta(\gamma P)(x, F \cap \hat{G}) = 0$$

où $\delta(\gamma P)$ *est ici un germe en* $(\underline{x}, \underline{F})$.

2.2. On en déduit les **propositions** suivantes

$1 \le k \le q-1$. *Si le* k*-plan* $F_{1\cdots(l_1-1)(l_1+1)\cdots(l_k-1)(l_k+1)\cdots q}$ *est permis et tangent à une composante non développable* V'_x *de* V_x *pour* x *permis, en un point* τ *permis, il y a appui sur* V:

$$(\Delta_k)_{\hat{l}_1\cdots\hat{l}_k}(x, F_{1\cdots(l_1-1)(l_1+1)\cdots(l_k-1)(l_k+1)\cdots q}) = 0.$$

On procède par une récurrence que nous résumerons. On obtient d'abord le résultat pour $k=1$. On le suppose vrai pour $k-1$. Supposons alors le k-plan permis:

$$\underline{F}_{1\cdots(l_1-1)(l_1+1)\cdots(l_k-1)(l_k+1)\cdots q}$$

tangent à $V'_{\underline{x}}$ en un point permis avec un contact que l'on supposera d'abord quadratique; le $k+1$-plan permis $\underline{F}_{1\cdots(l_1-2)(l_1+1)\cdots(l_k-1)(l_k+1)\cdots q}$ transverse à $V'_{\underline{x}}$ a une intersection de dimension k avec $V'_{\underline{x}}$ dont $\underline{\tau}$ est point quadratique avec pour espace tangent le k-plan précédent qui vérifie donc l'équation locale de la variété polaire de

$$\underline{F}_{1\cdots(l_1-2)(l_1+1)\cdots(l_k-1)(l_k+1)\cdots q} \cap V'_{\underline{x}},$$

soit:

$$P_k(\underline{x}, \underline{F}_{1\,2\cdots(l_1-1)(l_1+1)\cdots(l_k-1)(l_k+1)\cdots q}) = 0.$$

Si maintenant $\underline{F}_{1\cdots(l_2-1)(l_2+1)\cdots(l_k-1)(l_k+1)\cdots q}$ est un $k-1$-plan permis du k-plan précédent avec un contact quadratique, les $k-1$-plans voisins

$F_{1\cdots(l_2-1)(l_2+1)\cdots(l_k-1)(l_k+1)\cdots q}$ tangents à V'_x voisin sont, d'après le § 2.1, les "points" du germe d'hypersurface irréductible au point

$$(\underline{x}, \underline{F}_{1\cdots(l_2-1)(l_2+1)\cdots(l_k-1)(l_k+1)\cdots q}) \in C^{n+1} \times \Gamma^{k-1}$$

défini par:

$$\gamma P_k(x, F_{1\cdots(l_2-1)(l_2+1)\cdots(l_k-1)(l_k+1)\cdots q}) = 0,$$

tel que:

$$\delta(\gamma P_k)(\underline{x}, \underline{F}_{1\cdots(l_1-1)(l_1+1)\cdots(l_k-1)(l_k+1)\cdots q} \cap \hat{F}_{l_1}) = 0;$$

mais de l'hypothèse de récurrence résulte que les points de ce germe: $\gamma P_k = 0$ appartiennent nécessairement au germe défini au même point par:

$$(\Delta_{k-1})_{\hat{l}_2\cdots\hat{l}_k}(x, F_{1\cdots(l_2-1)(l_2+1)\cdots(l_k-1)(l_k+1)\cdots q}) = 0;$$

on en déduit (cf. § 1.4) que:

$$(\Delta_k)_{\hat{l}_1\cdots\hat{l}_k}(x, \underline{F}_{1\cdots(l_1-1)(l_1+1)\cdots(l_k-1)(l_k+1)\cdots q}) = 0.$$

Si le contact du k-plan n'est pas quadratique, par densité et continuité, $(x, F_{1\cdots(l_1-1)(l_1+1)\cdots(l_k-1)(l_k+1)\cdots q})$ vérifie l'équation *précédente*.

On obtient de même

$1 \leq k \leq q-1$. *Si le k-plan* $F_{1\cdots(l_1-1)(l_1+1)\cdots(l_{k+1}-1)(l_{k+1}+1)\cdots q0}$ *est permis et tangent à une composante non développable* V'_x *de* V_x *en un point permis, il y a appui sur* V, *c'est-à-dire que*:

$$(\Delta_k)_{\hat{l}_1\cdots\hat{l}_k}(x, F_{1\cdots(l_1-1)(l_1+1)\cdots(l_{k+1}-1)(l_{k+1}+1)\cdots q0}) = 0.$$

On a, enfin la **proposition** suivante.

On suppose que, pour tout x, V_x *n'a aucune composante développable. Soit* x *permis et* $\underline{F}_{1\cdots(l_1-1)(l_1+1)\cdots(l_k-1)(l_k+1)\cdots q}$ *un k-plan permis*; *on suppose que* $V_{\underline{x}}$ *a un point singulier permis, par où passe*: $\underline{F}_{1\cdots(l_1-1)(l_1+1)\cdots(l_k-1)(l_k+1)\cdots q}$, *alors*:

$$(\Delta_k)_{\hat{l}_1\cdots\hat{l}_k}(\underline{x}, \underline{F}_{1\cdots(l_1-1)(l_1+1)\cdots(l_k-1)(l_k+1)\cdots q}) = 0.$$

Il résulte, par exemple du théorème de Sard, la densité des x tels que V_x soit lisse et le k-plan précédent peut être approché par un k-plan tangent à V'_x lisse non développable; par continuité on obtient le résultat. On a la proposition analogue avec $F_{1\cdots(l_1-1)(l_1+1)\cdots(l_{k+1}-1)(l_{k+1}+1)\cdots q0}$.

2.3. On donnera d'abord une **proposition** qui est en un certain sens une réciproque de celle du 2.1

$\underline{x} \in \Omega_1$, Ω_1' est un voisinage ouvert de $\underline{x}$ tel que, pour $x \in \Omega_1'$, chaque V_x *est une hypersurface lisse*; $P(x, a, \sigma)$ est un polynôme de Weierstrass en σ, réduit, défini dans un voisinage de Weierstrass de $(\underline{x}, \underline{F}) \in \boldsymbol{C}^{n+1} \times \boldsymbol{C}^q$ (plus précisément le voisinage $\Omega_1' \times \mathscr{W}$). La partie $V_x \cap \Omega_2'$ de V_x formée des points de V_x, $x \in \Omega_1'$, dont le plan tangent F appartient à $\mathscr{W}$ est ouverte. On note P_x l'ensemble des $F \in \mathscr{W}$ tels que $P(x, F)=0$ et on suppose que P_x *est l'ensemble polaire de* $V_x \cap \Omega_2'$, alors:

$$\delta P(\underline{x}, \hat{F}) \neq 0.$$

On démontre cette proposition à l'aide du lemme suivant [1], [2] que nous avons un peu adapté.

Lemme. *Soit* σ *un zéro de* $P(\underline{x}, \underline{a}, \sigma)=0$; *l'hyperplan* $\underline{F} \simeq (\underline{a}, \underline{\sigma})$ *est tangent à* V_x *et sa direction est définie par* $\underline{a}$.

Pour tout voisinage assez petit de $\underline{a}$, *il existe* a *appartenant à ce voisinage, tel que chaque hyperplan tangent de direction* a *soit tangent à* $V_{\underline{x}}$ *en un seul point (c'est-à-dire que les* σ *correspondants sont distincts), avec un contact quadratique.*

2.4. On obtient ensuite les **propositions** suivantes.

On suppose que, pour tout x *permis,* V_x *n'a aucune composante développable.*

$1 \leq k \leq q-1$. *Si* $(\Delta_k)_{\hat{l}_1 \ldots \hat{l}_k}(x, F_{1 \ldots (l_1-1)(l_1+1) \ldots (l_k-1)(l_k+1) \ldots q})=0$ *alors, ou bien le k-plan permis* $F_{1 \ldots (l_1-1)(l_1+1) \ldots (l_k-1)(l_k+1) \ldots q})$ *est tangent à* V_x *en un point permis ou bien il contient un point singulier permis de* V_x.

$k=q$. *Si*: $(\Delta_q)_{\hat{1} \ldots \hat{q}}(x)=0$, V_x *a un point singulier permis.*

On procède par récurrence. Le résultat est évident pour $k=1$. On le suppose vrai pour $k-1$ et on veut démontrer alors que, si

$$F_{1 \ldots (l_1-1)(l_1+1) \ldots (l_k-1)(l_k+1) \ldots q}$$

permis n'est pas tangent à V_x en un point permis et ne contient pas de point singulier permis de V_x, alors:

2.4 (1)
$$(\Delta_k)_{\hat{l}_1 \ldots \hat{l}_k}(x, F_{1 \ldots (l_1-1)(l_1+1) \ldots (l_k-1)(l_k+1) \ldots q}) \equiv \delta(\Delta_{k-1})_{\hat{l}_2 \ldots \hat{l}_k}(x, F_{1 \ldots (l_1-1)(l_1+1) \ldots (l_k-1)(l_k+1)} \cap \hat{F}_{l_1}) \neq 0.$$

On remarque d'abord que:

$$V_x \cap F_{1 \ldots (l_1-1)(l_1+1) \ldots (l_k-1)(l_k+1) \ldots q}$$

est lisse dans l'ouvert permis et de dimension $k-1$; elle contient une partie ouverte dont les $k-1$ plans tangents sont permis et vérifient d'après le § 2.2, dont les hypothèses sont réalisées, l'équation suivante:

2.4 (2) $$(\Delta_{k-1})_{\hat{l}_2 \ldots \hat{l}_k}(x, F_{1\ldots(l_2-1)(l_2+1)\cdots(l_k-1)(l_k+1)\cdots q})=0;$$

réciproquement, si un tel $k-1$ plan vérifie cette équation, il résulte de l'hypothèse de récurrence qu'il est tangent à V_x, donc à l'intersection ci-dessus.

2.4 (2) est donc la polaire de la partie permise de l'intersection. Les hypothèse du § 2.3 sont réalisées, en remplaçant x par

$$(x, F_{1\ldots(l_1-1)(l_1+1)\cdots(l_k-1)(l_k+1)\cdots q})$$

et F par F_{l_1} et on a l'inégalité 2.4 (1).

On obtient de même:

On suppose que, pour tout x permis, V_x n'a aucune composante développable, $1 \leq k \leq q-1$. *Si:*

$$(\Delta_k)_{\hat{l}_1 \ldots \hat{l}_k}(x, F_{1\ldots(l_1-1)(l_1+1)\cdots(l_{k+1}-1)(l_{k+1}+1)\cdots q0})=0,$$

alors, ou bien le k-plan permis ci-dessus est tangent à V_x en un point permis ou bien il contient un point singulier permis de V_x.

Les résultats du § 2.2 et les précédents, montrent que, si V_x n'a aucune composante développable, l'appui est équivalent au contact ou à l'existence d'un point singulier.

On a les *conséquences* suivantes.

i) $0 \leq j \leq q-1$, $1 \leq k \leq q-j$, $l_1 > j+1$. Si:

$$\delta(\Delta_{k-1})_{\hat{l}_1 \ldots \hat{l}_{k-1}}(x, F_{[j]} \cap T_{(j+2)\cdots(l_1-1)(l_1+1)\cdots(l_{k-1}-1)(l_{k-1}+1)\cdots q} \cap \hat{F}_{j+1})=0$$

alors:

$$(\Delta_k)_{\widehat{j+1}\hat{l}_1 \ldots \hat{l}_{k-1}}(x, F_{[j]} \cap T_{(j+2)\cdots(l_1-1)(l_1+1)\cdots(l_{k-1}-1)(l_{k-1}+1)\cdots q})$$
$$\equiv (\Delta_k)_{\widehat{j+1}\hat{l}_1 \ldots \hat{l}_{k-1}}(x, A_j A_j^{j+1} \cdots A_j^{l_1} \cdots A_j^{l_k-1})=0.$$

ii) De même, $0 \leq j \leq q-2$, $1 \leq k \leq q-j-1$, $l_1 > j+1$, si:

$$\delta(\Delta_{k-1})_{\hat{l}_1 \ldots \hat{l}_{k-1}}(x, F_{[j]} \cap T_{(j+2)\cdots(l_1-1)(l_1+1)\cdots(l_k-1)(l_k+1)\cdots q0} \cap \hat{F}_{j+1})=0$$

alors:

$$(\Delta_k)_{\widehat{j+1}\hat{l}_1 \ldots \hat{l}_{k-1}}(x, F_{[j]} \cap T_{(j+2)\cdots(l_1-1)(l_1+1)\cdots(l_k-1)(l_k+1)\cdots q0})$$
$$\equiv (\Delta_k) \equiv_{\widehat{j+1}\hat{l}_1 \ldots \hat{l}_{k-1}}(x, A_j^{j+1} A_j^{l_1} \cdots A_j^{l_k})=0.$$

Pour obtenir i), on utilise à nouveau le § 2.3, et la remarque que si, $A_j A_j^{j+1} A_j^{l_1} \cdots A_j^{l_k-1}$ n'est pas tangent à V_x et ne contient pas de point singulier, alors la polaire de la partie permise de son intersection avec V_x est:

$$(\Delta_{k-1})_{\hat{l}_1 \ldots \hat{l}_{k-1}}(x, A_{j+1} A_{j+1}^{l_1} \cdots A_{j+1}^{l_k-1})=0,$$

d'après un raisonnement précédent (cf. formule 2.4 (2)). On obtient ii) de façon analogue.

iii) *Supposons que pour tout x permis, V_x n'a aucune composante développable. Si:*

$$\Delta_{S_q}(x)=0,$$

alors, ou bien une des arêtes de $S_q(x)$ est tangente à V_x en un point permis, ou bien V_x a un point singulier permis.

2.5. On se propose d'abord d'étudier le discriminant en F_{j+1} de $\Delta_{S_{q-j-1}}(x, F_{[j+1]})$.

On suppose que pour tout x permis, V_x n'a aucune composante développable.

Définition. $\Delta^{\vee}_{S_{q-j}}(x, F_{[j]})$ est le germe au point $(0, T_{[j]})$ obtenu en effectuant le produit des germes d'appuis des arêtes distinctes de S_{q-j} issues de A_j^{j+1}.

On obtient alors le

Lemme. $0 \leq j \leq q-1$. *La condition:*

$$\delta\Delta_{S_{q-j-1}}(x, F_{[j]} \cap \hat{F}_{j+1})=0$$

implique:

$$\Delta^{\vee}_{S_{q-j}}(x, F_{[j]})=0.$$

Il résulte des conséquences de la fin du § 2.4 et des définitions que l'hypothèse implique que, ou bien:

$$\Delta^{\vee}_{S_{q-j}}(x, F_{[j]})=0$$

ou bien $(x, F_{[j]})$ est tel qu'il existe un couple de conditions d'appuis de deux arêtes distinctes de S_{q-j-1} tel que les deux polynômes de Weierstrass en σ_{j+1} qu'elles définissent aient un facteur commun de degré ≥ 1 en σ_{j+1}.

Il suffit donc de démontrer que, si:

$$\Delta^{\vee}_{S_{q-j}}(x, F_{[j]}) \neq 0,$$

alors pour tout couple de conditions d'appuis de deux arêtes distinctes de S_{q-j-1}, les deux polynômes de Weierstrass en σ_{j+1} qu'elles définissent n'ont pas de facteur commun de degré ≥ 1 en σ_{j+1}.

L'inégalité précédente implique que les arêtes de dimension ≥ 1 du $q-j$-èdre de sommet A_j^{j+1} du simplexe S_{q-j} ne sont pas tangentes à V_x en un point permis et ne passent pas par un point singulier permis de V_x.

Les deux conditions d'appuis des arêtes sont de la forme (notation du § 1.3):

$$(\Delta'_{k-1})_{\hat{l}_1\cdots\hat{l}_{k-1}}[x, F_{[j]}\cap T_{(j+2)\cdots(l_1-1)(l_1+1)\cdots(l_{k-1}-1)(l_{k-1}+1)\cdots q},$$
$$(a_{j+1}^{j+2}, \cdots, a_{j+1}^{q}, \sigma_{j+1})]=0$$

ou une condition analogue contenant F_0.

Il résulte des démonstrations du § 2.4 que ce sont des polaires d'intersections lisses de V_x avec des arêtes issues de A_j^{j+1} du simplexe S_{q-j}. On en déduit à l'aide du lemme 2.3 que leurs hessiens respectifs ne sont identiquement nuls sur aucune de leurs composantes. La considération de la différence de leurs dimensions implique alors que les polynômes de Weierstrass en σ_{j+1} qu'elles définissent ne peuvent avoir de facteur commun de degré ≥ 1.

On obtient encore par des considérations de dimension de la variété polaire le deuxième **lemme.**

Si $\Delta_{S_{q-j-1}}(x, F_{[j]}\cap F_{j+1})=0$, pour tout hypreplan F_{j+1} permis passant par A_j^{j+1}, alors:

$$\Delta^{\vee}_{S_{q-j}}(x, F_{[j]})=0.$$

§ 3. Prolongements analytiques

3.1. Nous utiliserons le théorème de prolongement d'Hartogs, la Proposition 5.5 de [8] (p. 439) et le théorème sur la ramification en dimension 1 [6], [8] que nous rappellerons.

Théorème. *$f(u, \sigma)$ est un germe de fonction holomorphe à l'origine de $\boldsymbol{C}^l\times\boldsymbol{C}$ tel que: $f(0, \sigma)\not\equiv 0$; Ω est le polydisque de $\boldsymbol{C}^l$ de centre* 0, *de "rayon" R_1 et $\Omega_2=\{\sigma\in\boldsymbol{C}; |\sigma|<R_2\}$; $\Omega_1\times\Omega_2$ est un voisinage de Weierstrass de f et δf est holomorphe dans Ω_1; ici* (cf. § 1.2):

$$\delta f(u, \hat{\sigma})=(\mathrm{discr}_\sigma\, \pi_r(u, \hat{\sigma}))_r$$

θ: $\Omega_1\mapsto\Omega_2$ est une fonction holomorphe telle que $\theta(u_0)=0$, $u_0\in\Omega_1$. $\mathscr{U}$ est un germe de fonction holomorphe au point $(u_0, 0)\in\Omega_1\times\Omega_2$ se prolongeant analytiquement au revêtement simplement connexe de

$$\{(u, \sigma)\in\Omega_1\times\Omega_2; f(u, \sigma)\neq 0\}.$$

Alors l'intégrale, (*intégrale dans le plan complexe $\boldsymbol{C}_\sigma$ le long de segment joignant* 0 *et $\theta(u)$, pour u assez voisin de u_0*):

$$I(u)=\int_0^{\theta(u)}\mathscr{U}(u, \sigma)d\sigma$$

définit un germe de fonction holomorphe au point u_0, qui se prolonge *analytiquement au revêtement simplement connexe de:*

$$\{u \in \Omega_1; f(u, 0)\cdot f(u, \theta(u))\cdot \delta f(u, \hat{\sigma})\neq 0\}.$$

Dans notre étude u_0 pourra être choisi aussi voisin qu'on veut de 0 et $\theta(0)=0$.

Nous obtiendrons ensuite la:

Proposition. *$f(x, a, \sigma)$ est maintenant un germe de fonction holomorphe à l'origine de $\boldsymbol{C}^{n+1}\times\boldsymbol{C}^{q-1}\times\boldsymbol{C}$ tel que:*

$$f(0, 0, \sigma)\not\equiv 0, \qquad f(x, 0, 0)\not\equiv 0;$$

$\tilde{f}(x, \alpha, \sigma)=f(x, \alpha\sigma, \sigma)$ est holomorphe dans le voisinage de Weierstrass: $\Omega_1\times\tilde{\mathscr{A}}\times\mathscr{S}$ et $\delta f(x, \hat{F})$ est holomorphe dans Ω_1 (§ 1). θ: $\Omega_1\mapsto\mathscr{S}$ est une fonction holomorphe telle que $\theta(y)=\theta(0)=0$, $y\neq 0$, $y\in\Omega_1$; on suppose qu'il existe une fonction holomorphe $h(x)$ non identiquement nulle telle que:

$$\{x \in \Omega_1; f(x, a, \theta(x))=0, \forall a \text{ petit}, a \in \boldsymbol{C}\}\subset\{x \in \Omega_1; h(x)=0\}.$$

$\mathscr{U}(x, \alpha, \sigma)$ est un germe de fonction holomorphe au point:

$$(y, 0, 0) \in \Omega_1\times\tilde{\mathscr{A}}\times\mathscr{S},$$

se prolongeant analytiquement au revêtement simplement connexe de:

$$\{(x, \alpha, \sigma) \in \Omega_1\times\tilde{\mathscr{A}}\times\mathscr{S}; f(x, \alpha\sigma, \sigma)\neq 0\}.$$

L'intégrale:

$$I(x, \alpha)=\int_0^{\theta(x)} \mathscr{U}(x, \alpha, \sigma)d\sigma$$

définit un germe de fonction holomorphe au point $(y, 0)$.

On la suppose indépendante de α:

$$I(x, \alpha)=I(x).$$

Alors $I(x)$ se prolonge analytiquement au revêtement simplement connexe de:

$$\{x \in \Omega_1; f(x, 0, 0)\cdot h(x)\cdot \delta f(x, \hat{F})\neq 0\}.$$

En effet, il résulte du théorème précédent que $I(x, \alpha)$ se prolonge analytiquement au revêtement simplement connexe de:

$$\{(x, \alpha) \in \Omega_1\times\tilde{\mathscr{A}}; f(x, 0, 0)\cdot f(x, \alpha\theta(x), \theta(x))\cdot \operatorname{discr}_\sigma Q_r(x, \alpha, \hat{\sigma})\neq 0\}$$

où Q_r désigne le polynôme de Weierstrass du germe réduit de $\tilde{f}(x, \alpha, \sigma)$.

La proposition rappelée précédemment [8] (p. 439), le théorème d'Hartogs, le dernier lemme du § 1.2 impliquent que $I(x)$ se prolonge analytiquement au revêtement simplement connexe de:

$$\{x \in \Omega_1; f(x, 0, 0) \cdot h(x) \cdot \delta f(x, \hat{F}) \neq 0\}.$$

3.2. φ a été définie au § 1.4. $\mathcal{U}$ est un germe de fonction holomorphe au point $(y, 0) \in \Omega_1 \times \Omega_2$, $(y^0 = 0, y^1 \neq 0)$, se prolongeant analytiquement au revêtement simplement connexe de:

$$\{(x, \tau) \in \Omega_1 \times \Omega_2;\ \varphi(x, \tau) \neq 0\}.$$

On considère l'intégrale:

$$I(x) = \int_0^{x^0} d\tau_1 \cdots \int_0^{x^0 - \Sigma_{1 \leq k \leq j-1} \tau_k} d\tau_j \cdots \int_0^{x^0 - \Sigma_{1 \leq k \leq q-1} \tau_k} \mathcal{U}(x, \tau) d\tau_q;$$

elle définit un germe de fonction holomorphe au point y dont il s'agit d'étudier la ramification, (c'est-à-dire le prolongement analytique). Remarquons, que cette intégrale peut être considérée comme une intégrale sur un cycle singulier relatif à la réunion des faces du simplexe $S_q(x)$, de sorte qu'on la notera aussi:

$$I(x) = \int_{S_q(x)} \mathcal{U}(u, \tau) d\tau.$$

Les équations:

$$\tau_j + \sum_{j+1 \leq k \leq q} a_j^k \tau_k + \sigma_j = 0, \qquad 1 \leq j \leq q,$$

définissent pour des a assez petits une application holomorphe s_q:

$$(x, a_1, \cdots, a_{q-1}, \sigma_1, \cdots, \sigma_q) \longmapsto (x, \tau_1, \cdots, \tau_q).$$

On notera:

$$\mathcal{U}_q = \mathcal{U} \circ s_q;$$

c'est un germe de fonction holomorphe au point $(x = y,\ a = \sigma = 0)$; on considérera aussi:

$$\mathcal{U}_q(x, \alpha\sigma, \sigma)$$

obtenue en remplaçant chaque a_j^k par $\alpha_j^k \sigma_j$. On obtient la *formule de "changement de variables"*:

$$I(x) = \int_0^{-x^0} d\sigma_1 \cdots \int_0^{-\theta_j} d\sigma_j \cdots \int_0^{-\theta_q} \mathcal{U}_q(x, \alpha\sigma, \sigma) i_q(\alpha, \sigma) d\sigma_q,$$

pour α assez petit, où i_q est holomorphe voisin de $(-1)^q$ et où

$$\theta_j(x^0, a_1, \cdots, a_{j-1}, \sigma_1, \cdots, \sigma_{j-1})$$

a été défini au § 1.5.

De plus
$$\int_0^{-\theta_j} d\sigma_j \cdots \int_0^{-\theta_q} \mathcal{U}_q i_q d\sigma_q$$

ne dépend pas des α_k, $k \geq j$ et définit un germe holomorphe au point: $(x=y, \alpha=0, \sigma=0)$.

On obtient cette formule, soit en utilisant directement la formule de dérivation de l'intégrale de Leray [6], soit en la redémontrant comme dans [8].

Elle nous permettra d'obtenir le

3.3. Théorème. *On fait les hypothèse suivantes:*

H1 *Vx, V_x n'a aucune composante développable.*

H2 *Les conditions de Weierstrass permettant de définir les appuis sont satisfaites* (cf. § 1.4).

H3 $\Delta_{S_q}(x) \not\equiv 0$ *(sinon le résultat est vide, remarquons aussi que* $\Delta_q(x) \not\equiv 0$).

On obtient alors:

i) *il existe un voisinage* $\mathcal{V}_1$ *de* 0 *dans* $\boldsymbol{C}^{n+1}$ *tel que* $I(x)$ *se ramifie autour de l'hypersurface de* $\mathcal{V}_1$ *définie par:*

$$\Delta_{S_q}(x)=0,$$

c'est-à-dire formée des x *tels qu'un couple* (x, *une arête du simplexe* $S_q(x)$) *s'appuie sur* V: $\varphi(x, \tau)=0$;

ii) *de plus, il existe un voisinage* $\mathcal{V}_2$ *de* 0 *dans* $\boldsymbol{C}^q$ *tel que*: $\{\Delta_{S_q}(x) \equiv 0\} \subset \{x \in \mathcal{V}_1$; *une des arêtes de* $S_q(x)$ *est tangente à* V_x *dans* $\mathcal{V}_2$ *ou bien* V_x *a un point singulier dans* $\mathcal{V}_2\}$.

i) on procède par récurrence. On obtient d'abord, à l'aide du théorème du § 3.1, compte tenu de la condition:

$$\varphi(0, T_{[q-1]} \cap F_q) \neq 0,$$

que l'intégrale:

$$\int_0^{-\theta_q} \mathcal{U}_q i_q d\sigma_q$$

définit un germe holomorphe en $(y, 0)$ qui se prolonge au revêtement simplement connexe de:

$$\{x, \alpha_i, \sigma_i, 1 \leq i \leq q-1; \varDelta_{S_1}(x, F_{[q-1]}) \neq 0\}.$$

Ensuite, on remarque que:

$\varDelta_{S_{q-1}}(x, F_{[j]})$ satisfait la condition de Weierstrass:

$$\varDelta_{S_{q-j}}(0, T_{[j-1]} \dot{\cap} \dot{F}_j) \not\equiv 0,$$

que $\theta_j(x, \cdots, \alpha_l \sigma_l, \sigma_l, \cdots, \alpha_{j-1}\sigma_{j-1}, \sigma_{j-1})$ est tel que: $\theta_j(y, 0, 0) = 0$, et que de plus, d'après le § 2.5:

$$\{x, \cdots, \alpha_l, \sigma_l, \cdots, \alpha_{j-1}, \sigma_{j-1}; \varDelta_{S_{q-j}}(x, F_{[j]}) = 0, \forall F_j \text{ passant par } A^j_{j-1}\}$$
$$\subset \{x, \cdots, \alpha_l, \sigma_l, \cdots, \alpha_{j-1}, \sigma_{j-1}; \varDelta^{\vee}_{S_{q-j+1}}(x, F_{[j-1]}) = 0\}.$$

Enfin on remarque que l'intégrale:

$$\int_0^{-\theta_{j+1}} d\sigma_{j+1} \cdots \int_0^{-\theta_q} \mathscr{U}_q(x, \cdots, \alpha_l \sigma_l, \sigma_l, \cdots, \sigma_q) i_q(\ ) d\sigma_q$$

ne dépend pas des α_l, $l \geq j+1$ et définit un germe de fonction holomorphe au point $(x=y, \cdots, \alpha_l = \sigma_l = 0)$, $l \leq j$.

Supposons qu'elle se prolonge analytiquement au revêtement simplement connexe de:

$$\{x, \cdots, \alpha_l, \sigma_l, \cdots, \alpha_j, \sigma_j; \varDelta_{S_{q-j}}(x, F_{[j]}) \neq 0\}.$$

L'intégrale suivante:

$$\int_0^{-\theta_j} d\sigma_j \int_0^{-\theta_{j+1}} d\sigma_{j+1} \cdots \int_0^{-\theta_q} \mathscr{U}_q i_q d\sigma_q$$

ne dépend pas des α_l, $l \geq j$ et définit un germe holomorphe au point $(x=y, \cdots, \alpha_{j-1} = \sigma_{j-1} = 0)$.

On est dans les conditions de la proposition du § 3.1 et cette intégrale se prolonge au revêtement simplement connexe de

$$\{x, \cdots, \alpha_{j-1}, \sigma_{j-1}; \varDelta_{S_{q-j}}(x, F_{[j-1]} \cap T_j) \cdot \varDelta^{\vee}_{S_{q-j+1}}(x, F_{[j-1]}) \neq 0\}$$

compte tenu du § 1.3 et des résultats de la fin du § 2.4 et du § 2.5; c'est-à-dire encore au revêtement simplement connexe de:

$$\{x, \cdots, \alpha_{j-1}, \sigma_{j-1}; \varDelta_{S_{q-j+1}}(x, F_{[j-1]}) \neq 0\}.$$

La récurrence est donc obtenue et finalement $I(x)$ se ramifie autour de:

$$\{x \in \mathscr{V}_1; \varDelta_{S_q}(x) = 0\},$$

où $\mathscr{V}_1$ est le dernier voisinage permis en x obtenu.

ii) est immédiat, compte tenu de la définition du § 1.5 et des résultats du § 2.4.

Remarque. On peut remplacer l'hypothèse de non développabilité par une hypothèse d'équidimension des variétés polaires comme dans [6].

Bibliographie

[1] J. Briancon et J. P. Speder, Thèse, Nice, 1976.
[2] T. Fukuda et T. Kobayashi, A local isotopy theorem, Tokyo J. Math., **5**, (1982), 31–36.
[3] Y. Hamada and G. Nakamura, On the singularities of the solution of the Cauchy problem . . . , Ann. Scuola Norm. Sup. Pisa, **4** (1977), 725–755.
[4] J. P. Henry, M. Merle et C. Sabbah, Sur la condition de Thom stricte pour un morphisme analytique complexe, Publ. Centre Math. Ecole Polytech., Palaiseau, 1982.
[5] T. Kobayashi, On the singularities of the solution to the Cauchy problem . . . , Math. Ann. (à paraître) (1982).
[6] J. Leray, Un complément au théorème de N. Nilsson sur les intégrales . . . , Bull. Soc. Math. France, **95** (1967), 313–374.
[7] D. Schiltz, J. Vaillant et C. Wagschal, Problème de Cauchy ramifié à caractéristiques multiples en involution, C. R. Acad. Sci. Paris, **291** (1980), 659–662.
[8] ——, Problème de Cauchy ramifié: racine caractéristique double ou triple en involution. J. Math. Pures Appl., **4** (1982), 423–443.
[9] J. Vaillant, Intégrales singulières holomorphes (Journées E.D.P. Saint Jean de Monts, 1983).

UNIVERSITÉ PIERRE ET MARIE CURIE (PARIS VI)
MATHÉMATIQUES, TOUR 45-46, 5ÈME ÉTAGE
UNITÉ ASSOCIÉE AU CNRS 761, 4 PLACE JUSSIEU
75252 PARIS CEDEX 05

Taniguchi Symp. HERT
Kyoto 1984, pp. 415–423

Generalized Hamilton Flows and Singularities of Solutions of the Hyperbolic Cauchy Problem

Seiichiro WAKABAYASHI

§ 1. Introduction

Singularities of solutions of the hyperbolic Cauchy problem have been investigated by many authors. In these works, Hamilton flows (null bicharacteristics) played a key role. However, in general, Hamilton flows can not be defined meaningly unless the characteristic roots are smooth. To study singularities of solutions, one must generalize Hamilton flows. One may expect intuitively that results on singularities of solutions can be obtained from results on well-posedness of the Cauchy problem. To make it possible, Hamilton flows must be generalized suitably.

In this note we shall give a definition of generalized Hamilton flows and investigate the wave front sets of solutions of the hyperbolic Cauchy problem in the framework of C^∞, Gevrey classes or the space of real analytic functions, microlocalizing results on well-posedness and using generalized Hamilton flows.

§ 2. Microhyperbolicity

Let Ω be an open subset of $\boldsymbol{R}^N$ and $p(z) \in C^\infty(\Omega)$.

Definition 2.1. Let $z^0 \in \Omega$ and $\nu \in T_{z^0}\Omega \cong \boldsymbol{R}^N$. We say that $p(z)$ is microhyperbolic at z^0 with respect to ν if there are a neighborhood U of z^0 in Ω, $\delta > 0$, a non-negative integer k, $a_j(z) \in C^\infty(U)$ $(1 \leqq j \leqq k)$ and $e(z, s) \in C^\infty(U \times (-\delta, \delta))$ such that $a_j(z^0) = 0$ $(1 \leqq j \leqq k)$, $e(z, s) \neq 0$ for $(z, s) \in U \times (-\delta, \delta)$,

$$p(z+s\nu) = e(z, s)g(z, s) \qquad \text{for } (z, s) \in U \times (-\delta, \delta) \tag{2.1}$$

and $g(z, s) \neq 0$ for $z \in U$ and $s \in \boldsymbol{C}$ with $\operatorname{Im} s < 0$, where $g(z, s) = s^k + a_1(z)s^{k-1} + \cdots + a_k(z)$.

Remark. (i) If $p(z^0 + s\nu) = s^k(c + o(1))$ as $s \to 0$ for some $c \neq 0$, then, by Malgrange's preparation theorem, $p(z+s\nu)$ can be written in the form (2.1). Although $g(z, s)$ can not be determined uniquely, the above definition does not depend on the choice of $g(z, s)$ (see Lemma 2.4).

(ii) When $p(z)$ is real analytic, Kashiwara and Kawai [11] (and also Sjöstrand [15]) gave the same definition as the above.

Definition 2.2. Let $z^0 \in \Omega$. We define the localization polynomial $p_{z^0}(\delta z)$ of $p(z)$ at z^0 by

$$p(z^0+s\delta z)=s^{\mu}(p_{z^0}(\delta z)+o(1)) \quad \text{as } s\to 0,$$
$$p_{z^0}(\delta z)\not\equiv 0 \qquad \text{in } \delta z \in T_{z^0}\Omega,$$

if there is a derivative of p which does not vanish at z^0. Moreover, we define $p_{z^0}(\delta z)\equiv 0$ and $\mu=\infty$ if any derivatives vanish at z^0. We call μ the multiplicity of z^0 relative to p.

Remark. When $p(z)$ is a polynomial of z, the localization polynomial $p_{z^0}(\delta z)$ was defined by Hörmander [5] and Atiyah, Bott and Gårding [1] in the study of singularities of solutions of partial differential operators with constant coefficients.

The following lemma easily follows from the definition of microhyperbolicity (see [6], [8] and Lemma 8.7.3 in [7]).

Lemma 2.3. *If $p(z)$ is microhyperbolic at z^0 ($\in \Omega$) with respect to ν ($\in T_{z^0}\Omega$), then $p_{z^0}(\delta z)$ is a homogeneous polynomial of degree μ and hyperbolic with respect to ν*, i.e.,

$$p_{z^0}(\delta z-it\nu)\neq 0 \qquad \textit{for } \delta z \in T_{z^0}\Omega \textit{ and } t>0,$$

where μ is the multiplicity of z^0 relative to p.

Lemm 2.4. *Assume that $p(z)$ is microhyperbolic at z^0 ($\in \Omega$) with respect to ν ($\in T_{z^0}\Omega$), and let $p(z, s)$ be an almost analytic extension of $p(z+s\nu)$ with respect to s near $z=z^0$ and $s=0$, i.e., there are a neighborhood U of z^0 in Ω and $\delta>0$ such that $p(z, s) \in C^\infty(U\times\{s \in \boldsymbol{C};\ |s|<\delta\})$, $p(z, s)=p(z+s\nu)$ for $z \in U$ and $s \in \boldsymbol{R}$ with $|s|<\delta$ and $(\partial/\partial\bar{s})p(z, s)$ vanishes to infinite order on $U\times\{s \in \boldsymbol{R};\ |s|<\delta\}$ (see* [4], [12]). *Then the multiplicity μ of z^0 relative to p is finite and there are a neighborhood U_1 of z^0 and positive constants δ_1 and c such that*

$$|p(z, s)|\geqq c|\operatorname{Im} s|^{\mu} \qquad \textit{if } z \in U_1,\ s \in \boldsymbol{C},\ |s|<\delta_1 \textit{ and } \operatorname{Im} s\leqq 0.$$

Conversely if, for an almost analytic extension $p(z, s)$ of $p(z+s\nu)$ with respect to s, there are a non-negative integer k, a neighborhood U of z^0 and positive constants δ and c such that

$$|p(z, -is)|\geqq cs^k \qquad \textit{if } z \in U \textit{ and } 0\leqq s<\delta, \tag{2.2}$$

then $p(z)$ is microhyperbolic at z^0 with respect to ν.

Proof. If $p(z)$ is microhyperbolic at z^0 with respect to ν, then μ is finite and we can write $p(z+s\nu)=e(z, s)g(z, s)$ as in (2.1), where $g(z, s)$ is a polynomial of s of degree μ and $e(z^0, 0)\neq 0$. Microhyperbolicity of p implies that $|g(z, s)|\geqq|\mathrm{Im}\, s|^\mu$ for $z \in U$, $s \in \boldsymbol{C}$ with $\mathrm{Im}\, s\leqq 0$. Let $\tilde{e}(z, s)$ be an almost analytic extension of $e(z, s)$ with resepct to s near $z=z^0$ and $s=0$. Then there are a neighborhood U_0 of z^0 and positive constants δ' and c' such that

$$|\tilde{e}(z, s)g(z, s)|\geqq c'|\mathrm{Im}\, s|^\mu \qquad \text{if } z \in U_0,\ s \in \boldsymbol{C},\ \mathrm{Im}\, s\leqq 0 \text{ and } |s|<\delta'.$$

On the other hand, we have

$$(2.3) \qquad |p(z, s)-\tilde{e}(z, s)g(z, s)|\leqq C_l|\mathrm{Im}\, s|^l,$$

if $z \in U_1$, $s \in \boldsymbol{C}$, $|s|<\delta''$ and $l=0, 1, 2, \cdots$, where U_1 is a neighborhood of z^0 and δ'' is a positive constant. This proves the first part of the lemma. Next we assume that (2.2) is valid for an almost analytic extension $p(z, s)$. It is obvious that there is a non-negative integre k' such that $k'\leqq k$ and $p(z^0+s\nu)=s^{k'}(c+o(1))$ as $s\to 0$ for some $c\neq 0$. So we can write $p(z+s\nu)=e(z, s)g(z, s)$ as in (2.1). Then (2.3) is valid. From (2.2) it follows that there are a neighborhood U_2 of z^0 and $\delta_1>0$ such that $g(z, s)\neq 0$ for $z \in U_2$, $s \in \boldsymbol{C}$ with $|s|<\delta_1$ and $\mathrm{Im}\, s<0$. On the other hand, the roots of $g(z, s)=0$ in s are contained in the set $\{s \in \boldsymbol{C};\ |s|<\delta_1\}$ for $z \in U_3$ if U_3 is a sufficiently small neighborhood of z^0. This implies that $p(z)$ is microhyperbolic at z^0 with respect to ν. Q.E.D.

Lemma 2.5. *Assume that $p(z)$ is microhyperbolic at z^0 ($\in \Omega$) with respect to ν ($\in T_{z^0}\Omega$). If M is a connected subset of $T_{z^0}\Omega$, $\nu \in M$ and if $p(z)$ is microhyperbolic at z^0 with respect to any $\tilde{\nu} \in M$, then $M\subset\Gamma(p_{z^0}, \nu)$, where $\Gamma(p_{z^0}, \nu)$ is the connected component of the set $\{\delta z \in T_{z^0}\Omega;\ p_{z^0}(\delta z)\neq 0\}$ which contains ν.*

Proof. From Lemma 2.3 it follows that $p_{z^0}(\tilde{\nu})\neq 0$ for $\tilde{\nu} \in M$, which proves the lemma. Q.E.D.

Lemma 2.6 ([19]). *Assume that $p(z)$ is microhyperbolic at z^0 ($\in \Omega$) with respect to ν ($\in T_{z^0}\Omega$), and let $\tilde{p}(z)$ be an almost analytic extension of $p(z)$ with respect to z near $z=z^0$. Then, for any compact subset M of $\Gamma(p_{z^0}, \nu)$, there are a neighborhood U of z^0 and positive constants s_0 and c such that*

$$|\tilde{p}(z-is\tilde{\nu})|\geqq cs^\mu \qquad \textit{if } z \in U,\ 0\leqq s<s_0 \textit{ and } \tilde{\nu} \in M,$$

where μ is the multiplicity of z^0 relative to p.

Corollary. *If $p(z)$ is microhyperbolic at z^0 with respect to ν, then $\Gamma(p_{z^0}, \nu)$ is the connected components of the set $\{\tilde{\nu} \in T_{z^0}\Omega;\ p(z)$ is microhyperbolic at z^0 with respect to $\tilde{\nu}\}$ which contains ν.*

Theorem 2.7 ([19]). *If $p(z)$ is microhyperbolic at z^0 with respect to ν, then for any compact subset M of $\Gamma(p_{z^0}, \nu)$ there is a neighborhood U of z^0 such that p is microhyperbolic at z with respect to $\tilde{\nu}$ for $z \in U$ and $\tilde{\nu} \in M$ and $M \subset \Gamma(p_z, \nu)$ for $z \in U$.*

Proof. Lemmas 2.4 and 2.6 imply that there is a neighborhood U of z^0 such that p is microhyperbolic at z with respect to $\tilde{\nu}$ for $z \in U$ and $\tilde{\nu} \in M$. We may assume that M is connected (or convex) and that $\nu \in M$. Then from Lemma 2.5 it follows that $M \subset \Gamma(p_z, \nu)$ for $z \in U$. Q.E.D.

§ 3. Generalized Hamilton flows

In this section we assume that $p(x, \xi) = \sum_{j=0}^{m} p_j(x, \xi')\xi_1^{m-j}$, where $(x, \xi) \in T^*\boldsymbol{R}^n \cong \boldsymbol{R}^n \times \boldsymbol{R}^n$, $x = (x_1, x') = (x_1, x_2, \cdots, x_n) \in \boldsymbol{R}^n$, $\xi = (\xi_1, \xi') = (\xi_1, \xi_2, \cdots, \xi_n) \in \boldsymbol{R}^n$, $p_0(x, \xi') = 1$, $p_j(x, \xi') \in C^\infty(\boldsymbol{R} \times (T^*\boldsymbol{R}^{n-1} \setminus 0))$ and $p_j(x, \xi')$ is positively homogeneous of degree j in ξ'. Moreover, we assume that p is microhyperbolic at z with respect to $(0, \mathscr{I}) \in T_z(T^*\boldsymbol{R}^n) \cong \boldsymbol{R}^{2n}$ if $z = (x, \xi)$ and $\xi' \neq 0$, where $\mathscr{I} = (1, 0, \cdots, 0) \in \boldsymbol{R}^n$. Define

$$\Gamma_z = \Gamma(p_z, (0, \mathscr{I})) \quad \text{if } z = (x, \xi) \text{ and } \xi' \neq 0,$$

$$\Gamma_{(x, \xi_1, 0)} = \begin{cases} \boldsymbol{R}^{2n} & \text{if } \xi_1 \neq 0, \\ \boldsymbol{R}^n \times \Gamma_x & \text{if } \xi_1 = 0, \end{cases}$$

where $\Gamma_x = \bigcap_{\xi \in \boldsymbol{R}^n, |\xi|=1} \{\delta\xi;\ (0, \delta\xi) \in \Gamma_{(x,\xi)}\}$. It is easy to see that $\Gamma_{(x,0)} = \Gamma(p_{(x,0)}, (0, \mathscr{I}))$ if p is a polynomial of ξ and microhyperbolic at $(x, 0)$ with respect to $(0, \mathscr{I})$ for any $x \in \boldsymbol{R}^n$ (see [19]). Then Theorem 2.7 gives the following

Lemma 3.1 ([19]). *Let $z^0 \in T^*\boldsymbol{R}^n$, and let M be a compact subset of Γ_{z^0}. Then there is a neighborhood U of z^0 in $T^*\boldsymbol{R}^n$ such that $M \subset \Gamma_z$ for $z \in U$.*

Definition 3.2 ([17], [19]). Let $z \in T^*\boldsymbol{R}^n$ and $A \subset T_z(T^*\boldsymbol{R}^n)$, and define the dual cone of A with respect to the symplectic form σ on $T^*\boldsymbol{R}^n$ by

$$A^\sigma = \{(\delta x, \delta\xi) \in T_z(T^*\boldsymbol{R}^n);\ \delta x \cdot \delta\eta - \delta y \cdot \delta\xi\ (= \sigma((\delta y, \delta\eta), (\delta x, \delta\xi))) \geqq 0 \quad \textit{for any } (\delta y, \delta\eta) \in A\},$$

and the generalized Hamilton flows $K_z^\pm$ for p by

$$K_z^\pm=\{z(t)\in T^*\boldsymbol{R}^n;\ \pm t\geqq 0,\ \textit{and}\ \{z(t)\}\ \textit{is a Lipschitz continuous curve satisfying}\ (d/dt)z(t)\in\Gamma^\sigma_{z(t)}\ (\textit{a.e.}\ t)\ \textit{and}\ z(0)=z\}.$$

Remark. (i) For general microhyperbolic functions one can define similarly their generalized Hamilton flows. (ii) One can write $K^\pm_{(x,0)}=K_x^\pm\times\{0\}$, where $K_x^\pm=\{x(t)\in\boldsymbol{R}^n;\ \pm t\geqq 0$ and $\{x(t)\}$ is a Lipschitz continuous curve satisfying $(d/dt)x(t)\in\Gamma_x^*$ (a.e. t) and $x(0)=x\}$ and $\Gamma^*=\{\delta x;\ \delta x\cdot\delta\xi\geqq 0$ for any $\delta\xi\in\Gamma\}$.

Lemma 3.3. (i) *The definition of $K_z^\pm$ does not depend on the choice of the canonical coordinates.* (ii) *If $z\in T^*\boldsymbol{R}^n\setminus 0$, then $K_z^\pm\subset T^*\boldsymbol{R}^n\setminus 0$.* (iii) *If $p(z)\neq 0$, then $K_z^\pm=\{z\}$.* (iv) *If $z\in T^*\boldsymbol{R}^n\setminus 0$ and $p(x,\xi)=\prod_{j=1}^r p_j(x,\xi)^{\nu_j}$ has involutive characteristics, that is, if $\{p_j,p_k\}=a_{jk}p_j+b_{jk}p_k$ for $1\leqq j,k\leqq r$, then $K_z^\pm$ are the unions of the broken null bicharacteristics of p_j $(1\leqq j\leqq r)$ issuing from z along which $\pm x_1$ increase, where $p_j(x,\xi)=\xi_1-\lambda_j(x,\xi')$, and the $\lambda_j(x,\xi')$, the $a_{jk}(x,\xi)$ and the $b_{jk}(x,\xi)$ are smooth for $\xi'\neq 0$. $K_z^\pm$ are conoids with their vertexes at z.*

Theorem 3.4 ([17]). *Assume that $K_x^-\cap\{x_1\geqq 0\}$ is bounded for each $x\in\boldsymbol{R}^n$, and let $z^0=(x^0,\xi^0)\in T^*\boldsymbol{R}^n$. Then $K_{z^0}^-\cap\{x_1\geqq 0\}$ is compact, and $K_{z^0}^+$ is closed if $x_1^0\geqq 0$. Moreover, for any neighborhood U of $K_{z^0}^-\cap\{x_1\geqq 0\}$, there is a neighborhood U_1 of z^0 such that $K_z^-\cap\{x_1\geqq 0\}\subset U$ for $z\in U_1$.*

Theorem 3.5 ([19]). *Let Σ be the set $\{z\in T^*\boldsymbol{R}^n\setminus 0;\ p(z)=0\}$, and let Σ_1 be the set of singular points of Σ. If $T_z\Sigma_1\cap\Gamma_z^\sigma=\{0\}$ for any $z\in\Sigma_1$, then $K_z^\pm$ are the unions of the broken null bicharacteristics issuing from z along which $\pm x_1$ increase. Here $T_z\Sigma_1$ denotes the set $\{\delta z\in T_z(T^*\boldsymbol{R}^n);\ \delta z=\lim_{j\to\infty}a_j(z^j-z)$ for some sequences $\{a_j\}\subset\boldsymbol{R}$ and $\{z^j\}\subset\Sigma_1$ with $z^j\to z$ as $j\to\infty\}$.*

Remark. If p is effectively hyperbolic, then $p(x,\xi)$ can be reduced microlocally to $\xi_1^2-a(x,\xi')x_1^2-b(x,\xi')$, where $a(x,\xi')>0$ and $b(x,\xi')\geqq 0$ (see [13]). Thus, since $\Sigma_1\subset\{\xi_1=0,\ x_1=0,\ b(x,\xi')=0\}\subset\{x_1=0\}$, one can apply Theorem 3.5 to effectively hyperbolic operators.

We refer to [17] and [19] for further properties of $K_z^\pm$.

§ 4. Singularities of solutions

In this section we shall consider the Cauchy problem for general hyperbolic operators in Gevrey classes and that for symmetric hyperbolic systems in C^∞ together.

Definition 4.1. (i) For $1 \leqq \kappa < \infty$ we say that $f \in \mathcal{E}^{\{\kappa\}}$ if $f \in C^\infty$ and if for any compact subset K of $\boldsymbol{R}^n$ there are positive constants h and C such that

$$|D^\alpha f(x)| \leqq C h^{|\alpha|} (|\alpha|!)^\kappa \quad \text{for any } x \in K \text{ and any multi-indexes } \alpha,$$

where $D=(D_1, D')=-i(\partial/\partial x_1, \partial/\partial x_2, \cdots, \partial/\partial x_n)$. We write $\mathcal{E}^{\{\infty\}}=C^\infty$ (formally) and $\mathcal{D}^{\{\kappa\}}=\mathcal{E}^{\{\kappa\}} \cap C_0^\infty$ for $1<\kappa \leqq \infty$. We introduce usual locally convex topologies in these spaces. (ii) We denote by $\mathcal{D}^{\{\kappa\}\prime}$ the strong dual space of $\mathcal{D}^{\{\kappa\}}$ for $1<\kappa \leqq \infty$, and by $\mathcal{D}^{\{1\}\prime}$ the space of all hyperfunctions on $\boldsymbol{R}^n$. (iii) For $f \in \mathcal{D}^{\{\kappa\}\prime}$, $WF_{\{\kappa\}}(f)$ $(\subset T^*\boldsymbol{R}^n \diagdown 0)$ is defined as follows: $(x, \xi) \notin WF_{\{\kappa\}}(f)$ if $f \in \mathcal{E}^{\{\kappa\}}$ microlocally at (x, ξ).

Remark. (i) In the above definition, $WF_{\{\infty\}}(f)=WF(f)$ and $WF_{\{1\}}(f)=\{(x, \xi) \in T^*\boldsymbol{R}^n \diagdown 0;\ (x, i\xi\infty) \in \text{S.S.}f\}$, where S.S.$f$ denotes the singular spectrum of f (see [14]). (ii) For $f \in \mathcal{D}^{\{\kappa\}\prime}$ $(1<\kappa \leqq \infty)$ we can also define $WF_{\{\kappa_1\}}(f)$ $(1 \leqq \kappa_1 \leqq \infty)$.

Let us consider the Cauchy problems

$$(\text{CP})_\kappa \qquad \begin{cases} P(x, D)u=f, \\ \operatorname{supp} u \subset \{x_1 \geqq 0\}, \end{cases}$$

where $1 \leqq \kappa \leqq \infty$ and $f \in \mathcal{D}^{\{\kappa\}\prime}$ with $\operatorname{supp} f \subset \{x_1 \geqq 0\}$. We assume that for $1 \leqq \kappa < \infty$

(A–1)$_\kappa$ $P(x, \xi)$ is a hyperbolic polynomial with resepct to $\mathcal{I}$ for every $x \in \boldsymbol{R}^n$, i.e.,

$$p(x, \xi - it\mathcal{I}) \neq 0 \qquad \text{if } x \in \boldsymbol{R}^n,\ \xi \in \boldsymbol{R}^n \text{ and } t>0,$$

and the coefficients of $P(x, \xi)$ belong to $\mathcal{E}^{\{\kappa\}}$, where $p(x, \xi)$ is the principal part of $P(x, \xi)$.

And we assume that for $\kappa=\infty$

(A–1)$_\infty$ $P(x, D)=D_1-A(x, D')+B(x, D)$ and $A(x, \xi')^*=A(x, \xi')$, where $A(x, \xi') \equiv (A_{ij}(x, \xi'))$ is an $m \times m$ matrix-valued symbol whose entries are positively homogeneous of degree 1 in ξ' and belong to $C^\infty(\boldsymbol{R} \times (T^*\boldsymbol{R}^{n-1} \diagdown 0))$, $A(x, D')$ is a properly supported pseudo-differential operator with symbol $A(x, \xi')$ and $B(x, D)$ is a properly supported pseudo-differential operator of order 0.

For $\kappa=\infty$ we put $p(x, \xi)=\det(\xi_1 I_m - A(x, \xi'))$, where I_m denotes the identity matrix of order m. To state the results globally, we assume that

(A$-$2) $K_x^- \cap \{x_1 \geqq 0\}$ is bounded for every $x \in \boldsymbol{R}^n$, where K_x^- is defined for p as in Section 3.

Theorem 4.2 ([17], [18], [19]). *Assume that* (A–1)$_\kappa$ *and* (A–2) *are satisfied, and that* $1 \leqq \kappa < r/(r-1)$ *or* $\kappa = \infty$, *where* r ($\geqq 2$) *is more than or equal to the multiplicities of the roots of the equation* $p(x, \xi_1, \xi') = 0$ *in* ξ_1 *for* $x \in \boldsymbol{R}^n$ *and* $\xi' \in \boldsymbol{R}^{n-1} \setminus \{0\}$. *If* $u \in \mathscr{D}^{\{\kappa\}\prime}$ *satisfies the Cauchy problem* (CP)$_\kappa$, *then*

$$WF_{\{\kappa\}}(u) \subset \{z \in T^*\boldsymbol{R}^n \setminus 0;\ z \in K_w^+ \text{ for some } w \in WF_{\{\kappa\}}(f)\}.$$

Remark. (i) If $P(x, D)$ is a partial differential operator, then we have

$$\operatorname{supp} u \subset \{x \in \boldsymbol{R}^n : x \in K_y^+ \text{ for some } y \in \operatorname{supp} f\}.$$

(ii) Theorem 4.2 is also valid when $f \in \mathscr{D}^{\{\kappa\}\prime}$, $WF_{\{\kappa\}}(f) \subset \{x_1 \geqq 0\}$ and the condition $\operatorname{supp} u \subset \{x_1 \geqq 0\}$ is replaced by the condition $WF_{\{\kappa\}}(u) \subset \{x_1 \geqq 0\}$ in the Cauchy problem (CP)$_\kappa$.

(iii) For $\kappa = \infty$ we have the same results as in Theorem 4.2, concerning the wave front sets of solutions with respect to Sobolev spaces (see [19]).

Theorem 4.2 can be proved, using the following proposition and some properties of flows $K_z^\pm$ which are obtained from Theorem 2.7 or Lemma 3.1.

Proposition 4.3. *Assume that* (A–1)$_\kappa$ *and* (A–2) *are valid and that* $1 \leqq \kappa < r/(r-1)$ *or* $\kappa = \infty$. *If* $u \in \mathscr{D}^{\{\kappa\}\prime}$ *and* $z^0 = (x^0, \xi^0) \in WF_{\{\kappa\}}(u) \setminus WF_{\{\kappa\}}(Pu)$, *then for any compact subset* M *of* Γ_{z^0} *there is* $t_0 > 0$ *such that*

$$WF_{\{\kappa\}}(u) \cap \{z^0 - M^\sigma\} \cap \{x_1 = x_1^0 - t\} \neq \emptyset \qquad \text{for } 0 \leqq t \leqq t_0.$$

Remark. When $\kappa = 1$, Kashiwara and Kawai [11] (and, also, Sjöstrand [15]) proved Proposition 4.3, which is a microlocalization of Bony and Schapira [2]. For $1 < \kappa < r/(r-1)$, Proposition 4.3 is a microlocalization of Bronshtein [3] (see [17]). When $\kappa = \infty$, Proposition 4.3 can be proved, combining Ivrii's result [9], which is a microlocal version of Holmgren's uniqueness theorem, with the method of sweeping out due to John [10].

§ 5. Some remarks

We first note that generalized Hamilton flows are useful to study analytic hypoellipticity. Assume that $P(x, D)$ is a partial differential operator with real analytic coefficients and that its principal symbol $p(x, \xi)$ is microhyperbolic at any $z \in T^*\boldsymbol{R}^n \setminus 0$. Then we can define locally (or microlocally) the flows $K_z^\pm$ for p, choosing a microhyperbolic direction

$\nu \in R^{2n}$. From Proposition 4.3 ([11]) it follows that $P(x, D)$ is analytic hypoelliptic if $K_z^- = \{z\}$ for any $z \in T^*R^n \setminus 0$. For example, if $P(x, D) = D_1 + ix_1^{2k} D_2$ $(k=1, 2, \cdots)$ and $n=2$, then $P(x, D)$ is analytic hypoelliptic.

Next let us consider the Cauchy problem for hyperbolic operators $P(x, D)$ in C^∞. If the coefficients of $P(x, D)$ are real analytic (or in some Gevrey classes), and if the Cauchy problem is C^∞ well-posed, then Theorem 4.2 is valid even if $\kappa = \infty$. Assume that the coefficients of $P(x, D)$ belong to C^∞. Let $z^0 \in T^*R^n \setminus 0$, and let $\phi(x, \xi)$ be a real-valued positively homogeneous function of degree 0 defined in a conic neighborhood Γ of z^0 such that $\phi \in C^\infty(\Gamma)$, $\phi(z^0)=0$ and $-H_\phi(z) \in \Gamma_z$ for $z \in \Gamma$, where Γ_z is defined for the principal symbol $p(x, \xi)$ of P and $H_\phi = \sum_{j=1}^n \{(\partial\phi/\partial\xi_j)(\partial/\partial x_j) - (\partial\phi/\partial x_j)(\partial/\partial\xi_j)\}$. Let $\chi: \Gamma \ni (x, \xi) \mapsto (y(x, \xi), \eta(x, \xi)) \in \tilde{\Gamma}$ be a homogeneous canonical transformation satisfying $\chi(z^0) = (0, \eta^0)$ and $y_1(x, \xi) = \phi(x, \xi)$, where $\tilde{\Gamma}$ is a conic neighborhood of $(0, \eta^0)$. Then there are (classical) Fourier integral operators F_1 and F_2 corresponding to χ and χ^{-1} with principal symbols different from 0 at z^0 and $(0, \eta^0)$, respectively. For the definition of microhyperbolicity and Malgrange's division theorem it follows that there are a pseudo-differential operator $e(x, D)$ and a operator $Q(x, D)$ such that $Q(x, D) = \sum_{j=0}^{m'} q_j(x, D') D_1^{m'-j}$, $q_0(x, \xi') = 1$, $q_j(0, \eta^{0\prime}) = 0$ $(1 \leq j \leq m)$, $q(x, \xi)$ is hyperbolic with respect to $(0, \mathscr{I})$ and $F_1 P(x, D) F_2 \equiv e(x, D) Q(x, D)$ in a conic neighborhood of $(0, \eta^0)$, where $q_j(x, \xi')$ is the symbol of $q_j(x, D')$ and $q(x, \xi)$ is the principal symbol of Q (see [16]). We assume that for any $z^0 \in T^*R^n \setminus 0$ and any ϕ satisfying the above properties the Cauchy problem for $Q(x, D)$ is C^∞ well-posed. Then, using microlocal Holmgren's transformation, we can prove a microlocal version of Holmgren's uniqueness theorem, that is, if $z^0 \notin WF(Pu)$, $-H_\phi(z^0) \in \Gamma_{z^0}$ and $WF(u) \cap \{z \in \Gamma; \phi(z) < 0\} = \emptyset$, then $z^0 \notin WF(u)$. From this the method of sweeping out in [10] gives Theorem 4.2 with $\kappa = \infty$ for such hyperbolic operators (see [19]).

References

[1] M. F. Atiyah, R. Bott and L. Gårding, Lacunas for hyperbolic differential operators with constant coefficients I, Acta Math. **124** (1970), 109–189.

[2] J. M. Bony and P. Schapira, Solutions hyperfonctions du problème de Cauchy, Hyperfunctions and Pseudo-differential Equations, Lecture Notes in Math. **287**, Springer, 1973, 82–98.

[3] M. D. Bronshtein, The Cauchy problem for hyperbolic operators with variable multiple characteristics, Trudy Moskov. Mat. Obšč **41** (1980), 83–99.

[4] L. Hörmander, Lecture Notes at the Nordic Summer School of Math., 1969.

[5] ——, On the singularities of solutions of partial differential equations, International Conference of Functional Analysis and Related Topics, Tokyo, 1969.

[6] ——, The Cauchy problem for differential equations with double characteristics, J. Analyse Math. **32** (1977), 118–196.

[7] L. Hörmander, The Analysis of Linear Partial Differential Operators **I**, Springer, Berlin-Heidelberg-New York-Tokyo, 1983.
[8] V. Ya. Ivrii and V. M. Petkov, Necessary conditions for the Cauchy problem for non-strictly hyperbolic equations to be well-posed, Uspehi Mat. Nauk. **29** (1974), 3–70.
[9] V. Ya. Ivrii, Wave fronts of solutions of symmetric pseudo-differential systems, Sibirsk. Mat. Zh. **20** (1979). 557–578.
[10] F. John, On linear partial differential equations with analytic coefficients, unique continuation of data, Comm. Pure Appl. Math. **2** (1949), 209–254.
[11] M. Kashiwara and T. Kawai, Micro-hyperbolic pseudo-differential operators I, J. Math. Soc. Japan **27** (1975), 359–404.
[12] A. Melin and J. Sjöstrand, Fourier integral operators with complex valued phase functions, Lecture Notes in Math. **459**, Springer, 1974, 120–223.
[13] R. Melrose, The Cauchy problem for effectively hyperbolic operators, Hokkaido Math. J. **12** (1983), 371–391.
[14] M. Sato, T. Kawai and M. Kashiwara, Microfunctions and pseudo-differential equations, Lecture Notes in Math. **287**, Springer, 1973, 265–529.
[15] J. Sjöstrand, Singularités analytiques microlocales, Astérisque, **95** (1982), 1–166.
[16] F. Treves, Introduction to Pseudo-differential and Fourier Integral Operators II, Plenum Press, New York-London, 1980.
[17] S. Wakabayashi, Singularities of solutions of the Cauchy problem for hyperbolic systems in Gevrey classes, Japan. J. Math. **11** (1985), 157–201.
[18] ——, Analytic singularities of solutions of the hyperbolic Cauchy problem, Proc. Japan Acad. **59** (1983), 449–452.
[19] ——, Singularities of solutions of the Cauchy problem for symmetric hyperbolic systems, Comm. Partial Differential Equations **9** (1984), 1147–1177.

INSTITUTE OF MATHEMATICS
THE UNIVERSITY OF TSUKUBA
IBARAKI 305, JAPAN